MOLECULAR BEAM EPITAXY

A historic photograph of an early custom-built MBE machine. This was the first machine built at the Mullard Research Laboratories in Redhill (circa 1975) and owes a great deal to the design skills of Jim Neave and the technical abilities of the MRL Engineering Division. It was used for mass-spectrometry studies of MBE growth of GaAs, as described in chapter 4 (section 4.2).

Molecular Beam Epitaxy

A Short History

John Orton and Tom Foxon

OXFORD
UNIVERSITY PRESS

OXFORD
UNIVERSITY PRESS

Great Clarendon Street, Oxford, OX2 6DP,
United Kingdom

Oxford University Press is a department of the University of Oxford.
It furthers the University's objective of excellence in research, scholarship,
and education by publishing worldwide. Oxford is a registered trade mark of
Oxford University Press in the UK and in certain other countries

© John Orton and Tom Foxon 2015

The moral rights of the authors have been asserted

First Edition published in 2015

Impression: 1

Published in the United States of America by Oxford University Press
198 Madison Avenue, New York, NY 10016, United States of America

British Library Cataloguing in Publication Data
Data available

Library of Congress Control Number: 2014958219

ISBN 978–0–19–969582–9

Printed and bound by
CPI Group (UK) Ltd, Croydon, CR0 4YY

Preface

Molecular beam epitaxy, as a method of growing thin films of semiconductor materials, originated during the 1960s and, though it is difficult to define a precise birth date, one might reasonably suggest that the technique, now some 50 years old, has reached mature middle age. It therefore seemed to us that this might be an appropriate time to look back over the past five decades with a view to summarising not only the origins of MBE but also its undoubted achievements. Roughly one decade ago, thanks again to the support of Oxford University Press, one of us (J. O.) took the first faltering steps towards writing a history of semiconductor science and technology, a venture which proved successful enough to encourage this further attempt. Let us be clear, we have tried to write a history of the subject, rather than aiming at 'textbook' precision. Nor do we make any pretension to completeness—our intension has been to present snapshots of MBE taken at appropriate intervals so as to illustrate its development as a sophisticated method of material deposition, together with its contribution to those three interlinked activities: the understanding of crystal growth, the furtherance of semiconductor science and technology and the application to commercial success. Such an approach must inevitably leave gaps and we apologise in advance to those scientists whose contributions we may have overlooked. Such omissions by no means imply any judgement of inferior quality—we would simply suggest that any attempt at completeness would lead to an account of indigestible length. In similar vein, we might add that, though we are aware of MBE's contribution in other branches of materials science, we have felt obliged to restrict our presentation to the field with which we are personally familiar—that of semiconductors.

Our involvement with MBE began during our respective sojourns at the Mullard (later Philips) Research Laboratories, near Redhill in Surrey (an edifice now very sadly lying in desolate ruins). There, we were extremely fortunate in being able to work with a group of colleagues whose contribution to the fields of crystal growth, semiconductor physics and semiconductor technology was second to none and whose expertise and enthusiasm greatly stimulated our own attempts to understand and contribute to these subjects. We should like to record our deeply felt appreciation of their help during many years of successful research endeavour. They were exciting years and developed in us a love of MBE growth and semiconductor science without which the writing of this book would certainly have been impossible. In 1991 we both moved to Nottingham University, where we continued our romance with MBE, this time in the form of Group III nitride materials, and we must also acknowledge a debt of gratitude to the university for encouraging us in this stimulating activity. Once again, we were privileged to work with helpful, friendly colleagues, whom we should like to thank for their excellent collaboration. In 1996 one of us (J. O.) retired to concentrate on book writing, while the other (T. F.—only semi-retired) continues to contribute on a part-time basis to the work of the Department of Physics.

In common with the majority of MBE-based activities, our work with Philips was concerned entirely with the growth and understanding of compound semiconductors such as GaAs, AlGaAs, InP, InGaP, etc. and their application in the fields of high-frequency electronics and optoelectronics. Without doubt, the stimulus provided by the development of low-dimensional structures during the 1970s (at the IBM, Yorktown and Bell, Murray Hill Laboratories) led to a remarkable expansion in MBE growth facilities, worldwide, when it became clear that MBE was an ideal method for the growth of thin semiconductor films with monolayer resolution. This was just what the PhD scientist ordered for the production of accurately controlled quantum wells and two-dimensional electron gas structures, which served to drive research in microwave and optical devices from that decade onwards into the present era. At the same time, it also represented the ideal technique for studying fine details of crystal growth—one could watch the incorporation of semiconductor material almost literally atom by atom, an ability made possible by the incorporation into the MBE machine of a RHEED (reflection high energy electron diffraction) facility. Indeed, the (accidental!) discovery of RHEED oscillations in the Redhill laboratory at the end of the 1970s allowed crystal growers to measure growth rates by literally counting monolayers. This, together with the use of mass spectrometry to monitor atomic species desorbed from the growing surface, revolutionised the study of crystal growth and provided a detailed understanding of the relationship between surface structure and the incorporation of atoms arriving at, and leaving from, such a surface. This was heady stuff indeed. Combined with sophisticated developments in X-ray diffraction measurements and the invention of the scanning tunnelling microscope in 1981, allied with even greater sophistication in theoretical modelling methods, it provided materials scientists with a range of techniques enabling them to understand crystal growth at the atomic level to an altogether unprecedented degree. All this was made possible by MBE.

What was it about MBE that was different? The answer has to be that it represented an example of crystal growth in an ultra-high vacuum environment, where background impurity levels were reduced to negligible proportions and where sophisticated surface studies on *clean* surfaces could therefore be combined with observations of growth phenomena. This, in turn, demanded a stainless steel UHV apparatus with appropriate pumping facility. Needless to say, it involved a deal of complexity and no little monetary expense (in round figures, the early MBE machine might cost something approaching a million US dollars) so it was of vital importance for the ultimate take-up of MBE as a practical growth technique that the material quality and corresponding device structures should be as good as any available. In this regard, the work of Al Cho at Murray Hill was of considerable importance. As early as 1970, he was able to show that various electronic and optoelectronic devices grown by MBE could perform as well as those produced by other methods. This probably helped to stimulate the development of commercial MBE machines, not only in the USA but in England, France and Japan, from which point MBE has never looked back. However, it was certainly the development of low-dimensional structures which stimulated the rapid advancement of MBE during the 1980s. The number of MBE research groups increased tenfold between 1978 and

1988 and the number of MBE-based publications leaped correspondingly—in 1978 it was roughly 50 per annum, while in 1988 it had reached the dizzy heights of 2000 per annum.

MBE was clearly here to stay, one consequence being its application to the growth of an increasingly wide range of different compounds. The III-V compound spectrum widened to include not only arsenides but phosphides, nitrides and antimonides of aluminium, gallium, and indium. A range of II-VI compounds was soon added to the list and even silicon and germanium succumbed to its obvious allure. Needless to say, the understanding of growth mechanisms expanded correspondingly, an interesting and potentially application-worthy example being concerned with the growth of self-organised quantum dots, first in the form of InGaAs/GaAs and Ge/Si structures and later as a range of other material combinations. Fascinating new alloy properties were uncovered when studies began on the so-called highly mismatched alloys, which incorporated unusually small (in the case of nitrogen) or unusually large (in the case of bismuth) atoms. At the same time, the range of complex device structures grown by MBE techniques followed a similar upward curve, and Al Cho's innovative work finally came to fruition when a number of such devices came to be manufactured by MBE in preference to alternative growth methods—a prime example of this being the successful manufacture, by the Japanese company Rohm, of double heterostructure laser diodes for use in compact disc players. Not only, therefore, was MBE an ideal means for studying basic material properties but it could also be turned to good effect in the highly competitive field of device manufacture.

We have attempted to provide an overview of all these aspects of the subject, together with brief accounts of the background theory of growth mechanisms, RHEED studies, material and device structural properties, etc. as an aid to understanding. Needless to say, such theoretical aspects are covered at a relatively simplistic level—as we said earlier, our book makes no pretension to being an advanced textbook for specialist use but rather a history of an exciting era of semiconductor technology, seen from the particular viewpoint of the MBE grower. Uniquely, perhaps, we have also tried to provide an account of the origins of the subject, including the early use of molecular beams in quite different areas of physics and of their early use by our colleague Bruce Joyce, when at the Plessey, Caswell, laboratory as a means of studying the deposition of silicon atoms on the surface of a silicon crystal, following the thermal decomposition of silane. This activity, in the period 1966–1968, may be seen as one of the first applications of molecular beams in the study of semiconductor surfaces, coinciding rather closely with that of John Arthur's work on GaAs at Bell Labs. We make no attempt to decide who was 'first'—these were two quite independent activities which just happened to occur roughly together, the reason for such coincidence probably having to do with the (then) state of development of the essential UHV technology. Interestingly, they were both concerned with fundamental studies of surface physics (chemistry?)—thus 'proving' the somewhat controversial thesis that, left to its own devices, scientific research will always lead to some 'useful' outcome! We say no more here—the evidence is collected in the following pages. In particular, we might emphasise the wealth of device applications summarised in our last two chapters, which demonstrate just how effective MBE has been in generating commercial outcomes.

Having said all this, it must, by now, be clear that our book should be of interest to a wide range of readers. These obviously include material scientists, semiconductor scientists and technologists as well as those interested primarily in theories of crystal growth. It should also be of interest to device engineers who can benefit from a deeper understanding of the complexities of the semiconductor structures developed to solve their device problems. At the same time, there is no reason why such a book should not be useful as general reading for third-year students in physics, electrical engineering and materials science—the text makes no serious demands on mathematical knowledge, most of the mathematical detail being presented in 'boxes' which sit alongside the appropriate text. These may be read separately or together with the text, or simply ignored(!), the text being complete in itself. It represents a presentational technique used in the earlier book referred to above and found helpful by many readers. Finally, we should not overlook the possible interest for readers concerned with understanding the history of science and technology. We have tried to comment not only on what happened but on the how and why of its happening and, again, we should stress that the level of presentation is such that the non-specialist should have little difficulty in following the argument.

Finally, we must acknowledge the considerable help we have received from colleagues worldwide. Any attempt to write a book covering such a wide span of activity inevitably means that its authors are far from expert in many aspects of the subject and we have had to rely on advice and assistance from many friends and acquaintances. No request was ignored, help came quickly and willingly from all, and we should very much like to take the opportunity of thanking them all most sincerely (while accepting, of course, full responsibility for any errors and oversights which still remain). The list is a long one:

Yasuhiko Arakawa, Richard Ares, John Arthur, Tim Ashley, Alex Ayanselm, Seth Bank, Brian Bennett, Peter Blood, Pierre Bouchaib, Fernando Briones, David Brock, Philip Buckle, Louise Buckle, Richard Campion, Peter Capper, Andres Cantarer, Catherine Chaix, Noel Chase, Al Cho, Philip Cohen, Stuart Comber, Gavin Conibeer, Lutz Daeweritz, Grahem Davies, Phil Dawson, Peter Dobson, Ravi Droopad, Geoff Duggan, Laurence Eaves, Robin Farrow, Eric Friedman, Paul Fewster, Art Gossard, Nicholas Grandjean, Robert Grey, Shun-ichi Gonda, Jon Gowers, James Harris, Jeff Harris, Roger Heckingbottom, Rick Kiehl, Stewart Hooper, Seth Hughes, Don Hurle, Ayahiko Ichimiya, Sergey Ivanov, Jan-Theodoor Janssen, Mark Johnson, Bruce Joyce, Hiroshi Kamimura, Max Lagally, Eric Larkins, Amy Liu, John Lockley, Trevor Martin, Jean Massies, Eva Monroy, David Mowbray, Thomas Myers, Aneesh Nainani, Jim Neave, Sergei Novikov, Brad Orr, Mike Pashley, Loren Pfeiffer, Klaus Ploog, Kevin Prior, David Ritchie, John Roberts, Jaeger Roland, Darell Schlom, Alexey Shkolnik, Shri Ram Shukla, Jim Speck, Gunther Springholz, Stephen Sweeney, Steve Tabreham, Kiyoshi Takahashi, Ted Thrush, Richard Thomson, Eric Tournie, Brian Tuck, Vladimir Umansky, Tom Vandervelde, John Walling, Wladek Walukiewicz, Werner Wegscheider, Gunter Weimann, David Williams, Colin Wood, Homen Yuen, Jing Zhang, and Hans Zogg.

When the discovery of the fractional quantum Hall effect was announced in 1982, there were three names on the published 'Physical Review Letter'. In 1998 two of the authors were honoured by the award of a Nobel Prize while the third name was omitted—that of the MBE grower, Art Gossard. Needless to say, no reason has ever been given but one is left with the uneasy feeling that it was because Gossard '*only* provided the crystals'—if true, a shameful lack of appreciation of the vital contribution made by crystal growers to a wide range of condensed matter physics and the device engineering which depends upon it. We therefore dedicate our book to all those other crystal growers who 'only provide the crystals'—whether they do so using Czochralski, Bridgman, MBE, MOVPE, LPE, VPE or whatever other esoteric technique. May they not only continue to flourish but may their essential contribution to the advancement of solid state physics, surface chemistry, materials science, device innovation, etc. be adequately recognised for the vital role which it clearly plays.

Nottingham
February 2015

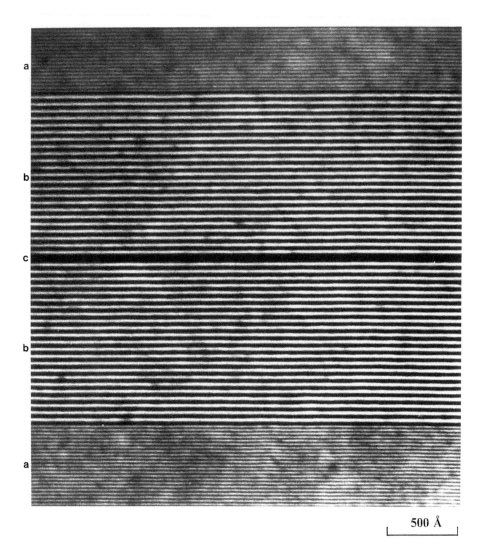

500 Å

Cross-sectional TEM micrograph of the GaAs quantum well and GaAs and AlAs superlattices (SL) in an all SL laser diode. At (a), (b) and (c) the layer thickness in monolayers (2.8Å) and period number are $(4+4)\times663$, $(8+8)\times23$ and $(20)\times1$.

This TEM micrograph taken by Jon Gowers is an excellent example of what a commercial MBE machine can achieve in growing superlattices composed of ultra-thin layers of AlGaAs and GaAs. The two distinct layers are 4 and 8 monolayers thick, respectively. The structure was grown at Philips Research Laboratories during the 1980s as part of a programme to develop semiconductor lasers.

Contents

1

Setting the Scene

1.1 Importance of Single Crystal Growth

Molecular beam epitaxy was developed as a technique for growing single crystal films of semiconductor materials during the 1960s and 1970s and has gone from strength to strength during the past 40 years and more. Its particular virtues can be seen to be its ability to produce well-controlled, extremely thin films (down to atomic or molecular dimensions), together with precise monitoring of thickness and structural characteristics by way of in situ measuring facilities. By far, its greatest success has been in relation to the so-called low-dimensional structures (LDS), which took semiconductor physics and technology by storm during the 1970s and 1980s, and it has gone on to make a major contribution to both physics and device technology right up to the present day. We shall discuss these aspects in detail later—for the moment, it will be sufficient to set MBE in the general context of semiconductor technology as it existed during the 1970s and, for this, we need to examine a little semiconductor history. This chapter, therefore, will provide a brief overview of the state of semiconductor material technology as it existed during the early years of the 1970s, when MBE was in process of taking its first youthful steps in the world of solid state electronics.

The importance of using single crystal samples in semiconductor measurements came to be appreciated surprisingly early (see Orton 2004 and Orton 2009 for more detailed accounts). Michael Faraday's pioneering experiments in 1833 on 'sulphuret of silver' (silver sulphide, Ag_2S) may have been performed on fused powder samples but, when Carl Ferdinand Braun discovered the phenomenon of rectification in 1874 while working at the University of Wurzburg, he was making use of naturally occurring crystals such as that of galena (lead sulphide, PbS) to study the electrical characteristics of the contact between a semiconductor and a metal. What is more, this was a deliberate choice, based on his earlier unsatisfactory experiments with fused samples of various salts (Kurylo and Susskind 1981). Likewise, when the one-time assistant to Marconi, the American-based Englishman Henry Joseph Round first reported observing electroluminescence in 1907, he too was working with natural crystals, in this case, of silicon carbide (SiC) (Orton 2009, p. 196). Again, while many early studies of photoconductivity in semiconductors were based on powder or fused samples, a significant number also made use of 'geological' single crystal specimens (Bube 1960). Perhaps the first use of artificial single crystals (during the 1920s) was concerned with the alkali halides, on the grounds that they were easy to prepare as large crystals from aqueous solution (Pohl 1937). Though not strictly semiconductors, these materials certainly helped to bring home the need for well-controlled samples if their solid state properties were to be properly understood.

The modern history of semiconductors began with the application of quantum theory by A. H. Wilson in 1931 (Wilson 1931a, b) to the understanding of their electrical properties and was consolidated during the Second World War by the application of silicon and germanium in making reliable 'cat's whisker' detectors for microwave radar systems. It was here that the combination of single crystal samples with careful resistivity and Hall-effect measurements first allowed a satisfactory theoretical understanding to emerge. Prior to this, semiconductor physics had acquired a singularly doubtful image, with many well-known figures openly deriding it as a danger to scientific reputations on account of its propensity for throwing up ill-understood (and apparently random) experimental data. It was only with the recognition that well-controlled samples were essential for reproducible experimental results that such an ill-favoured response could be adequately countered and it was significant that the breakthrough should occur in the United States, rather than in Europe. The history of scientific progress in Europe had been dominated by the love of science for science's sake, creating a strong emphasis on pure research and contrasted markedly with the American tendency towards empiricism and the search for practical outcomes (Thomas Edison being, perhaps, the prime practitioner). In the instance of semiconductor research, however, it turned out that progress in pure research depended strongly on the availability of both structurally perfect and chemically pure samples—in other words, on semiconductor technology—and this lay closer to the nature of American science.

Wilson's quantum approach to semiconductor physics had shown the essential importance of the valence and conduction bands in determining electrical conductivity and he had also predicted the behaviour of donor and acceptor impurity atoms in providing n- and p-type conductivity. This made clear the distinction between intrinsic and extrinsic conduction, together with their very different dependences on temperature. Thus, the basis for a proper understanding of electrical behaviour was available in 1931 but practical realisation was held up by the lack of adequate samples. Most of the semiconductor crystals available to pre-war experimentalists were far too impure to reveal distinct extrinsic and intrinsic behaviour, a feature well illustrated by Wilson's classification of silicon as a metal, even in notionally 'undoped' samples, free electron densities being so high as to force the Fermi level well into the silicon conduction band. It was ironic that the resolution of this apparent impasse should emerge from the very practical, urgent and, initially, largely empirical task of developing cat's whisker detectors for microwave radar systems. Various semiconductor materials were tried but the elemental semiconductors silicon and germanium showed most promise and attention was therefore focussed on attempts to control their respective electrical properties. The work had been begun in England but, following the famous Tizard visit to the United States in September 1940, it was taken up with greater intensity by MIT, Purdue University, the University of Pennsylvania and by Bell Telephone Laboratories. Bell, in particular, had already acquired a degree of competence in dealing with the vicissitudes of silicon but sample preparation was still in a very primitive state—nominally pure silicon was melted in a suitable boat and cooled slowly to produce a polycrystalline mass from which small pseudo-crystals could be disentangled. 'Pure' in those days meant 'chemically pure' rather than 'semiconductor pure' as we have come to recognise it today. Impurities might be present at the parts-per-million level, rather than the parts-per-billion or better required for modern integrated circuit manufacture. The Bell scientists were certainly aware of the importance of purity and had taken to using a vacuum furnace to minimise the take-up of impurities from the atmosphere (Riordan and Hoddeson 1997) but there were still problems inherent in the inevitable contact between their molten silicon and its container.

Progress in improving material quality came first in the guise of germanium detectors, being driven particularly by the goal of achieving the so-called high back-voltage devices. Initially, many detector crystals were found to burn out on being confronted with no more than a small fraction of the outgoing radar pulse, and considerable effort at Purdue was concentrated on trying to solve this problem (Lark-Horowitz 1954). It soon became clear that there was a need for diodes with large reverse breakdown voltages, obtained from germanium samples with particularly low impurity levels, an observation consistent with the Schottky barrier model of a metal–semiconductor contact. Note that the barrier model had been proposed only just prior to the start of the war (Rhoderick 1978), and the Purdue team, under Karl Lark-Horowitz, were probably first to apply it to a practical problem. The crucial skill was, of course, that of producing germanium with the necessary degree of purity and this required some concerted effort to improve the conditions under which samples were prepared—choice of material for the boat, control of the immediate environment, etc. Eventually, samples were grown of sufficient purity that they might then be back-doped to obtain the desired performance, a success which was soon repeated at Pennsylvania University by Frederick Seitz, with silicon diodes. Bell Labs were given responsibility for turning this basic research into production devices, and the skills they acquired were to prove of tremendous importance for the post-war development of the transistor. However, for our immediate purpose, it is perhaps more important that these high-quality semiconductor samples allowed, for the first time, unequivocal confirmation (by the use of Hall-effect measurements over a wide temperature range) of both intrinsic and extrinsic conductivity. For the future of semiconductor physics, this probably ranked in importance with the practical boost given to wartime radar development. Here was proof positive—silicon was no longer a metal.

Purity was clearly of fundamental importance but the long-term future of semiconductors, both scientifically and commercially, was to depend upon the development of appropriate crystal growth methods. The 'invention' of the transistor may have relied solely on samples of germanium gleaned by Walter Brattain from the Bell Labs wartime activity in radar development but it soon became clear to those responsible for its commercialisation that future progress demanded large single crystals. The grain boundaries present in even the best-quality material available to Brattain were always going to be fatal to minority holes, or electrons diffusing across the base region. The breakthrough came in somewhat bizarre fashion due to the sheer, dogged determination of Gordon Teal to defy his managers and apply the Czochralski method (already well established for the growth of metal crystals), first of all, to germanium and, later, to silicon (Teal and Little 1950; Teal and Buehler 1952). Not only did his success result in structurally superior material but also in much improved purity, a consequence of favourable impurity segregation between melt and solid phases. Minority carrier lifetimes in germanium were immediately measured to be some ten times longer than those appropriate to the high back-voltage material used by Brattain, an improvement which finally sold Teal's radical approach to those around him. Even William Shockley was impressed (Riordan and Hoddeson 1997, p. 179).

Molten germanium was contained in a crucible made from high-purity graphite, which allowed the use of induction heating, a seed crystal was dipped into the melt and slowly withdrawn, being rotated to improve radial uniformity of the grown crystal while a clean environment was maintained with the use of either flowing hydrogen or helium (see Figure 1.1). In the case of silicon, which tends to react strongly with graphite, a pure silica crucible replaced the graphite (though it could be placed within a graphite liner in the interest of induction heating). The relatively high-purity material produced allowed Teal and his collaborators to dope the growing crystal by adding suitable donor or acceptor species to the melt, a technique which later led to the demonstration of the first junction transistor. An n-type germanium crystal was first doped with gallium to form a p-type base region and then counter-doped with antimony to produce the n-type collector. The resulting transistors were relatively slow in operation on account of their rather large base widths but it was of crucial importance to prove the principal of transistor action in such structures—it spelled the demise of the point-contact device and set the solid state electronics industry on course for its future dominance.

Silicon, with its higher melting temperature and greater chemical reactivity, took rather longer to tame but the grown-junction silicon transistor eventually emerged in 1954. Of greater long-term significance, perhaps, was the development of the zone refining technique by William Pfann in 1952 (Pfann 1957). In the case of germanium, this involved the simultaneous passage of several molten zones along a germanium crystal contained in a horizontal boat (Figure 1.2), impurity atoms being swept to one end of the sample, whence they could be removed by the simple act of cutting the end off. Once again, silicon proved less co-operative, because of its tendency to interact with the silica boat, until its recalcitrance was brilliantly overcome by the introduction of the floating zone technique. A vertically mounted silicon crystal was clamped at both ends while a single molten zone, held in place by surface tension, was passed along its length. Several passes may have been necessary to obtain ultimate purity but the fact that the crystal no longer required any container had removed what originally seemed like a fundamental limit to crystal quality.

By the middle of the 1950s, then, both silicon and germanium were available, with quality far surpassing anything dreamt of prior to the Second World War, and silicon was already well on the way to becoming the material par excellence for transistor making—the first silicon bipolar devices having been produced at Texas Instruments (once again by the energetic Gordon Teal) in 1954. The year 1959 saw the emergence of the first integrated circuits at Texas and at Fairchild—silicon, once again, rapidly taking over as the material of choice. Particularly significant was the fact that both transistor and integrated

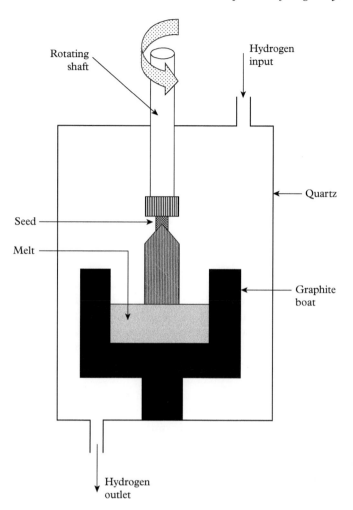

Rotating
shaft

Hydrogen
input

Quartz

Seed

Melt

Graphite
boat

Hydrogen
outlet

Figure 1.1 *Schematic drawing of a typical Czochralski crystal puller for growing bulk single crystals of germanium. The germanium is melted by RF induction heating of the graphite crucible, a seed crystal is dipped into the melt and slowly withdrawn while being rotated to improve radial uniformity. The diameter of the pulled crystal depends on the rate of withdrawal. A continuous flow of hydrogen prevents any take-up of impurities from the ambient. (From Orton 2004.)*

circuit manufacture were based on bulk single crystal material, purity and structural perfection being more than adequate for the purpose. In fact, from the strictly materials viewpoint, the important follow-up to this early success lay simply in the development of ever larger crystal boules. From the original 2-inch diameter samples, the technology has now advanced to the point of growing 12-inch monster crystals weighing as much

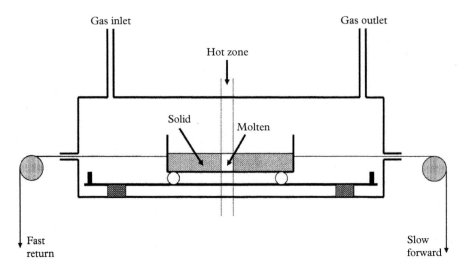

Figure 1.2 *Schematic diagram of a typical zone refiner for the purification of silicon or germanium. A molten zone is moved along the semiconductor sample from end to end, sweeping impurities with it. The resulting high-concentration end of the crystal is then removed. Oxidation of the material is prevented by flowing an inert gas through the quartz tube, surrounding the sample. In practice, several zones are swept along the sample simultaneously to speed up the process. (From Orton 2004.)*

as 50 kg while maintaining the same standards of purity and structural perfection that made possible the initial breakthrough. Finally, while the integrated circuit may dominate most people's perception of semiconductor electronics, we should not overlook the importance of the physically very much larger power devices which control so much of our electric power-generation and utilisation. To mention only a few such applications: from the small motor powering a typical kitchen gadget to the megawatt motors driving modern electric trains, speed control is effected by some form of semiconductor switch, the output from every wind turbine is matched to the electricity grid by similar devices while the efficient functioning of the average automobile also depends critically on power electronics. In many instances, these devices demand even greater purity and uniformity than those required by the integrated circuit industry, background impurity levels of 10^{18} m^{-3} (approximately 1 part in 10^{11}) and doping uniformity better than 1% across a whole wafer setting the ultimate standard. Nor should we forget the ever-burgeoning use of silicon in the manufacture of photovoltaic solar cells.

While this tremendous improvement in semiconductor materials quality was obviously stimulated by the rapidly developing electronic industry, it was also of enormous importance for the parallel development of semiconductor physics, the availability of high-quality, well-controlled samples being a prerequisite for obtaining high-quality, well-controlled semiconductor data. In its turn, reliable data furthered a concomitant growth in the understanding of semiconductor physics, a truism which we shall continue to meet at various stages of our studies. We have already referred to the use of resistivity and Hall-effect measurements over a wide temperature range as a means for studying extrinsic and intrinsic conductivity. Some of the best and most reliable of such data were obtained on samples of germanium and silicon (see e.g. Putley 1960), providing detailed information concerning electron and hole mobilities, invaluable in fostering theoretical understanding of free carrier properties. Cyclotron resonance measurements and studies of magnetoresistance provided accurate values of the anisotropic effective masses in these materials, confirming theoretical ideas of their band structure and of their optical properties. Studies of low-temperature photoluminescence provided new understanding of free and bound exciton properties (Haynes 1960). High-purity silicon, containing a low level of phosphorus-doping atoms, was used to demonstrate population inversion at microwave frequencies in the region of 10 GHz during the development of the microwave maser in the early 1960s (Orton 2004, p. 140) and, of far greater long-term interest, the quantum Hall effect was observed during measurements on silicon MOS structures (von Klitzing, Dorda and Pepper 1980).

In strictly commercial terms, of course, silicon came to dominate mainstream semiconductor electronics but demand for a multitude of special functions such as those of laser action, visible light emission, infrared photodetection, microwave generation and amplification, photovoltaic generation, photoelectric emission, etc., led to a host of other semiconductor materials being called into play, with need for similar material quality. As early as the mid-1950s, development work was in full flow in the case of several III-V compounds (GaAs, GaP, InP, InSb, AlSb), a group of IV-VI compounds (PbS, PbSe, PbTe) and a clutch of II-VI compounds (CdS, CdSe, CdTe, ZnS, ZnSe, ZnTe). While much of this early work was driven by scientific curiosity (see, for example the pioneering III-V work of Heinrich Welker in 1952—reviewed by him in Welker 1976), rather than by market forces, very soon there was growing enthusiasm for the commercial possibilities of gallium arsenide as an alternative to silicon in high-speed bipolar transistors, a hope soon to be dashed by the realisation that minority carrier lifetimes in GaAs were to be measured in nanoseconds, rather than microseconds. Nevertheless, material research was undertaken in a number of government and industrial laboratories in an effort to improve crystal quality and, in spite of considerable scepticism from various sources, this was to pay off in a number of unexpected directions. During the early 1960s came the GaAs light-emitting diode, the semiconductor laser and the microwave Gunn diode. Towards the end of the decade came the GaAs MESFET, with operating frequencies approaching 100 GHz. The sceptics may have been right in one sense but the era of the non-silicon niche market had arrived—and it brought with it a tremendous stimulus to semiconductor material development.

It was natural, of course, for crystal growers to apply similar methods to the growth of GaAs as had recently been applied successfully to silicon, and the use of the Czochralski method was demonstrated as early as 1955 by Gremmelmaier in Germany. On account of the high vapour pressure (approximately 1 atm) of arsenic at the GaAs melting point (1238 °C), he employed a sealed tube to prevent loss of arsenic from the growing crystal, together with magnetic coupling to provide the necessary pulling and rotational movements. A somewhat simpler alternative melt-growth technique used in the early days of GaAs development was the horizontal Bridgman method (see Figure 1.3) in

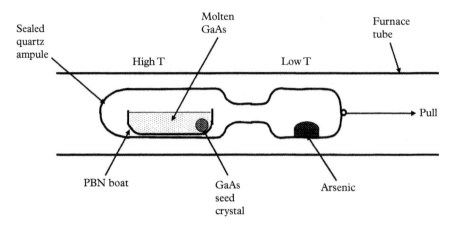

Figure 1.3 *Schematic diagram of a typical horizontal Bridgman apparatus used for the growth of bulk GaAs crystals. GaAs is melted in a PBN crucible contained within a sealed quartz tube, which is slowly withdrawn from the furnace so as to cool the GaAs from the seed end. A separate section of the ampoule holds a charge of arsenic, kept at a constant temperature of 617 °C to maintain a constant pressure of As vapour over the GaAs. (From Orton 2004.)*

which an arsenic vapour pressure was maintained over the growing crystal by means of a separate arsenic source kept at a constant temperature of 617 °C. The boat in this case was made from pyrolytic boron nitride (PBN), rather than silica (which tended to dope the GaAs n-type with silicon donors). Later developments saw the introduction of the boric oxide liquid encapsulation (LEC) method of Czochralski growth at RSRE Malvern in 1965 and the vertical Bridgman method in the 1980s. Structural quality in the resulting crystals was generally found acceptable but there were serious difficulties with regard to purity and (during the 1960s and early 1970s) this led to a major new development, in the shape of epitaxial growth. Bulk crystals, grown by either Bridgman or Czochralski methods, were used as substrates while the active device structures were deposited epitaxially—only thus was it possible to produce material with adequately low background doping levels or with the desired control of uniformity. Many different approaches have been adopted but they all have one feature in common, namely, much lower growth temperature (typically ranging from 500 °C to 800 °C) than the 1238 °C appropriate to bulk growth. There can be little doubt that this held a major influence over material purity and made possible the success of the GaAs devices referred to above, not to mention a wide range of devices based on other semiconductor material systems. What is more, it can also be seen to have had a similar influence on the development of semiconductor physics, not least the emergence of low-dimensional structures (LDS) during the 1980s. We are now in a position to look at epitaxial methods as a whole in order to set our particular topic of MBE in context.

1.2 VPE (Vapour Phase Epitaxy)

The word 'epitaxy' comes from the Greek, meaning 'arranged upon', and is usually taken to imply a single crystal film oriented according to the crystal plane of the substrate. This concept is straightforward when applied to 'homoepitaxy', for example, a film of GaAs grown on a GaAs substrate, but becomes somewhat more complex in the case of 'heteroepitaxy', where film and substrate differ from one another. Crucial parameters in this latter case are the differences between crystal structure, lattice parameter and thermal expansion coefficient of the two constituents, leading to possible strain or structural defects in the grown film. We shall say no more of this for the present, though it will figure large in our later detailed accounts of specific material systems. For the moment, we should merely note that heteroepitaxy made an early entrance on the semiconductor scene (*c.*1960) in the shape of films of GaAs grown on germanium substrates (their respective lattice parameters being closely similar), germanium films grown on GaAs substrates and GaAsP films grown on GaAs substrates.

VPE of semiconductor films originated within the context of increasing sophistication in the semiconductor device field. Following the introduction of the junction transistor at Bell Labs, during the 1950s, various technologies were explored in the hope of simplifying its manufacture and improving control over the base width (in the interest of high frequency operation). Alloy junctions were superseded by diffusion, leading to the all-important planar technology, which rapidly dominated the development of integrated circuits during the 1960s. All this was based on bulk germanium or silicon crystals. However, those scientists responsible for detailed improvement in device design saw a difficulty—it was not possible, by diffusion, to produce a lightly doped layer on a heavily doped substrate and this led, at IBM and at Bell Labs (see Grove 1967) to the deposition of lightly doped films of both germanium and silicon using the vapour phase reduction of germanium or silicon tetrachlorides by hydrogen; for example:

$$SiCl_4 + 2H_2 = Si + 4HCl \qquad (1.1)$$

Appropriate reactors are illustrated in Figure 1.4. Growth temperatures (in the case of silicon) were typically well above 1000 °C, and growth rates were about 1 μm per minute, appropriate to the production of films with thicknesses of 10 μm or more. These rather high growth temperatures were primarily a consequence of complications introduced by the presence of small amounts of water vapour in the hydrogen, which led to the formation of stacking faults but also to the need to make silicon surface atoms mobile enough to find appropriate lattice sites. Lowering the deposition temperature resulted in increasingly defective films. We shall see the importance of such considerations in many forms of epitaxy, the necessary growth temperature being intimately related to the melting point of the particular material (1412 °C in the case of silicon), a measure of the strength of the Si–Si chemical bond. As we shall see later, silicon epitaxy could also be effected by the pyrolysis of silane (SiH_4) or by reduction of hydrochlorides, such as $SiHCl_3$.

Interestingly, vapour phase epitaxial growth of III-V compound semiconductors was reported at almost the same time as that of silicon. It, too, was based on the use of halides

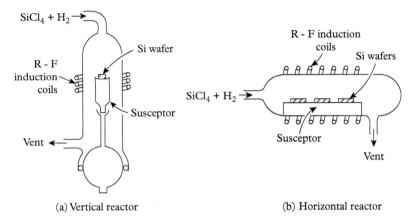

Figure 1.4 *Examples of vertical and horizontal reactors used for silicon epitaxy by the hydrogen-reduction of silicon tetrachloride. The substrate is mounted on a graphite susceptor to allow RF heating while keeping the quartz container relatively cool. (From Grove 1967.) Reproduced with permission of Wiley & Sons.*

but, because of the involvement of both Group III and Group V elements, the overall process was slightly more complicated. Indeed, a surprisingly large number of variants were studied before a definitive technique would emerge. Epitaxy can be traced back to earlier work involving vapour deposition of bulk crystals, in an attempt to grow material at much lower temperatures than was possible for melt growth. One of the first attempts was reported by Antell and Effer (1959) at the Metropolitan-Vickers Research Laboratory in Manchester, UK. They produced small crystals of InAs, InP, GaAs and GaP in an evacuated closed-tube system, using either iodides or chlorides of the appropriate metal. In fact, they described two slightly different approaches, but both based on the disproportionation reaction

$$3GaCl + 1/2As_4 = GaCl_3 + 2GaAs \tag{1.2}$$

In the first of these, the monochloride and arsenic were heated together at a uniform temperature and, as the temperature was lowered, the reaction moved from left to right, crystals of GaAs being deposited on the reaction tube. In the second process, the reaction tube was loaded with GaAs and $GaCl_3$ and placed in a furnace with a temperature gradient, the GaAs being at the hot end. Then, as the overall temperature was lowered, GaAs crystals were deposited at the cooler end, it being possible to transport all the original GaAs to this end of the tube. Similar reactions were proposed with In substituted for Ga, and P for As. Particularly noteworthy is the fact that a very similar method was used by Holonyak (Holonyak and Bevaqua 1962) to grow the GaAsP crystals with which he demonstrated laser action at visible wavelengths. He had originally acquired an interest in this material as a source of improved tunnel diode characteristics but made a rapid volte-face when the race to demonstrate the first semiconductor laser came alive in the early months of 1962. Incidentally, his work was of general interest in so far as it demonstrated, for the first time, that alloy materials behaved as 'proper' semiconductors, several doubts having been expressed that the random nature of an alloy might compromise its semiconducting behaviour. A similar approach was also adopted by Prior (1961) in growing small crystals of lead selenide, of interest as a far-infrared photodetector.

In this early work no attempt was made to grow material on a suitable crystalline substrate, though in 1961 Antell (now working for AEI Research) did try to grow GaP epitaxially on a seed crystal (Antell 1961). Unfortunately, he succeeded only in etching the seed away but the concept of VPE was clearly established—several other people also had the same idea. Within the next few years, epitaxial growth of GaAs and GaP was reported by workers at RCA (Somerville), Bell Labs, General Electric, Texas Instruments, Plessey, Philips and RCA (Princeton), the preponderance of industrial researchers emphasising just how important these III-V compounds were seen to be in the context of semiconductor electronics. While no one seriously pretended that these compounds would rival silicon as the electronic workhorse, it was already becoming clear that they might offer special properties not available to silicon, such as light emission, microwave generation and microwave amplification.

Important progress towards the development of a commercially practical growth system was reported from RCA by Newman and Goldsmith (1961). Rather than accepting the physical limitation of working with an evacuated closed-reaction tube, they chose an open-tube system, constantly flushed with pure hydrogen, in which GaAs feed material was transported from a hot zone ($c.1000$ °C) to a cooler zone ($c.900$ °C) containing lapped and polished GaAs seed crystals. The hot zone was loaded with crushed Bridgman GaAs crystals, together with 'five-nines' arsenic and the transport effected by way of a stream of HCl. The chemistry can be represented by the pair of equations:

$$3\text{GaAs} + 3\text{HCl} = 3\text{GaCl} + 3\text{As} + 3/2\text{H}_2 \tag{1.3}$$

$$3\text{GaCl} + 2\text{As} = 2\text{GaAs} + \text{GaCl}_3 \tag{1.4}$$

As can be seen, the first reaction uses GaAs and HCl to form the monochloride of gallium within the system, rather than supplying it from an external source, while the second is the disproportionation reaction referred to above. Thus, Newman and Goldsmith were able to study the effects of substrate orientation and substrate preparation, together with some electrical measurements which indicated the presence of p-type impurities in notionally undoped samples. They also demonstrated the possibility of n-type doping with tellurium and tin, thus allowing them to make epitaxially grown p–n junctions showing well-behaved current–voltage characteristics. Although ultimate purity still left something to be desired, VPE was already showing promise in the struggle to obtain 'silicon quality' III-V materials.

Research in other laboratories soon showed that smooth epitaxial films could be grown at significantly lower temperatures, typically between 700 °C and 800 °C, but the growth of high-purity films was dependent on the nature of the source material. The use of GaAs was always going to be problematic in so far as Bridgman crystals were known to be relatively impure and there was little likelihood of any significant improvement in this respect—after all, the chief motivation for the growth of epitaxial films was to circumvent this very weakness in bulk material. The key to achieving high purity lay with the choice of starting materials and this pointed towards the use of elemental gallium (available to five-nines or even better purity), rather than the GaAs with which most of the earlier work had been performed. Ing and Minden (1962), at GE Syracuse, were amongst the first to follow this approach. They used the open-tube method to grow GaAs from gallium, arsenic and hydrogen chloride and obtained n-type films with free carrier densities as low as 10^{23} m^{-3} and electron mobilities approaching 0.3 m^2 V^{-1} s^{-1}. (Note that the theoretical mobility for lightly doped n-type GaAs approximates to 0.9 m^2 V^{-1} s^{-1}.) This was undoubtedly a step in the right direction, though even better results could be obtained by using arsenic trichloride as the source of arsenic, rather than elemental arsenic. This could be supplied to the reaction tube by bubbling pure hydrogen through AsCl$_3$ liquid at room temperature, as demonstrated by Finch and Mehal (1964) in the growth of GaAsP films. Thus was a consensus reached as to the optimum method of growing high-purity GaAs films.

A typical reactor, employing a three-zone furnace, is shown schematically in Figure 1.5. The gallium source is located in the second zone at a temperature of 800 °C while the GaAs substrate is placed further downstream at a temperature of 750 °C. In the first zone, at a temperature of about 550 °C, the AsCl$_3$ reacts with hydrogen according to

$$2\text{AsCl}_3 + 3\text{H}_2 = 6\text{HCl} + 1/2\text{As}_4 \qquad (1.5)$$

The arsenic and HCl thus produced flow over the gallium boat to form a skin of GaAs which is then transported according to

$$4\text{GaAs} + 4\text{HCl} = 4\text{GaCl} + \text{As}_4 + 2\text{H}_2 \qquad (1.6)$$

Finally, in the third zone GaAs is deposited epitaxially on the substrate following the reaction

$$6\text{GaCl} + \text{As}_4 = 4\text{GaAs} + 2\text{GaCl}_3 \qquad (1.7)$$

GaCl$_3$ is deposited further downstream on the cooler parts of the apparatus. (Note that all these reactions are intended as a rough guide and should not be taken too literally.)

Using this approach, with palladium-diffused hydrogen, pure gallium and AsCl$_3$, Knight, Effer and co-workers at the Plessey, Caswell, laboratory were first to report electron mobilities approaching the theoretical limit (Effer 1965; Knight et al. 1965). They confirmed the advantage to be derived from the supply of arsenic as AsCl$_3$, rather than in the elemental form—free electron densities were close to 10^{21} m^{-3} in the former case and 10^{22} m^{-3} in the latter. Even though the gallium and AsCl$_3$ sources were of five-nines purity, they further purified the AsCl$_3$ by a gradient-freeze method while, at the same time, demonstrating the use of vapour-etched substrates to avoid the incorporation of impurity atoms from the chemically cleaned surface. This was achieved by

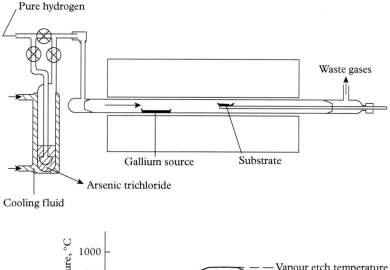

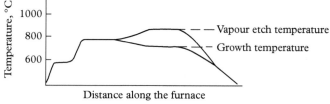

Figure 1.5 *Apparatus for the growth of pure epitaxial layers of GaAs by the arsenic trichloride VPE method. Pure hydrogen is bubbled through liquid $AsCl_3$ before passing over a boat containing liquid gallium at a temperature of 800 °C. A polished GaAs substrate is held further downstream at a temperature of 750 °C. Reproduced by permission of The Electrochemical Society.*

reversing the temperature gradient in the reactor, thus raising the substrate temperature to 900 °C. Then, lowering this to 750 °C allowed growth on the newly etched surface. Such careful attention to detail resulted in films with room temperature mobilities as high as 0.88 m^2 V^{-1} s^{-1} and a mobility at 77 K of 3.8 m^2 V^{-1} s^{-1}, the highest value recorded at that time. (In fact, this represents material with a total impurity concentration of 10^{22} m^{-3} and a compensation ratio N_A:N_D of about 0.8—see e.g. Blood and Orton 1992, p. 103). Experiments using both selenium and tin dopants resulted in n-type doping levels of up to 3×10^{24} m^{-3}, illustrating the versatility of this growth method for future device applications. At that time, Plessey workers were interested in growing material for high-voltage diodes and for FETs, both of which were to come to fruition, but the subsequent impact of the Gunn diode was to introduce yet another demand on crystal growers, as we shall see in a moment. However, before proceeding with this, we should mention one further perturbation to the progress of VPE.

Just one year later, in 1966, Tietjen and Amick (1966), at RCA, reported an alternative approach to the VPE growth of GaAs and GaP. (In fact, they grew samples of GaAs$_{1-x}$P$_x$ covering the range x = 0 to x = 0.7.) Instead of using AsCl$_3$ (or

PCl_3) to supply arsenic (or phosphorus), they made use of the corresponding hydrides, AsH_3 or PH_3, gallium being transported by passing HCl over a heated gallium boat. The hydrides were obtained from commercial sources, with no further purification, yet the resulting GaAs films showed characteristics superior even to those obtained by the Plessey workers, the lowest free electron concentration being about $3 \times 10^{20} \, m^{-3}$ with a 77 K mobility just under $6.0 \, m^2 \, V^{-1} \, s^{-1}$. It thus became apparent that either chloride or hydride VPE could be employed to grow high-quality III-V compounds, and these techniques were to be put to a severe test by the requirements of several device applications.

In a book on epitaxy it is clearly impracticable to describe each and every application for the resulting epitaxial films. Instead, we shall simply refer to a number of typical examples and, in the VPE context, it is convenient to concentrate attention on the microwave Gunn diode, which challenged the practitioners of VPE growth in the mid-1960s. Gunn (1963) discovered microwave oscillations in 1963 as a by-product of his studies of high-speed effects in several GaAs and InP samples. These were very thin bulk samples with electrical contacts on either surface, and the oscillations appeared when a modest DC voltage was applied between them. Gunn noted that the oscillation frequency varied inversely with sample thickness but was unable to explain the origin of the effect. Nevertheless, in practical terms, such a simple device created considerable excitement—current microwave generators were large, expensive, vacuum-tube devices requiring high-voltage power supplies—and research into the new effect was immediately taken up in several laboratories. Surprisingly, it took two years before the controversy concerning the origin of the effect was settled in favour of the 'transferred electron effect'.

The transferred electron effect had been predicted theoretically by Ridley and Watkins (1961) and by Hilsum (1962) as a method of obtaining a negative resistance in a microwave circuit, thus making possible the development of a high frequency oscillator (for a simplified account, see Orton 1971). Free electrons in the gamma conduction band minimum, accelerated by an applied field, may gain sufficient energy to transfer into adjacent X minima (see Figure 1.6), where they abruptly slow down because the electron effective mass in X minima is much larger than in gamma. Thus, an increase in applied field (or applied voltage) results in a *decrease* of electron velocity (or current), implying a negative differential resistance. Under such circumstances, Ridley then showed that a uniform electric field distribution would no longer be stable, resulting in a narrow high-field region (a high-field 'domain') at one end of the sample, with a low-field region over the rest of the sample. The domain would drift through the sample, disappear at the anode end, then reform at the cathode, and this repetitive process would result in an oscillatory current in an external circuit, the oscillation frequency being determined by the time taken for the domain to drift between cathode and anode. In GaAs the domain drifts at a velocity $v_D = 10^5 \, ms^{-1}$, which implies that, for operation at 10 GHz, the sample length should be 10 μm. So much for sample length—can we say anything about the required doping level? This is related to the time taken for the domain to form, which depends on the sample resistivity and, therefore, on doping level. For optimum efficiency in converting DC input power into microwave power, it can be shown that

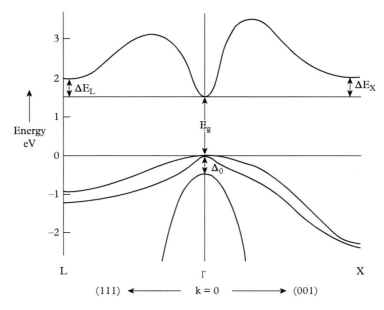

Figure 1.6 *Band structure of GaAs, showing the relative positions of the Γ, X and L conduction band minima. Electrons in the central minimum are characterised by a small effective mass whereas those in the X and L minima have much larger masses, so electrons transferring from Γ to X or L are slowed down significantly, resulting in a negative differential resistance. (From Orton 2004.)*

this formation time should be approximately one-third of the transit time and, finally, we arrive at an admirably simple criterion that can be expressed as

$$n_0 l = 10^{16} \, \text{m}^{-2} \qquad (1.8)$$

where n_0 is the required doping level. For the case of our 10 GHz oscillator, this implies a value of $n_0 = 10^{21} \, \text{m}^{-3}$ and this, the reader will recall, is close to the lowest value achieved by VPE under optimum growth conditions. It was clearly going to present a challenge to crystal growers when asked to produce uniformly doped epilayers with controlled thickness on a reproducible basis. However, assuming this could be done successfully, the Gunn device was simplicity itself, consisting of a low-doped epilayer grown on an n^+ substrate (the back contact) while a top metal contact was defined photolithographically. A few volts DC applied between the two contacts would stimulate the (almost magical!) generation of microwave power in a suitable microwave circuit, such as a small copper resonator. A new era in microwave technology was clearly in view—but, as ever, the crystal grower was once again in the hot seat. Could he deliver the goods?

In principle, the task sounded relatively straightforward, once it had been established that material of the required n-type doping could be produced and, as the initial thrust was for diodes to operate at a frequency of 10 GHz (X-band), this meant a free electron concentration of 10^{21} m^{-3}. The requirement was therefore for a 10 μm thick, uniformly doped n⁻ layer grown on an n⁺ substrate and it was necessary to control both doping level and layer thickness to an accuracy of a few per cent. The first decision concerned the substrate—it turned out that silicon was the preferred dopant because the other serious contender, tellurium, was a fast diffuser and prone to contaminate the n⁻ layer during growth. This choice, however, brought with it its own difficulty on account of the amphoteric doping behaviour of silicon in the III-V compounds—n-type when substituting for the metal atom, p-type for the Group V atom. It then became critically important to avoid the incorporation of arsenic vacancies in the region of the substrate-layer interface, because they tended to force silicon to occupy arsenic sites and produce a heavily compensated region, which came to be widely known as the 'interface dip' (i.e. a dip in the doping profile). It represented a serious problem because it produced a high-resistance region near the anode which prevented the field at the cathode rising to the necessary threshold value. Only a sensibly flat doping profile could be accepted by the device engineer and this caused much heart-searching amongst the crystal-growing fraternity.

No doubt, several proprietary techniques were employed, around the world, to minimise the problem but we can describe with confidence the one developed in our own (Philips) laboratory. It should be clear that a serious difficulty was presented by the important vapour-etching part of the growth sequence. In dropping the substrate temperature from its etching value of 900 °C to the growth temperature of 750 °C, it was extremely difficult to maintain the necessary stoichiometric balance, so the Philips crystal growers adopted an alternative procedure (Easton et al. 1971). Etching of the substrate was performed at the growth temperature by using a stream of AsCl$_3$ which had not been passed over the gallium boat, then switching to a stream which had reacted with gallium to initiate growth of the active layer. Care was also necessary, while heating the substrate to the growth temperature, to do so in an atmosphere of excess arsenic. It soon became clear that commercial success would be critically dependent on adequate control of the timing of the required sequence of operations and this led to the design of an automatically programmed equipment, which did, indeed, allow the growth of reproducible device structures.

In summary, the successful development of Gunn diodes, GaAs FETs and GaAs detector diodes, made possible by the mastery of vapour phase epitaxy, effected a dramatic transformation of microwave circuitry and allowed microwave techniques to offer solutions to a wide range of new problems. With this in mind, therefore, it might be well to emphasise the importance of VPE in providing n-type epitaxial layers, grown on semi-insulating substrates, for the development of microwave MESFETs. Initially, the requirement was for layers of approximately 5 μm thickness, doped at 10^{21} m^{-3} but as gate lengths were reduced, in the interest of higher operating frequencies, thicknesses came down to 1 μm or less, with doping levels up to 10^{23} m^{-3}. The accurate control of thickness and doping level at these levels was essential to the successful development of microwave integrated circuits towards the end of the 1970s.

This may sound like an unqualified vote of confidence in VPE, and, indeed, there can be no doubting its success but, before leaving the topic, we must note the introduction of a modified version of the technique that has proved of tremendous importance in its longer-term development. At the end of the 1960s, Manasevit, at North American Rockwell Corporation, reported his innovative work on 'The Use of Metal-Organics in the Preparation of Semiconductor Materials' (Manasevit and Simpson 1969). This paper described the use of triethyl gallium, $Ga(C_2H_5)_3$, and trimethyl gallium, $Ga(CH_3)_3$, in the growth of GaAs, GaP, GaAsP and GaAsSb epitaxial films on GaAs and a number of insulating substrates, the Group V species being provided by an excess flow of the appropriate hydride. The equipment used is shown schematically in Figure 1.7. A particular advantage of MOVPE was the fact that it required only one temperature zone, rather than the two or more employed in earlier VPE systems. It also turned out that growth rate was independent of this temperature over the range 600 °C–800 °C, which meant that there was no need for accurate temperature control. Furthermore, while in conventional VPE the balance between deposition and etching depended critically on the temperature of the substrate, in MOVPE the etching reaction occurred at a negligible rate, making the process very much easier to control. It was also a straightforward matter to dope the films either n- or p-type using, on the one hand hydrogen sulphide

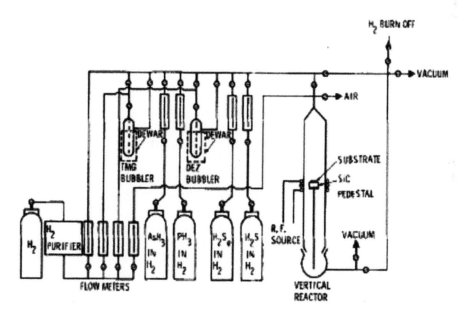

Figure 1.7 *Outline of the MOVPE apparatus used by Manasavit and Simpson to grow epitaxial layers of GaAs and GaP. Arsenic and phosphorus are supplied as hydrides, and gallium as trimethyl gallium. Dopants take the form of hydrogen sulphide or hydrogen selenide (n-type) or diethyl zinc (p-type). (From Manasevit and Simpson 1969.) Reproduced by permission of The Electrochemical Society.*

or selenide, and, on the other, diethyl zinc. Manasevit and Simpson succeeded in producing free electron densities up to 6×10^{24} m^{-3} and hole densities up to 3×10^{25} m^{-3}, their background doping levels being in the mid-10^{22} m^{-3} range.

The metal-organic compounds are liquids at room temperature and were transported to the reaction chamber by bubbling pure hydrogen gas through sealed containers of the appropriate liquid, and the gas mixed with the necessary hydride, before being presented to the substrate, mounted on a silicon carbide-lined graphite pedestal, heated by an RF source to a temperature of 700 °C. This method of heating ensured that the reaction chamber, itself, remained relatively cool, thus minimising any possible impurity incorporation. Growth rates were directly proportional to the flow rate of the metal-organic and lay in the range between 0.2 and 0.8 μm/min, considerably lower than used in conventional VPE, making this method more suitable for the growth of very thin films, a matter of vital importance, as we shall see later, for the growth of low-dimensional structures. Also of great significance was the fact that, when (in the 1980s) a second source was added, containing trimethyl aluminium, it was possible to grow films of AlGaAs with excellent control over the Ga/Al ratio. GaAs lasers, solar cells and a multiplicity of low-dimensional structures were to benefit over the succeeding years.

Returning, however, to the overall state of epitaxy when MBE was first introduced, we should simply note that, by 1975, some moderately high-quality GaAs had already been grown by MOVPE. The Rockwell group reported room temperature mobilities of about 0.5 m^2 V^{-1} s^{-1}, with free electron densities in the low 10^{21} m^{-3} range, though they observed some slightly puzzling p-type behaviour in thin undoped films (Thorsen and Manasevit 1971). A group at NEC in Japan (Ito et al. 1973; Seki et al. 1975) then demonstrated the importance of the arsine/organometallic ratio in determining doping type, their explanation depending on the amphoteric behaviour of carbon or silicon impurities, which could be switched from p- to n-type by increasing this ratio. They obtained corroborative evidence from low-temperature photoluminescence data which indicated the presence of two acceptor levels in n-type material. In their second paper they reported electron mobilities a little over 10 m^2 V^{-1} s^{-1} at 77 K, still some way short of the material quality achieved by VPE and LPE at this time, but showing considerable improvement over earlier work. Clearly, MOVPE was already a very promising growth method for the III-V compounds and we shall meet it again in the following chapters as a serious rival to MBE.

1.3 LPE (Liquid Phase Epitaxy)

The urgent need for improved crystal quality in III-V materials was emphasised by the fact that, at more or less the same time as VPE was under development, the alternative technique of liquid phase epitaxy also appeared on the semiconductor scene. Solution growth of crystals had, of course, been well known for many years so it was hardly surprising that someone should apply it to the growth of GaAs, and, as so often seems to happen, two quite different approaches were proposed, more or less simultaneously. Of particular subtlety was the 'travelling solvent' method described by Mlavsky and

Weinstein of the Tyco Laboratory in Waltham, Mass, while a conceptually much simpler method was demonstrated by Nelson, of RCA, Princeton, who used it to grow films of both GaAs and germanium on GaAs substrates.

Mlavsky and Weinstein (1963) adopted an earlier suggestion by Pfann, the inventor of the zone melting technique for sample purification. Two flat GaAs samples were arranged to be separated by a thin film of molten gallium (saturated with GaAs) which could be passed through the upper GaAs crystal along a temperature gradient, thus recrystallising it, as GaAs was precipitated at the lower liquid–solid interface. The epilayers so formed, at growth temperatures of about 900 °C, showed significantly improved purity, compared to the starting materials, with n-type doping levels close to 10^{22} m^{-3} being obtained from bulk samples characterised by free carrier densities of 2×10^{23} m^{-3}. Room temperature electron mobilities as high as 0.56 m^2 V^{-1} s^{-1} were recorded, and well-behaved p–n junctions were also fabricated by the addition of suitable acceptor species to the gallium film prior to the growth stage. Comparison with concurrent results from VPE experiments suggested this to be a promising alternative, though, in the event, it appears not to have been further developed, leaving the field to Nelson and his subsequent rivals.

The so-called tipping method, described by Nelson (1963), is shown in Figure 1.8. This made use of a horizontal furnace tube, mounted on a gimbal which allowed a saturated solution of GaAs in molten tin to be tipped onto a GaAs substrate. The furnace temperature was then slowly reduced, causing GaAs to precipitate onto the substrate until, the required thickness of epitaxial film having been achieved, the melt was tipped back to its original position. The growth temperature was close to 600 °C. Nelson's use of tin as solvent was certainly successful in demonstrating the possibility for growing smooth epitaxial layers but the fact that tin acts as a donor in GaAs meant that the films were doped n-type at a level of about 5×10^{23} m^{-3}, implying a serious loss of flexibility in the process. However, he went on to perform similar experiments using gallium as solvent, which required a somewhat higher growth temperature (in the region of 800 °C), and he then showed that tellurium could be added to the melt in order to achieve n-type doping, or zinc for p-type. Films, typically about 100 μm thick, were used for studies of laser diodes, work which, interestingly, was reported in June 1961 (Nelson 1961), roughly a year before the first successful lasers, which all depended on bulk material. Nelson chose to grow these films on a (100) GaAs crystal plane, which allowed laser mirrors to be made by cleaving along (110) cleavage planes. Threshold current densities reported in his 1963 paper, based on grown, rather than diffused, junctions were almost an order of magnitude lower than those measured on the first successful laser diodes, suggesting a significant improvement in material quality over the earlier bulk crystals. This was, perhaps, a foretaste of better things to come, as we shall see in a moment when we look at the development of the double heterostructure laser which eventually led to the first important commercial application for a semiconductor laser. However, in making comparison with VPE, it was important to consider ultimate purity of the grown layers and, in this context, we should remember that the standard was set originally by Knight and Effer (1965) with samples having n = 2×10^{21} m^{-3} and mobility at 77 K of 3.8 m^2 V^{-1} s^{-1}.

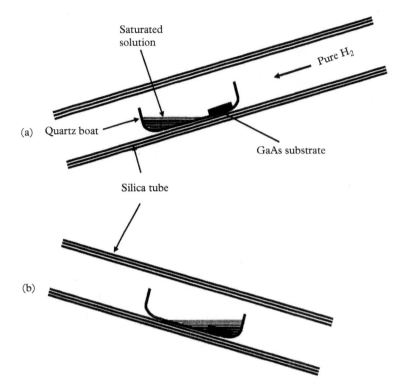

Saturated
solution

Pure H$_2$

(a)   Quartz boat

GaAs substrate

Silica tube

(b)

Figure 1.8 *Illustration of the 'tipping' method of LPE first demonstrated by Nelson of RCA. A saturated solution of GaAs in molten tin, at one end of the boat, is tipped over a GaAs substrate at the opposite end, the temperature is slowly lowered and then the melt is tipped back again when the required thickness of GaAs has been deposited. (From Orton 2004.)*

It took several years before LPE material could compete with this but in 1967 Kang and Greene (1967) from Hewlett-Packard, using the tipping method described by Nelson, went one better, reporting the growth of material with free carrier concentrations in the 10^{20} m^{-3} range and 77 K mobilities between 5.3 and 9.5 m^2 V^{-1} s^{-1}. However, the race was really on—VPE had already moved ahead, Bolger et al. (1967), from the STC laboratory in Harlow, Essex, measuring n = 6.6×10^{20} m^{-3} and μ_{77} = 10.6 m^2 V^{-1} s^{-1}. Interestingly, STL had also invested in LPE, Goodwin et al.(1968) reporting growth of reproducible samples with free electron concentrations in the 10^{21} m^{-3} range, and one 'best' sample with n = 1.7×10^{20} m^{-3} and 77 K mobility of 5.2 m^2 V^{-1} s^{-1}. (We might note, in passing, that Goodwin et al. produced the necessary Ga/GaAs melt by passing AsCl$_3$-H$_2$ over molten GaAs at 800 °C–900 °C.) A year later, Hicks and Manley (1969), also at STL, reported the growth of a reproducible sequence of layers with background doping levels in the 10^{19} m^{-3} range and 77 K mobilities as high as 17.5 m^2 V^{-1} s^{-1}. Harris and Snyder (1969), at Stanford, also achieved a background level of 10^{19} m^{-3} and successfully doped their material with tin to make Gunn effect samples with $n = 10^{21}$ m^{-3} and $\mu_{77} = 4.6$ m^2 V^{-1} s^{-1}. Compensation ratios as low as 0.4 were measured on this material. Obviously, the situation was moving rapidly and, at the Aachen conference on gallium arsenide and related compounds in 1970, Wolfe and Stillman (1970) were able to summarise a considerable volume of data from around the world on both VPE and LPE samples, covering a range of free carrier concentrations from 10^{19} m^{-3} to 10^{21} m^{-3}. Their final conclusion was that there was no essential difference in the quality of material grown by the two techniques, though samples with the lowest values of compensation ($N_A:N_D \sim 0.2$) appeared to be mainly those grown by VPE.

In common with growth by VPE, the principal source of unwanted impurity in LPE was probably leaks to the atmosphere, and material with low background doping level was the result of much care in constructing the growth apparatus. The number of joints had to be minimised and highly pure hydrogen flushed through at all times. The possibility of contamination from the boat containing the solution was reduced by using pyrolytic carbon or high-purity quartz. In most cases, low values of impurity segregation coefficient between the solid and solution reduce the detrimental effect of impure GaAs source material. Finally, also in common with VPE growth, an influence on impurity take-up by substrate orientation, possibly through changes in segregation coefficient, was observed by several workers. Needless to say, not all these effects were well understood but a reasonable degree of care would allow routine growth of material with background doping levels below 10^{20} m^{-3}, more than adequate for Gunn diode, varactor diode or MESFET requirements. On the other hand, factors such as layer thickness control and surface quality were of more than passing significance. LPE was in its element when layers of several tens or even hundreds of microns were demanded but was somewhat stretched to produce layers of a micron or less, such as might be required for GaAs MESFETs. Surface quality was compromised to some extent by the need to remove any remnants of gallium from the layer (by mechanical scraping and chemical dissolution) and there developed a consensus to the effect that VPE had the upper hand in situations demanding smooth surfaces. Finally, there were problems associated with the layer–substrate interface. Common practice was to arrange for the starting growth temperature to be such that there was an initial etching action before film growth commenced and this could cause difficulty in controlling both the thickness of thin layers and the doping profile near the interface. As we shall see later, it also led to problems when growing multilayer structures, particularly for the case of low-dimensional structures. It would, however, be altogether too negative to overemphasise these difficulties—the successful development of heterostructure lasers illustrates the degree to which such difficulties could be overcome.

In contrast to the need for high-purity material in the development of microwave devices, the other main application of the LPE process was concerned with electroluminescent diodes, which demanded relatively high doping levels, both n- and p-type. There were two important aspects, one concerned with visible light-emitting diodes and one with efficient GaAs lasers which could be operated CW at room temperature. Both of these applications depended crucially on the growth of $Al_xGa_{1-x}As$ alloy films, for which LPE turned out to be particularly convenient. The first report of alloy growth by LPE came from IBM, Yorktown Heights, in the persons of Ruprecht, Woodall and Pettit (1967), who demonstrated the possibility of efficient LEDs operating in the red part of the spectrum but the technique was rapidly taken up elsewhere and applied to the development of heterostructure lasers. We shall follow the laser trail first.

The very first laser was made by Maiman at Hughes Research Labs in May 1960, using appropriate energy levels in a crystal of chromium-doped alumina (ruby). It emitted in the red part of the spectrum. Javan at Bell Labs followed shortly afterwards with the first gas (HeNe) laser, and the world of science was excited beyond measure. Nor was semiconductor science to be excluded—a little over two years later no less than four separate groups had submitted papers to American Physical Society journals, describing their observation of laser action in semiconductor diodes. Three of these referred to GaAs diodes (emitting in the infrared) and one to a GaAsP diode (emitting in the red). All were based on n-type bulk crystals counter-doped by the in-diffusion of zinc acceptors and, because they were hopelessly inefficient, operated only in short pulses at a temperature of 77 K. (For a more detailed account, see e.g. Casey and Panish 1978, Part A.) All this represented an exciting breakthrough and established an important existence theorem but there was clearly much work to be done before semiconductor lasers could be regarded as commercially viable devices—in fact, the first serious application, in the CD player, had to wait until 1978, 16 long years after these pioneering efforts. In a few brief words, the material quality was inadequate, injected minority carriers were not confined close to the junction region and light was not confined close to the recombination region; and it was vital to rectify these defects before the goal of room temperature CW operation could be reached. It was here that LPE was to make its first major contribution by producing GaAs of sufficient purity to allow efficient radiative recombination at room temperature and films of $Al_xGa_{1-x}As$ with values of x appropriate to both carrier and optical confinement.

As we have noted already, Nelson, in first describing his 'tipping' apparatus for LPE growth, applied it to the construction of simple GaAs homojunction lasers and obtained very significant reduction in threshold current, compared to those measured on bulk samples. Typical room temperature threshold current densities for diffused-junction lasers were of order 100 kA/cm^2, whereas LPE grown-junction devices were characterised by values of order 20 kA/cm^2. Ruprecht (1966), at IBM, reported very similar results using a vertical, 'dipping' version of LPE. The grown-junctions differed somewhat in their degree of abruptness but the important difference was almost certainly the improved radiative efficiency of the LPE material, deep recombination levels in the bulk material being largely absent from epitaxial layers. Evidence from photoluminescence studies provided strong support for this thesis and, interestingly, such deep levels were also seen in VPE layers—it would appear that LPE was much the preferred process for making lasers. However, the problems associated with free carrier confinement and optical waveguiding had yet to be solved.

Laser light is generated when minority carriers, injected across the p–n junction, recombine and, in the interest of building up a high density of photons to emphasise stimulated, rather than spontaneous, emission, it is essential that this recombination be concentrated close to the junction. As late as 1967, most workers were still reliant on bulk GaAs, diffused with zinc to form the p–n junction and in these homojunction devices minority carriers were free to diffuse away from the junction plane, thus spreading the recombination over several microns. Progress towards reduced threshold current was correspondingly slow. First attempts to improve laser design depended on the introduction of a single heterojunction step on the p-side of the junction, designed to reflect diffusing minority electrons back towards the junction. Alferov and co-workers at the Ioffe Institute in Leningrad (as it then was) had been thinking of using a GaAs/GaAsP combination but were frustrated by the lattice mismatch between GaAs and GaP. It was only when Woodall at IBM demonstrated how to grow AlGaAs by LPE that the way was clear to a practical heterostructure. The idea was taken up by Kressel and Nelson (1969) and by Hayashi and Panish (1970), both of whom grew p-type (Zn-doped) layers of AlGaAs on an n-type GaAs substrate, followed by out-diffusion of zinc into the GaAs to form a 2 μm-thick p-layer in which the recombination would occur. Both groups were able to claim worthwhile reductions in threshold current density, as a result. It was suggested that the concomitant step in refractive index might be responsible for a degree of optical confinement but the dominant effect was surely the confinement of minority carriers close to the junction.

The next step was (of course!) to extend the idea to include a second heterojunction with the aim of confining holes as well as electrons, thus forming a double heterostructure (DH) laser, and that was precisely what the Russian group did, reporting their results marginally before their American counterparts at Bell Labs. However, though it sounds easy enough in principle, practical realisation placed heavy demands on the crystal growers as it involved the growth of three distinct epilayers; n-type $Al_{0.3}Ga_{0.7}As$, p-type GaAs and p-type $Al_{0.3}Ga_{0.7}As$. To do this with the simple LPE equipment hitherto available was nigh-on impossible and required a major rethink of crystal-growing strategy. The answer took the form of the, by now well-known, sliding well technique, a typical example being shown in Figure 1.9. Appropriate solutions are contained in a series of wells which can be brought over the GaAs substrate in turn to grow the desired multilayer structure. Careful calculations of the required solution compositions are necessary in order to ensure they are all at the necessary saturation point when brought into use. Even minor inaccuracies could lead to dissolution rather than deposition, with the possibility of a previously deposited layer being substantially removed before the new one could be grown and, recognising that the optimum thickness of the central GaAs film was a mere 0.2 μm, one immediately realises how easily such a catastrophe might occur. It says much for the growers' skill and ingenuity that successful DH lasers met the criteria for low-threshold, CW operation at room temperature. It had taken no less than eight years since the original pioneering work, a measure, perhaps, of the many problems inherent in such an endeavour.

This was undoubtedly an important breakthrough but it was not quite the end of the story. The double heterostructure effectively solved the carrier confinement problem but efficient optical waveguiding demanded a yet more complex structure—an outer pair of AlGaAs layers with 50% aluminium content provided a step in refractive index appropriate to the desired optical confinement (see Figure 1.10). This innovation appears to have originated from the STC laboratory in Harlow in the persons of Thomson and Kirkby

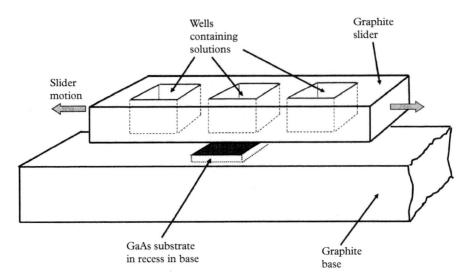

Figure 1.9 *An example of sliding LPE equipment for growing multiple layer structures of AlGaAs and GaAs. The substrate is held in a small declivity in the graphite base while appropriate solutions are contained in a sequence of wells which can be brought over the substrate in turn. (From Orton 2004.)*

(1973), who used yet another version of multilayer LPE based on rotation rather than sliding and, independently, from Hayashi (1972) at Bell Labs (see Panish et al. 1973). Threshold current densities, having started at some 100 kA/cm², were now down to a few hundred amps per square centimetre, and the GaAs laser was, at last, a commercially viable entity. There was more work to be done in controlling optical beam properties but the main challenge facing the LPE grower had successfully been met. However, as we have already indicated, yet another light-emitting device was placing parallel demands on the crystal growers' skills.

GaAs electroluminescent diodes had been under investigation from the early days of III-V materials development, based almost entirely on bulk n-type samples, back-doped by the diffusion of zinc. As we commented above in relation to GaAs lasers, it was natural to explore the virtues of LPE material in so far as its improved purity might result in greater luminescence efficiency but Ruprecht (1966) of IBM demonstrated yet another example of LPE's virtues by making use of the amphoteric doping behaviour of silicon in GaAs. In stoichiometric material, silicon substitutes largely on the gallium site and behaves as a donor while, under arsenic-deficient conditions, it substitutes for arsenic and behaves as an acceptor. In gallium solution growth, as GaAs is deposited, the solution becomes gradually depleted of arsenic so, while initially silicon may act as a donor, later in the process it gradually takes on an acceptor preference, forming a relatively wide, compensated p-type region. Electron injection into this compensated layer gives rise to radiation at an energy some 0.1 eV below the band gap energy, a difference great enough to minimise reabsorption. Ruprecht grew these structures on a

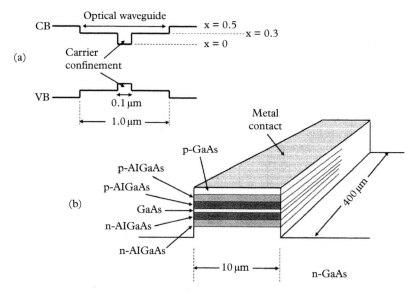

Figure 1.10 *Energy band and structural outline of a mesa stripe, separate confinement, double heterostructure gallium arsenide laser. The outer $Al_xGa_{1-x}As$ layers (x = 0.5) form an optical waveguide to confine emitted photons close to the central region, where carriers recombine. The inner $Al_xGa_{1-x}As$ layers (x = 0.3) serve to confine the recombining free carriers. Light is emitted through the cleaved end facets, that is, normal to the plane of the diagram. (From Orton 2004.)*

silicon-doped substrate, starting growth at a temperature of 935 °C and cooling at 0.5 °C per minute. The change to p-type doping occurred after roughly an hour and a half and produced a 50 μm-wide recombination region. The overall effect was to increase the external quantum efficiency of such diodes by a factor of, typically, four, compared with conventional zinc-diffused diodes.

It may seem obvious that a similar approach could be adopted to make visible LEDs in the AlGaAs alloy system, though it was first necessary to develop a capability for growing the alloy by LPE. The growth of aluminium-containing films by any process is always complicated by aluminium's propensity for combining with oxygen, making the need for oxygen-free surroundings even more important than was the case when growing GaAs. In fact, this approach was, indeed, adopted by Kressel et al. (1969) but it turned out to be less than ideal for the very reason that proved virtuous in the case of GaAs diodes. Because the silicon acceptor level is relatively deep in GaAs, it results in light emission at photon energy well below the band edge, and this tendency only grows greater with the addition of aluminium. Whereas the maximum direct energy gap in the AlGaAs system is about 1.9 eV, it transpires that the maximum emission photon energy from amphoterically doped LEDs is little more than 1.5 eV, too small to be seen by the human eye. Because the eye sensitivity varies steeply in the red spectral region, it is vital, in the interest of achieving optimum brightness, to select the maximum possible photon energy and this means using a shallow acceptor such as zinc. Ruprecht et al. (1967), using their 'dipping' LPE method and doping first with tellurium to produce an n-type layer, then counter-doping with zinc, were able to make diodes emitting at 1.70 eV (730 nm), which provided brightness adequate for initial commercialisation. Further improvements were made later and AlGaAs red LEDs could claim commercial success for a number of years, until the alloy system AlGaInP finally took over the role of chief purveyor of efficient red/orange/yellow light emission; but this was during the 1990s, well beyond our immediate period of interest.

1.4   A Wide Range of Materials

In setting out the background to the emergence of MBE as a practical growth technique, we have concentrated largely on two materials, GaAs and AlGaAs, on the basis that the early phase of MBE work was particularly concerned with these two compounds. However, it would be misleading to imply that they were the only compounds of interest to the electronics and optoelectronics communities. Indeed, one need only pick up Otfried Madelung's excellent book 'Physics of III-V Compounds' (Madelung 1964) or the II-VI Compound Semiconductors conference proceedings (Thomas 1967) to realise that the range of materials under investigation in the early 1960s was quite dauntingly extensive. In this section, therefore, while we obviously cannot hope to cover more than a small fraction of the total, we offer a brief summary of other fields of practical interest in an attempt to restore a modicum of balance to our discourse. Broadly speaking, in addition to the ongoing work on Si epitaxy, three groups of compound semiconductors were under active investigation during the 1960s and early 1970s: III-V compounds, II-VI compounds and IV-VI compounds. It probably makes good sense to discuss them in this order.

Of the III-V materials, those most strongly in contention were indium antimonide, indium phosphide, gallium phosphide and gallium arsenide-phosphide. InSb is a narrow gap semiconductor of interest for detecting far-infrared radiation, with application to thermal imaging. Because it is a highly stable compound with a fairly modest melting temperature (530 °C), at which the vapour pressure of antimony is relatively low, crystal growth from the melt was straightforward, and high-quality bulk samples became available as early as the 1950s. Thus, by the early 1960s the electrical and optical properties of InSb had been thoroughly explored. Its unusually small electron effective mass implied unusually large electron mobilities, and values as high as 7.7 m^2 V^{-1} s^{-1} at room temperature and over 60 m^2 V^{-1} s^{-1} at 77 K had been recorded. In consequence, it became the material of choice for a number of Hall-effect devices such as multipliers, power meters and magnetic flux meters. It also played a serious role in the early development of far-infrared detectors but all this in the form of bulk crystal—it seems that epitaxy only made an appearance much later when research made demands for multilayer structures such as the GaInSb/InAs superlattice for application to infrared imaging.

Indium phosphide first came to prominence in 1970 as a contender for the prize of microwave Gunn diode supremo. As we saw above, GaAs had dominated the early stages of Gunn oscillator development but it was then proposed that InP might have a significant advantage in allowing greater DC-to-microwave conversion efficiencies and this stimulated material scientists to develop both bulk crystal growth, in the form of the liquid encapsulation Czochralski method, and epitaxy, largely in the shape of LPE from indium solution. Within a matter of two years InP became available in excellent quality, sufficient for the experimental testing of these predictions. A somewhat complex story can, perhaps, be oversimplified by simply noting that InP has proved of considerable value in the development of millimetre-wave devices operating in the 100–300 GHz frequency band (for more detail see Orton 2004, p. 207). While this, one may say, represented no more than a small fraction of the overall semiconductor microwave market, the inherent materials technology, nevertheless, proved of inestimable value when fibre-optic communications required semiconductor lasers operating at wavelengths close to 1.5 μm. During the latter half of the 1970s, InP-based materials were to come very much to the fore in the development of both laser sources and photodetectors.

Gallium phosphide and GaAsP both came to early prominence as materials for visible light emitters. We have already met Holonyak's work on the GaAsP red laser in 1962 but the commercial future of the material was to lie with the humble LED. In 1962 General Electric offered red LEDs based on bulk-grown GaAsP but ultimate commercial success came only with the development of epitaxy—VPE-grown films of the alloy material deposited on GaAs substrates led both Hewlett-Packard and Monsanto (in 1968) to market cheap red emitters, which found application in a wide range of products. They were neither particularly efficient nor particularly bright and were later to be rapidly overtaken by devices based on AlGaAs and InGaP; but they well demonstrated the possibilities for semiconductor LEDs to make inroads into the market for indicator lamps, displays and, ultimately, general illumination. Two technical problems limited the brightness that could be achieved. First, GaAsP maintained a direct band gap only for phosphorus fractions up to 45%, at which point the band gap was 1.95 eV, while maximum radiative efficiency occurred at a slightly lower energy, corresponding to a wavelength of 650 nm. At this wavelength the human eye is about ten times less sensitive than at its peak performance, in the green part of the visible spectrum. Second, the lattice mismatch of 4% between GaAs and GaP meant that acceptable structural perfection required the growth of a graded buffer layer between the substrate and the active GaAsP layer, though, at best, this was no more than partially successful in producing material with adequate radiative efficiency. In spite of all this, however, GaAsP devices still satisfy a market for cheap, unpretentious red emitters, where price is of greater significance than performance.

Gallium phosphide was one of the first visible-light-emitting materials to be investigated, primitive red-emitting devices being reported as early as 1955 and the first 'proper' LEDs in the early 1960s. The band gap of GaP (2.25 eV) was large enough to allow both red and green emission, though, needless to say, these initial results represented extremely low efficiencies. It was only with the development of high-quality GaP substrates by the LEC Czochralski process followed by epitaxial deposition of even higher-quality active material that practical devices could be obtained. In 1969 Bell Labs reported external efficiencies of 7% for red LEDs while in 1972 Fairchild achieved 15%.

Green emitters were much less efficient, though roughly equal in brightness. The really interesting feature of GaP light emission concerned the nature of the recombination processes—GaP, being an indirect gap material, should have been totally unsuitable for efficient LED manufacture but was 'saved' by the peculiar properties of so-called iso-electronic centres. Green light was generated by the recombination of an exciton bound to a nitrogen impurity in GaP while red emission involved an exciton bound to a (Zn-O) centre. Because electrons are bound very strongly by these centres, their wavefunctions involve components of momentum from right across the Brillouin zone, thus circumventing the limitation imposed by the indirect energy gap of the host crystal. Thus, during the 1970s GaP could compete effectively as a red-emitter and was unique in its ability to produce green light with modest efficiency. It was not until the 1990s that this dominance was to be seriously challenged, first by AlGaInP, then, more dramatically, by InGaN.

Several II-VI compounds were also studied with a view to the possible development of visible-light-emitting devices. In particular, CdS (2.50 eV), ZnS (3.68 eV), ZnSe (2.82 eV) and ZnTe (2.39 eV) were characterised by suitably wide direct band gaps, and efficient visible photoluminescence was recorded from all of these during the very early days of semiconductor materials research. Polycrystalline and powder samples were capable of bright-light emission and this led to their widespread application as phosphors, though the hoped-for emergence of efficient LEDs was frustrated by the apparent impossibility of achieving both n-type and p-type conductivity. CdS, ZnS and ZnSe remained resolutely n-type while ZnTe could only be doped p-type. Only CdTe could readily be doped both n- and p-type but its band gap of 1.5 eV was far too small to permit visible emission. As we might reasonably expect, the first single crystal material resulted from melt growth (Bridgman) methods which required high temperatures (well over 1000 °C in all cases) and required careful compensation for the high vapour pressure of the chalcogenide species. Under such conditions, thermodynamics postulated the inevitable existence of significant densities of crystalline point defects which might well be responsible for blocking the movement of the Fermi level towards conduction or valence band and, for many years the accepted wisdom had it that this was, indeed, the case. Indeed, for a long time, there appeared little hope that p–n homojunctions would ever be produced in these wide gap materials.

Efforts to overcome this problem (during the early 1970s) were many and various, including the use of ion implantation to effect type conversion, the use of MIS structures and reverse-biased metal–semiconductor contacts, and the ingenious introduction of heterojunctions (for details, see Bergh and Dean 1976, p. 245 et seq.). Ion implantation of the lithium acceptor into ZnSe was successful in producing p-type conduction, though with inconveniently large resistivity, while Cl implantation into ZnTe, followed by a 450 °C anneal, resulted in n-type conductivity but again with high resistivity; p–n junction luminescence was obtained in a number of cases but with no more than modest efficiency. Similar comments can be attached to attempts at ZnTe MIS electroluminescence, where the light is probably generated by recombination of avalanche-generated electrons with holes in the bulk ZnTe. Finally, we should make mention of the interesting concept of a heterojunction between p-ZnTe and n-ZnSe (semiconductors having nearly equal band gaps) and which generated green electroluminescence but again with only moderate efficiency.

This last structure relied on the use of liquid epitaxy (from zinc or bismuth solution) to grow a film of ZnTe on a bulk ZnSe crystal surface while similar structures were also reported by the VPE growth of CdS on a crystal of CuGaS. We might fairly comment on the fact that epitaxy came rather late on the II-VI scene (1973 rather than 1963). This is, perhaps, surprising in so far as epitaxial growth takes place at significantly reduced temperatures compared with melt growth and might therefore be expected to provide different point-defect concentrations. Indeed, as we shall discuss in more detail later, it was the introduction of low-temperature MBE growth that eventually led to successful p–n junction luminescence in ZnSe.

To complete this brief survey of the early work on semiconductor materials, we must finally say something about the narrow gap compounds that formed the basis of the militarily important infrared detector market. The excitement began during the First World War with the 'thalofide cell', based on thallium sulphide, but this material was sensitive at wavelengths only as far into the infrared as 1 μm. The first serious contenders to satisfying the need for thermal imagers covering the 3–5 μm and 8–14 μm atmospheric windows were the lead chalcogenides, PbS, PbSe and PbTe, which, when cooled to 90 °C, showed sensitivities out to about 9 μm (though, somewhat incongruously, their success emerged only *after* the Second World War!). In their original incarnation as photoconductors, they took the form of polycrystalline thin films and, though bulk (Bridgman) single crystals were first grown at RRE Malvern in the early 1950s, it took until about 1970 before the quality was adequate for their electro-optic properties to be fully established. In the meantime relatively perfect crystals of InSb (sensitive to about 5.5 μm) offered a serious challenge, at least for operation in the 3–5 μm window, and considerable success was obtained with such detectors when applied to mechanically scanned imaging systems. However, further progress towards realising full coverage of the desired IR wavelength range and developing large arrays of detectors for electronically scanned imagers demanded both new detector materials and more flexible growth methods. The 1960s saw two important materials innovations in the shape of the English-born mercury cadmium telluride HgCdTe vying with the American-born lead tin telluride PbSnTe. Both material systems came with the important bonuses of variable (and controllable) energy gaps and the (later) capability for epitaxial growth. The world of thermal imaging was to be taken by storm (though, to be fair, a storm which took many years to develop its full intensity!).

Single crystal MCT was, again, the brainchild of the RRE Malvern group under W. D. Lawson (Lawson et al. 1959). They were able to demonstrate a continuously variable direct band gap (at low temperatures—see Figure 1.11) from the 1.6 eV of CdTe to the –0.3 eV of HgTe (the latter being a semimetal), implying a zero gap at a composition of approximately 15% CdTe (x = 0.15 in $Hg_{1-x}Cd_xTe$). Of particular interest for infrared detector technology was the realisation of band gaps of 0.1 eV (at x = 0.2) and of 0.25 eV (at x = 0.3), corresponding to the 8–14 μm and 3–5 μm windows, respectively. This early promise resulted in several other groups around the world taking up the challenge to develop satisfactory device material. The early work was all

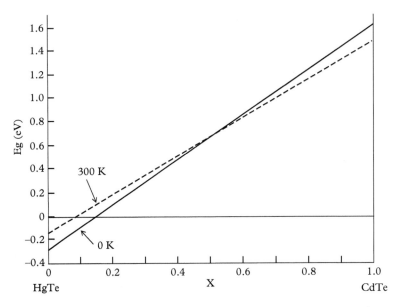

Figure 1.11 *Plot of the energy gap of the alloy Hg$_{1-x}$Cd$_x$Te as a function of composition, at room temperature and at low temperature. Values of x = 0.2 and x = 0.3 result in band gaps of 0.1 eV and 0.25 eV, respectively, appropriate to the 8–14 μm and 3–5 μm windows. (Following Kruse 1981.)*

based on some form of bulk crystal growth, involving careful melting of an appropriate mixture, either of the elements or of the compounds CdTe and HgTe. It was important to homogenise the mixture thoroughly by annealing for many hours at a temperature of about 850 °C—high vapour pressures of unreacted mercury had a tendency to cause explosion—followed either by slow cooling or, alternatively, by a rapid cool and long anneal to obtain satisfactory uniformity of composition in the final product. MCT probably represented a far more serious materials challenge than any so far met with and, in spite of the strength of the military interest in its successful development, progress was unusually slow. As the demand for larger, more uniform samples appropriate to detector arrays became more insistent, it became more and more obvious that some form of epitaxy was essential. First attempts took the form of close-spaced epitaxy, pioneered during the late 1960s at the CNRS laboratory in Bellevue, France. Polycrystalline MCT was heated in close proximity to a CdTe substrate in a vacuum furnace, resulting in a slow build-up of an epitaxial film on the surface of the CdTe. Suitable films might take as long as 100 h to grow and they suffered from poor uniformity in the direction normal to the film, as a result of mercury diffusion from film to substrate so during the 1970s first VPE and then LPE techniques were explored in the hope of developing more practical alternatives. Minimising interdiffusion clearly required faster growth and lower growth temperatures, and VPE offered the possibility of some improvement, using deposition temperatures on the order of 450 °C rather than the 550 °C typical of close-spaced epitaxy. On the other hand, LPE from either mercury or tellurium solution required temperatures close to 500 °C and still suffered significantly from the diffusion problem. In developing the materials technology of MCT, nothing was easy and it is a reflection of the difficulties involved that right up until 1980 most of the current device work still depended on Bridgman growth.

Lead tin telluride was 'invented' at the MIT Lincoln Laboratory in the mid-1960s (see Melngailis and Harman 1970). Like MCT, this system shows an approximately linear variation of band gap from the 0.2 eV of PbTe to the –0.3 eV of SnTe, though, in this case, both materials are semiconductors, the 'inversion' resulting from an interchange between a pair of bands. In practical terms, band gaps between 0 and 0.2 eV can be obtained with alloys represented by values of x between 0.4 and 0 (in $Pb_{1-x}Sn_xTe$). A value of x = 0.25 corresponds to an absorption edge at a wavelength of about 15 μm, appropriate to the 8–14 μm window. In effect, the two material systems show very similar behaviour, and one might suppose them to be readily interchangeable, apart from the fact that PbSnTe is characterised by an extremely large value of dielectric constant (typically several hundred), a factor of importance in attempts to design high-speed photodiodes. The difficulty lies in the unusually large junction capacitance, which leads to long RC time constants. Single crystals of PbSnTe were produced by both vapour growth and by the Bridgman method but required lengthy annealing at temperatures in the range of 500–750 °C in order to reduce the free carrier concentrations to acceptable values (not particularly desirable in the context of commercial applications). As is also true for MCT, stoichiometry can have a major effect on free carrier levels, a feature which makes for additional complication in attempts to produce device-quality material. Liquid phase epitaxy was introduced in the early 1970s, using metal-rich solutions at temperatures in the range 400 °C–700 °C, significantly below the 825 °C appropriate to Bridgman growth. The advantages were clear—not only was it possible to grow device-quality material without the need for annealing but the greater flexibility of the epitaxial process allowed much more sophisticated structures to be realised. An important example was the development during the 1970s of double heterostructure long-wavelength lasers, which showed much superior performance to the earlier diffused-junction devices. As we shall see, though MCT gradually came to dominate the market for infrared detectors, LPE and MBE growth methods for IV-VI compounds were to cross swords over the development of long-wavelength lasers in the struggle to achieve room temperature CW operation (shades of the earlier DH GaAs laser battle!).

This brings us to the end of our scene-setting survey of semiconductor growth processes as they existed at the time when MBE was born, in the 1970s. Clearly, a great deal of work had been done to develop sophisticated growth techniques over a wide range of materials and material systems. Various forms of epitaxy had already reached an advanced state of development in the interests of producing well-controlled thin films of material, characterised by thicknesses in the order of 0.1 to 100 μm, with free carrier concentrations as low as 10^{20} m^{-3} and minority carrier lifetimes appropriate to efficient light-emitting devices. The use of heterostructures was well established as a direct result of the introduction of epitaxial growth methods and these had become an essential component of laser, LED and photodetector devices. Not all semiconductor problems had been solved, of course—the fact that the potentially important II-VI compounds could still not be doped both n- and p-type was a significant example and, though it was not yet appreciated just how important would be the use of very much thinner films, the idea of the semiconductor superlattice was already beginning to surface. Though the crystal-growing community could feel a fair degree of satisfaction at the progress already made, there were certainly no grounds for complacency. New demands from the physics and device entrepreneurs could be confidently anticipated.

REFERENCES

Antell, G R (1961) Investigation of a method of growing crystals of GaP and GaAs from the vapour phase. Brit J Appl Phys 12, 687.

Antell, G R and Effer, D (1959) Preparation of crystals of InAs, InP, GaAs, and GaP by a vapor phase reaction. J Electrochem Soc 106, 509.

Bergh A A and Dean P J (1976) 'Light Emitting Diodes', Clarendon Press, Oxford.

Blood, P and Orton J W (1992) 'The Electrical Characterisation of Semiconductors: Majority Carriers and Electron States', Academic Press, London.

Bolger, D E, Franks, J, Gordon, J and Whitaker, J (1967) in 'Proceedings of the International Symposium on Gallium Arsenide, Reading', IPPS, London, p 16.

Bube, R H (1960) 'Photoconductivity of Solids', Wiley, New York.

Casey, H C Jr and Panish, M B (1978) 'Heterostructure Lasers', Academic Press, New York.

Easton, B C, Fisher, C A and Wilkinson, J A (1971) Philips Research Laboratories Annual Review 1971, p 78.

Effer, D (1965) Epitaxial growth of doped and pure GaAs in an open flow system. J Electrochem Soc 112, 1020.

Finch, W F and Mehal, E W (1964) Preparation of $GaAs_xP_{1-x}$ by vapor phase reaction. J Electrochem Soc 111, 814.

Goodwin, A R, Gordon, J and Dobson, C D (1968) High-mobility gallium arsenide grown by liquid-phase epitaxy. J Phys D: Appl Phys 1, 115.

Grove, A S (1967) 'Physics and Technology of Semiconductor Devices' Wiley, New York.

Gunn, J B (1963) Microwave oscillation of current in III-V semiconductors. Sol St Commun 1, 88.

Harris, J S and Snyder, W L (1969) Homogeneous solution grown epitaxial GaAs by tin doping. Sol St Electron 12, 337.

Hayashi, I (1972) Double heterostructure laser diodes. US Patent 3691476.

Hayashi, I and Panish, M B (1970) $GaAs–Ga_xAl_{1-x}As$ heterostructure injection lasers which exhibit low thresholds at room temperature. J Appl Phys 41, 150.

Haynes, J R (1960) Experimental proof of the existence of a new electronic complex in silicon. Phys Rev Lett 4, 361.

Hicks, H G B and Manley, D F (1969) High purity GaAs by liquid phase epitaxy. Sol St Commun 7, 1463.

Hilsum, C (1962) Transferred electron amplifiers and oscillators. Proc IRE (New York) 50, 185.

Holonyak, N and Bevacqua, S F (1962) Coherent (visible) light emission from $Ga(As_{1-x}P_x)$ junctions. Appl Phys Lett 1, 82.

Ing, S W and Minden, H T (1962) Open tube epitaxial synthesis of GaAs and GaP. J Electrochem Soc 109, 995.

Ito, S, Shinohara, T and Seki, Y (1973) Properties of epitaxial gallium arsenide from trimethyl-gallium and arsin. J Electrochem Soc 120, 1419.

Kang, C S and Greene, P E (1967) Preparation and properties of high-purity epitaxial GaAs grown from Ga solution. Appl Phys Lett 11, 171.

Knight, J R, Effer, D and Evans, P R (1965) The preparation of high purity gallium arsenide by vapour phase epitaxial growth. Sol St Electron 8, 178.

Kressel, H, Hawrylo, F Z and Almeleh, N (1969) Properties of efficient silicon-compensated $Al_xGa_{1-x}As$ electroluminescent diodes. J Appl Phys 40, 2248.

Kressel, H and Nelson, H (1969) Close-confinement gallium arsenide pn junction lasers with reduced optical loss at room temperature. RCA Rev 30, 106.

Kruse, P W (1981) 'The Emergence of $Hg_{1-x}Cd_xTe$ as a Modern Infrared Sensitive Material' in 'Semiconductors and Semimetals', Vol 18, (ed R K Willardson and A C Beer), Academic Press, New York, p 1.

Kurylo, F and Susskind, C (1981) 'Ferdinand Braun', MIT Press, Cambridge, MA.

Lark-Horovitz, K (1954) 'The New Electronics' in 'The Present State of Physics' (ed F S Bracket), American Association for the Advancement of Science, Washington, DC, p 57.

Lawson, W D, Nielsen, S, Putley, E H and Young, A S (1959) Preparation and properties of HgTe and mixed crystals of HgTe-CdTe. J Phys Chem Solids 9, 325.

Madelung, O (1964) 'Physics of III-V Compounds', Wiley, New York.

Manasevit, H M and Simpson, W I (1969) The use of metal-organics in the preparation of semiconductor materials I. Epitaxial gallium-V compounds. J Electrochem Soc 116, 1725.

Melngailis, I and Harman, T C (1970) 'Single-Crystal Lead-Tin Chalcogenides' in 'Semiconductors and Semimetals', Vol 5 (ed R K Willardson and A C Beer), Academic Press, New York, p 111.

Mlavsky, A I and Weinstein, M (1963) Crystal growth of GaAs from Ga by a traveling solvent method. J Appl Phys 34, 2885.

Nelson, H (1961) 'Epitaxial Crystal Growth from the Liquid Phase', Solid State Device Conference, Stanford University, June 26, 1961.

Nelson, H (1963) Epitaxial growth from the liquid state and its application to the fabrication of tunnel and laser diodes. RCA Rev 24, 603.

Newman, R L and Goldsmith, N (1961) Vapor growth of gallium arsenide. J Electrochem Soc 108, 1127.

Orton, J W (1971) 'Material for the Gunn Effect', Mills and Boon, London.

Orton, J W (2004) 'The Story of Semiconductors', Oxford University Press, Oxford.

Orton, J W (2009) 'Semiconductors and the Information Revolution', Academic Press, Amsterdam.

Panish, M B, Casey H C Jr, Sumski, S and Foy, P W (1973) Reduction of threshold current density in GaAs–$Al_xGa_{1-x}As$ heterostructure lasers by separate optical and carrier confinement. Appl Phys Lett 22, 590.

Pfann, W G (1957) 'Techniques of zone melting and crystal growing' in 'Solid State Physics: Advances in Research and Applications', Vol 4, (ed F Seitz and D Turnbull), Academic Press, New York, p 423.

Pohl, R W (1937) Electron conductivity and photochemical processes in alkali-halide crystals. Proc Phys Soc 43, 3.

Prior, A C (1961) Growth from the vapor of large single crystals of lead selenide of controlled composition. J Electrochem Soc 107, 82.

Putley, E H (1960) 'The Hall Effect and Related Phenomena', Butterworth, London.

Rhoderick, E H (1978) 'Metal-Semiconductor Contacts', Oxford University Press, Oxford.

Ridley, B K and Watkins, T B (1961) The possibility of negative resistance effects in semiconductors. Proc Phys Soc 78, 293.

Riordan, M and Hoddeson, L (1997) 'Crystal Fire: The Birth of the Information Age', Norton and Company, New York.

Ruprecht, H (1966) 'New Aspects of Solution Regrowth in the Device Technology of Gallium Arsenide' in 'Proceedings of the International Symposium on Gallium Arsenide, Reading', IPPS, London, p 57.

Ruprecht, H, Woodall, J M and Pettit, G D (1967) Efficient visible electroluminescence at 300 °K from $Ga_{1-x}Al_xAs$ p-n junctions grown by liquid-phase epitaxy. Appl Phys Lett 11, 81.

Seki, Y, Tanno, K Iida, K and Ichika, E (1975) In situ mass-spectrometric evaluation of impurities in trimethylgallium. J Electrochem Soc 122, 1108.

Teal G K and Buehler, E (1952) Growth of silicon single crystals and of single crystal silicon pn junctions. Phys Rev 87, 190.

Teal, G K and Little, J B (1950) Growth of germanium single crystals. Phys Rev 78, 647.

Thomas, D G (ed) (1967) 'II-VI Compound Semiconductors: Proceedings of the International Conference', Benjamin, New York.

Thompson, G H B and Kirkby, P A (1973) (GaAl)As lasers with a heterostructure for optical confinement and additional heterojunctions for extreme carrier confinement. IEEE J Quantum Electron QE-9, 311.

Thorsen, A C and Manasevit, H M (1971) Heteroepitaxial GaAs on aluminum oxide: electrical properties of undoped films. J Appl Phys 42, 2519.

Tietjen, J J and Amick, J A (1966) The preparation and properties of vapor-deposited epitaxial $GaAs_{1-x}P_x$ using arsine and phosphine. J Electrochem Soc 113, 724.

Von Klitzing, K, Dorda, G and Pepper, M (1980) Realization of a resistance standard based on fundamental constants. Phys Rev Lett 45, 494.

Welker, H J (1976) Discovery and development of III-V compounds. IEEE Trans Electron Devices ED-23, 664.

Wilson, A H (1931a) The theory of electronic semi-conductors. Proc Roy Soc A 133, 458.

Wilson, A H (1931b) The theory of electronic semi-conductors. II. Proc Roy Soc A 134, 277.

Wolfe, C M and Stillman, G E (1970) 'High Purity GaAs' in 'Proceedings of the International Symposium on Gallium Arsenide and Related Compounds, Aachen', IPPS, London, p 3.

2

MBE—The Early History

2.1 Origins

MBE is usually seen as originating towards the end of the 1960s, and this chapter will be concerned with exploring its several origins and understanding exactly where it came from and why. However, in attempting to do this, we are immediately faced with a problem of definition—what, exactly, should we understand by the term 'molecular beam epitaxy'? How can it best be defined in relation to the other forms of epitaxial deposition we met in Chapter 1? As has been pointed out on many occasions, MBE is nothing more than a sophisticated form of vacuum evaporation and we should not be surprised to discover that a number of such techniques were explored well before the modern version of MBE came into being. Vacuum evaporation in a simple bell-jar system, pumped by a standard rotary pump, had been used over many years for depositing thin films of metals and insulators, so it was natural for materials scientists to explore the possibility of growing epitaxial films of semiconductors in a similar manner. To choose an example at random, we might note that the growth of epitaxial films of PbS on NaCl substrates was reported as early as 1948 (Elleman and Willman 1948). What is more, because the lead chalcogenides evaporate as diatomic molecules, this might be seen as a very early example of MBE!

Two caveats can immediately be raised against this. First, one must question the quality of the vacuum (10^{-3} Torr in this instance) and its influence on material quality—modern MBE systems employ UHV conditions, with background pressures of 10^{-10} Torr or better, in the interest of minimising impurity incorporation in the growing film. Second, we should note that the original definition of a molecular beam implied beam conditions such that molecules suffered no collisions, conditions that demanded a Knudsen cell with an aperture that was small compared with the mean-free-path of molecules within the cell (see Box 2.1). In many applications, this was of major importance and, though it was also vital to some pioneering MBE studies, one can question whether it has any relevance to crystal growth, in general. The objective must surely be to supply appropriate atoms or molecules to the substrate surface and leave surface diffusion, surface reactions and inevitable desorption from the surface to play their various roles in generating an epitaxial film. In most cases, the precise details of molecular velocities within the beam are unlikely to be of major significance and, if we accept this, the nature of the molecular source is only of secondary importance. Indeed, modern MBE sources cannot, in general, claim to be Knudsen cells—in the interest of increased growth rate, their apertures have been enlarged to the point of being significantly greater than the molecular mean-free-path. In this respect, many MBE machines differ very little from some of the earlier evaporation equipment. Against this, however, we should point out that some modern MBE machines utilise gas sources, rather than effusion cells, which not only marks them off from typical evaporators but adds further to the problems of finding a precise definition of MBE.

Another important feature of the modern MBE process is the use of heated substrates to allow the diffusion of adsorbed species over the substrate surface in order to find suitable sites for necessary chemical reactions. For example, an As_4 molecule must break down into four As atoms in order to be incorporated into a GaAs surface. Similarly, a silane molecule must react with surface silicon atoms when breaking down thermally to form a silicon film. Some of the early evaporation work may have failed to recognise this point, with consequent deterioration in crystal quality but there is, nevertheless, plenty of published work in which substrate heating was seen as essential. In any case, it might seem perverse to suggest that MBE should be defined by such a criterion.

Finally, we must note another important aspect of MBE growth, in so far as most depositions involve more than one source. An obvious example is the use of separate gallium and arsenic sources when growing GaAs films, while the growth of alloy films requires three or more sources and doping at least two more, in order to achieve both n- and p-type doping. Thus, a modern MBE machine will typically be equipped with at least 8, and in many cases, 14 or more source cells, which certainly distinguishes it from a simple evaporator growing films of a lead chalcogenide. On the other hand, the protagonist of simple evaporation can claim that he alone is using *a* molecular beam. It is also true that GaAs films were grown by evaporation using separate gallium and arsenic sources in 1964 (Davey and Pankey 1964). The quality of vacuum used in these experiments (10^{-7} Torr) was certainly less than that available in a modern MBE machine but it would be splitting hairs to imply that this disqualified the work from being molecular beam epitaxy.

Clearly, we are faced with a dilemma in trying to define exactly what is and what is not MBE. The fact of the matter is that the technique developed gradually from a number of quite distinct roots, and the following discussion is intended to acknowledge that fact. It is convenient to look at MBE as currently practised and adopt as a working definition 'MBE is the growth of epitaxial films on a crystalline substrate by appropriate chemical reaction of one or more molecular beams impinging on a heated substrate surface. It usually takes place in UHV equipment to ensure that the arrival rate of unwanted impurity atoms from the vacuum ambient is negligible compared with that from the beam(s) (though we should accept that growth actually takes place under pressures as high as 10^{-6} Torr). It also usually involves the use of UHV-compatible surface characterisation techniques.' Note that this definition makes no attempt to specify the precise nature of the beams, though we shall, nevertheless, use this as a criterion in making our presentation. Thus, we begin by describing work which depended on the use of collision-free beams, then consider the history of relevant evaporation processes. The two approaches can be seen to have come together to realise modern MBE practice, which is our ultim-ate target. Perhaps, in closing this introductory section, we should also make the point that, in most cases, the ultimate criterion for judging any deposition technique is the quality of the resulting material, in relation to its application. It would be unwise to over-look the possibility that a relatively basic method might fulfil the demands of many end users. On the other hand, MBE has been tested to the very highest level in numerous de-manding applications and we must therefore acknowledge the genuine need for a highly sophisticated version of the method—it is this, of course, which forms the subject of our book. Finally, we should also make the point that modern MBE is rather more than just a growth method in so far as it usually incorporates sophisticated surface characterisation measurements which allow its practitioners to gain some valuable insight into the growth process, while simultaneously monitoring the quality of the growing film. In this respect, MBE is probably unique amongst epitaxial growth methods.

2.2   A History of Molecular Beams

Atomic and molecular beams had already acquired some considerable importance long before they were taken up as a basis for thin film crystal growth. While making no pretence at complete coverage of their various contributions to physics, chemistry and materials science, we shall examine their significance and illustrate their versatility by reference to three distinctive applications, namely the Stern–Gerlach experiment, the development of nuclear spin resonance and the invention of the maser. (For much of the material appearing in this section, we are indebted to the splendid little monograph 'Molecular Beams' by K. F. Smith (1955).)

In all these contributions, the essential property of the beam was the absence of inter-actions between the constituent atoms or molecules, a feature which, as we shall see later, was also vital to some early experiments on crystal growth. In a word, it allowed the pos-sibility for measurements on isolated atoms or molecules, rather than on considerably more complex interactive systems, which would be impossibly difficult to model the-oretically. Thus, the Stern–Gerlach experiment confirmed the quantisation of angular momentum (as predicted by early quantum theory) in isolated atoms, nuclear reson-ance measured the hyperfine splitting of energy levels in isolated molecules, while the first observation of maser action employed the energy levels of isolated ammonia mol-ecules. Consequently, the common feature of all these experiments was an appropriate method of generating a collimated beam of molecules with intensity high enough to permit meaningful experimental data while low enough to avoid significant numbers of intermolecular collisions.

A typical experimental set-up for the generation of a molecular beam is illustrated in Figure 2.1. It consists essentially of a Knudsen cell: that is, an oven containing the ap-propriate source material and including a small aperture or slit through which the beam is to emerge. The source material may take the form of a gas or, more frequently, a liquid or solid in thermal equilibrium with its vapour. The oven is situated within a sec-ond evacuated chamber so as to avoid the beam molecules being scattered by oxygen or

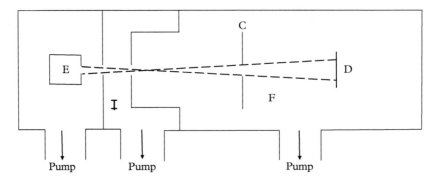

Figure 2.1 *Basic apparatus used to form a molecular beam. The oven E contains the gas molecules which effuse through a small aperture into an isolating chamber I. The observation chamber F contains a second small aperture C which defines the beam and a detector D. (Following Smith 1955).*

nitrogen molecules from the air. As is apparent in Figure 2.1, the vapour in the interior of the oven is effectively pumped by the vacuum pump used to evacuate the outer chamber; so, in the interest of maintaining the necessary pressure differential between the two chambers, the oven aperture should, ideally, be small compared with the mean-free-path of the gas molecules within the oven. Of even greater importance, however, is the fact that this same condition ensures that the exit beam is almost free of intermolecular collisions, as use of a larger opening would result in turbulent flow rather than the desired collision-free, rectilinear motion of the emitted molecules.

The beam is collimated by means of one, or more, further apertures placed in line with the first and, assuming molecular flow, it is straightforward to derive an expression for the intensity of the beam as a function of distance from the source aperture. This we do in Box 2.1, for two specific cases, corresponding to the Stern–Gerlach experiment and to MBE growth, respectively. In this section we shall be concerned with the well-collimated, narrow beams used in precise experimental work, and we begin by considering the experiment performed in Frankfurt in 1921 by Stern and Gerlach, when a beam of silver atoms was used to observe the quantisation of angular momentum predicted by early quantum theory.

The effect of the magnetic moment associated with the angular momentum of atoms was first observed in atomic spectroscopy, in the form of the Zeeman effect. However, it was not immediately possible to confirm the predictions of quantum theory concerning the quantisation of angular momentum, because classical theory was also able to mount a plausible explanation. However, the experiment proposed by Stern provided a straightforward and direct method of distinguishing between the predictions of classical and quantum theory and, as such, was of considerable importance in establishing the provenance of the new (and still controversial) quantum theory. It depended on the use of a highly non-uniform magnetic field in deflecting an atomic beam. An outline of the experimental arrangement is shown in Figure 2.2, where a narrow, bar-shaped beam of silver atoms passed through a magnetic field and impinged on a

Box 2.1 The Knudsen source

In order to arrive at an 'order-of-magnitude' understanding of the properties of a molecular beam, we can use the kinetic theory of gases to calculate beam intensity at a specified distance from a Knudsen source—an oven containing the relevant gas with a small aperture through which the gas effuses. To simplify the calculation, we shall assume an ideal aperture and we shall neglect the complicating fact that molecular velocities follow the Maxwell distribution, our purpose being merely to establish appropriate orders of magnitude.

Our starting point is the well-known equation for the number of molecules striking a small area of the oven wall (and therefore the number passing through the aperture) in unit time:

$$N = (1/4)\ ncA \tag{B2.1}$$

where n is the number density of molecules within the oven, c is the molecular velocity, and A the area of the aperture. Because the effective aperture seen by a molecule emerging at an angle θ varies as $\cos\theta$, it is easy to show that the flux of the exit beam follows a Lambertian cosine law:

$$F(\theta) = F_0\cos\theta \tag{B2.2}$$

where θ is the angle with the normal to the aperture. Thus, the number of molecules passing through the elementary area $2\pi r\sin\theta \cdot r\,d\theta$ at distance r from the aperture in unit time is

$$dN = F_0\cos\theta \cdot 2\pi r^2\sin\theta\,d\theta \tag{B2.3}$$

Integrating this expression and equating it to the total number of molecules effusing through the aperture N, we obtain an expression for the beam flux at distance r from the aperture:

$$F(r,\theta) = ncA\cos\theta/4\pi r^2 \tag{B2.4}$$

As we are usually interested in the intensity normal to the aperture, we can conveniently put $\cos\theta$ equal to unity.

In order to relate this to parameters under our experimental control, we must refer to three other equations involving the vapour pressure p, temperature T and mean-free-path λ of a molecule within the oven. Thus,

$$p = (1/3)nmc^2 \tag{B2.5}$$

$$c = \{3kT/m\}^{1/2} \tag{B2.6}$$

$$\lambda = \{n\pi\sigma^2\}^{-1} \tag{B2.7}$$

where m is the mass and σ is the diameter of a molecule. Combining (B2.5) and (B2.6) yields the simple result that

continued

Box 2.1 *continued*

$$p = nkT \tag{B2.8}$$

Note that the product λp is independent of the molecular density:

$$\lambda p = kT/\pi\sigma^2 \tag{B2.9}$$

emphasising that the mean-free-path is inversely proportional to the vapour pressure within the oven. This is important in deciding one's choice of aperture size, bearing in mind that an exit beam shows true molecular flow only when the aperture is small compared with λ. Note that, if we take $\sigma = 10^{-9}$ m and $T = 300$ K, equation (B2.9) gives us the useful rule-of-thumb relation that, when $p = 0.1$ mm of mercury (0.1 Torr), then $\lambda = 0.1$ mm.

Using equations (B2.4), (B2.6) and (B2.8), we arrive at an expression for the flux in terms of the vapour pressure and temperature within the oven:

$$F = Ap/\left[4\pi r^2\{mkT/3\}^{1/2}\right]$$

$$= 0.138Ap/\left[r^2\{mkT\}^{1/2}\right] \tag{B2.10}$$

$$= 9.09 \times 10^{23} Ap/\left[r^2\{MT\}^{1/2}\right] \text{mol/m}^2\text{/s}$$

where M is the molecular weight of the molecule concerned, and the pressure is in Pascals. (Bearing in mind that pressures in MBE work are customarily expressed in Torr, we should note that 1 Torr = 133 Pa.)

In the following discussion, we distinguish two situations which we shall consider separately. The first of these concerns experiments designed to measure specific properties of the molecules in the beam, such as the confirmation of angular momentum quantisation measured by the Stern–Gerlach experiment. The second refers to epitaxial deposition of molecules on a crystal substrate, which will, of course, be our chief concern in this book.

The Stern–Gerlach experiment

In this experiment the beam was defined by a pair of narrow slits so that the detector 'saw' a beam profile some 2 mm in length but only 0.01 mm wide. We can imagine the target area to be 'illuminated' from a 2 mm line of square apertures, each 0.01×0.01 mm^2, which results in a much enhanced signal—by a factor of $2/0.01 = 200$ times—compared with that generated by a single aperture. The slit width $b = 0.01$ mm was chosen so as to make $b \sim \lambda$ in order to avoid intermolecular scattering normal to the plane of the slits and this enables us to express the flux at the target (distance r from the source) in terms of the product λp, which, as we noted above, depends only on the source temperature T. Thus,

$$F = l \times \lambda p/\left[4\pi r^2\{mkT/3\}^{1/2}\right] \tag{B2.11}$$

where l is the length of the slits. Substituting from equation (B2.9) then gives

$$F = l \times \{3kT/m\}^{1/2}/4\pi^2\sigma^2r^2$$

$$= 4.0 \times 10^{18} \times \left[l/r^2\right] \times \{T/M\}^{1/2} \text{mol/m}^2\text{/s} \tag{B2.12}$$

where we have used a value of $\sigma = 10^{-9}$ m for the effective diameter of a molecule.

It is now straightforward to substitute values of $l = 2 \times 10^{-3}$ m, $r = 1$ m, $T = 1000$K and $M = 47$ (appropriate to silver) to obtain an estimate of the flux at the detector. The answer of 3.7×10^{16} mol/m^2/s implies a time of approximately 300 s (or 5 min) for deposition of a single monolayer of silver atoms (approximately 10^{19} molecules per square metre). Given that a single monolayer is visible to the eye, this constitutes an acceptable time scale for the experiment.

Epitaxial deposition

The geometry in this case differs from the above in that we must use a circular aperture of radius a, aligned axially with a circular substrate of very much larger radius—say 1 cm or more. Again, we must choose the diameter of the aperture to be no greater than the molecular mean-free-path, so we take $a \sim \lambda$ and write the beam flux as

$$F = \pi a^2 p/4\pi r^2 \{mkT/3\}^{1/2}$$

$$= (\lambda p)^2/4pr^2 \{mkT/3\}^{1/2}$$

$$= (kT)^{3/2}/4\pi^2\sigma^4 r^2 p\{m/3\}^{1/2} \tag{B2.13}$$

$$= 5.52 \times 10^{13}\ T^{3/2}/r^2 pM^{1/2}\ \text{mols/m}^2\text{/s}$$

Inserting values of $T = 1000$K, $r = 30$ cm, $p = 10^{-3}$ Torr (0.133 Pa) and $M = 30$ results in a flux of approximately 10^{19} mols/m^2/s, or 1 monolayer per second, which is typical of a practical MBE growth system.

Why, one may ask, is this intensity so very much larger than that calculated for the Stern–Gerlach experiment above? The reason is easy to understand when we calculate the mean-free-path (and therefore the size of the aperture), using equation (B2.9). Thus,

$$\lambda = kT/\pi p\sigma^2$$

$$= 3.3 \times 10^{-2}\ \text{m}$$

In other words, the aperture in this case has a radius of 33 mm, some 3000 times larger than the 0.01 mm slit width used previously. Note that, because the flux is inversely proportional to the vapour pressure in the oven (equation (B2.13)), the intensity is optimised by employing a low vapour pressure, such as we assume here. While it is vital in the Stern–Gerlach experiment that the beam must be narrow, in the interest of detecting a relatively small deflection, this is far from the case in MBE, where a wide beam is perfectly acceptable.

Finally, we should consider the importance of having a low background pressure in the growth chamber. Referring again to equations (B2.1), (B2.6) and (B2.8), we see that the number of molecules impinging on unit area of the growth chamber (and therefore on the substrate) is given by

$$N = p/4\{mkT/3\}^{1/2} \tag{B2.14}$$

from which we calculate, for a background pressure of 10^{-10} Torr $(1.33 \times 10^{-8}$ Pa$)$ and at room temperature, a value of $N = 4 \times 10^{14}$ mol/m^2/s—roughly 10^{-5} monolayers per second—compared with 1ML/s for the growing film. Sticking coefficients are usually much less than unity, so this provides an upper limit to the level of impurity atoms likely to be incorporated in the growing film.

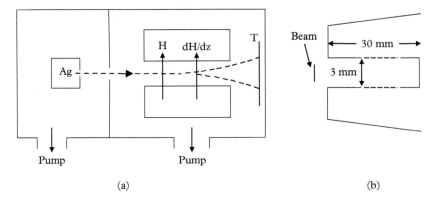

Figure 2.2 *Schematic arrangement used by Stern and Gerlach to confirm the quantisation of angular momentum as predicted by quantum theory. Figure 2.2(a) shows the overall layout of the experiment, in which the beam of silver atoms passes through a magnetic field with a large field gradient normal to the direction of the beam. T represents the detector screen on which the silver is condensed. Figure 2.2(b) shows the configuration of the magnet pole-pieces used to obtain the necessary field gradient. (Following Smith 1955.)*

condensation target serving as detector. The magnetic field was provided by an electro-magnet with specially designed pole-pieces which generated a field gradient (dH/dz) of some 10^3 Tm^{-1} in the direction of the field (see Figure 2.2(b)). Whereas in a uniform field the magnetic moment associated with the silver atoms would simply experience a couple, causing them to precess about the field axis, the presence of the field gradient results in a transverse force, bending the beam away from the original beam axis, and it was this deflection that the experiment was designed to measure.

Silver atoms are characterised by an electron configuration having a single unpaired 5s electron with ground state $^2S_{1/2}$, for which the corresponding magnetic moment is given by

$$\mu = g\beta M_s \tag{2.1}$$

where β is the Bohr magneton, $g = 2$, and $M_s = \pm \tfrac{1}{2}$ Therefore, according to quantum theory, silver atoms in the beam should be equally divided in number between those having moment +1 Bohr magneton and those having −1, the former being deflected in the direction of the field gradient, the latter in the opposite sense. With typical experimental geometry, the magnitude of the deflection was about 1 mm, much larger than the width of the beam (of order 0.02 mm) and, indeed, much larger than any blurring effects due to the statistical spread in atomic velocities, which we have neglected in our desire to present a simplified account. As must be clear, the deposit on the detector took the form of a pair of lines from whose separation an estimate of the silver atom g-factor could be made. The result gave $g = 2$, within experimental error, thus confirming the prediction of the quantum theory. (Note that the observation of such sharp lines depends crucially on the absence of any component of velocity normal to the beam direction and therefore on a lack of scattering within the beam.) By contrast, classical theory assumed that all possible orientations of the silver atom magnetic moment were equally probable and, as the measured deflection would be proportional to the component of magnetisation normal to the field axis, all deflections between zero and a limiting value would also be equally probable. In other words, the detector would register a uniform smear of deflected atoms, rather than the two well-separated lines actually observed.

The choice of silver was doubtless made because it represented the simplest possible case, that of pure electron spin. The experiment was later extended to a range of other atoms whose ground-state configurations included orbital contributions, leading to more complex deflection schemes. However, we are concerned here only with the concept of using an atomic beam to provide an appropriately non-interactive system in the interest of theoretical understanding; so, we shall not explore this topic in any greater detail but move on to consider a second important innovation which emerged some 20 years later, somewhat bizarrely, in the early stages of the Second World War. It came to be known as magnetic resonance and was, later, to lead to some exciting practical applications in the fields of medicine, communications and optoelectronics.

Expressed in quantum mechanical terms, the basis for the resonance method is relatively straightforward. If we have a molecular sample which shows no electronic moment but has a nuclear moment, then, to first approximation its magnetic energy can be written as

$$E_m = mg_I\beta_I H \tag{2.2}$$

where g_I is the nuclear g-factor, β_I is the nuclear magneton, and H is the (uniform) applied magnetic field. The magnetic quantum number m takes integral (or half-integral) values between $-I$ and $+I$ in the usual manner (where I is the appropriate nuclear spin), all $(2I + 1)$ states being equally occupied at normal temperatures. For pedagogical simplicity, we may consider the case of $I = 1/2$, for which there are only two levels. If the sample in the static magnetic field H is also subjected to an oscillating field at an appropriate RF frequency ν such that

$$\begin{aligned} h\nu &= E_{1/2} - E_{-1/2} \\ &= g_I\beta_I H \end{aligned} \tag{2.3}$$

the RF field will induce transitions between the two states. Detecting this resonant condition in terms of the field H and frequency ν clearly allows one to measure the nuclear g-factor, and the original nuclear magnetic resonance experiments made use of molecular beam methods to effect such detection.

An outline of the apparatus used for these experiments is shown in Figure 2.3. On the left was an oven to provide the beam of molecules, followed by a section containing three magnets and a collimating slit. The outer pair of magnets was designed to provide high field gradients so as to deflect the beam, in the manner already described for the Stern–Gerlach experiment, while the centre magnet provided a highly uniform field within which was an RF coil through which the beam must pass. It was here that the resonance transitions were induced to change the state of molecules, before the second deflection stage and their final arrival at the detector. Note that the deflector stages have oppositely directed field gradients so as to deflect the beam in opposite senses. In this case the

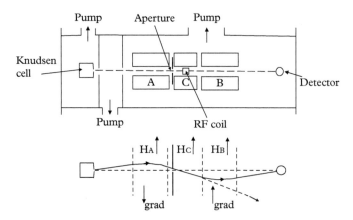

Figure 2.3 *Schematic layout of the apparatus used by Rabi to detect nuclear magnetic resonance. A beam of alkali halide molecules passes through three regions of magnetic field, A, B and C. H_A and H_B are non-uniform but with field gradients in opposite directions, so that, in the absence of an RF field, molecules deflected in region A will be deflected back again in B and still reach the detector. The addition of an RF field via a small coil in the uniform field region C induces transitions between magnetic states so that the balance between the two deflections in A and B is disturbed and the beam intensity at the detector is significantly reduced. (Following Smith 1955.)*

detector was a small-area device chosen so as to be sensitive to very small deflections of the beam—that is, if the beam moved only a small distance, the output signal dropped sharply.

The apparatus was first lined up so that, in the absence of deflecting fields, the beam passed axially through, to strike the detector and produce maximum signal (as suggested by the dotted line). On introduction of the two deflector fields, molecules were arranged to follow curved paths but, provided the two contra-deflections were in balance, and molecules remained throughout in the *same* energy state, the beam would still reach the detector, with only minimal reduction in intensity. (Once again, note the importance of minimal scattering, in this instance in the interest of ensuring that molecules are not flipped spontaneously between the two configurations.) Of course, the selection of molecules had to change because only those molecules making a small angle with the axis could pass through the central slit—however, this was of no consequence in practice because the oven produced a wide beam with only a slow variation in intensity with angle from the normal. Note that there are two symmetrical curved paths, one above and one below the axis, corresponding to molecules in the two states m = $\pm\frac{1}{2}$. The crux of the experiment was, of course, to adjust the uniform magnetic field in the central magnet, and the RF frequency so as to flip molecules in the beam between their two stable states, as they passed through the central section. This resulted in the deflection in the final section being in the wrong sense, causing the beam to move off the detector, reducing the signal by perhaps 10% of the original intensity. The whole complex of slits and deflecting magnets was no more than a means for detecting the resonance transitions induced by the RF field.

The first spectra were observed by Rabi and his collaborators at Columbia University, New York, in the early 1940s. They chose to study alkali halide molecules which have no electronic magnetic moment, an essential requirement for this experiment because the large electron Bohr magneton would result in huge deflections and thus mask the nuclear effect being sought. An interesting detail concerned the fact that both nuclei involved (the alkali and the halogen) possessed nuclear spins, which, in the absence of a magnetic field coupled together to form a pair of spins—it was only at high values of magnetic field that this coupling was broken down, allowing the separate nuclei to be studied individually. The technique was soon taken up by others and developed to measure nuclear electric quadrupole effects, which result in small splittings in the nuclear magnetic resonance spectrum. Another interesting study made by the molecular beam resonance method related to the three hydrogen molecules, H_2, HD and D_2. Proton resonance took on a special significance when it came to be used in magnetic resonance imaging but, by then, the resonance condition was detected by measuring the absorption of energy from (or emission into) the RF field, rather than by the much more elaborate molecular beam method, and further consideration would take us very far from our current interest. It is time to move on to consider our third example of molecular beam work: the demonstration of maser action in ammonia molecules.

The concept of maser action appeared in various guises during the 1950s. Based on the much earlier discussion of Einstein (1917), who pointed out the importance of stimulated emission of radiation from an excited state of any atomic or molecular system, several scientists began thinking that it might be possible to use stimulated emission to make an amplifier or oscillator. First to do so, however, was the group at Columbia University under Charles Townes, who has written a graphic account of the exciting sequence of events leading to their success (Townes 1999). As we saw above, Columbia was the home of molecular beam studies, under the direction of I. I. Rabi, so it was, perhaps, appropriate that Townes should make use of a molecular beam of ammonia. He had previously studied the microwave spectrum of ammonia in some detail so was well placed to capitalise on it to make a microwave oscillator. The structure of the ammonia molecule is characterised by a plane of hydrogen atoms, with a nitrogen atom placed symmetrically above or below the plane. These two configurations are energetically equivalent, and the nitrogen atom is able to tunnel between them, giving rise to an 'inversion' spectrum, the excited and ground states of the system being separated by approximately 25 GHz. In practice, these states are further split by interaction with the various rotational and nuclear spin states of the molecule, giving rise to a complex set of spectral lines but, from our point of view, it is sufficient to recognise that the most intense of these occurs at a frequency close to 24 GHz. In a similar manner to that employed in the Stern–Gerlach experiment and in the development of nuclear magnetic resonance, Townes made use of a beam-deflection method to separate ground-state from excited-state molecules (see Figure 2.4). Because of the lop-sided shape of the ammonia molecule, the two states possess different electric dipole moments and this allowed the separation to be achieved by sending the beam along the axis of a quadrupolar electrostatic field. Ground-state molecules were bent away from the axis,

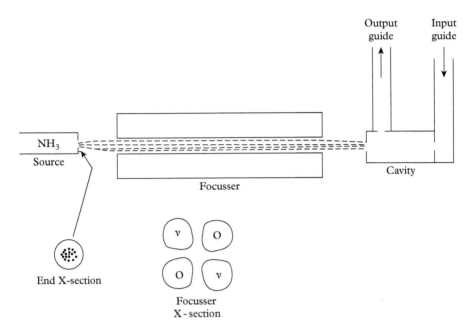

Figure 2.4 *Block diagram of the apparatus used by Townes et al. to observe maser action in the microwave spectrum of ammonia molecules. The electric quadrupole field focuses molecules in the upper energy state on a small hole in the microwave cavity, while deflecting away molecules in the lower energy state. Microwave radiation coupled into the cavity stimulates downward transitions in the excited molecules when its frequency satisfies the resonance condition. When the resulting emission is sufficient to overcome cavity losses, self-sustained oscillations occur. (From Gordon et al. 1954.) Reprinted with permission from Gordon, J P, Zeiger, H J and Townes, C H (1954) Phys Rev 95, 282. Copyright 1954, American Physical Society.*

while excited-state molecules were focussed towards the axis, thus allowing a beam of excited-state molecules to be directed through a small hole into a microwave cavity tuned to the resonant frequency of 24 GHz. Downward transitions were stimulated in the cavity by means of a microwave input from a klystron oscillator, thus producing a degree of amplification in the signal. Further to this, oscillations were produced from the noise background, when the beam intensity reached approximately 10^{13} molecules per second. It was at this point that the acronym **M**icrowave **A**mplification by **S**timulated **E**mission of **R**adiation was coined.

The narrowness of the output spectrum and the constancy of its frequency were confirmed by beating the outputs of a pair of masers together, and the device was provisionally adopted as an 'atomic clock'. However, the full import of the new invention was only recognised gradually. Two distinct developments ensued: the solid state maser, based on the energy levels of the Cr^{3+} ion in ruby, was employed as low-noise microwave amplifier in a variety of satellite communication and radio astronomy experiments, and the same material (coincidentally) was used by Theodore Maiman in demonstrating the first optical maser, or laser, in 1960. The laser, itself, may have taken off rather slowly but there can be no doubt that it now fulfils a quite bewildering range of tasks in both scientific and domestic applications. It may have been a long time between Einstein's original contribution and the practical realisation of maser action but, following Townes' initial success, progress has certainly been rapid. Molecular beams may no longer play a role in today's devices but, without their contribution, today's devices may not even exist. We must now turn our attention to their use in growing even more of today's devices. Molecular beam epitaxy can be seen to enter the fray sometime in the middle of the 1960s.

2.3　Molecular Beams Applied to Nucleation Studies

So far as we are aware, the first application of collision-free molecular beams to crystal growth was the work performed by Bruce Joyce and Bob Bradley at the Plessey Allen Clarke Research Centre, Caswell. Together with their collaborators, Brian Watts at Plessey and Roger Booker at the University of Cambridge, they published a series of papers between 1966 and 1968 concerned with the nucleation of silicon thin films grown by the pyrolysis of silane (Booker and Joyce 1966; Joyce and Bradley 1966; Joyce et al. 1967; Joyce 1968; Watts et al. 1968). The reader will remember from Chapter 1 that silicon epitaxy was developed during the early 1960s as a means of depositing lightly doped films of silicon on heavily doped substrates and that a variety of techniques were employed, based on the hydrogen reduction of silicon chlorides or the thermal decomposition of silane on a heated silicon substrate. The Plessey company had been involved in the development of silicon integrated circuits from the beginning of the 1960s and had established an interest in silicon epitaxy from 1962 onwards. Understandably, then, their research staff became interested in the mechanism whereby silane reacted with a hot silicon surface to deposit an epitaxial film; that is, on the nucleation of the first few atomic layers (it being reasonable to believe that the ultimate quality of the film might depend strongly on the details of the nucleation process). Their first attempts to study nucleation were made in the basic gas-flow system used for practical film growth and this imposed severe restrictions on the interpretation of experimental data. In his review article, Joyce (1968) lists three problem areas: (a) the rate-limiting step is frequently one of gaseous diffusion to and from the surface, rather than the kinetics of surface nucleation, (b) the length of time taken for the gas-flow system to reach equilibrium is usually longer than the time for which nucleation can be observed, making it impossible to obtain reproducible behaviour and (c) in the gas (typically at atmospheric pressure), there is a likelihood of homogeneous, gas phase chemical reactions occurring, which confuse the understanding of any silane–silicon surface interactions. It was with a view to overcoming these difficulties that they introduced a fundamentally different approach—the use of a molecular beam system.

From the viewpoint of gaining a fundamental understanding of the growth process, this confers a number of important advantages. The collision-free nature of a molecular beam ensures that there will be no gas phase reactions—nucleation results entirely from surface reactions between silane molecules and silicon surface atoms. Likewise, gas phase diffusion plays no role. The beam can be switched on and off by a shutter, almost instantaneously, which allows the study of any predetermined phase of the growth process. As we see from Box 2.1, it is straightforward to calculate the intensity of the beam as it strikes the silicon surface, and measuring the temperature of the Knudsen source from which the beam emerges defines the energy of the impinging molecules (following the Maxwell–Boltzmann distribution). This contrasts with the situation in a gas-flow system, where silane molecules are heated variously by gas phase collisions as they approach the silicon surface. Finally, the use of an ultra-high vacuum system means that the nucleation process can be studied without interference from background impurity molecules. There can be little doubt that this Plessey work represented a major step forward in the understanding of epitaxial growth and could clearly be adapted to the study of other epitaxial processes.

The apparatus used by Joyce et al. is illustrated in Figure 2.5. Its construction was constrained by the need to study atomically clean surfaces, which implied the use of a stainless steel UHV system, pumped by a pair of oil-diffusion pumps: one for the source region and one for the deposition chamber. Both pumps were provided with liquid nitrogen cold traps and molecular sieve traps. The whole chamber could be baked at 300 °C to outgas the various components. Background pressures, measured by ionisation gauges, were 5×10^{-7} Torr in the source region and 3×10^{-9} Torr in the

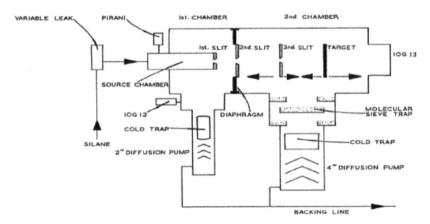

Figure 2.5 *Diagram of the stainless steel molecular beam system used by Joyce et al. to study the nucleation of silicon atoms on a silicon substrate, by the pyrolytic decomposition of silane. Oil diffusion pumps with cold traps and molecular sieve traps achieved background pressures of 3×10^{-9} Torr in the deposition chamber and 5×10^{-7} Torr in the beam formation chamber. Pressures were measured by Bayard–Alpert ionisation gauges. Silane gas was introduced into the Knudsen source by way of a controlled leak and its pressure measured by a Pirani gauge to permit calibration of beam intensity. Silicon growth rates were of order 1 Å per minute. (From Joyce and Bradley 1966, courtesy of Taylor and Francis.)*

deposition chamber. Two mechanical feed-throughs were used for adjusting the system geometry, shutting off the beam and beam chopping. The source gas, silane, was leaked into a Knudsen cell and kept at room temperature, its pressure being monitored by a calibrated Pirani gauge. The silicon substrate (or target) was electrically heated to temperatures in the range 800 °C–1200 °C, its value being measured with an optical pyrometer. Considerable care was taken to measure the beam intensity and shape at the target in order to confirm calculated values, based on source pressure and temperature, and system geometry. Good agreement between the two encouraged confidence in the interpretation of subsequent experiments. In passing, we might note the similarity with later MBE equipment—the use of stainless steel UHV chambers, cold traps, ionisation gauges, Knudsen sources, heated substrates, optical pyrometry, mechanical shutters, etc., all being typical of modern MBE practice.

A typical experiment began by pumping down to the appropriate background pressures and then heat-cleaning the substrate at a temperature of 900 °C before reducing the temperature to the value required for film growth, usually about 850 °C. (Substrates were pre-cleaned by chemical polishing in an HF-HNO_3-CH_3COOH etching solution.) The silane beam was turned on and, when relevant parameters had reached stable values, the beam shutter was opened and growth commenced. The beam intensity was typically 7×10^{20} molecules m^{-2} s^{-1}, resulting in a growth rate of about 1 Å per minute (10^{-2} monolayers per second). This corresponded to about 0.01% of silane molecules decomposing to form silicon surface atoms, the very first example of molecular beam epitaxy. The object was not, of course, to grow practical films but to study the nature of the growth process by applying TEM and other surface measurement techniques. The TEM replica micrograph of a silicon (100) surface, shown in Figure 2.6 (Joyce 1968), demonstrated the growth of surface nucleation centres, which grew in extent with continuing beam application time. An interesting feature was the observation of an induction period prior to the formation of the first nucleation sites, ascribed to the time taken to remove silicon oxide from the surface via the reaction of silicon with silicon oxide, according to

$$Si + SiO_2 = 2SiO \tag{2.4}$$

forming the volatile SiO molecule, which rapidly desorbed. Once this oxide had thus been removed, leaving a clean Si (7×7) reconstructed surface, the density of nucleation sites grew rapidly to reach a saturation value and remained at this level until coalescence of these centres caused it to drop again. This saturation level was found to increase with increasing beam intensity and to decrease exponentially with increasing temperature. The data were consistent with a kinetic model in which a critical cluster of atoms was essential to the formation of stable nuclei, the cluster size being 2 for a (111) and 3 for a (100) surface, respectively. The measured 'activation energy' (NB the density of nuclei was not strictly thermally activated but was proportional to $\exp\{E_A/kT\}$) was associated with a combination of energies: the energy gained from the formation of the cluster, the activation energy for the decomposition of silane and the activation energy for surface diffusion. Growth on *clean* surfaces was therefore shown to proceed

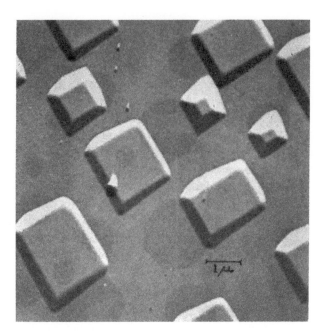

Figure 2.6 *Replica electron micrograph of a silicon (100) surface, during MBE deposition, showing the formation of individual nucleation centres. Electron diffraction measurements showed that, on a clean surface, each centre was free of defects. (From Joyce 1968, courtesy Elsevier.)*

in this manner, rather than by lateral propagation of surface steps but, if deliberate carbon contamination was introduced, step propagation might well dominate the growth process.

However, our immediate purpose is not to present a discussion of crystal growth mechanisms but to recognise the origins of MBE, as illustrated by this pioneering work of Bruce Joyce and colleagues in the second half of the 1960s. Joyce presented a review paper summarising their work at the Second International Conference on Crystal Growth (ICCG2) in Birmingham, in1968, following which the British Association for Crystal Growth (BACG) was formed. History was well and truly in the making, though it must be seen as significant that this first foray into MBE made absolutely no pretension towards being a practical growth method—growth rates of a micron per week were hardly likely to set the semiconductor device world buzzing! As we shall see in the next section, the essential steps towards the establishment of MBE as a practical growth method occurred on the opposite side of the Atlantic. But, as we shall also see, Bruce Joyce's involvement with MBE was far from over. He was to make a considerable name for himself in a later incarnation, within the confines of the Mullard (later Philips) Research Laboratory near Redhill in Surrey, and later at Imperial College, London.

2.4   Early Evaporation Methods

As we saw in Section 2.1, thermal evaporation of semiconductor films had been studied for many years before the arrival of MBE in the 1970s and it would be futile to try
and draw a sharp dividing line between the two. Perhaps, the growth of IV-VI films for
infrared devices best illustrates the development of what was originally called 'evaporation' into what came to be called 'molecular beam epitaxy'. The binary compounds
PbS, PbSe and PbTe were seen as promising materials for thermal imaging applications
during the run-up to the Second World War, when polycrystalline films were deposited
on glass substrates. Indeed, such films served as the basis for most of the far-infrared
detector developments until the early 1950s, when artificial bulk single crystals first became available. Even this advance, however, was found to be less than totally satisfactory
from the device viewpoint, due to problems of structural perfection and chemical purity
(the fact that the band gaps of these materials were not accurately determined until the
1970s illustrates this point rather well—see Orton 2004, p. 402). There was still an urgent need for single crystal epitaxial films. There was also an urgent need for a material
system allowing choice of band gap in the range 0.1 eV to 0.4 eV, covering the infrared
windows between 3–5 μm and 8–14 μm, and this requirement was satisfied, initially at
least, by the pseudo-binary alloy PbSnTe, variation of the ratio of tin to lead effecting
the desired variation of band gap. The first far-infrared lasers were made in 1964, using
bulk single crystals of PbSe and PbTe but the need for heterostructure devices (to allow
CW operation at temperatures of 77 K or greater) called for epitaxy, and MBE came
first to the rescue. PbSnTe lasers operating CW at 65 K were reported by the middle
of the 1970s and it is interesting to look at the growth technology then in use. This
has been well reviewed by Holloway and Walpole (1980) of the Ford Motor Company,
Dearborn, Michigan, and the Lincoln Laboratory of MIT, respectively, who have been
major contributors to the field.

We pointed out in Section 2.1 that the lead chalcogenides evaporate in molecular
form, so the growth of these films on a suitable crystalline substrate is relatively straightforward. Early work at the US Naval Ordnance Laboratory in Maryland (Schoolar and
Zemel 1964; Zemel et al. 1965) made use of NaCl or KCl substrates, which have the
same crystal structure as the lead salts, and the evaporations took place in a liquid-
nitrogen-trapped, diffusion-pumped vacuum system (see Figure 2.7). Finely pulverised
PbS (later PbSe and PbTe) was loaded into a quartz oven and evaporated onto a
substrate, heated to approximately 300 °C, the whole being pre-baked with a shutter
between source and substrate. Once a steady-state condition had been established, the
shutter was removed and PbS deposited at a rate of 100 Å per minute (very roughly
a monolayer per second, typical of modern MBE practice). The authors omit mention
of the background pressure attainable in this ion-pumped vacuum system but quote a
value of approximately 10^{-5} Torr during deposition. The resulting films showed optical
properties closely similar to those of bulk single crystals and electron mobilities within a
small factor of those measured on the best bulk material. However, the low-temperature
data suggested that grain-boundary scattering was probably significant—in other words,
these samples were still polycrystalline. In a later paper (Schoolar and Lowney 1971),

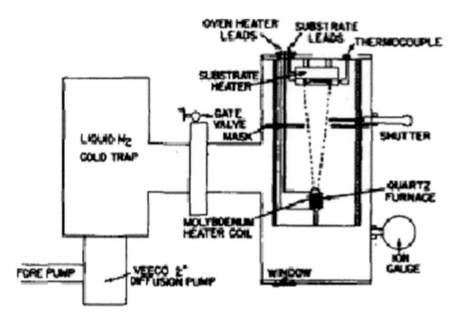

Figure 2.7 *Schematic diagram of the evaporation apparatus used by Schoolar and Zemel (1964) to grow epitaxial films of PbS on NaCl substrates. Finely pulverised galena was placed inside a quartz oven, heated by a molybdenum coil. Substrates were heated to 300 °C and deposition rates were about 100 Å per minute. The pressure, during evaporation, was in the region of 10^{-5} Torr. (From Schoolar and Zemel 1964.) Reprinted with permission from Schoolar, R B and Zemel, J N (1964) Jnl Appl Phys 35, 1848. Copyright 1964, American Institute of Physics.*

films of PbSe were used to make infrared photoconductive detectors, though it was found necessary to remove the films from their substrates on account of problems with excessive strain, on cooling the devices to their operating temperature.

The next important step in improving material quality came with the use of lead chalcogenide, SrF_2 or BaF_2 substrates (Holloway and Walpole 1980). The lattice match between the lead salts and the fluorides is better than that with the alkali halides, and material quality was correspondingly improved, though the use of an improved vacuum system may also have contributed—the system was ion pumped and achieved pressures during deposition of about 10^{-7} Torr (one can only guess that background pressures were in the region of 10^{-9} Torr, and therefore within the UHV range). For example, low-temperature electron mobility in evaporated PbTe-on-BaF_2 films reached 50 m^2 V^{-1} s^{-1}, indicative of much improved crystalline perfection. Such films were also used for making photodiodes with useful detectivities at temperatures attainable by thermoelectric cooling. Perhaps of greater significance in practical terms was the deposition of films of PbSnTe appropriate to both the 3–5 μm and 8–14 μm windows. Care was necessary to employ a homogeneous sample of the alloy (by prolonged annealing) and to use only a modest fraction of the source material during deposition. As evaporation was not strictly congruent, its composition changed slowly during the period of evaporation. It was also necessary to use an open quartz crucible, rather than a Knudsen cell. Given these precautions, satisfactorily uniform films were obtained and sensitive photodiodes made. Considerable success was also achieved in making far-infrared heterostructure lasers from films grown on chalcogenide substrates, and significantly lower threshold currents were obtained, compared with those obtained using bulk crystal devices. In fact, MBE was to play a major role in the longer-term future of far-infrared lasers but to discuss this now would take us well outside our present time frame. It is important, however, to recognise just how close this deposition system (assembled about 1970) was to a modern MBE system. It employed ion pumping, UHV background pressures, substrate heating, boron nitride crucibles for source material, and growth rates of a few Ångstroms per second. It lacked only multiple sources and surface analysis equipment.

Needless to say, the IV-VI semiconductors were not the only ones being grown by vacuum evaporation during the 1960s and early 1970s. One of the earliest examples was the work of Davey and Pankey (1964, 1968) at the Naval Research Laboratory in Washington, DC, who described the growth of GaAs films by the so-called 'three-temperature' method. This technique was developed at Siemens during the early 1960s for the growth of polycrystalline thin films of InSb and InAs for Hall effect devices and has been well reviewed by Freller and Gunter (1982). It is based on the principle that, in evaporating a compound which decomposes into species with significantly different vapour pressures, a regime can be found in which the stoichiometric compound is deposited on the substrate—this by using two sources and controlling both their temperatures and that of the substrate. Davey and Pankey's first paper referred to deposition on glass substrates and so could hardly be described as epitaxy; however, the later paper referred to growth on GaAs and Ge substrates and showed evidence of single crystal epitaxial growth. As for many compounds, GaAs does not evaporate congruently and it was therefore necessary to employ separate sources for the constituent atoms, the three temperatures referred to being those of the gallium, the arsenic and the substrate zones, respectively. A schematic diagram of the apparatus is shown in Figure 2.8—it was designed to allow a sequence of substrates to be moved into the substrate oven. The method was based on the principle that, if arsenic was present at the substrate in excess, the growth rate would be determined by the arrival rate of gallium atoms. Thus, the temperature of the gallium cell was crucial in controlling growth rate and needed to be optimised for crystal quality—too high a rate, and free gallium appeared on the film surface. Typical temperatures were 910 °C for the gallium, 295 °C for the arsenic and 450 °C for the substrate, at which temperature films were found to be essentially single crystals. Substrates were cleaned in situ by argon ion bombardment, which was followed by thermal annealing, the whole apparatus was baked and pumped to a background pressure of 10^{-7} Torr, growth rates were held below 2 Å/s and film quality was monitored (ex situ) by reflection electron diffraction. Once again, the similarity to modern MBE growth of GaAs is striking.

Perhaps of even greater significance was the work of Ito and Takahashi (1968) at the Tokyo Institute of Technology, where they reported the growth of Ge epitaxial films on silicon substrates (the optoelectronic properties of heterojunctions were growing rapidly

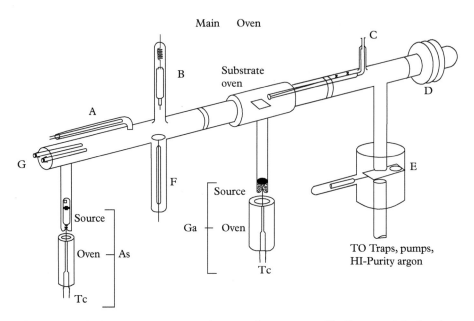

Figure 2.8 *The 'three-temperature-zone' evaporation system used by Davey and Pankey for the growth of GaAs epitaxial films on GaAs and Ge substrates. The gallium cell was held at a temperature of 910 °C, the arsenic cell at 295 °C and substrate temperatures were varied between 175 °C and 450 °C. Deposition rates for optimum film quality were of order 100 Å per minute. The quartz tubing was pumped to a base pressure of 10^{-7} Torr and baked, the substrate cleaned by argon ion bombardment and annealed at 500 °C and then the substrate temperature adjusted to the appropriate value for deposition. (From Davey and Pankey 1968.) Reprinted with permission from Davey, J E and Pankey, T (1968) Jnl Appl phys 39, 1941. Copyright 1968, American Institute of Physics.*

in importance at the time). They made use of a stainless steel vacuum system, pumped by an ion pump and a titanium sublimation pump; background pressure, after baking, reaching 10^{-9} Torr. Substrates were heat-cleaned in situ at 1250 °C and Ge films evaporated from an electrically heated tungsten spiral. Substrate temperatures during growth ranged from 330 °C to 930 °C and were measured either by a thermocouple or by an optical pyrometer. Many of the features of a modern MBE machine were already in place—could Japan lay claim to being first to develop MBE? Or, perhaps we should see this as an important step along the way? Kiyoshi Takahashi certainly played a major role in furthering the growth of a variety of III-V compound materials for application as FETs, graded base transistors, etc., by what he called a 'normal MBE system' but this was several years later, when MBE was already achieving international fame (see e.g. Naganuma et al. 1975 and Tateishi et al. 1976). Exactly what contribution he made to the development of such normal MBE systems is not clear. Returning, for a moment, to the growth of Ge films, we should also note examples of work in Russia (Datsenko et al. 1971). Films were deposited under vacuum conditions of 3×10^{-6} Torr at substrate temperatures between 240 °C and 800 °C and their structural quality investigated by X-ray topography. Films ranged from polycrystalline to 'mosaic-monocrystalline'. The authors appear ignorant of the earlier paper of Ito and Takahashi and their work added little of merit.

To complete this brief survey of evaporation precursors to modern MBE, we shall simply note that various other materials were grown in this way. Examples include the growth of CdS on various substrates by Shimaoka (1971) and of ZnTe on Ge by Calow et al. (1972). The II-VI compounds evaporate congruently so it was necessary only to use a single source. The vacuum used by Shimaoka was 5×10^{-5} Torr and substrate temperatures ranged from 23 °C to 500 °C. Calow et al. were careful to explore the use of three different vacuum regions, ranging from 10^{-5} Torr to 10^{-8} Torr, and deposition temperatures from 250 °C to 450 °C. Substrates were prepared by chemical polishing but were not heat-cleaned in situ. Oxygen contamination was clearly a problem, as was the purity of their starting material, and film quality left much to be desired. The lack of in situ surface analysis made it difficult to draw reliable conclusions but this, we have to admit, represents a somewhat unfair criticism, based, as it is, on considerable hindsight! It would certainly be unreasonable to dismiss this early work as unimportant, bearing in mind that the resulting films, however imperfect, were probably superior to those grown by any alternative method at the time. The point that we should take from these evaporation experiments is that they were feeling the way towards modern MBE technology and certainly acted as a worthwhile stimulus to the development of the sophisticated equipment we now take for granted. Many of the difficulties experienced led, gradually, to the appreciation that high vacuum, high purity of materials and in situ characterisation were of crucial importance to the growth of high-quality epilayers, all of which were very soon to come together in the development of commercial MBE machines. For readers unfamiliar with UHV technology, we include a brief summary in Box 2.2.

Box 2.2 The evolution of UHV

Standard practice in modern MBE systems involves the use of a stainless steel vacuum enclosure with pumps able to evacuate it, after bakeout, to background pressures of 10^{-10} Torr (roughly 10^{-8} Pa) or better. Such vacuum systems are now taken largely for granted but it is interesting to take a brief look at their evolution.

Vacuum evaporation of both metal and insulator films has a long history but was performed in glass vessels, employing rubber vacuum seals and characterised by base pressures of, typically, 10^{-5} Torr, achieved with a combination of a water-cooled, oil-diffusion pump backed by a standard rotary pump. As we saw in Box 2.1, such a vacuum would imply approximately one monolayer of impurity atoms striking the substrate surface every second, making it quite impossible to deposit high-purity films. However, by employing relatively large evaporation rates, adequate films for most purposes could be obtained, typical examples, within the confines of semiconductor technology, being the ohmic and Schottky barrier contacts formed by evaporation of various metals, and the formation of transparent contacts by evaporation of

continued

Box 2.2 *continued*

indium tin oxide. The motivation for improved vacuum systems came initially from studies of surface science, where it was important to be able to work with atomically clean surfaces. Again, to quote an example from semiconductor experience, the development of infrared photocathodes made an important step forward when it was discovered that a single mono-layer of caesium on a clean GaAs surface could result in negative electron affinity (Scheer and van Laar 1965). A perfectly clean surface was achieved by cleaving a GaAs crystal in the vacuum system but, clearly, it was essential that this surface should remain clean through the time taken to deposit the caesium film. Experimental studies on clean surfaces could obviously demand that a surface should remain uncontaminated for even longer periods—much improved vaccua were therefore a vital component of such work, the study by John Arthur of the behaviour of Ga and As atoms on the surface of GaAs, during the latter half of the 1960s, being particularly relevant.

Early UHV systems, developed during the second half of the 1950s, were made in glass or quartz and pumped with sorption pumps and ion pumps. The very possibility of working at such low pressures depended on an interesting development in the ion gauges used to measure them. For a long time, it was believed that the lowest pressure achievable lay in the region of 10^{-6} Torr, until it was discovered that this was an artefact of the way in which the gauge worked. Electrons from a heated cathode were attracted to a positively charged grid and, in transit, ionised residual gas atoms, which were then attracted to a negatively charged collector. This ion current was taken as a measure of the gas pressure and it worked very well down to pressures of about 10^{-8} Torr but this appeared to be a lower limit until it was recognised that, when electrons impacted on the grid, they released soft X-rays which, in turn, ejected photoelectrons from the collector, thus producing a current which was indistinguishable from the ion current. This secondary component was independent of gas pressure and appeared to imply a limit to the lowest pressures that could be measured. The solution, in the shape of the Bayard–Alpert gauge, was to reduce the area of the collector by making it in the form of a fine wire. This had the effect of minimising the number of X-ray photons intercepted and thus the spurious collector current. From 1950 onwards, it became possible to realise background pressures as low as 10^{-10} Torr, and UHV was effectively born.

Glass enclosures may have been considerably cheaper than stainless steel but they were certainly less convenient, requiring to be resealed for each new experiment. They were also more fragile. Exchanging samples or introducing new measuring devices could be tricky and, furthermore, it was far from easy to mount auxiliary equipment such as an Auger electron analyser or mass spectrometer. Most people interested in clean surface studies therefore tended to choose the stainless steel option, in spite of its greater cost and, let it be said, the inconvenient fact that stainless steel contains a considerable quantity of hydrogen, which outgases throughout the lifetime of the system. Of particular significance was the development by the Varian Company in the USA of the 'conflat flange', which provided a relatively easy and totally reliable method of connecting different pieces of the apparatus together. It involved a pair of stainless steel knife edges, biting into a copper gasket which distorted sufficiently to seal any minor imperfections in the knife edges. Two pumping systems were available: one using suitably trapped diffusion pumps, the other based on a combination of sorption, sublimation and ion pumps (and therefore oil free).

In the oil-free design, the primary pumping uses two or three sorption pumps to reduce the pressure to the point where the sputter ion pump can begin operation. Such pumps are filled with 'molecular sieve' and are activated by heating to about 200 °C, which removes gas previously pumped. They are then cooled to 77 K and used sequentially. The first pump might reach about 10^{-3} Torr, and the second an ultimate pressure of 10^{-5} Torr, at which point the sputter ion pump can take over. This reduces the pressure to about 10^{-8} Torr, when the system is baked at 180 °C–200 °C to remove all water vapour. The resulting pressure of about 10^{-10} Torr consists mainly of H_2, CO and CO_2. A final pumping stage typically uses a titanium sublimation pump to achieve the ultimate vacuum. The main advantage of this system is that it is completely oil free; the main disadvantage is that it does not easily handle a gas load. It would, for example, be quite inappropriate for the plasma-assisted MBE growth of nitride materials, where nitrogen gas is continuously generated during growth. In modern designs the pumping may be augmented by a closed-cycle helium cryopump, which does have the advantage of pumping gases such as nitrogen efficiently. It is also important in MBE growth of ultra-high-purity material (as we shall see, e.g. in Chapter 6).

Let us look briefly at the modus operandi of these various pumps. The sublimation pump takes the form of a titanium filament, which can be heated by passing a high current through it. Titanium is sublimed so as to coat a water- or liquid-nitrogen-cooled surface within the chamber with a thin film of clean metal, which reacts strongly with many residual gas molecules (though not with noble gas molecules). This active surface coating gradually becomes contaminated, so it must be renewed by further sublimation—the pump therefore operates in a series of short bursts. The ion pump works on a rather complicated set of principles. First, residual gas molecules are ionised by interaction with an electron cloud; second, these ions are attracted by a potential of several kilovolts to a cathode (usually titanium); third, they sputter atoms from its surface to deposit a fresh layer of material on the anode; and, fourthly, this newly deposited material absorbs gas atoms by either chemisorption or physisorption, depending on the chemical reactivity of the gas species. The whole process can be likened to a constantly replenished titanium sublimation pump, though this is a slight oversimplification because some pumping action does occur at the cathode as well. An important feature of pump design involves the enhancement of the electron–gas molecule interaction by the addition of a 'crossed' magnetic field to induce spiral electron orbits within open-box-like anode cells known as 'Penning cells'. The cryopump, as its name implies, works by condensing gas molecules on a cold surface. Pumping nitrogen, hydrogen or helium demands a temperature as low as 10 K, which is achieved by using compressed helium in a 'helium-liquefier-like' closed system. Its pumping efficiency decreases slowly during use, as the surface becomes saturated with gas molecules, so it must be regenerated from time to time by raising the panel to room temperature or above, conveniently during bakeout of the whole vacuum system.

The second pumping option uses a properly trapped diffusion pump and, with the development of low-vapour-pressure silicone and fomblin oils, is capable of achieving very similar ultimate pressures. The English company Vacuum Generators (later to become VG Semicon) pioneered the use of a 'CCT' trap, which combined a liquid nitrogen trap with an integral titanium sublimation pump. The advantages of diffusion pumping are reduced cost and the ability to handle a gas load. A minor disadvantage, for some applications, is the modest amount of vibration associated with the backing pump. A more serious objection is the possibility of

continued

Box 2.2 *continued*

---

contaminating the UHV system with high-vapour-pressure oil from this same backing pump in the event of system failure (or human error!). We have already discussed the sublimation pump—the diffusion pump operates by boiling the oil to produce a jet of oil molecules which are directed through a system of baffles to form a jet of vapour. Residual gas molecules are swept from the vacuum system by the force of this jet and transferred to the backing pump. The oil vapour condenses and is returned to the boiler.

One final comment on UHV pumping systems—many modern apparatuses make use of a turbo-molecular pump, which was introduced in 1958 by W. Becker and is a development of the original Gaede molecular-drag pump. This pump operates by fast-moving rotor blades 'hitting' molecules and thereby imparting momentum to them, directing them into the pump exhaust. The rotors must rotate at very high speeds, and bearing design has been crucial to effective working. Early models used oil bearings, later replaced by low-vapour-pressure grease and, ultimately, by magnetically levitated bearings. In the latter case, such pumps, usually operating into a conventional backing pump, are capable of achieving system pressures of 10^{-10} Torr, thus competing effectively with the alternatives described above.

2.5   The Bell Labs Involvement

The Bell Telephone Laboratory at Murray Hill, NJ was built in 1941, and played a significant role in the US scientific war effort. However, its real purpose was clearly one of applying basic scientific methods to the post-war development of the communications industry, an objective well illustrated in 1947 by the discovery of transistor action in a small piece of germanium left over from the wartime radar programme (Riordan and Hoddeson 1997; Orton 2004). Such was the success of the transistor programme that basic science in an industrial context appeared secure for evermore, and Bell scientists were renowned for the quality (and quantity) of their research output. A particular activity of relevance to our interest here was the successful development of a room-temperature, CW GaAs laser in 1970 by Morton Panish and Izuo Hayashi (Casey and Panish 1978). This was based on their development of LPE growth of GaAs and AlGaAs, referred to in Chapter 1, and led to considerable emphasis being placed at Murray Hill on the crystal growth of III-V materials. The development of MBE very soon came to be one of the more important parts of this programme and, in true Bell Labs style, practical and fundamental aspects were pursued with equal vigour. Two young, aspiring scientists, John Arthur and Alfred Cho, joined Panish's group during the 1960s and, in establishing international reputations for themselves, played important roles in developing MBE into a reliable, practical growth method. What is more, according to Arthur, the very name for the process, 'molecular beam epitaxy' was coined by Panish. Clearly, the subject owes a great deal to Bell Labs.

John Arthur joined the Murray Hill lab in 1961, with a background in the study of catalysis on metal surfaces, and decided to apply his expertise in the kinetics of surface reactions to the materials of the moment—semiconductors. Being a member of the group that had raised LPE of III-V compounds to a new level of sophistication, it was natural for him to concentrate initially on GaAs and related materials. It was also helpful that surface science techniques were developing rapidly at that time—UHV conditions were essential to the study of clean surfaces, the Auger electron spectrometer had recently been invented and quadrupole mass spectrometers were available for analysing species desorbing from a surface. LEED (low energy electron diffraction) was also available for studying surface reconstruction, post epitaxial film growth. Some work had already been done on germanium and silicon surfaces, showing evidence for complex reconstruction (the surface arrangement of atoms differing significantly from that in the bulk) but little was known about the surface structure or surface kinetics of the III-V materials. The field was wide open and the Bell management philosophy of the time encouraged its graduate staff to explore whatever interested them. It was, perhaps, the period of maximum confidence for the electronics industry, and research could do little wrong! In Europe, too, the philosophy was typified within the Philips 'empire' by the Casimir doctrine that research, left to itself, would produce practical good, no matter what! In the case of MBE, it seemed to work.

Arthur set up a bakeable stainless steel UHV system with background pressure of 10^{-10} Torr, an argon sputter gun for substrate cleaning, and a facility to perform LEED and Auger spectroscopy (see Figure 2.9). The molecular beam sources took the shape of pyrolytic BN cells, and the beams could be chopped, while the desorbed species were measured with a mass spectrometer. He proceeded to analyse the vapour pressures of Ga, As_2 and As_4 over GaAs, proving that, at typical cell temperatures of $850\,°C$–$900\,°C$, the dominant arsenic species is As_2, and the gallium arrival rate at the substrate is close to a monolayer per second, while the arsenic rate is roughly an order of magnitude greater. As the growth rate, under excess arsenic, is determined by the gallium arrival rate, this ensures a net growth rate of one monolayer per second, typical of normal MBE practice. He went on to study the growth mechanism of GaAs by measuring the surface lifetimes of Ga and As_2 (i.e. the mean residence times of a Ga or As atom on the GaAs surface), using chopped beams and mass spectrometric detection of the desorbed species. In the absence of gallium atoms on the surface, the arsenic lifetime was found to be extremely short but, on a Ga-covered surface, the arsenic lifetime was of order seconds, similar to the values appropriate for gallium. These results, in fact, led to the proposal that adsorbed gallium is necessary for arsenic to adsorb on a GaAs surface at elevated temperatures (and, thus, that growth rate is determined by the arrival rate of gallium). We shall look at Arthur's work in more detail in Chapter 4 (Section 4.2). In passing, we might note that this work of Arthur's took place at almost exactly the same time as (though quite independently of) Bruce Joyce's work on silicon epitaxy at Plessey.

As Arthur's research progressed, it became clear that MBE could be used to grow 'real' epilayers of III-V compound semiconductors—indeed, he himself demonstrated this by growing high-quality films of GaAs, GaP and GaAsP on GaAs and GaP substrates (Arthur and LePore 1969). It also became clear that the technique had a distinct advantage over LPE for growing DH laser structures in so far as the layer thickness could be much better controlled, a fact with obvious attraction for laboratory management. The upshot was a decision in 1968 to recruit a new member of staff, Al Cho, who had recently completed a PhD degree from the University of Illinois, studying

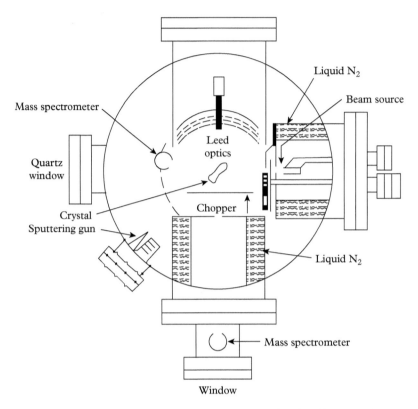

Figure 2.9 *Top view of the stainless steel MBE system developed by John Arthur at Bell Labs for studying the kinetics of molecular beam-surface interactions. The mass spectrometer was arranged so that it could be used to analyse both incident and scattered beams, the incident beam being modulated by a mechanical chopper. Surface structure could be monitored by the LEED system. The chamber was ion pumped and could achieve a background pressure of 10^{-10} Torr, after an 8 h bakeout. (From Arthur and Brown 1975.) Reprinted with permission from Arthur, J R and Brown, T R (1975) J Vac Sci Technol 12, 200. Copyright 1975, American Institute of Physics.*

metal surfaces and the effect of monolayer additions on their work functions. (Incidentally, Cho claims to have built the first quadrupole mass spectrometer as part of his doctoral research.) He immediately set about converting an existing UHV system to III-V MBE use and equipped it with Auger electron spectroscopy (AES) and reflection high energy electron diffraction (RHEED) capabilities, together with an argon ion gun for substrate cleaning (see Figure 2.10). In a series of papers published between 1970 and 1975 (summarised in the review article by Cho and Arthur 1975), Cho not only made important contributions to the understanding of III-V film growth but demonstrated their application to a wide range of device structures. He realised that, for

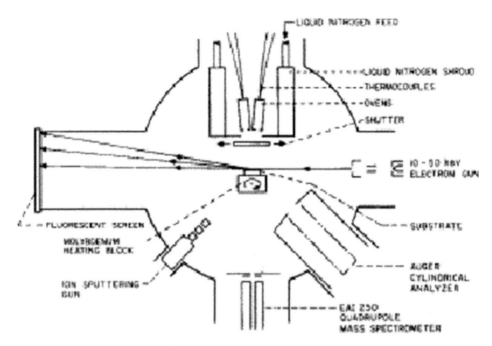

Figure 2.10 *Overview of the MBE system used by Cho to grow GaAs, AlGaAs and GaP films for a range of device structures. It included doping sources as well as Ga, Al, As and P sources, therefore requiring an increased number of effusion cells. Surface structure of the films could be monitored by the glancing-incidence RHEED system. Substrates could be cleaned by ion bombardment, and surface cleanliness monitored by the Auger electron system. (From Cho and Casey 1974.) Reprinted with permission from Cho, A Y and Casey, H C (1974) Jnl Appl Phys 45, 1258. Copyright 1974, American Institute of Physics.*

MBE to be accepted as a viable growth method, it would be necessary for such devices to perform at least as well as those produced by the well-established methods, and his dramatic success in this endeavour changed attitudes towards MBE for ever. No longer was it seen simply as an interesting 'blue skies' research tool but now as a practical technique for preparing challenging new devices, with clear potential for future application in industrial development and production facilities. In particular, Cho's work provided a strong stimulus for the development of the commercial MBE machines which appeared in the mid 1970s and set the scene for a quite remarkable surge of international activity.

Use of the AES facility revealed valuable information concerning the relationship between substrate cleanliness and film growth, its sensitivity to surface impurity atoms being typically 0.1% of a monolayer. Very briefly, AES is performed by irradiating the substrate surface with a beam of electrons with energies in the range 3–5 keV and analysing the energies of the electrons re-emitted. Ionised core levels in surface atoms are filled by means of a two-electron Auger process in which one of the electrons gains sufficient energy to allow it to leave the surface, and the energies of these electrons are characteristic of the atoms involved. Chemically polished substrates showed strong peaks representing carbon and oxygen, the latter being readily removed by heat treatment at 540 °C. Carbon, on the other hand, was unaffected by heating and, if present at more than about 10% ML, had to be removed by argon ion cleaning, otherwise, the resulting film suffered from faceting and twinned growth. Such studies led, therefore, to a reliable method of substrate preparation involving chemical polishing with (in the case of GaAs) a solution of bromine in methanol, followed by washing in distilled water to leave an oxidised surface. This oxide could then be removed simply by heating in the UHV chamber (though the precise temperature required has been the subject of some dispute!).

RHEED proved extremely convenient for monitoring surface structure as it could be used at glancing incidence, while the beams arrived normally, thus allowing measurements to be made during growth (see Figure 2.10). Diffraction patterns from heat-cleaned substrates showed three-dimensional 'spots' indicating rough surfaces, the electron beam penetrating through asperities and being diffracted as though through bulk material. After the growth of one micron of GaAs, RHEED patterns had been replaced entirely by 'streaks' which were taken to represent diffraction from a flat surface (though, as we explain in Chapter 4, this was not strictly correct). The details of these streak patterns revealed the existence of a considerable variety of surface structures and provided rich pickings for anyone interested in the physics/chemistry of surface reconstruction. For the moment, we simply note that two particular structures on the GaAs (100) surface were associated with specific growth conditions, the $C(2 \times 8)$ structure corresponding to 'arsenic-stabilised' growth (i.e. growth with a large excess of arsenic in the beam), while the $C(8 \times 2)$ structure corresponded to gallium-stabilised conditions. The appearance and disappearance of these various structures as a function of substrate temperature were also to prove valuable as they provided a method of estimating growth temperatures. We cannot overemphasise the importance of this pioneering work on surface structural analysis, as it came into universal use in future MBE development.

Discussion of doping in MBE growth of GaAs will be taken up later but we should recognise at this point that several n- and p-type dopants were studied by the Bell group, who reached the conclusion that silicon or tin (n-type) and magnesium (p-type) were probably the most convenient. They also found that doping levels in nominally undoped GaAs were significantly lower when separate Ga and As sources were used, rather than a single source of polycrystalline GaAs. This was presumed to be due to the relatively high concentration of silicon impurities in commercially available GaAs. The essential point to note here is that Cho was able to grow GaAs, AlGaAs and GaP layers doped to whatever level was required by the demands of the device structures he was concerned to investigate.

Cho went on to demonstrate that MBE could grow several kinds of device structure which performed at least as well as those grown by other methods. They included varactor diodes, mixer diodes, IMPATT (impact ionisation avalanche transit time) diodes, laser diodes, Schottky barrier FETs and optical waveguides. In all cases, success depended on the ease with which MBE was able to grow well-controlled thin layers with similarly well-controlled doping profiles. He showed results for Schottky barrier varactor diodes with a range of capacitance–voltage characteristics, made microwave mixer diodes suitable for beam–lead technology, demonstrated an SB GaAs FET, operating at 4 GHz, with hysteresis-free current–voltage characteristics (indicative of low trap density in the active layer), and grew excellent 'low-high-low' IMPATT structures which oscillated at 11.7 GHz with 18% efficiency and a low noise factor ('low', that is, in IMPATT terms—the IMPATT is very much noisier than the Gunn diode). As an example of doping control, we show the IMPATT doping profile in Figure 2.11. The accurate control of the 0.1 μm-wide doping spike resulted in remarkably reproducible characteristics from diodes made over a single GaAs slice. But, perhaps of more immediate interest for the Panish research group, was the work done on double heterostructure lasers. The early years of the 1970s were witnessing the first stirrings of the optical fibre revolution, and Bell were very keen to obtain an efficient GaAs laser source for fibre communication links. Initially, LPE had been successful in demonstrating room-temperature CW operation with threshold current densities of about 2 kA/cm^2 but was struggling to achieve the reduced active layer thickness required for even lower thresholds. Clearly, MBE had the capability for growing thin layers, and Cho was not slow to produce structures with active layer thicknesses down to 0.2 μm, some two-to-three times smaller than the best LPE

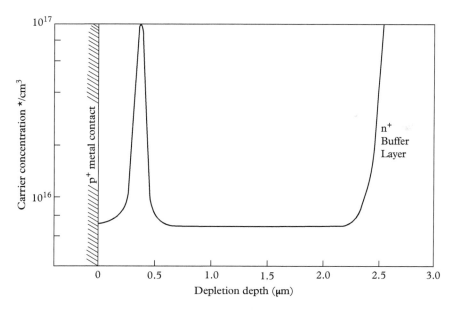

Figure 2.11 *An example of the doping control made possible by MBE growth. This is a typical carrier profile for a low-high-low GaAs IMPATT diode. (From Cho et al. 1974.) Reproduced with permission from Cho, A Y, Dunn, C N, Kuvas, R L and Schroeder, W E (1974) Appl Phys Lett 25, 224. Copyright 1974, AIP Publishing LLC.*

layers. Threshold currents were disappointing at first but came close to emulating those on devices grown by LPE, following annealing treatment. The problem seemed to lie with the Mg used as p-type dopant—its strong affinity for oxygen appeared to be introducing non-radiative recombination centres into the active region. Modified structures, with n-type active layers immediately improved matters, threshold current densities of 2.5 kA/cm^2 rapidly emerging from the earlier confusion. The long-term saga of MBE and the DH laser will form an engrossing aspect of our later chapters but the immediate impact on Bell Labs was clear enough—here was a technology they could surely exploit in an area of interest close to their company's heart. It is probably fair to say that Cho's initiatives were very largely responsible for the commercial future of MBE as a growth method, a fact clearly recognised in the higher echelons—Al Cho was awarded the US National Medal for Science in 1993 and the US Medal for Technology in 2007 for his pioneering work on MBE.

2.6   Superlattices and Low-Dimensional Structures

MBE is strongly associated with 'low-dimensional structures' on account of its ability to deposit ultra-thin layers, down to the single monolayer level, so it is interesting to examine the manner in which this came about. The idea for the superlattice came first—it was proposed in 1970 in a theoretical paper by Esaki and Tsu (1970) from the IBM Thomas J Watson Research Centre in Yorktown Heights, NJ. Their idea was to grow a large number of alternating ultra-thin layers of AlGaAs and GaAs in which electrons could tunnel through the wider gap layers and interact with the periodic potential of the superlattice in much the same way they might interact with the periodic potential of a uniform semiconductor crystal lattice. The very different nature of the superlattice potential was predicted to produce a current–voltage characteristic showing novel features, in particular, a region of negative resistance, electrons slowing down as they approach the boundary of the superlattice mini-zone (compare the similar effect we discussed in relation to the transferred electron effect in Section 1.2). Leo Esaki had done his PhD work at the University of Tokyo and then worked for the Sony Company, where he discovered the negative resistance associated with the tunnel diode (a p–n junction diode made with heavily doped material). Moving to IBM in 1960, he pursued the holy grail of negative resistance in the shape of the artificial superlattice and was awarded the Nobel Prize in 1973 for his research into tunnelling phenomena.

Initially, Esaki's group tried to grow superlattice structures by LPE but soon found this method to be far from appropriate, requiring, as they did, layers of thickness measured in Ångstroms rather than microns. Learning from Cho, in 1970, the virtues of MBE, Esaki immediately switched to the new technique and built his own version of the necessary growth machine. It was pumped by a series of ion pumps and a titanium sublimation pump with a liquid nitrogen shroud, the background pressure, after baking, being 10^{-10} Torr; it featured six effusion cells, a quadrupole mass spectrometer (for measuring the flux of atoms directed at the substrate) and a scanning RHEED system; also, of particular significance, layer thicknesses could be controlled by a dedicated computer. In this regard, Esaki's being at IBM was certainly a considerable advantage—at that time the digital computer was still very much in its infancy (the microprocessor was invented only in 1971) and necessary skills were far from widely available. The first IBM paper to describe the growth of a superlattice appeared in 1973 (Chang et al. 1973a—see also the review by Chang and Esaki 1980). It referred to structures of a hundred periods of 60 Å GaAs and 10 Å $Al_{0.5}Ga_{0.5}As$ layers, which did, indeed, show evidence of a negative resistance region in their current-voltage characteristics, as predicted by Chang and Esaki's earlier theory. The technique of thinning samples so as to allow transmission electron microscope images of such structures to be made was apparently not available to them, so Chang et al. (1973b) showed, instead, a scanning electron microscope picture of a structure with a period of about 1000 Å (see Figure 2.12). We should note, however, that Cho had, himself, grown similar multilayer structures as early as 1971 (Cho 1971). In the paper, he showed pictures of layers some 2000 Å thick

Figure 2.12 *A scanning electron micrograph, showing the cross-section of a multilayer GaAs-GaAlAs structure grown by Esaki and co-workers at IBM. The sample had been cleaved and etched, contrast resulting from different etch rates for the two constituents. The dark stripes represent GaAs, light stripes AlGaAs. The period of the structure is about 1000 Å. (From Chang et al. 1973b.) Reprinted with permission from Chang, L L, Esaki, L, Howard, W E, Ludeke, R and Schul, G (1973b) J Vac Sci Technol 10, 655. Copyright 1973, American Institute of Physics.*

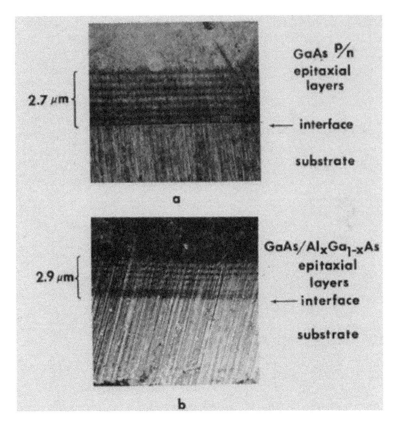

Figure 2.13 *A scanning electron micrograph of a cleaved section through a GaAs/AlGaAs structure grown by Cho at Bell Labs in 1971. The alternating layers are about 2000 Å thick. Cho had also grown periodic structures with layer thicknesses down to a few atomic layers but he had no method of imaging such structures at that time. (From Cho 1971.) Reprinted with permission from Cho, A Y (1971) Appl Phys Lett 19, 467. Copyright 1971, American Institute of Physics.*

(Figure 2.13) on the same grounds that he had (at the time) no means of imaging very thin layers. We shall come across a number of alternative applications for superlattices in later discussion, mostly rather more mundane than the sophistication of Esaki's work, but we shall leave the subject now for the moment.

As the negative resistance observed in superlattice structures was never great enough to lead to any useful devices, Esaki moved on to the study of resonant tunnelling in simple structures, consisting of a single GaAs well, sandwiched between a pair of AlGaAs barrier layers (Chang et al. 1974). As the voltage across such a structure increases, current rises sharply when the energy of electrons outside the barrier resonates with a confined energy state in the well; then, as it increases further, current drops again, giving rise to yet another type of negative resistance. Much interest was generated in this as a possible very high frequency oscillator, because the tunnelling effect on which it depends is extremely fast. Again, no useful devices have materialised but there can be no questioning the quality of the scientific insight. Of much greater long-term significance was the measurement of optical absorption spectra in quantum wells first reported from Bell Labs by Dingle and co-workers (Dingle et al. 1974, 1975). This will form the subject of detailed discussion in a later chapter so we shall say no more about it here but it is important to set it in context—it represented yet another 'scoop' for MBE growth in the period up to the mid 1970s, when MBE was to come of age with the development of commercial machines.

2.7   1975: Current Status

Having looked briefly at the various strands of MBE development, we shall bring this chapter to a close by presenting a snap-shot of the overall state of affairs at the middle of the 1970s, a date chosen as marking the beginning of commercial MBE machine availability. It seems clear that, if industrial firms were prepared to invest very large sums of money in designing and constructing elaborate pieces of high-tech equipment, MBE had reached the stage of being internationally accepted as a growth method with a considerable future. From this point onwards, one could obviously anticipate a considerable burgeoning of activity in the MBE growth of a wide range of semiconductor structures for both purely scientific and device-oriented motives (though one must also recognise the futility of trying to draw sharp dividing lines between these two intimately related aspects!). Consequently, from this point onwards, our presentation needs to change from one of general coverage to one of more specific topicality. But, first, let us set out the situation as we might have found it in 1975.

The first point to make is that all the MBE activities under way in 1975 depended on equipment that was largely home-made and therefore personal to the individual research group. Commercial firms such as Varian, in the USA, Vacuum Generators, in the UK, and Riber, in France, were marketing UHV equipment for surface studies and it was possible to buy a vacuum system, together with certain monitoring tools such as the quadrupole mass spectrometer and this is what most workers would naturally do. However, there was obviously an element of customisation in any individual MBE system. In particular, effusion cells to generate the molecular beams were not generally available, and each group was obliged to design its own version. Another important feature of machines yet to be developed was the use of load locks for transferring substrates and sample films to and from the growth chamber. The need to open the chamber to air after each growth (or a small number of growth runs) was, to say the least, tedious and, more importantly, was detrimental to ultimate purity of the grown films. (The problem of oxygen incorporation in films of AlGaAs was a particularly critical example of this.) Certainly, future experience was to show, beyond doubt, that UHV systems continued to 'clean up' for a surprisingly long time when kept continuously under vacuum.

It seems convenient that we consider the situation of MBE in 1975 country by country and there can be no apology for beginning with the USA, where the strongest moves towards practical growth of films for new devices and new physics were taking place. We have already referred to the work of Cho at Bell Labs. His publications on GaAs and related alloys at this time were concerned with an impressive range of activities, including photoluminescence, impurity doping, superlattices, heterostructure lasers and various other devices. One of Bell's principal interests at this time was, of course, that of optical communications and the need for suitable lasers to act as sources. As we have already mentioned, double heterostructure GaAs lasers had been grown by LPE in Panish's group with threshold current densities as low as 2 kA/cm^2 and it was hoped to improve on this by further reducing the thickness of the active layer. However, the control of thickness in LPE layers thinner than about 0.5 µm was difficult, and MBE was showing promise as an alternative growth method. Cho had demonstrated his ability to grow suitable DH structures but threshold currents were an order of magnitude too high. The secret probably lay with the quality of the AlGaAs cladding layers, as illustrated by their poor luminescence efficiency, and this, in turn, stemmed from the incorporation of oxygen from the relatively high concentration of CO in the MBE vacuum system. Careful removal of CO by the repeated operation of titanium sublimation pumps led to dramatic improvement in luminescence and corresponding reduction in laser threshold current. A secondary problem associated with the use of magnesium as p-type dopant was also related to oxygen incorporation and was similarly much improved by removal of CO. Cho was responsible for much of the early work on dopant selection, particularly for n-type doping. He made comparison between silicon, germanium and tin for producing sharp doping profiles in GaAs films and demonstrated the importance of surface segregation in the case of tin. Broad, asymmetric profiles resulted from the accumulation of tin on the growing surface, whereas silicon and germanium gave much sharper, symmetric profiles, limited only by the ability of the Schottky barrier capacitance–voltage technique to resolve them. Also, at this time Cho reported on the use of MBE deposition of GaAs to make primitive integrated circuits. Silicon oxide was deposited on a semi-insulating GaAs substrate, holes etched in it, and GaAs grown by MBE over the whole surface. Within the holes the deposition was monocrystalline, while on the oxide it was fine-grain polycrystalline. This polycrystalline material was highly resistive, providing excellent isolation between the crystalline islands, whose electrical properties, he determined, were identical to those of conventional epilayers grown under the same conditions. Finally, we might give brief mention to the fact that Cho was working on varactor diodes, microwave mixer diodes, IMPATT diodes and Schottky barrier FETs (Cho and Arthur 1975). As we have already remarked, the long-term future of MBE technology owes a very great deal to Al Cho's pioneering efforts.

In contrast to Cho's extensive foray into the growth of films for practical applications, Arthur was continuing his basic studies of the behaviour of gallium and arsenic on clean surfaces of GaAs (see e.g. Arthur and Brown 1975), while two new MBE activities had appeared within the Murray Hill laboratory. The first of these represented an expansion of the activity in Panish's group, involving another newcomer, Marc Ilegems, the second a new MBE development in the group run by Ray Dingle. Ilegems became interested in studying phase diagrams, such as that of the quaternary system Ga-Al-As-P (Ilegems and Panish 1974), which were of relevance for both LPE and MBE growth, played an important role in the development of GaN vapour epitaxy (Ilegems 1972; Ilegems and Montgomery 1973), took an interest in electrical conduction in the relaxation regime (Ilegems and Queisser 1975) and also found time to grow films of AlGaAs by MBE. Of particular interest was his work (Van der Ziel and Ilegems 1976) involving MBE growth of AlAs/GaAs multilayer structures for the study of non-linear optical phenomena, where radiation propagated in the plane of the layers. (It followed from some earlier work of Cho's in which GaP waveguides were grown on a CaF_2 substrate (Cho and Chen 1970).) Of much longer-term significance was Ilegems' initiative in exploring the use of beryllium as a p-type dopant in GaAs, grown by MBE (Ilegems 1976) but we shall discuss this in detail later. Similar MBE multilayers, grown by Art Gossard (Dingle et al. 1975), also formed part of the interest of the Dingle group and this was shortly to lead to the burgeoning development of low-dimensional structures which form the subject of Chapter 6. Bell Laboratories were ever the scene of frenzied activity—but there can never have been a time more exciting than the mid 1970s, when MBE was rapidly coming of age.

Elsewhere in the USA, we are already familiar with the superlattice work of Esaki at IBM and of the MBE growth of IV-VI materials for infrared devices by Holloway at the Ford Motor Company in Dearborn, Michigan and by Walpole at the MIT Lincoln Laboratory. Conveniently for our purpose, Esaki summarised the position his group had established in a review paper presented at the Third International Conference on Thin Films, in Budapest, during August 1975 (Esaki and Chang 1976). He entitled it 'Semiconductor *Superfine* Structures by Computer-Controlled Molecular Beam Epitaxy' (our italics) and in it proposed an interesting alternative to the usual 'superlattice structures'. The fact that it seems not to have caught on need scarcely concern us—the work, itself, clearly represented an important aspect of MBE growth and the use of computer-control was to be taken up by many a future devotee of the technique. The films referred to consisted of alternating layers of AlAs and GaAs (or their alloys) with thicknesses in the range 10–100 Å, and the first challenge was one of confirming experimentally the accuracy of these dimensions in practical structures. As we have already mentioned, the use of transmission electron microscopy to image such fine scale structures was not available to Esaki so he therefore relied on small-angle X-ray scattering to demonstrate the coherence of the layers on an atomic scale. The principle is similar to that of optical interference in studying thin films and depends on the fact that the refractive index for X-rays of wavelength $\lambda = 1.54$ Å differs significantly between AlAs and GaAs. Theoretical modelling of the resulting interference patterns showed excellent agreement with their experimental data (Chang et al. 1976), confirming that MBE could indeed produce smooth, coherent layers with such small dimensions. The motivation for the work was, of course, to study the electrical and optical properties of superlattice and double barrier structures—suffice it here to say that negative resistance was observed in superlattices, resonant tunnelling in double barriers, and evidence for confined electron states in photocurrent measurements on superlattices (Tsu et al. 1975). The era of low-dimensional structures and devices was clearly imminent even before the arrival of commercial MBE machines. With regard to Holloway and Walpole's work on IV-VI compounds, we can simply note that, by 1975, good progress had already been made in developing far-infrared detectors, optical waveguides and DH lasers—the even more sophisticated distributed feedback laser was to appear in 1976. Their version of MBE may have been less advanced than the state-of-the-art systems available at Bell Labs but there can be no doubt as to its efficacy in producing working devices.

The take-up of MBE growth outside the USA may have seemed somewhat slower—nevertheless, in 1975 we can evidence several activities in Japan, those of Kiyoshi Takahashi, in Tokyo, referred to above, of Shun-ichi Gonda of the Electro-technical Laboratory, also in Tokyo, and of S. Hiyamizu at Fujitsu, H. Sugiura at NTT and S. Nagata at Matsushita. Indeed, these constitute rather more different *laboratories* working on MBE growth than anywhere else in the world. Materials studied included Si, Ge, GaAs, AlGaAs, GaAsP, AlN, ZnSe, ZnTe and ZnS and it is interesting to note this very early application of MBE to the growth of II-VI compounds—they were also being grown by Perkin Elmer in the USA at much the same time (see Chapter 8—Section 8.1). In England Bruce Joyce and Tom Foxon at the Mullard Research Laboratory Redhill were studying GaAs growth mechanisms, while at RRE Malvern Robin Farrow was look-ing into the MBE growth of InP. In continental Europe, Klaus Ploog was in process of setting up equipment for growing doping superlattices at Stuttgart, while in France, Jean Massies, having joined Thomson CSF in 1974, was already planning an MBE activity but little of substance had materialised in 1975. That sums up the overall situation—let us look briefly at some of the details.

Takahashi had been in the MBE business for some time, having published his first MBE paper in 1968, concerned with the growth of germanium films on silicon sub-strates. In 1975 his interests had shifted, in line with those elsewhere, towards the III-V compounds. In addition to reporting the fabrication of a GaAs FET on MBE mater-ial (Naganuma et al. 1975), he reported the use of ionised zinc as a method of p-type doping of GaAs (Naganuma and Takahashi 1975a). At that point, no really satisfactory p-type dopant had been found—the Bell workers settled for magnesium but it proved to be difficult to localise, being a rapid diffuser—while zinc had demonstrated an extremely low sticking coefficient on GaAs. By using zinc ions, it was possible to accelerate them into the growing film, a process essentially like ion implantation, and doping levels as high as 3×10^{25} m^{-3} were obtained, a considerable improvement on anything achieved previously. Had not beryllium come on the scene, Takahashi's ionised zinc may well have achieved universal application. These same authors reported the growth of GaAs, GaP and GaAsP films using graphite effusion cells for the various sources (Naganuma and Takahashi 1975b) and shortly afterwards, applied this capability to graded band-gap material for application to both transistors and solar cells (Tateishi et al. 1976). Gonda came much later to the field but published two papers at this time (Gonda et al. 1975; Gonda and Matsushima 1976), concerned with MBE growth of GaAs and GaAsP using an 'up-to-date' MBE machine (graphite effusion cells, background pressure 10^{-10} Torr, quadrupole mass spectrometer, RHEED analysis, etc.), achieving material quality com-parable with that appropriate to other growth methods. Already a sound base for MBE growth had clearly been established in Japan.

In England, the MBE scene had shifted as a result of Bruce Joyce's moving from Plessey to Mullard with a brief to study the Si/SiO$_2$ system, which had become of special significance with the development of the MOS transistor at Bell and at RCA and its application to memory circuits at Intel in the early 1970s. Almost as a sideline, he and Tom Foxon began a project to use modulated beam techniques to study surface interactions of Ga and As$_4$ on GaAs, at the time, a very basic surface science programme but one that was to lead to a major growth programme later in the decade. Their original motivation was unrelated to MBE growth, being based, rather, on the need for a better understanding of the conventional VPE process but experimentally the programme relied on the use of UHV conditions to maintain atomically clean surfaces and employed many of the standard features of MBE growth. The already well-established use of modulated beams in conjunction with mass spectroscopic analysis had been applied to GaAs by Arthur (1968) as a means of distinguishing between signals related to desorption from the substrate and those due to background species. Foxon and Joyce (Foxon et al. 1974; Foxon and Joyce 1975; Joyce and Foxon 1975) extended the method by refining techniques of signal analysis and studied the detailed kinetic behaviour of the As$_4$ molecule on clean GaAs surfaces as a function of Ga flux and of temperature over the range 300 K to 900 K. This, together with Cho's and Arthur's work, provided a sound basis for the rapidly growing understanding of exactly how Ga and As atoms were incorporated in films of GaAs (by way of a physisorbed precursor state and subsequent chemisorption) and its relation to surface structure. Robin Farrow's research at RRE Malvern was (innovatively) concerned with InP, which he demonstrated could be grown on InP substrates by MBE from indium and InP source cells (Farrow 1974). At the time, InP was not particularly high on most research agendas but a burst of interest had been generated by a theoretical paper from Hilsum and Rees of RRE which predicted enhanced performance from InP as a material for microwave Gunn diodes (see Orton 2004, p. 207). In consequence, some priority was given to the growth of InP in several UK laboratories, and Farrow's activity can be seen as a part of this. In the event, InP was to achieve technological fame in the quite different realm of optical communications and this early stimulus certainly proved to be of considerable value.

In summary, we can see that MBE had achieved international status, and the growth of III-V epitaxial films had already reached a sophisticated level by the mid 1970s. Several different materials, GaAs, GaP, GaAsP, AlAs, AlGaAs and InP, had been grown successfully and both n- and p-type doping were available (though the advent of beryllium as the preferred p-type dopant lay still in the future). Of even greater significance was the proven ability of MBE to generate uniform, multilayer, ultra-thin films with smooth interfaces, a capability which was to find application in an amazingly wide range of activities during the next 35 years. All this, of course, was still very much a matter of basic research—MBE's direct application to industrial production was dependent on the development of commercial growth equipment but that development was only a year or two away and progress would then be even more impressive, as we shall see. As a final comment, we should simply make the point that 1975 represented an MBE watershed in so far as its position had been well summarised in the excellent review by Cho and Arthur (1975). Anyone looking for more detail on the early development of III-V

MBE growth will find it therein. Finally, let us not overlook the Japanese work on II-VI compounds—MBE was to play a vital part in their subsequent development. Nor can we leave this introductory chapter without reference to the parallel development of Si MBE growth, together with corresponding studies of Si crystal growth which also date back to the early 1970s. We take up the MBE growth of II-VI compounds and of Si and Si/Ge (which also dates from the mid 1960s) in Chapter 8.

....

REFERENCES

Arthur, J R (1968) Interaction of Ga and As$_2$ molecular beams with GaAs surfaces. J Appl Phys 39, 4032.
Arthur, J R and Brown, T R (1975) Velocity distributions of As2 and As4 scattered from GaAs. J Vac Sci Technol 12, 200.
Arthur, J R and LePore (1969) GaAs, GaP, and GaAs$_x$P$_{1-x}$ epitaxial films grown by molecular beam deposition. J Vac Sci Technol 6, 545.
Booker, G R and Joyce, B A (1966) A study of nucleation in chemically grown epitaxial silicon films using molecular beam techniques: II. Initial growth behaviour on clean and carbon-contaminated silicon substrates. Phil Mag 14, 301.
Calow, J T, Kirk, D L and Owen S J T (1972) The growth of epitaxial ZnSe upon germanium substrates. Thin Solid Films 9, 409.
Casey, H C Jr and Panish, M B (1978) 'Heterostructure Lasers', Academic Press, New York.
Chang, L L and Esaki, L (1980) 'Semiconductor Superlattices by MBE and Their Characterization' in 'Molecular Beam Epitaxy' (ed B R Pamplin), Pergamon Press, Oxford, p 3.
Chang, L L, Esaki, L, Howard, W E and Ludeke, R (1973a) The growth of a GaAs–GaAlAs superlattice. J Vac Sci Technol 10, 11.
Chang, L L, Esaki, L, Howard, W E, Ludeke, R and Schul, G (1973b) Structures grown by molecular beam epitaxy. J Vac Sci Technol 10, 655.
Chang, L L, Esaki L and Tsu, R (1974) Resonant tunneling in semiconductor double barriers. Appl Phys Lett 24, 593.
Chang, L L, Segmüller, A and Esaki, L (1976) Smooth and coherent layers of GaAs and AlAs grown by molecular beam epitaxy. Appl Phys Lett 28, 39.
Cho, A Y (1971) Growth of periodic structures by the molecular-beam method. Appl Phys Lett 19, 467.
Cho, A Y and Arthur, J R (1975) Molecular beam epitaxy. Prog Sol St Chem 10, 157.
Cho, A Y and Casey, H C Jr (1974) GaAs–Al$_x$Ga$_{1-x}$ As double-heterostructure lasers prepared by molecular-beam epitaxy. J Appl Phys 45, 1258.
Cho, A Y and Chen, Y S (1970) Epitaxial growth and optical evaluation of gallium phosphide and gallium arsenide thin films on calcium fluoride substrate. Sol St Commun 8, 377.
Cho, A Y, Dunn, C N, Kuvas, R L and Schroeder, W E (1974) GaAs IMPATT diodes prepared by molecular beam epitaxy. Appl Phys Lett 25, 224.
Datsenko, L I, Gureev, A N, Korotkevich, N F, Soldatenko, N N and Tkhorik, Yu A (1971) Crystalline structure of germanium films on silicon substrates: I. Investigation of the perfection of germanium heteroepitaxial films on silicon by X-ray diffraction methods. Thin Solid Films 7, 117.

Davey, J E and Pankey, T (1964) Structural and optical characteristics of thin GaAs films. J Appl Phys 35, 2203.

Davey, J E and Pankey, T (1968) Epitaxial GaAs films deposited by vacuum evaporation. J Appl Phys 39, 1941.

Dingle, R, Gossard, A C and Wiegmann, W (1975) Direct observation of superlattice formation in a semiconductor heterostructure. Phys Rev Lett 34, 1327.

Dingle, R, Wiegmann, W and Henry, C H (1974) Quantum states of confined carriers in very thin Al_xGa_{1-x} As-GaAs-Al_xGa_{1-x} As heterostructures. Phys Rev Lett 33, 827.

Einstein, A (1917) Zur Quantum Theorie der Strahlung. Phys Z. 18, 121.

Elleman, A J and Willman, H (1948) The structure and growth of PbS deposits on rocksalt substrates. Proc Phys Soc 61, 164.

Esaki, L and Chang, L L (1976) Semiconductor superfine structures by computer-controlled molecular beam epitaxy. Thin Solid Films 36, 285.

Esaki, L and Tsu, R (1970) Superlattice and negative differential conductivity in semiconductors. IBM J Res Dev 14, 61.

Farrow, R F C (1974) Growth of indium phosphide films from In and P_2 beams in ultra-high vacuum. J Phys D: Appl Phys 7, L121.

Foxon, C T, Boudry, M R and Joyce, B A (1974) Evaluation of surface kinetic data by the transform analysis of modulated molecular beam measurements. Surf Sci 44, 69.

Foxon, C T and Joyce, B A (1975) Interaction kinetics of As_4 and Ga on {100} GaAs surfaces using a modulated molecular beam technique. Surf Sci 50, 434.

Freller, H and Gunter, K G (1982) Three-temperature method as an origin of molecular beam epitaxy. Thin Solid Films 88, 291.

Gonda, S and Matsushima, Y (1976) Effect of substrate temperature on composition ratio x in molecular-beam-epitaxial $GaAs_{1-x}P_x$. J Appl Phys 47, 4198.

Gonda, S, Matsushima, Y, Makita, Y and Mukai, S (1975) Characterization and substrate-temperature dependence of crystalline state of GaAs grown by molecular beam epitaxy. Jap J Appl Phys 14, 935.

Gordon, J P, Zeiger, H J and Townes, C H (1954) Molecular microwave oscillator and new hyperfine structure in the microwave spectrum of NH_3. Phys Rev 95, 282.

Holloway, H and Walpole, J N (1980) 'MBE Techniques for IV-VI Optoelectronic Devices' in 'Molecular Beam Epitaxy' (ed B R Pamplin), Pergamon Press, Oxford, p 49.

Ilegems, M (1972) Vapor epitaxy of gallium nitride. J Cryst Growth 13/14, 360.

Ilegems, M (1976) [Abstract]. J Electronic Matls 5, 445.

Ilegems, M and Montgomery, H C (1973) Electrical properties of n-type vapor-grown gallium nitride. J Phys Chem Solids 34, 885.

Ilegems, M and Panish M B (1974) Phase equilibria in III–V quaternary systems—Application to Al-Ga-P-As. J Phys Chem Solids 35, 409.

Ilegems, M and Queisser, H J (1975) Current transport in relaxation-case GaAs. Phys Rev B 12, 1443.

Ito, K and Takahashi, K (1968) Epitaxial growth of Ge layers on Si substrates by vacuum evaporation. Jap J Appl Phys 7, 821.

Joyce, B A (1968) Growth and perfection of chemically-deposited epitaxial layers of Si and GaAs. J Crystal Growth 3, 43.

Joyce, B A and Bradley, R R (1966) A study of nucleation in chemically grown epitaxial silicon films using molecular beam techniques I.—experimental methods. Phil Mag 14, 289.

Joyce, B A, Bradley, R R and Booker G R (1967) A study of nucleation in chemically grown epitaxial silicon films using molecular beam techniques III. Nucleation rate measurements and the effect of oxygen on initial growth behaviour. Phil Mag 15, 1167.

Joyce, B A and Foxon, C T (1975) Kinetic studies of the growth of III–V compounds using modulated molecular beam techniques. J Cryst Growth 31, 122.

Naganuma, M, Kamimura, K, Takahashi, K and Sakai, Y (1975) GaAs FET prepared with molecular beam epitaxial films. Jap J Appl Phys 14, 581.

Naganuma, M and Takahashi, K (1975a) Ionized Zn doping of GaAs molecular beam epitaxial films. Appl Phys Lett 27, 342.

Naganuma, M and Takahashi, K (1975b) GaAs, GaP, and $GaAs_{1-x}P_x$ films deposited by molecular beam epitaxy. Phys Stat Sol (a) 31, 187.

Orton, J W (2004) 'The Story of Semiconductors', Oxford University Press, Oxford.

Riordan, M and Hoddeson L (1997) 'Crystal Fire', Norton, New York.

Scheer, J J and van Laar, J (1965) GaAs-Cs: a new type of photoemitter. Sol St Commun 3, 189.

Schoolar, R B and Lowney, J R (1971) Photoconductive PbSe epitaxial films. J Vac Sci Technol 8, 224.

Schoolar, R B and Zemel, J N (1964) Preparation of single-crystal films of PbS. J Appl Phys 35, 1848.

Shimaoka, G (1971) Structure and epitaxy of evaporated cadmium sulfide films. Thin Solid Films 7, 405.

Smith, K F (1955) 'Molecular Beams', Methuen Monograph, London.

Tateishi, K, Naganuma, M and Takahashi, K (1976) Graded-bandgap III-V ternary compound films by molecular beam epitaxy. Jap J Appl Phys 15, 785.

Townes, C H (1999) 'How the Laser Happened', Oxford University Press, Oxford.

Tsu, R, Chang, L L, Sai-Halasz, G A and Esaki, L (1975) Effects of quantum states on the photocurrent in a "superlattice". Phys Rev Lett 34, 1509.

Van der Ziel, J P and Ilegems, M (1976) Second harmonic generation in a thin AlAs-GaAs multilayer structure with wave propagation in the plane of the layers. Appl Phys Lett 29, 200.

Watts, B E, Bradley, R R, Joyce, B A and Booker, G R (1968) A study of nucleation in chemically grown epitaxial silicon films using molecular beam techniques. IV. Additional confirmation of the induction period and nucleation mechanisms. Phil Mag 17, 1163.

Zemel, J N, Jensen, J D and Schoolar, R B (1965) Electrical and optical properties of epitaxial films of PbS, PbSe, PbTe, and SnTe. Phys Rev 140, A330.

3

Machine Technology

3.1 Overview

The historical account given in Chapter 2 has already provided an introduction to the most important design features of an MBE machine but we must now examine them in greater detail, emphasising individual functions, design compromises, materials, dimensions, etc. This chapter will therefore be concerned with essentially practical aspects of machine design, while important theoretical aspects of MBE growth will be taken up in Chapter 4. As we have already remarked, commercial machines became available towards the end of the 1970s and these have dominated the field ever since, so we have no option but to base much of our discussion on these (with appropriate acknowledgement, as far as possible). The early development in the West was dominated by three companies: Varian in the USA, Vacuum Generators in England and Riber in France, their undoubted success being clearly demonstrated by the fact of their continuing existence right up to the present day, though VG is now part of the Riber portfolio and the Varian MBE activity now trades as VEECO. More recently, a number of other companies have entered the fray, notably Karl Eberl, Createc-Fischer and Omicron Nanotechnology in Germany, and DCA Instruments in Finland. Two Japanese companies, Anelva and EIKO Corporation, also took up the challenge in the 1970s and are still manufacturing complete MBE machines, very largely for sale in the Far East. Numerous other companies supply UHV-compatible components for customised MBE systems. A very rough estimate of the number of commercial MBE machines supplied to academic and industrial users round the world suggests this could now be approaching 3000, an impressive indicator of the overall importance of MBE as both a research and a production tool. We shall return to the question of the range of machine types available but first we need to analyse the function of the various components in a basic single-wafer MBE machine.

A typical schematic arrangement of the growth chamber of an MBE system is shown in Figure 3.1. With only a very few exceptions, the chamber is made from stainless steel, being bolted together with 'conflat' flanges and pumped to a background pressure (after bakeout) of 10^{-10} Torr, or better, with a combination of ion, turbo-molecular, titanium sublimation and cryopumps. (Each of these has been briefly described in the previous chapter—Box 2.2.) We should emphasise once again that such background

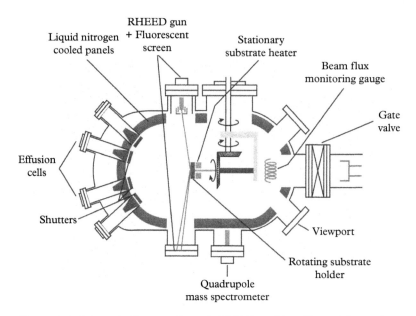

Figure 3.1 *Schematic diagram of a typical MBE machine. The stainless steel vacuum system can be pumped down to a pressure of 10^{-10} Torr or better. Molecular beams are produced by a series of shuttered effusion cells pointing towards the substrate, mounted on a rotating holder. Beam intensities can be calibrated with the ion gauge. A RHEED gun and fluorescent screen allow monitoring of the surface structure. Samples may be introduced and removed via the sample load lock, without letting the chamber up to atmospheric pressure.*

pressures are essential in the interest of ensuring the ultimate purity of deposited films, it being important that the arrival rate of unwanted species be much lower than those from the various effusion cells. As we showed in Box 2.1, at a background pressure of 10^{-10} Torr and growth rate of 1 monolayer per second, the ratio of these two quantities is about 10^{-5}, though this takes no account of impurity sticking coefficients which help considerably to enhance ultimate film purity. A brief comment on bakeout procedure is probably in order. The machine is surrounded by a removable enclosure made from electrically heated panels, which can be relatively rapidly constructed and, while any delicate external components are taken to a place of safety, the whole chamber is heated overnight to a temperature of typically 200 °C, while being continuously pumped.

It is standard practice to surround the substrate with a liquid-nitrogen-cooled cryopanel, which serves both to remove heat from the system and to trap impurity species which might otherwise find their way to the growing film. Furthermore, there is overwhelming evidence to show that layer purity improves considerably with time, following the initial pump-down and bakeout, so it is of vital importance that substrates can be introduced into the machine without letting the system up to air. The development of the vacuum interlock towards the end of the 1970s (see Section 3.5) has therefore played an essential role in securing an international role for MBE in the growth of high-purity semiconductor films. It also plays an important role in avoiding the oxidation of source materials, such as arsenic, phosphorus and aluminium, which would otherwise occur each time the machine was allowed up to atmosphere. We should also emphasise here just how important has been the design of reliable substrate handling. Not only has the substrate to be introduced into the chamber, it must be manipulated onto the substrate heater in the desired orientation, and such delicate manoeuvring is far from trivial.

Molecular beams are provided by a set of effusion cells (see Section 3.2) mounted on a source flange, each individual source being provided with a moveable shutter which can be used to interrupt the flow of molecules to the substrate. The flange will normally be provided with a second cryopanel to remove surplus heat from the sources and to prevent cross-contamination between the cells, any wandering molecules being condensed on the panel rather than infiltrating an alien cell. The individual cells have grown considerably in size over the years, both in the interest of allowing the long run times necessary to achieve ultimate film purity, and the demands of efficient commercial applications, and the source flange, carrying up to eight, ten or more cells, has grown in proportion, representing one of the largest features of the machine.

The substrate holder (see Section 3.4) represents a key feature of any MBE machine, it being important that substrates be heated uniformly to a temperature which can be reliably measured and controlled. Substrates are normally heated by radiation and are continuously rotated to smooth out non-uniformities due to the off-axis arrangement of the effusion cells (see Figure 3.1). Without rotation, the growth rate can vary by as much as a factor of 2 across a typical 3-inch wafer, so the design of the rotation mechanism is critical—note, in particular, that it is essential to avoid the use of ordinary lubricants in order to maintain the necessary high vacuum in the system.

One of the most important features of MBE growth is the possibility for in situ monitoring of both growth rate and structural perfection of the growing film. Structural analysis is usually performed by reflection high energy electron diffraction (RHEED; see Section 3.6), generally preferred to low-energy electron diffraction on account of RHEED's more convenient geometry. As shown in Figure 3.1, an electron gun is mounted on one side of the growth chamber so as to send a narrow beam of electrons at glancing incidence onto the surface of the film, the resulting diffraction pattern being detected on a phosphor screen mounted diametrically opposite the gun. Growth rate is ultimately measured by observation of RHEED oscillations, which we discuss in detail in Section 4.3, though in situ ellipsometry has also been used in a few instances (see e.g. Maracas et al. 1992). Direct monitoring of beam flux has been achieved by measuring optical absorption (Pinsukanjana et al. 2003) but is more usually dependent on the use of an ion gauge, which can be moved into the sample position (see Section 3.3). However, it is important to recognise that the ion gauge is actually a density detector, rather than a flux monitor (see Box 3.2). It also shows very different sensitivities for different molecules, and its absolute sensitivity can vary considerably as a result of the deposition of molecules on the collector. Without doubt, it is a very useful instrument but some degree of care is necessary when using it as a beam monitor.

For completeness, we should mention (again) the use of Auger electron spectroscopy for monitoring impurity atoms on the substrate surface (Cho and Arthur 1975) and the use of mass spectrometry for analysing impurities within the vacuum system but in modern day practice such techniques have, to a large extent, been made redundant by the high standard of vacuum system design and the reliable sample cleaning procedures which are now typical.

3.2   Cell Design

Perhaps the first point to be made concerning effusion cell design is that there is considerable variation, depending on the species to be evaporated and on the application—the needs of pure research differing significantly from those of production. Clearly, the essential requirements are, first, that the beam intensity should be sufficient to achieve the required growth rate (typically one monolayer per second—equivalent to approximately one micron per hour) and, second, that impurity levels within the beam should be low enough to allow the growth of acceptably pure films. Given that practical doping levels may, in extreme cases, be as low as 10^{18} m^{-3} (or roughly 1 part in 10^{10}), this places serious demands on the purity of the evaporant species. A well-known illustration of the importance of purity concerns the growth of GaAs. In the early days the arsenic beam was generated from a cell containing GaAs itself, the vapour pressure of arsenic being much greater than that of gallium, but it soon became clear that GaAs could not be obtained with anything like the desired level of purity. On the other hand, the elemental source arsenic could be so obtained and was very soon adopted by all growers.

Returning to the question of beam flux intensity, we should note that effusion cells have developed quite dramatically from the original Knudsen sources used in molecular beam experimentation, where it was of paramount importance that the beam was collision free (see Sections 2.2 and 2.3), to the 'monster' cells used in today's commercial MBE machines. Perhaps unsurprisingly, evolution was gradual. The first step consisted in opening up the source aperture to increase beam intensity, and an example of such early cells, taken from our own experience, is shown in Figure 3.2. Already the conditions which must be satisfied for a true Knudsen cell have been abandoned, namely, that the area of the aperture should be small compared to the internal surface area so that the molecules leaving the cell do not perturb the equilibrium vapour pressure within the cell and, second, that the aperture diameter should be small compared to the mean-free-path of molecules within the cell, so as to satisfy the collision-free condition. These 'home-made' cells were considerably smaller than modern commercial cells but nevertheless contained all the principal features. As shown in Figure 3.3, they consisted of a crucible to hold the evaporant and which was typically made from pyrolytic boron nitride (though graphite was acceptable in some applications), the important criterion being the need to avoid chemical reaction with the evaporant, to prevent cross-contamination. The crucible was surrounded by an electric heater, which had to be capable of reaching the

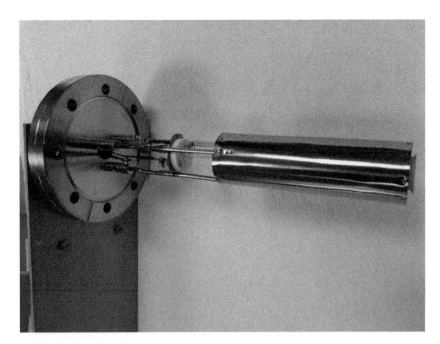

Figure 3.2 *An example of a simple home-made MBE effusion source. It takes the form of an open BN crucible within an electrically heated furnace, surrounded by radiation shields.*

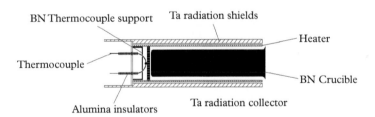

Figure 3.3 *Schematic diagram of the effusion cell shown in Figure 3.2. The thermocouple is arranged to be in intimate contact with the base of the crucible so as to measure the melt temperature as accurately as possible.*

temperature appropriate to the required beam flux while introducing the absolute minimum amount of impurity into the vacuum system. Therefore, the metal wire or tape usually took the form of tungsten or tantalum, suitably insulated, of course, to avoid short-circuiting. The final essential components of the cell were the radiation shields, which minimise the amount of heat lost from the cell and are important not only from the viewpoint of efficiency but also to reduce outgassing from the cell's surroundings, with consequent contamination of the growing film. As we have already noted, cells are usually surrounded by a liquid-nitrogen-cooled cryopanel, with this same purpose in view.

The source temperature is a particularly important parameter, determining, as it does, the beam flux, and its accurate measurement therefore takes on a special significance. The standard method involves using a tungsten/tungsten–rhenium thermocouple but it is vital that this should measure the temperature of the evaporant rather than that of the heater and this implies it should not 'see' radiation from the heater. In the design shown in Figure 3.3, this is achieved by the thermocouple being sandwiched between the crucible and a dummy crucible. We should note in passing that this condition is not always satisfied in some commercial cells—the user should be aware of this! It can be checked by measuring the apparent activation energy for evaporation. If the measured temperature T_m differs from the actual temperature T of the evaporant, a plot of log (flux) against T_m will have an incorrect slope, and comparison with standard heats of evaporation will then enable the ratio T/T_m to be determined. The thermocouple is used not only to measure the cell temperature but also to stabilise the emitted flux by acting as the sensor in a control loop. Because the flux depends exponentially on temperature, it is necessary to control temperature to a high degree of accuracy, typically to within $0.25 \,^\circ$C (a difficult task in the early days when control was effected with 12-bit logic but a simple enough matter now!).

Reference to commercial cells makes this an appropriate point at which to illustrate typical examples. Figure 3.4 shows modern cells by Riber and Veeco, respectively. These differ from the 'home-made' cells described above mainly in respect of size, being some 30–40 cm long by 15–20 cm diameter. Cell capacity is typically as great as 1700 cc, compared with roughly 40 cc for their forebears. They operate even further from equilibrium than their junior predecessors but are designed on the principle that each pump-down should last as long as possible, in the interest of 'production' efficiency, and that the enhanced purity of grown films is a function of time during the growth run. Total thickness of material deposited during a single 'commercial' run may be typically of order 3–4 mm over seven 6-inch substrates, as much as 25 000 wafers. Note, too, that the size of the charge needed to provide any desired run time may depend on the nature of the growth process. For example, when growing GaAs, it is standard practice to grow with excess arsenic, thus demanding a significantly larger arsenic charge than might at first seem likely. We shall discuss details of commercial machines in Section 3.7, so will say no more about cell design here, other than to note that the flux varies slowly with time due to the gradual diminution of the cell's contents. The flux distribution will also vary slightly during a run, being more or less forward focussed compared to that from an ideal Knudsen cell. As can be seen in Figure 3.5, the cell acts as a collimator with regard to flux distribution and, as the level of charge decreases, this collimating action correspondingly increases, the beam becoming narrower with growth time, and this has important consequences for uniformity in the growing film. We might, therefore, take the opportunity for a brief discussion of this aspect of cell design.

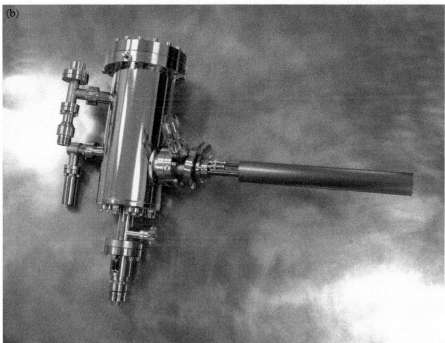

Figure 3.4 *Modern effusion cells manufactured by (a) Riber and (b) Veeco (courtesy of Riber and Veeco, respectively).*

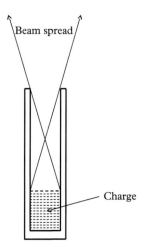

Figure 3.5 *Sketch to illustrate the collimating effect of the crucible in an open-cell MBE source. The width of the emerging beam clearly depends on the level of source material in the cell.*

Perhaps the first demonstration of highly uniform film growth in an MBE system was due to Cho and Cheng (1981), who made use of substrate rotation to achieve better than 1% uniformity of thickness, composition and doping in AlGaAs/GaAs structures grown over a 2-inch slice. Their emission cells were cylindrical in section with apertures of 2.5 cm and spaced 12 cm from the substrate. (Note that the somewhat oversimplified theory of flux uniformity presented in Box 3.1 gives results in quite good agreement with their experimental data.) Later computer modelling by Curless (1985), Saito et al. (1987), Yamashita et al. (1987) and Wasilewski et al. (1991a, b) clearly demonstrated the importance of cell design in determining flux uniformity. In particular, the use of conical cells with cone angles sufficient that all points within the cells can be 'seen' from all points on the substrate (see Figure 3.7) made considerable improvement. Saito et al. (1987) of the Fujitsu company used this cell design, together with an unusually large cell-substrate separation of 55 cm to achieve 1% uniformity over an 18 cm diameter substrate holder (while still maintaining growth rates of approximately 1 μm/h)—an important indicator for the future development of MBE machines for semiconductor device production. An important detail is the use of a secondary heater at the cell aperture to prevent the condensation of gallium droplets, which cause instability in the emission pattern.

Box 3.1 Flux uniformity

Generally speaking, the MBE grower is concerned with producing epitaxial films of uniform thickness, composition and doping level, so it is of interest to examine the uniformity of the beam flux over a typical substrate of, say, three inches in diameter. This is influenced by three distinct factors: first, the beam flux from an ideal Knudsen source varies with the angle θ between the beam direction and the normal to the aperture, and this leads to non-uniformity even when the source is aligned along the axis of the substrate; second, it is generally the case that sources are aligned in an off-axis geometry; third, practical sources usually diverge from the ideal, and the collimating effect of the cell tubular geometry, for example, results in forward-focussing of the beam. We shall discuss these three effects in turn.

Figure 3.6(a) illustrates the very simple geometry represented by an ideal point source aligned on the substrate axis. The distance r from source to an arbitrary point on the substrate is given by

$$r(\theta) = d/\cos\theta \tag{B3.1}$$

where d is the axial separation of aperture and substrate.

As the flux $F(\theta)$ depends on $1/r^2$, this leads to a variation in flux proportional to $\cos^2\theta$. In addition, as we showed in Box 2.1, the flux emerging from the cell aperture varies as $\cos\theta$ and, finally, the substrate area over which this segment of the beam is distributed increases with $\sec\theta$, leading to an overall dependence of flux:

$$F(\theta) \sim \cos^4\theta \tag{B3.2}$$

Taking typical dimensions of $d = 15\,\text{cm}$ and substrate diameter $D = 8\,\text{cm}$, we calculate a radial variation of approximately 13% from the axis to the perimeter of the substrate.

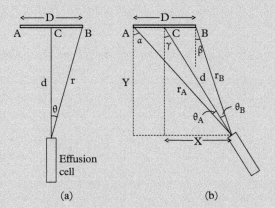

(a)   (b)

Figure 3.6 *Diagrams relevant to calculation of the uniformity of the molecular beam intensity over the substrate. In (a) the source is aligned with the axis of the substrate, while in (b) it is off-axis (the more usual condition).*

continued

Box 3.1 *continued*

Note that this remains unaffected by rotating the substrate about its axis. Indeed, a similar remark can be made concerning any axially symmetric arrangement even though the cell may be far from a point source.

A typical off-axis cell configuration is represented in Figure 3.6(b), where we assume the cell to be pointed directly at the centre of the substrate. The flux at point A is less than that at C because $r_A > d$, because the areas over which the fluxes are incident are proportional to $\sec \alpha$ and $\sec \Upsilon$, respectively, and, finally, because the flux from the source varies as $\cos \theta$. Combining these three factors allows us to write

$$F(A)/F(C) = (d/r_A)^2 \times (\cos \alpha / \cos \Upsilon) \times \cos(\alpha - \Upsilon) \qquad (B3.3)$$

From Figure 3.6(b) it is easy to see that the ratio (d/r_A) is given by $(\cos \alpha / \cos \Upsilon)$, so our flux ratio can be written as

$$F(A)/F(C) = (\cos \alpha / \cos \Upsilon)^3 \times \cos(\alpha - \Upsilon) \qquad (B3.4)$$

Similarly,

$$F(B)/F(C) = (\cos \beta / \cos \Upsilon)^3 \times \cos(\Upsilon - \beta) \qquad (B3.5)$$

where

$$\tan \alpha = (X + D/2)/Y \qquad (B3.6)$$

$$\tan \beta = (X - D/2)/Y \qquad (B3.7)$$

and

$$\tan \Upsilon = X/Y \qquad (B3.8)$$

the quantities X, Y and D being characteristic of the MBE machine configuration. Taking representative values of $D = 8$ cm, $Y = 15$ cm and $X = 8$ cm, we can evaluate the angles α, β and Υ as $38.7°$, $14.9°$ and $28.1°$, respectively, giving flux ratios of $F(A)/F(C) = 0.68$ and $F(B)/F(C) = 1.28$. In other words, for a three-inch substrate, the flux at B is almost double that at A, a rather serious discrepancy! Note, however, that the average of these two ratios is 0.98, pointing to the considerable advantage that can be realised by rotation of the substrate. However, this is not quite the whole story because rotation effectively averages the fluxes appropriate to all points round the perimeter and it is clear that, in the direction at right angles to the line AB, we are still faced with an edge-to-centre ratio to be calculated according to equation (B3.2), yielding a value of 0.90. Without getting bogged down in any more complex three-dimensional geometry, it is easy to see that the overall edge-to-centre ratio appropriate to a rotating substrate is likely to be of order 0.94—that is, some 6% variation from centre to edge of a 3-inch substrate. The precise numerical values should not be taken too seriously— our model, based on the Knudsen cell, is certainly not applicable to the 'open plan' cells used in most practical machines, nor have we made any attempt to optimise the cell orientation. Nevertheless, they do agree surprisingly well with the results of more sophisticated modelling and they certainly emphasise the need to rotate the substrate about its axis in order to achieve a reasonably uniform film thickness.

So far can one go with simple modelling, but a realistic theory of flux uniformity, applicable to real cells with bucket-like geometry, almost certainly requires assistance from the computer—and even then there is a question as to how to deal with molecules which bounce off the cell walls on their way to the substrate. (Should one assume specular reflection or thermal accommodation and re-emission?) An early example of such calculation was that of Curless (1985) who simulated the growth of GaAs from an open conical gallium cell coated with source material to an arbitrary depth and a cylindrical arsenic cell, the cells being placed off-axis, at 5 o'clock and 11 o'clock, respectively. The simulation showed a GaAs film growing under arsenic-stabilised conditions over most of its surface but under gallium-stabilised conditions over some 10% of the surface and showed variation of film thickness amounting to 50% over a 4 cm square, a result not inconsistent with that of our more naive calculation above. Curless' modelling has been further developed by Yamashita et al. (1987) to include the effect of liquid sources with sloping surfaces, though their simulations show a very similar degree of non-uniformity. Finally, we take note of the modelling work reported by Wasilewski et al. (1991a, b) which clearly demonstrates how cell design can influence flux uniformity across a three-inch diameter substrate. Use of a standard cylindrical cell resulted in an edge-to-centre ratio of 0.6, while a conical crucible improved this to about 0.75. A vital factor was found to be the extent to which all points within the crucible could be 'seen' from the substrate and this is clearly enhanced by an increase in the taper angle of a conical cell (see Figure 3.7). However, they also found that other details were important, such as the depletion of the liquid level in the gallium crucible and the precise orientation of the cell with respect to the substrate. Including substrate rotation and optimising the cell orientation suggested the possibility of obtaining uniformity to within less than 1% over a 3-inch wafer using a conical cell in a standard commercial MBE machine.

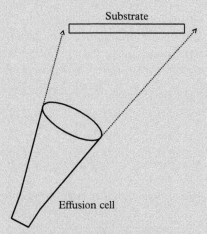

Substrate

Effusion cell

Figure 3.7 *Schematic illustration of the condition where all points within the cell are visible from all points on the substrate.*

Cells of the type illustrated are suitable for most applications but circumstances sometimes demand the design of special cells, such as the 'valved cracker' cell used to generate beams of the Group V species P_2, As_2 or Sb_2, the first such cell being designed by Jim Neave at PRL for the study of deep levels in GaAs (Neave et al. 1980). The cracker is designed to achieve two objectives: to produce a beam of dimers, rather than the tetramers which occur naturally, and to generate a beam flux which remains constant with time. All three Group V elements evaporate as molecules which must be decomposed in order for growth of the appropriate compound to occur, and all suffer from the fact that they sublime rather than melt, so the evaporation rate depends on the total surface area of an agglomeration of particulates forming the charge within the crucible. This, in turn, implies that, as the growth cycle continues and the evaporant is gradually depleted, the flux is correspondingly reduced. The cracker cell typically consists of a chamber or reservoir containing the relevant solid material, together with a second space where the emerging tetramer molecules can be dissociated. The reservoir chamber is heated to a temperature appropriate to producing a suitable vapour pressure of the element, the flux effusing into the cracking zone being regulated by a mechanical valve which maintains a constant flow of molecules. The temperature in the cracking zone is adjusted so that the tetramers emerging from the valve are decomposed into dimers. The first crackers were developed to decompose As_4 molecules into As_2, a reaction which occurs at a temperature of about 900 °C. Various reasons for preferring the dimer species have become apparent, one of the most important being concerned with the growth of double heterostructure lasers. This is probably associated with the fact that AlGaAs layers grown with As_4 show a marked tendency to roughness, which is completely avoided when growing with the dimer. On the other hand, AlGaAs/GaAs two-dimensional electron gas samples show superior characteristics when grown with As_4, a mystery not yet unravelled! More recently in the development of materials for 'spintronics', it has become clear that the low-temperature growth of GaMnAs requires As_2 in order to minimise the incorporation of arsenic anti-site defects. In such matters, generalisation is clearly impossible—each case must be treated on its own particular merits.

Other special cells include electron beam sources which are used to evaporate low-vapour-pressure materials such as carbon, silicon or tungsten. The source material is irradiated with high-energy electrons from a 25 kV electron gun, some 80% of the electron energy being converted into heat at the sample surface, resulting in the desired vapourisation of surface atoms. It is a process which has been used for a long time to deposit thin films of a wide variety of materials, and its adaptation to MBE growth was scarcely to be wondered at. Perhaps its most important application is to the evaporation of silicon when growing silicon/silicon–germanium structures—while a conventional cell may be adequate for doping III-V compounds with silicon donors, the available flux is far too small for the growth of silicon epilayers. (We take up the subject of Si MBE in Chapter 8.) Leaping now to the opposite extreme of the vapour pressure spectrum, we encounter beam molecules which already exist as gases. Examples include carbon tetrabromide CBr_4, arsine AsH_3 or phosphine PH_3. We leave this particular topic until Chapter 5, where we discuss gas source MBE in detail; nevertheless, given the ever-growing significance of the Group III nitride semiconductors, it is appropriate to emphasise here the importance of the special sources developed for providing active nitrogen or excited molecular nitrogen from nitrogen gas. Nitrogen is one of the most stable of gases and requires extremely high temperatures to decompose it so that it can react with a growing nitride film. However, a far more practical approach to the growth of nitride films is to use active nitrogen from a plasma source (Hoke et al. 1991). Nitrogen molecules are decomposed in a high-temperature RF plasma with active nitrogen atoms supplied to the film surface, which may be kept at a very much lower temperature (typically about 700 °C) compatible with avoiding gallium re-evaporation. Nitrogen gas is contained in a boron nitride crucible and heated directly by an RF coil to a temperature of about 6000 K, at which all the N_2 molecules are dissociated, though only some 3–5% of the active nitrogen actually reaches the film surface before recombining to form molecular nitrogen. An alternative approach to generating active nitrogen species makes use of an electron cyclotron resonance (ECR) source (Paisley et al. 1990; Lei et al. 1991) which operates at microwave frequencies, rather than the 13.56 MHz typical of RF excitation. Active hydrogen can also be produced using the RF plasma source, though a simpler arrangement is often used in which the gas is passed over a heated filament in order to decompose it into atomic hydrogen.

In summary, we should simply note that the important consideration is to supply just the desired species to the growing film and to avoid unwanted contaminants. This may often involve pretreating the source to decontaminate it, typically by heating it to a temperature well above its normal operating temperature both before and after loading the charge. In many cases it is also necessary to hold it at an elevated temperature when not in use, to avoid contamination from other gases in the MBE system. Precise details vary from one source to another, there being a deal of 'local' knowledge within the MBE community which must be acquired by newcomers to the 'trade'.

3.3 Beam Calibration

As the growth rate of epitaxial films depends on beam flux, it is clearly desirable to have some reliable method of measuring the flux from each of the cells, and various techniques have been applied over the years. In the original studies of surface chemistry and growth dynamics (Cho and Arthur 1975), a quadrupole mass spectrometer was used to analyse both the flux of atoms arriving at the substrate and that leaving it, a requirement which involved moving the spectrometer round an arc so it could sample both beams (see Figure 2.9). The group at IBM studying superlattice growth also calibrated their molecular beams with a mass spectrometer (Chang et al. 1973a), though in this case the geometry of their MBE machine was such as to allow some fraction of the beam fluxes to bypass the substrate and reach the QMS behind it. The modern arrangement with rotating substrate holder makes this either difficult or impossible—it is usually necessary to move the substrate holder out of the way in order to monitor beam flux.

By far the most commonly used method of measuring flux makes use of an in situ beam monitoring ion gauge (BMIG), which can be manipulated into a position coincident with the substrate holder. In Box 3.2 we emphasise that the ion gauge is actually sensitive to the density of molecules in the beam rather than to flux, so it is necessary to correct the reading to allow for this. Given that flux equals density times velocity, this implies that, to obtain a number proportional to the flux, we must multiply the reading by $T^{1/2}$, where T is the temperature of the gas molecules inside the cell, and this raises the question as to how accurately we can measure it. As we pointed out in the previous section, the temperature as measured by the cell thermocouple T_m may differ significantly from the true temperature T and it is necessary to check the slope of a log (flux) vs T^{-1} plot in order to correct such discrepancy. Knowing, then, the correct temperature, one might use equation (B2.13) (see Box 2.1) to compare calculated and measured fluxes, and this we actually did in the early days, with some success, but today's large effusion cells can hardly be expected to perform as Knudsen cells, and the simple fact is that no one does anything like this nowadays! Nevertheless, it should be clear from a brief reading of Box 3.2 that using an ion gauge to measure beam flux is far from straightforward, and reliable results require a certain amount of care. What is more, neither the QMS nor the BMIG are absolute gauges and must therefore be calibrated, a gallium beam being convenient for this. With the exception of the plasma-assisted growth of the nitrides (see below), the III-V compounds are all grown under Group V-rich conditions, the growth rate being determined purely by the arrival rate of the Group III species and this means that a measurement of growth rate of the appropriate compound can be used to calibrate the ion gauge. Probably the most convenient method of achieving this (at least in research) is based on RHEED oscillations (see Section 4.3) and this would typically be done on a very small sample, of dimension $5 \times 5 \, mm^2$, placed on the rotational axis of the substrate holder.

Box 3.2 Flux measurement with the ion gauge

It is common practice to monitor molecular beam flux by moving an ion gauge into the substrate position so as to intercept the beam. However, one must be careful to allow for the considerable variation in gauge sensitivity between different molecules and to bear in mind that the ion gauge actually measures density, rather than flux. Taking this second point first, we note that beam flux F and beam density n_B are related by

$$F = n_B \times c \qquad (B3.9)$$

where c is the molecular velocity, which depends on the molecular mass and the temperature of the source. For a Knudsen cell, c is given by equation (B2.6) and F by equation (B2.10). Combining these with equation (B2.8) leads to the simple relationship

$$n_B = An/4\pi r^2 \qquad (B3.10)$$

A being the area of the source aperture, and a few moments contemplation leads to the conclusion that this is obvious!

Ion gauge sensitivity has been studied extensively in the past (Dushman and Young 1945), the result being conveniently expressed in terms of the sensitivity to nitrogen at room temperature:

$$S = 0.038Z + 0.465 \qquad \text{(B3.11)}$$

where Z is the appropriate atomic number. (Note that equation (B3.11) yields $S = 0.997$ for nitrogen.) Combining this and equation (B3.9) allows us to calculate the net sensitivity for a range of molecules, the results being collected in Table 3.1. The fifth column shows the result of applying equation (B3.11), while the sixth column includes the correction for molecular velocity. The seventh column expresses the results in the sixth column in terms of the sensitivity for a gallium beam, while some experimental values appear in the eighth column for comparison. Clearly, there are considerable variations in sensitivity on account of both the above factors. Experimentally, there is reasonable agreement for metals, while the calculated values for the molecular species are somewhat underestimated. In part, this is due to the fact that the ionisation energy of a typical gauge (70 eV) is sufficient to fragment the parent molecule, as is evident from mass spectrometer studies. For example, in the case of As_2, mass peaks are seen with approximately equal intensities at 75 and 150, corresponding to As^+ and As_2^+, respectively, though, if the ionisation energy is reduced to 12 eV, the mass 75 peak disappears. The increased sensitivity which has been measured is therefore partly the result of the two contributions to the ion current, corresponding to the two species, both of which can be ionised.

Table 3.1 *Calculated and experimental ion gauge sensitivities for a range of molecules.*

Mol.	Z	Mol. wt	Flux T (°C)	S wrt N_2	Vel.	S wrt Ga	Expt
N_2	14	28	27	1.00	1.00	0.77	
Al	13	27	1000	0.96	0.46	0.35	0.5
Ga	31	70	930	1.64	1.3	1.00	1
In	49	115	832	2.33	2.46	1.9	1.4–1.7
P_4	60	124	650	2.75	3.3	2.54	
As_2	66	150	950	2.97	3.43	2.63	~5
As_4	132	300	500	5.48	11.2	8.6	~10
Sb_4	204	483	~650	8.22	19.5	15	
Bi_4	332	1036	~650	13.1	45.5	35	
Mn	25	55	867	1.42	1.02	0.79	0.82

In the case of a binary alloy such as AlGaAs, calibration of both Group III fluxes is equivalent to measuring both growth rate and composition. However, in many instances, the grower will be required to deposit layers with different (but specific) aluminium fractions—for example, the DH laser employs one composition for the optical waveguide and another for carrier confinement. For accurate control of composition, it is clearly necessary to calibrate the aluminium flux as a function of cell temperature, then to adjust the temperature to achieve any required composition. For AlGaAs the calibration is straightforward, using the same GaAs substrate for both gallium and aluminium deposition rates, but in other cases there may be subtleties to circumnavigate. For instance, in the case of InGaAs, it would be necessary to use a small InAs substrate to measure the indium flux, and a GaAs substrate for gallium, and to bear in mind that a correction must be made for the different lattice parameters of the two materials, when converting from growth rate to flux.

Unlike the arsenides and antimonides, the Group III nitrides are normally grown under Group III-rich conditions, where the growth rate is determined by the nitrogen flux. What is more, growth under excessively nitrogen-rich conditions results in an unusual form of columnar growth which cannot be relied on for accurate calibration purposes. This means that a calibration of the Group III flux involves a certain amount of dexterity. In particular, it requires that one must first establish the point at which the two fluxes are approximately equal and then grow under slightly nitrogen-rich conditions, where growth is determined by the Group III flux. Measurement of growth rate then provides a calibration of this aforementioned flux. However, one's problems do not terminate here, because it is extremely difficult to obtain RHEED oscillations in nitride growth (probably a result of growing on non-equivalent substrates); so, some alternative method of measuring growth rate is necessary. This is usually done by using optical interference from suitably thick films, though it should not be forgotten that this depends on accurate knowledge of the appropriate refractive indices.

An altogether different approach to flux measurement depends on measuring the optical absorbance of the beam atoms and this has been developed into a reliable practical technique for use in device production. It was first proposed by Kometani and Wiegmann (1975) at Bell Labs. They measured both gallium and aluminium fluxes during the growth of AlGaAs by transmitting an optical beam through windows on either side of their MBE chamber so that it sampled the molecular beams close to the substrate. Because each species shows strong absorption at its own characteristic wavelength, they made use of a hollow cathode lamp as source, filled with the appropriate metal, the absorbance being measured with a suitable monochrometer and photomultiplier, followed by a lock-in detector. Though the signal-to-noise ratios they obtained left something to be desired, they were able to demonstrate good correlations with mass spectrometer measurements and with cell temperature. They pointed out the importance of a technique such as this in providing continuous flux monitoring, noting that, over any significant time period, the flux from a cell will vary as the charge is consumed. In many cases, flux instabilities can also occur on much shorter time scales and, perhaps of even greater importance, flux changes as a result of a change in cell temperature when the beam shutter is opened or closed.

In the event, a number of workers have developed the technique, improving its sensitivity the while (e.g. see McClintock and Wilson 1987). At Sandia, Chalmers and Killeen (1993), who were concerned with the growth of AlGaAs vertical cavity surface emitting lasers (VCSELs), also applied it to measuring gallium and aluminium fluxes, using the resulting data for MBE shutter control when growing distributed Bragg reflectors (DBRs). They simplified the experimental arrangement by using narrow band filters instead of a monochrometer and improved sensitivity by incorporating a source stabilisation loop and heated quartz windows (the appropriate wavelengths being in the near UV). DBR layer thicknesses were thus controlled to an accuracy of 0.3%, compared with the earlier method based on measuring cell temperatures which achieved accuracies no better than 5%. More recently, Pinsukanjana et al. (2003) have refined the method by employing optical fibres for signal handling and by reflecting the optical beams back through the flux beams to double the absorption. They obtained reliable, clear-cut correlations between absorbance and growth rate for aluminium, gallium, and indium fluxes and used the method as a measure of the composition of InGaAs layers lattice-matched to InP substrates and strained layers in PHEMT structures. Of particular significance is the fact of their applying it for flux monitoring during the long time scales of HBT and PHEMT device production runs. As they make clear, in the context of production, it is far more convenient than the use of RHEED oscillations. Once calibrated, it provides a reliable, continuous record of the various beam fluxes, and this record can be used to maintain accurate control of device structures over a lengthy production campaign.

3.4   The Substrate Holder

Epitaxial growth presupposes the existence of a well-prepared substrate, and that, in turn, demands a suitable holder. Two requirements of the substrate holder in an MBE machine are that it should allow rotation of the substrate, to smooth out the inevitable flux non-uniformities arising from the off-axis arrangement of the sources, and that it should facilitate heating of the substrate to a temperature appropriate to surface diffusion of molecular species on their way to finding appropriate lattice sites. (According to application, such temperatures may lie anywhere within the range from 200 °C to 800 °C.) What is also of importance is the uniformity of this substrate temperature—diffusion, like so many physical phenomena, depends exponentially on temperature, which exaggerates the effect of relatively small temperature variations. Such considerations, as we shall see, make stringent demands on the details of sample mounting—not only must the substrate be held firmly in place and heated uniformly, it must be a relatively easy matter to transfer it to and from the growth position. What is more, in the interests of reproducibility, it must be possible to measure and control the temperature, this being particularly true with regard to commercial applications, where critical structural and doping parameters may be temperature sensitive.

For many years, in the early development of MBE (see Cho and Arthur 1975), substrates were effectively 'glued' onto molybdenum blocks by means of a film of liquid indium (or occasionally gallium). Surface tension held the substrate in place, while the excellent thermal contact meant that the substrate temperature was determined by that of the molybdenum, which could be heated by radiation from an electric filament placed close behind the block. Temperatures could readily be measured with an embedded tungsten/rhenium thermocouple. However, with the advent of substrate rotation (Cho and Cheng 1981), it became clear that physical contact between substrate and thermocouple was no longer possible and it became standard practice to place the thermocouple behind the heater, at a distance equal to the separation between heater and substrate block (a typical arrangement being shown in Figure 3.8). Implicit was the assumption that, in this geometry, the purely radiative heating of substrate and thermocouple would be identical. Such an assumption had to be accompanied by a certain degree of scepticism, and doubts were often expressed as to whether results from different laboratories could be reliably compared. One might say that 'substrate temperatures' were generally regarded as being contained within inverted commas! This scepticism was clearly demonstrated by the fact that attempts were made to standardise by reference to certain well-defined(?) 'fixed points' such as the temperature at which the oxide was thermally removed from GaAs. Early measurements suggested that this occurred at a temperature of approximately 530 °C (Cho and Arthur 1975), though this has later been amended to ~620 °C (see Chang et al. 1973b—it depends to some extent on the arsenic flux present during desorption). Some more recent measurements (Foxon and Harris 1986) quote 600 °C, while Hellman and Harris (1987) suggest a value closer to 660 °C, an indication in itself of the difficulty inherent in such procedures. Another calibration point made use of the congruent evaporation point for gallium arsenide. Raising the temperature above 637 °C results in the preferential evaporation of As_2, leaving gallium droplets on the surface of the GaAs. Finally (see Cho and Arthur 1975), it was found that the gallium arsenide surface restructured at well-defined temperatures, depending on the As-to-Ga flux ratio. For example, the transition between the gallium-stabilised C(8 × 2) and the arsenic-stabilised C(2 × 8) occurs at temperatures in the range 510 °C–590 °C for As_2/Ga ratios between 1 and 10. Similarly, at about 400 °C, there is a transition from

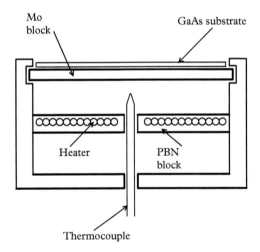

Figure 3.8 *A typical substrate holder employing indium solder to hold the sample against a heated molybdenum block. The heater is embedded in a PBN block to provide uniform heating. The thermocouple is mounted midway between the heater and the Mo block to sample the temperature of the thermal cavity formed between them.*

the arsenic-stabilised (2×4) structure to a $C(4 \times 4)$ structure (also dependent on the arsenic flux). None of these, however, provide absolute temperatures and they are, of course, particular to a specific material system—each new material required its own calibration points, not at all an ideal situation.

A partial solution to this measurement problem was provided by the use of infrared pyrometry. Typical of such instruments was the IRCON 2000 pyrometer, which measured the absolute intensity of thermally emitted radiation at wavelengths near 1 μm and which was applicable to GaAs at temperatures above about 500 °C, where the GaAs band gap is small enough to bring the measuring wavelength above the band edge. At lower temperatures, the pyrometer simply detects radiation emitted by the molybdenum block, which is transmitted through the GaAs. Even at 500 °C and above, the number of photons emitted by the GaAs is rather small, thus compromising the accuracy of the measured temperature, and there is always some slight uncertainty concerning the emissivity of the GaAs surface. This depends not only on surface preparation but also on the density of free carriers—in our own work we used values of 0.5 for semi-insulating GaAs and 0.62 for silicon-doped material but these values are only approximate and other workers certainly used somewhat different values! Another complication is introduced by the fact that the detected radiation must pass through a window in the vacuum system before detection, and windows in MBE systems tend to become coated with stray molecules such as As_2. This can be treated formally as a reduction in emissivity but, no matter how one regards it, it leads to an error in the measured temperature. Today this particular problem can be countered by using a heated window from which deposited films are very largely desorbed but it was not an option in the days when MBE was being developed. Hellman et al. (1986) estimate a minimum uncertainty of ± 10 °C in their temperature measurements as a consequence. Allowing for uncertainty in emissivity, the overall error in temperature might well be as much as 30 °C. Needless to say, temperatures of wider band-gap materials such as GaP and GaN could not be measured with infrared pyrometers unless they were grown on narrow-gap substrates or mounted on molybdenum back plates, and their needs have only recently been met by the introduction of band-gap monitoring, which we discuss below.

The next important development in substrate holder design appeared in the mid 1980s, based on the use of direct radiative heating. It was motivated by two considerations. First, the use of indium bonding resulted in a rough substrate back surface which was incompatible with mass production microlithography and, second, as substrates became progressively larger, there was increasing likelihood of uneven bonding over the whole area—with corresponding variation in temperature. Several different groups had explored methods of doing away with indium mounting—Wood (1976), as early as 1976, loosely attached his substrates to a fused silica block which had a tantalum heater filament deposited on its back surface, Palmateer et al. (1984) used a ring mount and sputtered a titanium/tungsten film on the back of the substrate to absorb radiation from an adjacent heater, Erickson et al. (1985) described a similar approach using sputtered molybdenum while Oe and Imamura (1985) contrived to grow successful 2DEG layers without the need for any sputtered backing film. The end product of these various developments was described by Hellman et al. (1986), their version being adopted by

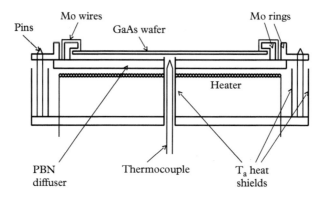

Figure 3.9 *An example of an indium-free substrate mounting system in which the sample is lightly held by molybdenum springs. The molybdenum rings which support the sample can be removed in order to exchange samples.*

Varian for the Gen II MBE machine. Doing away with indium bonding led to a complete redesign of the substrate holder, as illustrated in Figure 3.9. The three-inch substrate was mounted on a molybdenum ring with fine molybdenum wires which prevented intimate contact (and thus prevented conductive heating) but which introduced only minimal stress in the substrate. Highly uniform substrate heating was achieved by radiation from a PBN diffuser, and the temperature was measured with a thermocouple centred between substrate and diffuser. In effect, their two surfaces formed a radiation cavity whose well-defined temperature was measured by the thermocouple. Temperature stabilisation was effected by controlling the heater power so as to maintain a constant thermocouple reading. The molybdenum ring was mounted on a second, rotatable ring but could easily be removed in order to transfer the substrate to and from the load lock. Hellman et al. estimated that their substrate temperatures were uniform to within about 6 °C, which could be taken as justification for the new design, but their comparisons between temperatures measured with an IR pyrometer and those measured by the thermocouple showed just how difficult it is to arrive at reliable results—at 'thermocouple' temperatures of 700 °C they found a difference of some 50 °C between the two techniques.

Clearly, there was a need for a more reliable method of measuring substrate temperature, a need to be satisfactorily met, towards the end of the 1980s, by a completely new approach. It was based on the simple fact that the band gap of a semiconductor varies with temperature in a reproducible fashion, so a measurement of substrate band gap was effectively a measure of its temperature. Hellman and Harris (1987) demonstrated an in situ method of measuring the band gap of GaAs substrates, making use of the infrared radiation from the substrate heater to record the transmission spectrum on a conventional spectrometer, which was scanned over the appropriate range of wavelengths near the band edge. Recording the absorption coefficient as a function of wavelength allowed the authors to represent the band edge in terms of the wavelength at which the absorption coefficient took the value of 40 cm^{-1}, facilitating a direct comparison with literature data. They estimated the precision of the temperature measurement as being ±2 °C, a tremendous improvement over previous methods and pointing the way to the first totally reliable method of measuring substrate temperature on any material for which the necessary absorption data was available. The principal drawback was the inherent slowness of the measurement, which made it unsuitable for use in temperature control. However, it was perfectly possible to use a thermocouple reading to effect control, while relying on band-edge measurement for absolute accuracy. At last the problem of temperature measurement had been solved—application to other materials would swiftly follow. An important example was the work of Lyon et al. (1997), who applied it to the growth of InAs and HgCdTe films, problems with pyrometer measurements being particularly serious for such narrow-gap materials.

The final piece in this particular jigsaw puzzle was not put in place until 2003, with the development of a commercial band-edge detector as a replacement for the scanned spectrometer (see Farrer et al. 2007). This made use of an array of semiconductor diode detectors to record the intensity of individual segments of the dispersed spectrum, rather than relying on a slow scan across a single detector. The principle of the method remained unchanged, of course, but the measurement was more or less instantaneous and therefore far more convenient than that based on a conventional spectrometer. However, rather than using the transmitted radiation from the substrate heater to determine the band gap, it was found preferable to measure the diffuse reflection from the substrate, using a quartz halogen white-light source and this technique was found to work well for measurements on semi-insulating GaAs substrates over a temperature range from 200 °C to 600 °C. In the case of doped substrates, free-carrier absorption tends to smear out the band edge—nevertheless, this standard method could be applied on substrates doped up to about 1×10^{24} m^{-3}, only at p-type doping levels of 1×10^{25} m^{-3} did it become impossible to detect the absorption edge. Some special procedure, such as the use of the equipment in a pyrometer mode, was required in this case. Indium bonded substrates also demanded careful attention because, particularly at high substrate temperatures, radiation from the molybdenum holder tended to swamp the reflected light from the substrate surface. In this case, it was still possible to detect the band edge, provided a subtraction technique was employed to remove this interfering signal. In addition to measuring the temperature of GaAs substrates, the technique has also been applied successfully to InP and Si and, more recently, to GaN (Harris et al. 2007).

3.5   Load Locks

In the early days of MBE it was necessary to let the main chamber up to atmosphere in order to change samples after each growth run and this proved inconvenient for a number of reasons. Clearly, the chamber had to be re-baked and outgassed each time, and the time taken to reach an acceptable background pressure gradually increased as a result of the build-up on the chamber walls of volatile materials such as arsenic, which absorbed moisture and oxygen from the air. Worse still was the fact that the arsenic source material became oxidised and had to be heated and pumped to remove oxygen. Not only did each sample change take upwards of ten hours but, after one to two years of operation, it became essential to clean the whole chamber and to recondition or replace the pumps. It also became clear (as was well illustrated by the struggle to grow high mobility two-dimensional electron gas samples in the 1980s) that material quality improved gradually with time over a long sequence of growth runs, provided a clean vacuum could be continuously maintained. Thus, there was an obvious need to develop means for transferring samples into and out of the main chamber while keeping it under high vacuum and this became standard practice from about 1980 onwards.

An early example of a successful interlock system was reported by Robinson and Ilegems (1978). They described the use of a vertical transfer system which allowed the whole sample mount to be raised through a gate valve into a separate loading chamber, where it was moved horizontally through a demountable port to allow the substrate to be changed. The vertical motion was effected by virtue of a bellows system. This was in the days when substrates were mounted on a molybdenum block with indium 'solder', so it was necessary to move the whole block, together with its heater assembly, this arrangement having the advantage that a new substrate could be heated and degassed in the loading chamber. The sample change took no more than a matter of minutes, during which the pressure in the main chamber remained below 10^{-7} Torr (there was some slight leakage through the gate valve). The loading chamber was then evacuated before opening the gate valve and transferring the sample to the growth position. In about one hour, the system pressure had recovered to 3×10^{-9} Torr and a new growth run could be initiated. It was apparent too that the use of the interlock allowed the main pumps to be used for considerably longer than had previously been the case.

The advantages to be gained by the use of a load lock system were readily appreciated and commercial MBE machines adopted them from the beginning. Luscher and Collins (1980) describe a typical vacuum interlock as used in the Varian V360 machine, while Luscher (1981) outlines the system used on the Varian Gen II machine. Perhaps the most significant advance concerned the use of magnetic transfer rods which allowed movement by as much as a metre, avoiding the problems associated with UHV bellows. Additionally, such commercial load locks allowed the incorporation of an extra chamber in which samples could be stored and/or processed, prior to being introduced into the growth chamber. A separate ion pump for this subsidiary chamber meant that the pressure in the growth chamber could be maintained continuously at levels in the region of 10^{-9} Torr for as long as several months.

An improved transfer system was described by Andersson et al. (1983), based on the Varian V360 MBE machine (see Figure 3.10). By using two magnetic transfer rods, three gate valves and a carousel capable of holding six substrates, they were able to load several GaAs samples into an intermediate storage chamber, where they could be cleaned by heating in an arsenic atmosphere or by argon ion bombardment, prior to being transferred into the main growth chamber. There was also a facility for transferring processed samples into an independent analysis chamber. The loading chamber was pumped by an oil-free membrane pump and by sorption pumps, while the storage chamber was ion-pumped. Pressure in the latter chamber took about a half-hour to fall

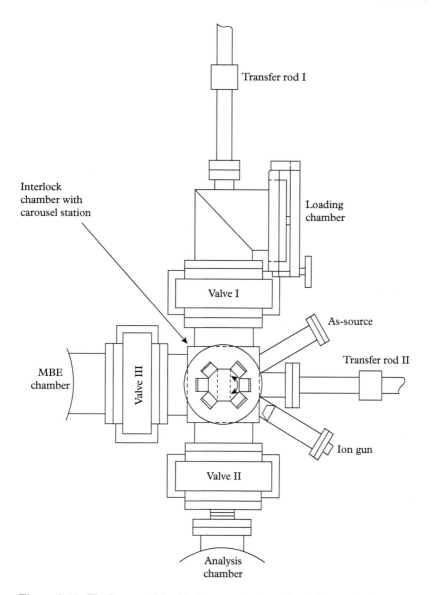

Figure 3.10 *The 'improved' load lock system developed by Andersson for the Varian V360 MBE machine. It employs magnetic transfer rods to facilitate movement of a cassette of substrates through a preparation chamber where they could be cleaned by argon ion bombardment and by heating. (From Andersson et al. 1983, courtesy Institute of Physics.)*

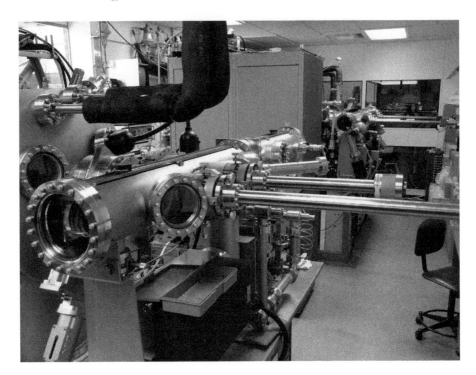

Figure 3.11 *Photograph of the load lock assembly used on the authors' MBE machine in Nottingham.*

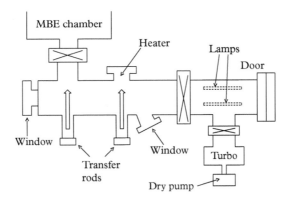

Figure 3.12 *A schematic diagram of the load lock arrangement shown in Figure 3.11. Samples are introduced by way of the end door and dried by the heating lamps, while the outer chamber is evacuated. They are then transferred to the preparation chamber, where they can be heat-cleaned, before being taken into the main growth chamber.*

to 10^{-8} Torr, at which point the valve between it and the main MBE chamber could be opened and the samples exchanged.

To give an impression of a typical load lock system as found on many modern research machines, we offer a photograph (Figure 3.11) and a corresponding diagram (Figure 3.12) of one of our own MBE machines in Nottingham. The loading chamber is pumped with a dry backing pump and small turbo-molecular pump. Samples are loaded through the end 'door' onto a trolley with 12 mounting positions, which can be moved magnetically on rails into the ion-pumped preparation chamber. Here, samples may be picked off and moved into a heating site for removal of water vapour before being transferred by a second transfer rod into the main growth chamber. There is a facility (not at present utilised) for adding arsenic or hydrogen sources.

3.6   In Situ Characterisation

One important feature of MBE film growth is the relative ease with which various characterisation measurements can be performed either within the MBE growth chamber or, only slightly less conveniently, in an adjacent chamber to which a simple vacuum transfer is possible. While Auger electron spectroscopy (AES) was used to monitor substrate cleanliness, these techniques have, for the most part, concentrated on measuring the dynamics of film growth or the structural properties of grown films. We know of no reports of in situ characterisation by either electrical or photoluminescence measurements on bulk films. In the first instance, the reason is obvious—electrical measurements require electrical contacts. In the second case, photoluminescence is generally performed at low temperatures and therefore demands a separate cryogenic system. This apart, it seems worthwhile, in discussing MBE machine technology, to include a brief account of those in situ techniques which have been applied successfully to studying MBE film growth.

It seems sensible to begin our discussion with the question of background pressure in the MBE chamber, it being essential, when growing high-quality material, that the arrival rate of unwanted species at the substrate is negligible compared to those of the molecular beams. Most systems will reach a an ultimate pressure of about 5×10^{-11} Torr after bakeout and overnight pumping unless there exists either a leak from outside or an internal leak from a trapped volume. In either event, it is extremely helpful to be able to identify the nature of the gaseous species responsible and this is most easily achieved by means of a residual gas analyser, usually in the shape of a quadrupole mass spectrometer with a mass range of 1–100 amu. This enables the user to use helium gas to check for external leaks, while also being able to analyse the mass spectrum from possible internal leaks. For example, oxygen and nitrogen might well characterise leakage from a trapped volume, and the ratio of mass 14 to mass 16 represents a valuable indicator of such problems, excess mass 14 coming from nitrogen. Typical mass analysers, equipped with an electron multiplier, achieve sensitivities to background pressures as low as 10^{-14} Torr (the less expensive versions employing a Faraday detector for only 10^{-12} Torr) but skill is required in either case to interpret the raw data. Experience leads to understanding of the species to be anticipated in any particular case and depends on the nature of the material being grown. There can be no simple recipe for success! For completeness, we should also note that mass spectrometers have been much used to study growth kinetics and the detailed chemistry of the growth process. This usually involves comparing the nature of the species arriving on the substrate and comparing this with what leaves. Such an instrument will usually have a much greater capability to detect higher mass numbers and will often make use of sophisticated modulation techniques to distinguish these differing signals. We shall discuss this aspect in greater detail in Chapter 4.

We continue our discussion with AES which, as we remarked in the previous chapter, was much used in the early days of MBE. The sample is excited by a focussed beam of primary electrons with energies in the range 3–10 keV, the resulting ionised atom core levels being filled by a two-electron Auger process, allowing one electron to be ejected from the crystal surface with an energy characteristic of the atom from which it came. By using an energy analyser in the shape of a retarding grid or cylindrical mirror analyser, it is therefore possible to identify the atom in question. The depth from which information can be obtained is determined by the very limited escape depth of the emitted electron, which is usually no more than a few atomic layers. AES is therefore ideal for detecting surface contamination of the substrate, prior to film growth, a seriously important requirement in early growth experiments. As we have already intimated, AES was an invaluable aid to checking substrate contamination prior to film growth, a requirement which, for GaAs, at least, has now receded into the technological distance with the availability of clean, lightly oxidised substrates which need little more than gentle heating to prepare them for film deposition. But AES, combined with argon ion sputtering, can also be used for depth profiling, in some instances an invaluable process but one that leads to contamination of the growth chamber. In consequence, it is usually performed in a separate chamber, to which the sample can be transferred, still under vacuum.

RHEED studies go back to the first stirrings of MBE as a process for growing practical films. As described by Cho and Arthur (1975), the first electron diffraction studies were performed by LEED, using electron energies in the range 50–150 eV, but the glancing incidence geometry of RHEED, using energies of 3–50 keV, proved much more convenient. Both systems involved electron penetration to no more than a few monolayers of atoms and so were ideal for studying surface structure, a subject that we shall explore in Chapter 4. For the present we shall merely take a brief look at one or two practical details. In some cases (C. T. Foxon, private communication), the Auger electron gun doubled as a RHEED gun; in other cases, black-and-white TV sets were raided for their electron guns, which were then mounted on UHV flanges; later, all three manufacturers of MBE machines (Varian, Riber and VG) developed their own guns; but, finally, guns from Staib became the industry standard for most commercial systems. The advantage offered by these latter guns was a facility for rocking the beam so as to adjust the angle of incidence, without moving the sample. The first RHEED screens were also taken from television sets, before tailor-made samples became available. A problem with these early systems was the gradual blackening as a result of contamination from the MBE beams. A partial solution was found in coating the screen with a thin metal film— blackening still occurs but at a more acceptable rate. It is also worth noting that the use of higher energy electrons (40 keV, rather than 4 keV) was beneficial in allowing the screen to be placed further from the sample surface and thus further from the sources of contamination. Because of its ability to provide surface structural information during MBE growth, RHEED has been widely used in research into the growth of many different semiconductor materials and is generally regarded as standard equipment in any MBE machine. Perhaps its most significant contribution to the understanding of film growth stems from the discovery in 1981 at the Philips Research Laboratory in Redhill, UK, of RHEED intensity oscillations, the period of the oscillation corresponding to the time taken to deposit a single monolayer of material (Harris et al. 1981a, b; Wood 1981). This facility for accurate determination of growth rate has found more or less universal application—yet another aspect of the technique that we shall discuss in the following chapter.

So much (for the moment) on electron diffraction—needless to say, other techniques for studying surface structure and growth rate also exist and have been applied to MBE. One naturally looks to optical methods, which are compatible with the presence of windows in all MBE systems. We have already referred to the use of optical absorption to measure the band gap of the growing film (Hellman and Harris 1987). This work was performed at Stanford—meanwhile a totally different approach was adopted at the Bellcore laboratory in New Jersey, based on measurement of optical reflection, using photon energies well above the semiconductor band gap (Aspnes et al. 1987; Harbison et al. 1988). This made use of the (then) newly discovered phenomenon of 'reflectance-difference spectroscopy' (RD for short) in which polarised light is reflected from the sample surface at approximately normal incidence. Studies on silicon and germanium had previously shown that the intensity of the reflected signal varied with the crystal orientation (typically by roughly 1%), and the RD signal represented the difference between intensities at a pair of mutually perpendicular directions. In the case of the (001) surface of GaAs and AlAs, for example, there was a readily detected difference for light polarised along the [110] and [−110] directions, respectively. Because the interaction with bulk material must be isotropic (for cubic crystals), the RD signal represented a purely surface effect and was later shown to be a measure of surface reconstruction (Kamiya et al. 1992). The experiment was performed using a Xenon lamp as the source, the reflected signal being dispersed in a grating monochrometer. The relevant photon energies lay in the range from approximately 2 to 4 eV. The particular significance for MBE growth arose from observation of RD signal changes in respect of changes in growth conditions, for example, on the difference between arsenic and gallium-stabilised conditions when growing GaAs. In fact, following initiation of GaAs growth, oscillations in RD intensity were detected, showing the same period as those seen in the RHEED signal (though considerably less well resolved). In this sense, the RD spectrum can be seen as a method of measuring instantaneous growth rate. However, a much easier optical method is to measure film thickness by recording interference between light reflected from the growing surface and that reflected from the film–substrate interface, as was first reported by SpringThorpe and Majeed (1990) and later by the group at Liverpool University (Armstrong et al. 1992), who were concerned with the growth of GaAs by chemical beam epitaxy (CBE). These measurements can be made either in transmission, using thermal radiation from the substrate heater together with a narrow band detector, or in reflection, using a HeNe laser (wavelength 632.8 nm) as light source. A minor inconvenience implicit in these proceedings is the necessity to grow a thin layer of AlAs on the substrate (to provide the necessary difference in refractive index) before growing the GaAs film.

Two other optical techniques used to study semiconductor surfaces in MBE systems are laser light scattering (at glancing incidence) and ellipsometry. In 1987 the group at RSRE Malvern applied LLS to study substrate cleaning and layer nucleation in silicon MBE (Robins et al. 1987). Later, the same group applied it to GaAs growth, measuring surface topography changes during substrate cleaning, film growth and subsequent thermal annealing (Smith et al. 1991). They point out that, using an argon ion laser source (wavelength 488 nm), the technique is sensitive to surface texture on a length scale of about one micron. Another interesting application of LLS was to the detection of lattice relaxation in InGaAs/GaAs strained-layer superlattices—the misfit dislocations occurring as the surface relaxes showing up as an increase in reflectivity in a specific [110] direction (Celii et al. 1993). The Aspnes group at Bellcore claim the honour of being first to apply ellipsometry to the study of GaAs and AlGaAs film growth (Aspnes et al. 1990). As they pointed out, ellipsometry actually measures the complex dielectric function of the material(s) in question, and interpretation of this data in terms of structural and chemical properties of what may be a complex multilayer structure demands detailed modelling. It also requires that the dielectric functions be accurately known in advance, in particular as a function of temperature over the temperature range used during crystal growth. They used it to monitor the oxide desorption process from GaAs, showing that it took place in a number of stages. They also demonstrated the possibility of measuring the aluminium fraction and thickness of very thin AlGaAs films. These ideas were taken up by Maracas et al. (1992), who emphasised some of the problems associated with in situ measurements in the MBE chamber. Because of the chamber geometry, the choice of incident and reflected angles is severely restricted and the desirable feature of using variable angles totally denied. The accuracy and reproducibility of the incident angle also suffers from lack of finesse in the construction of typical substrate holders, a degree of wobble being observed during the rotation essential to uniform film growth. (We might add that vibration from vacuum pumps and other extraneous equipment can also be a considerable nuisance!) They nevertheless contrived to apply the method of spectroscopic ellipsometry over a wavelength range of 250–1000 nm to the measurement of surface temperature (observing unexpectedly long stabilisation times), alloy composition and thickness of AlGaAs films and studies of the interface quality in AlGaAs/GaAs quantum wells.

As the sophistication of semiconductor device structures increased during the 1980s, and the use of mismatched substrate-layer combinations took on greater importance, it became more than ever important to develop techniques for monitoring the incidence of strain-relieving dislocations, and the application of in situ X-ray topography represented a significant innovation. The first such foray into technological futurism can be credited to a UK consortium involving RSRE Malvern, the Daresbury laboratory of SERC and the Universities of Durham and Hull. The design of the MBE growth facility, combined with in situ XRT is described by Whitehouse et al. (1992). Careful design of the sub-strate holder and minimisation of vibration were only two of the critical requirements but the system had been demonstrated to detect the onset of misfit dislocations in InGaAs layers grown on GaAs substrates. These were found to be confined to only one of the [110] directions in an (001) plane. Double crystal rocking curves were used to show a small variation in film thickness from centre to edge of the two-inch wafer, dislocations being confined to the thicker central zone and thus providing a precise determination of the critical thickness for strain relief. In a later example of the use of in situ X-ray diffraction, Braun and Ploog (2006) compared surface kinetics for (001) surfaces of GaAs, InAs and GaSb both during growth and in the so-called recovery stage following a growth interruption. Using X-ray wavelengths of approximately 1 Å incident at graz-ing incidence, they were able to follow the changes in island size over time scales of the order of a thousand seconds, their data showing remarkable differences in the detailed behaviour of these three very similar compounds. A particularly interesting observation was their discovery of oscillations in the scattered X-ray intensities which corresponded to the similar oscillations seen in RHEED intensities. As with RHEED, these oscillations start strongly as growth is commenced but decay in amplitude over times of the order of a hundred seconds (corresponding to the growth of about 20 monolayers). Their origin can be understood in terms very similar to those of their RHEED counterparts.

Finally, we should mention two other in situ characterisation techniques. Scanning electron microscopy (SEM) is an obvious method of looking at semiconductor surfaces and it is totally compatible with the UHV environment in MBE. Inoue (1991) first dem-onstrated the possibility of adding an SEM column to an MBE machine and obtaining real-time images of grown surfaces with resolution of the order of 10 nm. He used it to study the formation of gallium droplets on the Ga-stabilised, (4 × 2) reconstructed GaAs surface and showed that the migration length of gallium atoms was several orders of magnitude longer than on an arsenic-stabilised surface. Another vacuum-based ana-lysis method is that of scanning tunnelling microscopy (STM), which involves the use of a separate chamber connected to the MBE system (see e.g. Biegelsen et al. 1990). Krause et al. (2005) also described such an add-on and used it to examine the surface structure of both silicon and GaAs epitaxial layers. The ability to resolve single atoms makes this a particularly powerful tool for studying the clean surfaces produced in an MBE environment, though, once again, extreme care is necessary to eliminate vibra-tional interference. A practical point of interest in the Krause et al. paper is their use of liquid helium to cool the sample, opening the possibility for other low-temperature techniques to be applied similarly.

3.7 Commercial Machines

As we hope to have made clear in Chapter 2, MBE, as a method of growing practical semiconductor films, emerged from a slowly clearing mist of esoteric surface studies, and in 1970 there were probably very few participants in the struggle to master the vagar-ies of film growth who anticipated the possibility that the word 'commercial' might one day be uttered in the same breath as those of 'molecular', 'beam' and 'epitaxy'. In spite of the admirable pioneering work of Al Cho in demonstrating the possibility of growing semiconductor device structures by MBE, it was thought for a long time that MBE would remain for ever a research tool. The excellent control of film thickness, interface quality, doping profile, surface flatness, etc. proved invaluable in the design and development of esoteric new device structures but it was widely accepted that, when it came to 'production', MBE was itself too esoteric (and expensive!) a technique to be taken seriously. How wrong we all were! And how quickly were we to be disabused of our mistaken prejudices! The first commercial application of the technique emerged in the early 1980s, when the Japanese company Rohm undertook the production of AlGaAs/GaAs double heterostructure lasers by MBE. Following the launch of the compact disc player by Philips and Sony in 1980, it became clear that there was a potentially valuable market for any company able to manufacture reliable semiconductor lasers at a competitive price and, to most people's surprise, MBE, in the hands of Rohm, came to dominate that market for some considerable time (Tanaka and Mushiage 1991). More recently, it has become commonplace for MBE growth to take a central position in many commercial production facilities and no longer does anyone turn a hair at the idea of basing a major business venture on such an esoteric technology. The transition from fundamental research to large-scale production activity makes an interesting story and, in this section, we shall attempt to make clear some of its essential features.

The story began in the mid 1970s when an elite group of entrepreneurial companies recognised the commercial possibilities of manufacturing complete MBE machines to sell to the rapidly growing body of university and industrial researchers intent upon exploiting the obvious advantages of MBE as a means of making high-quality structures for fundamental research in both semiconductor science and semiconductor device development. Within the space of a single year Varian in the USA, Riber in France and Vacuum Generators (VG) in the UK each unveiled its version of what we shall call a 'commercial' MBE machine (to distinguish it from the numerous 'home-made' systems which had been assembled in research laboratories around the world) and that, to the Western World, was that. But, in fact, it wasn't! At more or less the same time, two Japanese companies, Anelva and Eiko, entered the market, though they appear to have exported their products outside Japan only to a very limited extent. All of these companies had been involved in some aspect of UHV technology and had supplied a variety of components to those 'home builders' who had pioneered the technique in the first place, and, scarcely surprisingly, there were several examples of collaboration between the two sides of the enterprise. Al Cho himself collaborated with Riber, Bruce Joyce advised VG, and Hellman et al. (1986) assisted Varian. Needless to say, these machines were all designed basically as research tools, there being no thought at the time of production possibilities. Varian gave their machine the title of 'V 360', Riber came up with MBE 500 and VG with '288' (two-inch substrate, eight cells, eight shutters!)—none of which titles could be seen as particularly romantic, this being a serious venture into serious scientific endeavour. Their Japanese counterparts were no more adventurous— Anelva chose 831, 430 and 620, Eiko EV-100 and EV-500, ULVAC MBC-300 and Daido Sanso VCE-S2020. Perhaps it was not surprising that all these machines showed more similarity than originality—they all came from the common source of scientific inventiveness. However, two notable differences could be discerned in the pumping arrangements—while the other companies opted for a combination of ion and titanium sublimation pumps, VG preferred diffusion pumps (though their later machines came with an option of ion pumping) and, while everyone else made use of liquid nitrogen for cryogenic pumping, Eiko preferred water cooling. There was yet another distinguishing feature—at this time the importance of using a load lock for sample transfer was only just beginning to be appreciated and Anelva's 1975 machine lacked such a facility—it was to be upgraded in 1977. We should also note that the concept of substrate rotation was missing at the time these early machines were designed—it was only incorporated in 1981 after Cho and Cheng's pioneering paper appeared. In terms of performance, there were no more than minor differences, though the cost of *running* a machine on cold water are significantly less than those employing liquid nitrogen, and it must be significant that Eiko have stayed loyal to their original gamble, even to the present day. (We are not aware of any detailed comparison of film quality in terms of cooling arrangement—could it be possible that, for many applications water cooling is perfectly adequate?)

Needless to say, the nearly 40 years that have rolled by since this initial burst of commercial activity have witnessed a number of changes in the competitive scene—some companies have gone and some newcomers have emerged. In the West, Riber has survived and appears to be flourishing, having taken over responsibility for marketing VG equipment since 2005, and new arrivals include three German companies, Omicron Nanotechnology, Dr Karl Eberl and Createc-Fischer (two of which represent spin-offs from the research activity of Klaus Ploog at the Max Planck Institute in Stuttgart), and one Finnish company, DCA Instruments. For a brief period in the 1980s Perkin Elmer was active in the MBE business, though they presumably decided their competitors were too strongly established and made a strategic withdrawal. Nevertheless, samples of the Perkin Elmer 430 MBE machine survive as irrefutable evidence. Varian's MBE activities were so successful that they were taken over in 1992 by Intervac, who passed them on to EPI in 1994. EPI (who had been in the business of effusion cell manufacture for some time) achieved nearly a decade of MBE machine production until, in 1994, another American company, Veeco, bought them out and have continued selling what are basically updated and enlarged Varian machines to the present day. In marked contrast to this flurry of Western commercial transactions, the two Japanese participants remain even-handed competitors within the Japanese mainland, though Anelva appear to have moved tentatively in the direction of developing a production machine while Eiko still rely on supplying the research market with machines that handle only a single two-inch wafer.

To a considerable degree, the demands of research have changed relatively little over the years, apart from the ever-widening range of materials to which MBE has been applied. The requirement for high-purity, high-quality interfaces, atomically flat surfaces, etc. were well satisfied by the first commercial machines, and the major significant improvement in machine performance has been the addition of a second UHV chamber for advanced material characterisation, as discussed in the previous section. It is worth noting that several commercial suppliers offer machines complete with SEM, STM,

ESCA or XPS analysis facilities. Indeed, companies such as Omicron, Createc-Fischer and Karl Eberl specialise in surface analysis techniques, and one might, perhaps, think of them as supplying MBE chambers as add-ons to their analysis equipment! Meanwhile, the availability of load lock wafer exchange facilities meant that the rate of sample production could easily keep pace with those of routine characterisation (for checking sample probity) and detailed analysis (for the pursuit of new scientific ideas). In short, the average university researcher was more than happy with his MBE machine—he had problems of a very different nature (such as where his next tranche of research funding might come from!). As more and more academics came to recognise the virtues of MBE for research, the market for basic machines appears to have remained surprisingly constant—'surprisingly' because MBE machines wear out only slowly and the market for replacements is correspondingly small. One significant statistic, here, is the fact that the V80 machine first marketed by VG in 1982 remained as their mainstay research instrument for no less than 23 years, over 200 being shipped during this time. It was a three-inch substrate machine and the only significant changes during its long life were concerned with improved reliability, rather than with a view to offering improved facilities.

In fact, progress in machine design has, perhaps inevitably, centred on the demands of those responsible for production of advanced semiconductor devices. At the beginning of the 1980s, it was already becoming clear that some of the new microwave and optoelectronic devices which depended on accurate control of very thin layers and precise doping profiles might viably be produced by MBE—question marks were being raised against such artefacts as running costs, throughputs, reliability, reproducibility, rather than mere technological capability—and, as we saw above, Rohm was already busy developing its MBE facility for the growth of DH lasers. In a significant paper published in 1981 (Luscher 1981), Paul Luscher of Varian Associates reviewed the position with regard to possible commercial production of devices by MBE, focussing attention on a number of relevant parameters such as machine uptime, layer thickness, device area, substrate size, growth rate, sample exchange time, etc. He pointed out that devices had already been grown by MBE with performance superior to those grown by alternative techniques and listed FETs, IMPATTs, mixer diodes, varactor diodes, Gunn diodes, light-emitting diodes, laser diodes, photodetectors, optical waveguides, integrated optical devices, integrated circuits and solar cells. In particular, he selected three devices for detailed consideration: FETs, IMPATT diodes and laser diodes. Assuming a (somewhat ambitious!) MBE growth rate of 5 μm/h, he estimated typical production costs of 1–4 cents per device, based on a current MBE machine handling two-inch substrates, figures which stimulated further interest in several production managers' minds and in MBE machine designers' imaginations. Nor had it escaped the notice of those concerned that computer-control of shutter operation offered an ideal facility within a production environment, as did the absence of toxic gases. The obvious question was: how could these figures be improved yet further by possible developments in machine technology?

Reliability was already a recognised feature of MBE growth, so attention was now focussed on the question of throughput. How many structures, over what area could be produced in a single pump-down and how long might that pump-down last? Relevant machine parameters were growth rate, cell size, substrate handling capacity and sample exchange rate. Ease of substrate cleaning was another important factor which could be solved outside the MBE machine. At least, in the case of GaAs, wafer technology advanced to the point that substrates could be reliably cleaned in an intermediate preparation chamber by the simple process of radiative heating to remove the very thin residual layer of oxide. The size of such substrates depended on developments in Czochralski crystal growth and, following the example of silicon for the integrated circuit market, boule diameters grew steadily from the 2-inch standard available in 1975 to the 6-inch standard available today (indeed, 8-inch boules also exist but cannot yet be regarded as standard). But this, in itself, did not represent the ultimate limit to total substrate area—substrate holders (or platens) capable of holding arrays of wafers became commonplace, to the point that a typical production machine may process as many as seven 6-inch wafers at a time—a total area of approximately 200 square inches—over 100,000 square millimetres. This latter figure not only sounds enormously greater but gives a more immediate idea of the number of devices involved—at roughly ten devices per square millimetre, this amounts to a million devices per platen. With a growth time of perhaps an hour, it represents something like 20 million devices per day (modern production machines run on a 24×7 basis). This is not, of course, quite the whole story—some time should obviously be added for dicing, contacting, cleaving, mounting, etc.—but it surely illustrates the capability of MBE to satisfy the demands for the epitaxial part of the production process, yields being typically as high as 95%.

Performing the calculation in this fashion ignores the question of machine downtime—just how long does a growth 'campaign' actually last? The answer is typically about nine months, with a further three months given over to cleaning and maintenance. To the research worker growing his few slices per week, this sounds like an eternity. To achieve it, however, required a considerable degree of machine development to provide the large amount of beam flux involved. For example, a modern production 'tool' growing GaAs-based devices employs as many as 11 cells, with perhaps 4 cells, each loaded with 10 kg of gallium—this compares with the original research machine cell, holding perhaps 250 g! Nor will it have escaped the reader's notice that a platen capable of holding seven 6-inch substrates has a diameter of no less than 18 inches and this has to pass through at least one gate valve with room to spare—not only cells have increased in size, the whole sample handling facility has been much enlarged and, of course, the whole process is computer controlled, from loading to unloading. Typically, some 50 substrates are loaded into a cassette and the cassette is then loaded into the machine. Everything else is automated. The result is a mind-boggling one billion (10^9) devices in a year (depending on the actual size of each device)—one can only wonder where they all go!

All this didn't happen in the blink of an eyelid—it took 25 years and quite a number of individual machine developments. For example, VG first marketed the 288 machine in 1979 with a capability of handling a single 2-inch substrate, cell sizes being 10 cc for the Group III materials and 40 cc for the Group V materials. Interestingly, two out of only four machines produced were sold in Japan. In 1982 the 288 was replaced by the long-lived V80 machine, which handled a 3-inch substrate and had larger cells, eventually 40 cc and 150 cc, respectively. VG's first production machine, shipped in 1989, was the V90, with 10 cell positions and a 500 cc (later 1 litre) Group V cell. Its initial application was in the growth of optoelectronic devices. Next came the V100 (first shipped in 1990 with application to microwave device production) with 12 sources, 70 cc Group III cells, 1.5 litre As cracker sources and a handling capacity of four 4-inch substrates. Finally, the largest VG machine, the V150, shipped in 2001, was designed for four 6-inch substrates with 150 cc Group III cells and 'big buckets' for the Group V species. Correspondingly, Riber marketed their first fully automated machine in 1991, the MBE 49, which handled four-inch wafers. In 1996, they brought out a new version, handling four such wafers, followed in 1998 by the MBE 6000, handling four 6-inch wafers, then in 2000, the MBE 7000, handling seven 6-inch wafers. Thickness, composition and doping of device layers are all claimed to be uniform to better than 1% over a single wafer and reproducible between different wafers. Combined with device yields of 95%, this constitutes a tempting basis for any business venture. It also provides Riber with an annual turnover of some €11 million. Similarly, Varian/Intervac/EPI/Veeco developed a whole spectrum of new machines. It began in 1981 with the Varian Gen II machine, which was intended to be used for production but which turned out to be a favourite with many research groups. It was designed for a single 2-inch wafer but the much later Gen 10 could handle a single 3-inch wafer, and the Gen 20 a single 4-inch wafer, while the Gen 200 could grow on seven 3-inch wafers and the Gen 2000 on seven 6-inch wafers. Thus, the Gen 2000 and the Riber 7000 are direct competitors, with comparable specifications. No doubt both companies will continue to match their machine capabilities to the size of available substrates and to introduce them to an ever-widening range of semiconductor materials as commercial demand dictates. Nor should we overlook the competition from some of the lesser-known companies— one, at least, DCA Instruments, offers machines designed to grow on eight-inch substrates.

....

REFERENCES

Andersson, T G, Nilsson, B, Svensson, S P and Flemming, E (1983) A combined vacuum interlock and preparation assembly for MBE and surface analysis. J Phys E: Sci Instrum 16, 364.
Armstrong, J V, Farrell, T, Joyce, T B, Kightley, P, Bullough, T J and Goodhew, P J (1992) Monitoring real-time CBE growth of GaAs and AlGaAs using dynamic optical reflectivity. J Cryst Growth 120, 84.
Aspnes, D E, Harbison, J P, Studna A A and Florez, L T (1987) Optical reflectance and electron diffraction studies of molecular-beam-epitaxy growth transients on GaAs(001). Phys Rev Lett 59, 1687.
Aspnes, D E, Quinn, W E and Gregory, S (1990) Application of ellipsometry to crystal growth by organometallic molecular beam epitaxy. Appl Phys Lett 56, 2569.
Biegelsen, D K, Swartz, L-E, and Bringans, R D (1990) GaAs epitaxy and heteroepitaxy—a scanning tunnelling microscopy study. J Vac Sci Technol A 8, 280.

Braun, W and Ploog, K H (2006) In situ studies of semiconductor growth by synchrotron X-ray diffraction. Nucl Inst Meth Phys Res B 246, 50.

Celii, F G, Beam, E A, Filessesler, L A, Liu, H Y and Kao, T Y C (1993) In situ detection of relaxation in InGaAs/GaAs strained layer superlattices using laser light scattering. Appl Phys Lett 62, 2705.

Chalmers, S A and Killeen, K P (1993) Real-time control of molecular beam epitaxy by optical-based flux monitoring. Appl Phys Lett 63, 3131.

Chang, L L, Esaki, L, Howard, W E and Ludeke, R (1973a) The growth of a GaAs–GaAlAs superlattice. J Vac Sci Technol 10, 11.

Chang, L L, Esaki, L, Howard, W E, Ludeke. R and Schul, G (1973b) Structures grown by molecular beam epitaxy. J Vac Sci Technol 10, 655.

Cho, A Y and Arthur, J R (1975) Epitaxial growth and optical evaluation of gallium phosphide and gallium arsenide thin films on calcium fluoride substrate. Prog Sol St Chem 10, 157.

Cho, A Y and Cheng, K Y (1981) Growth of extremely uniform layers by rotating substrate holder with molecular beam epitaxy for applications to electro-optic and microwave devices. Appl Phys Lett 38, 360.

Curless, J A (1985) Molecular beam epitaxy beam flux modelling. J Vac Sci Technol B 3, 531.

Dushman, S and Young, A H (1945) Calibration of ionization gauge for different gases. Phys Rev B 68, 278.

Erickson, L P, Carpenter, G L, Seibel, D D and Palmberg, P W (1985) MBE film growth by direct free substrate heating. J Vac Sci Technol B 3, 536.

Farrer, I, Harris, J J, Thomson, R, Barlett, D, Taylor, C A and Ritchie, D A (2007) Substrate temperature measurement using a commercial band-edge detection system. J Cryst Growth 301–302, 88.

Foxon, C T and Harris J J (1986) The growth of high purity III-V structures by molecular beam epitaxy. Philips J Res 41, 313.

Harbison, J P, Aspnes, D E, Studna, A A, Florez, L T and Kelly, M K (1988) Oscillations in the optical response of (001)GaAs and AlGaAs surfaces during crystal growth by molecular beam epitaxy. Appl Phys Lett 52, 2046.

Harris, J J, Joyce, B A and Dobson, P J (1981a) Oscillations in the surface structure of Sn-doped GaAs during growth by MBE. Surf Sci 103, L90.

Harris, J J, Joyce, B A and Dobson, P J (1981b) RED intensity oscillations during MBE of GaAs: comment. Surf Sci 108, L444.

Harris, J J, Thomson, R, Taylor, C, Barlett, D, Campion, R P, Grant, V A, Foxon, C T and Kappers, M J (2007) Evaluation of sapphire substrate heating behaviour using GaN band-gap thermometry. J Cryst Growth 300, 194.

Hellman, E S and Harris, J S Jr (1987) Infra-red transmission spectroscopy of GaAs during molecular beam epitaxy. J Cryst Growth 81, 38.

Hellman, E S, Pitner, P M, Harwit, A, Liu, D, Yoffe, G W, Harris J S Jr, Caffee, B and Hierl, T (1986) Molecular beam epitaxy of gallium arsenide using direct radiative substrate heating. J Vac Sci Technol B 4, 574.

Hoke, W E, Lemonias, P J and Weir, D G (1991) Evaluation of a new plasma source for molecular beam epitaxial growth of InN and GaN films. J Cryst Growth 111, 1024.

Inoue, N (1991) MBE monolayer growth control by in-situ electron microscopy. J Cryst Growth 111, 75.

Kamiya, I, Aspnes, D E, Florez, L T and Harbison J P (1992) Reflectance-difference spectroscopy of (001) GaAs surfaces in ultrahigh vacuum. Phys Rev B 46, 15894.

Kometani, T Y and Wiegmann, W (1975) Measurement of Ga and Al in a molecular-beam epitaxy chamber by atomic absorption spectrometry (AAS). J Vac Sci Technol 12, 933.

Krause, M, Stollenwerk, A, Awo-Affouda, C, Maclean, B and LaBella, V P (2005) Combined molecular beam epitaxy low temperature scanning tunneling microscopy system: enabling atomic scale characterization of semiconductor surfaces and interfaces. J Vac Sci Technol B 23, 1684.

Lei, T, Fanciulli, M, Molnar, R J, Moustakas, T D, Graham, R J and Scanlon, J (1991) Epitaxial growth of zinc blende and wurtzitic gallium nitride thin films on (001) silicon. Appl Phys Lett 59, 944.

Luscher, P E (1981) Molecular beam epitaxy: an emerging epitaxy technology. Thin Solid Films 83, 125.

Luscher, P E and Collins D M (1980) 'Design Considerations for Molecular Beam Epitaxy Systems' in 'Molecular Beam Epitaxy' (ed B R Pamplin), Pergamon Press, Oxford, p 15.

Lyon, T J, Roth, J A and Chow, D H (1997) Substrate temperature measurement by absorption-edge spectroscopy during molecular beam epitaxy of narrow-band gap semiconductor films. J Vac Sci Technol B 15, 329.

Maracas, G N, Edwards, J L, Shiralagi, K, Choi, K Y, Droopad, R, Johs, B and Woolam, J A (1992) In situ spectroscopic ellipsometry in molecular beam epitaxy. J Vac Sci Technol A 10, 1832.

McClintock, J A and Wilson, R A (1987) Optical measurement of Ga beam flux for MBE. J Cryst Growth 81, 177.

Neave, J H, Blood, P and Joyce, B A (1980) A correlation between electron traps and growth processes in n-GaAs prepared by molecular beam epitaxy. Appl Phys Lett 36, 311.

Oe, K and Imamura, Y (1985) Two-dimensional electron gas at n-AlGaAs/GaAs interface grown by molecular-beam epitaxy using direct-radiation substrate heating. Jap J Appl Phys 24, 779.

Paisley, M J, Sitar, Z, Yan, B and Davis, R F (1990) Growth of boron nitride films by gas molecular-beam epitaxy. J Vac Sci Technol B 8, 323.

Palmateer, S C, Lee, B R and Hwang, J C M (1984) Molecular beam epitaxial growth on radiation-heated substrates. J Electrochem Soc 131, 3028.

Pinsukanjana, P R, Marquis, J M, Hubbard, J, Trivedi, M A, Dickey, R F, Tsai, J M-S, Kuo, S P, Kao, P S and Kao, Y -C (2003) InGaAs composition monitoring for production MBE by in situ optical-based flux monitor (OFM). J Cryst Growth 251, 124.

Robins, D J, Pidduck, A J, Cullis, A G, Chew, N G, Hardeman, R W, Gasson, D B, Pickering, C, Daw, A C, Johnson, M and Jones, R (1987) In-situ light scattering studies of substrate cleaning and layer nucleation in silicon MBE. J Cryst Growth 81, 421.

Robinson, J W and Ilegems, M (1978) Vacuum interlock system for molecular beam epitaxy. Rev Sci Instrum 49, 205.

Saito, J, Igarashi, T, Nakamura, T, Kondo, K and Shibatomi, A (1987) Growth of highly uniform epitaxial layers over multiple substrates by molecular beam epitaxy. J Cryst Growth 81, 188.

Smith, G W, Pidduck, A J, Whitehouse, C R, Glasper, J L, Keir, A M and Pickering, C (1991) Surface topography changes during the growth of GaAs by molecular beam epitaxy. Appl Phys Lett 59, 3282.

SpringThorpe, A J and Majeed, A (1990) Epitaxial growth rate measurements during molecular beam epitaxy. J Vac Sci Technol B 8, 266.

Tanaka, H and Mushiage, M (1991) MBE as a production technology for AlGaAs lasers. J Cryst Growth 111, 1043.

Wasilewski, Z R, Aers, G C, SpringThorpe, A J and Miner, C J (1991a) Studies and modeling of growth uniformity in molecular beam epitaxy. J Vac Sci Technol B 9, 120.

Wasilewski, Z R, Aers, G C, SpringThorpe, A J and Miner, C J (1991b) Growth uniformity studies in molecular beam epitaxy. J Cryst Growth 111, 70.

Whitehouse, C R, Barnett, S J, Soley, D E, Quarrell, J, Aldridge, S J, Cullis, A G, Emeny, M T, Johnson, A D, Clarke, G F, Lamb, W, Tanner, B K, Cottrell, S, Lunn, B, Hogg C and Hagston, W (1992) An MBE growth facility for real-time in situ synchrotron x-ray topography studies of strained-layer III-V epitaxial materials. Rev Sci Instrum 63, 634.

Wood, C E C (1976) Molecular beam epitaxial GaAs layers for MESFET's. Appl Phys Lett 29, 746.

Wood, C E C (1981) RED intensity oscillations during MBE of GaAs. Surf Sci 108, L441.

Yamashita, T, Tomita, T and Sakurai, T (1987) Calculations of molecular beam flux from liquid source. Jap J Appl Phys 26, 1192.

4

Fundamentals

4.1 Introduction

In this chapter we take up the question of exactly how epitaxial films put themselves together in terms of the surface chemistry and growth dynamics relevant to the MBE method of crystal growth. However, before delving into the complexities of surface chemistry and growth dynamics, it may be useful to think briefly and qualitatively about the various processes which are essential to the incorporation of appropriate atoms into the growing crystal surface. We have already noted the need for effusion cells to generate the necessary beams of molecules, and the reader will have taken on board the notion that, while some species such as lead sulphide do evaporate as molecules, thereby simplifying the deposition process, many others evaporate dissociatively and many elements evaporate as dimers, tetramers, etc. rather than in atomic form. Clearly, this has importance in defining the nature of the surface chemistry involved but, from the cell viewpoint, we need merely consider the relation between vapour pressure and cell temperature for each molecular species. Such information was, of course, widely available before the needs of MBE began to be appreciated and the distinction between Knudsen cell and free evaporation (or Langmuir evaporation) was already well understood, in particular, the possibility that different species might be produced under the two different conditions. Note that this has particular relevance in so far as modern effusion cells operate under conditions a good deal closer to those associated with Langmuir than with Knudsen. Our account of cell characteristics in Section 3.2 also drew attention to the use of cracker cells to control the arsenic or phosphorus species: As_2 or P_2, rather than the corresponding tetramers. Knowledge of the relationship between evaporant vapour pressure and cell temperature is clearly an important requirement in understanding the performance of such molecular sources.

But this is only the beginning of the story. Once appropriate atomic or molecular species are supplied to the growing crystal surface, we must also ask how they come to be incorporated into the crystal, or, indeed, *whether* they will be incorporated. The possibility of Langmuir evaporation from the crystal surface depends in like manner on the temperature of the substrate and, as we shall see, on the composition and structure of the surface. In order to take up permanent residence on that surface, the arriving species must be chemically bonded to the outermost atoms and this implies that they find their way to appropriate lattice sites where chemical bonds are available to bind them. This, in turn, makes two further assumptions—first, that the species to be incorporated are present on the surface in an appropriate chemical form, for example, as atoms rather than molecules, and, second, that they are able to diffuse over the surface in order to reach the safe haven offered by those willing chemical bonds. Note that this may imply the existence of some intermediate state in which these atoms (or molecules) are unusually mobile—they are in some way attached to the surface but are nevertheless essentially free to wander. On the other hand, one must surely ask whether such migrants are all the more likely to leave the surface, never to return. If fully chemically bonded surface atoms are able to evaporate (as we know they certainly are!) is there, perhaps, an even greater possibility of loosely bound migrating species doing likewise? Such thoughts clearly raise the question of just *how* they are bonded to the surface in the first place? A moment's further deliberation then yields the question as to their precise nature—are they atoms or molecules? To be specific, does arsenic diffuse over a GaAs surface in the form of atoms, dimers or tetramers? And, given that it must be incorporated as atoms, what exactly goes on when an arsenic molecule does reach its appointed final resting place? If, for example, an As_2 molecule loses one atom to a hungry gallium bond, what exactly happens to its other half? Does this imply, for instance, that the sticking coefficient of As_2 can never be greater than a half? We should, of course, remember that both diffusion and evaporation are thermally activated processes, so substrate temperature plays an essential role in mediating the conflict between them to determine the ultimate fate of each and every migrating species.

So much (for the moment) regarding the incoming molecular flux—what, though, of the crystal surface? We have suggested that certain lattice sites are vital to the process of chemical bonding but have so far said nothing concerning their precise nature. Perhaps the first point to make is that the arrangement of crystal atoms on a surface depends on the crystallographic orientation of that surface and, by the same token, so does the arrangement of chemical bonds. We must therefore expect crystal growth to proceed differently on different surfaces. We must also take account of the possibility that some surfaces will reconstruct—the precise location of surface atoms being different from that predicted from knowledge of the bulk crystal structure—and that may have important consequences for the appropriate surface chemistry. However, it would be naive in the extreme to imagine that any real surface might exist as an ideal (singular) flat plane. Even if such a surface could be prepared, it would immediately be modified by the arrival of the first partial layer of incoming atoms. In practice, we must anticipate that the very process of crystal growth will introduce so-called step edges which result in a completely new configuration of chemical bonds. An obvious question then concerns the relative probability for an arriving atom to be captured by a step edge or a site on a flat region of the surface. The notion of an atom's being captured by a step edge implies, of course, that surface diffusion must play an important role. On the other hand, capture on a plane surface could, in principle, take place without the need for diffusion. However, it would imply, in itself, the generation of a projection atop that surface which again results in a special site to which an incoming atom may diffuse. Crystal growth is clearly a complex process which sets up its own dynamic. This apart, most real surfaces will differ to some extent from the planar ideal—in particular, the so-called vicinal surface, which is characterised by a small angular departure from the ideal, contains a more or less regular series of step edges whose separation depends on and can therefore be controlled by variations in this angle. As we shall see, the possibility of controlling step-edge distribution has proved extremely useful as a means of understanding mechanisms of crystal growth.

One final and obvious consideration in the case of multi-element compounds is the question of possible differences between the incorporation of different atomic species. For example, if a GaAs surface were to contain predominantly arsenic atoms, we might reasonably expect it to capture atoms from a gallium beam in preference to those from an arsenic beam, and vice versa. In this context it is important to remember that different principal crystal planes may show significant differences—growth on the {001}, {111} A and {111} B surfaces of GaAs showing their own particular characteristics. On the other hand, the fact that the grower can determine the relative arrival rates of the different molecular species by the simple expedient of controlling cell temperatures or activating a shutter mechanism allows him/her to influence the growth process in a manner not available to proponents of alternative epitaxial growth methods, a facility tuned to maximum usefulness in techniques such as migration-enhanced epitaxy, to be discussed in Chapter 5. The fact that GaAs films are generally grown under arsenic-rich conditions, growth rate being limited by the arrival rate of gallium atoms, is just one example of the flexibility inherent in the MBE process.

This is probably enough of an introduction to the many facets of MBE growth but before we dive into the depths of detailed growth mechanisms, it might be worthwhile saying a few words about the techniques used in learning their inner secrets. The above account will already have suggested the possibility of comparing measures of arrival rate at and desorption rate from a crystal surface. As noted earlier, desorption depends on temperature and must therefore be studied as a function of substrate temperature over the range appropriate to MBE growth but it must also be understood in relation to the other processes taking place on the surface—for example, the influence of gallium on the desorption rate of arsenic, during GaAs growth. An obvious method for identifying specific atomic or molecular species is that of mass spectrometry and this was taken up with considerable effect by the early investigators. It therefore forms the subject of Section 4.2. An alternative approach to studying epitaxial growth depends on measuring growth rate and this, as we have seen, can conveniently be done by measuring RHEED oscillations, an aspect which we take up in Section 4.3. However, RHEED is also an essential component in the study of surface structure, which, in turn, influences surface chemistry so we are obliged to examine these aspects too, if we are to understand epitaxial growth in all its complexity. During the early days of MBE, from its inception in the 1970s to its wide scale application in the 1980s, these two techniques played a major role in developing our understanding of growth mechanisms but we should certainly not overlook the importance of scanning tunnelling microscopy which came into prominence during the 1980s (Binnig et al. 1982; Pashley et al. 1988a, b) and saw application to in situ MBE measurements during the 2000s (Krause et al. 2005). This highly sophisticated technique allows resolution of individual atoms on a crystal surface and has proved particularly suited to studying surface reconstruction, as we describe in Section 4.4. Making a distinct change of direction, we then take up the very important topic of doping in Section 4.5. In several instances MBE has found it necessary to discover doping species which differ from those preferred by other growth methods and we shall examine the rationale behind such choices. As in most scientific studies, the interplay between experiment and theory has proved central to our understanding of crystal growth—indeed, theoretical modelling can well be said to have steered the experimental approach in appropriate directions. We therefore offer an account of modelling in Section 4.6. One final word of explanation is appropriate here—because the majority of relevant studies were initially concerned with the III-V compound semiconductors, we shall confine our attention in this chapter to these materials. They serve as excellent examples of the methods used to unravel crystal growth problems, while our commitment to providing a balanced account of the broader sweep of semiconductor materials will be met in Chapter 7, where we discuss a number of alternative materials which have been successfully grown by MBE.

4.2   Mass Spectrometer Studies

Much of the early work aimed at understanding surface chemistry and growth mechanisms was based on the use of a mass spectrometer to analyse the nature of molecular species desorbed from a semiconductor surface under various conditions of substrate temperature and incident beam intensities, the spectrometer typically having a mass range from 1 amu to about 350 amu. As we saw in Chapter 2, the first mass spectrometer surface studies on III-V compounds were undertaken by John Arthur at Bell Labs, even before MBE became recognised as a promising method of film growth. He had already recognised one significant problem—that the desired signal could easily be swamped by background molecules within the vacuum system (specifically As, P or Sb with high vapour pressures) reaching the spectrometer—and, to counter this, he used a shutter to switch the incident beam and looked to detect a sympathetic change in spectrometer signal. He also recognised the inherent variation in spectrometer sensitivity to different molecules and the complication due to molecular dissociation in the high energy ioniser. For example, with pure As_2 he saw peaks at mass numbers 75 (As^+) and 150 (As_2^+) (modern equipment is able to work at much smaller energy so the problem is correspondingly less). Clearly, one had to be rather careful when interpreting experimental observations—other similar artefacts were to emerge from later work.

   Arthur's first contribution (Arthur 1967) was concerned with the measurement of the vapour pressures of arsenic and gallium over GaAs and, to this end, he measured the species evaporating from a Knudsen cell containing GaAs over the temperature range 900 K–1500 K, taking pains to obtain a reliable measurement of cell temperature by using both a thermocouple and a calibrated pyrometer. Working back from the spectrometer signals, he could determine the equilibrium vapour pressures of Ga, As_2 and As_4 within the cell (see equation (B2.10) in Chapter 2). Over this range, the dominant arsenic species was seen to be As_2 (rather than As_4) and, at approximately 960 K

(690 °C), the vapour pressures of Ga and As_2 were found to be equal; but, above this temperature, that of As_2 predominated. Thus, at temperatures above 960 K, arsenic is lost from the GaAs, leaving liquid gallium on the surface, and equilibrium is rapidly established between solid GaAs, liquid gallium and the various vapour species. Because arsenic is relatively insoluble in liquid gallium, such a source can be relied on to provide beams of both gallium and arsenic until nearly all the arsenic is exhausted. While this certainly led to a much improved understanding of the Ga–As liquidus, the use of GaAs itself as a source for MBE growth has largely been superseded by separate gallium and arsenic cells which allow greater flexibility and give greater purity in the resulting epilayers.

As a brief aside, we should perhaps define the so-called congruent evaporation temperature T_c more precisely. At temperatures below T_c, equal numbers of *atoms* of Ga and As leave the surface of GaAs (the desorbing *fluxes* being related by $J_{Ga} = 2J_{As2}$) on the following basis: Ga evaporates, leaving As atoms on the surface, which evaporate as As_2 because an arsenic-rich surface is thermally unstable above 600 K. In other words, the rate of evaporation of the compound is determined by the desorption rate of gallium. Above T_c, arsenic desorbs more rapidly than gallium and Ga droplets form on the surface. Note that the above definition implies that T_c differs slightly from the temperature at which the two vapour pressures are equal. It is easy to show that, at the congruent evaporation temperature so defined, the ratio of vapour pressures is given by

$$p(Ga)/p(As_2) = 2\{m(Ga)/m(As_2)\}^{1/2}$$
$$= 1.37 \tag{4.1}$$

Thus $p(Ga) > p(As_2)$, and the corresponding congruent evaporation temperature is approximately 20 K lower than the crossover temperature measured by Arthur, that is, 940 K. (Later measurements (Foxon et al. 1973) suggest a more accurate value may be $T_c = 930$ K.)

Arthur was also greatly concerned with understanding the surface chemistry of the III-V semiconductors and made a major contribution to the early studies of epitaxial growth. In 1968 (Arthur 1968) he reported on the interaction between pulsed beams of Ga and As_2, impinging on GaAs surfaces. For example, by measuring the desorption of gallium and arsenic from a GaAs {111} surface, he was able to show that gallium atoms impinging on the surface were characterised by a temperature-dependent lifetime τ, given by

$$\tau = 2.5 \times 10^{-14} \exp\{2.6/kT\} \text{ s} \tag{4.2}$$

(where kT is measured in eV). Thus, over the temperature range 850 K to 1000 K (~580 °C to 730 °C), gallium atoms resided for mean times varying from roughly 50 s down to 0.3 s, extremely long lifetimes. In other words, the sticking coefficient of gallium on the {111} surfaces is essentially unity. On the other hand, the behaviour of As_2 molecules depended strongly on the nature of the surface. In the absence of a gallium beam, the sticking coefficient S_{As2} was effectively zero but, when a gallium beam was introduced, S_{As2} increased in proportion to the beam intensity. It was this result, of course, which led to the idea of growing GaAs films under excess arsenic, growth rate being determined simply by the arrival rate of gallium atoms. More recently (Arthur 1974) he updated his earlier studies with temperature-programmed thermal desorption measurements which led him to propose a model in which As_2 molecules are adsorbed into a weakly bound precursor state from which arsenic atoms may be chemisorbed onto the GaAs surface.

While Arthur's work stimulated a considerable number of further studies, it soon became clear that the acquisition of reliable data required some improved method of data analysis, and later work was based largely on the use of continuous beam modulation combined with synchronous detection of the spectrometer signals. Progress in this direction was led by the group at the Philips Research Laboratory in Redhill; they repeated much of Arthur's work and went on to propose detailed growth mechanisms for several III-V compounds and alloys. Bruce Joyce and Jim Neave transferred their affiliations from Plessey to Philips in 1969, with a brief to work on the silicon–silicon oxide system, it being at this time that the invention of the CMOS circuit obliged every serious-minded electronic company to concentrate research effort on the development of improved MOS devices. They were joined by Tom Foxon, who had been recruited to work on electron microscopy. In spite of these counter motivations, it was one of those odd quirks of the human experience that they were soon to make their joint names in the MBE growth of III-V compounds, an activity which was to grow rapidly during the 1970s with the arrival of Mike Boudry, Colin Wood, Karl Woodbridge, John Roberts and Jeff Harris.

In those early days, it was necessary of course to rely on home-built equipment, and the apparatus employed to study GaAs growth, shown in outline in Figure 4.1 (Foxon and Joyce 1981), owed much to the practical skills of Jim Neave. As can be seen, it provided a capability for modulating either incident beam or desorbed species by means of rotating chopper blades, which could be driven over a range of speeds by bakeable hysteresis motors. By suitable choice of blade geometry, the relevant beam was modulated in the form of a square wave with a fundamental frequency between 4 and 250 Hz. This allowed synchronous detection of the spectrometer signal, thus distinguishing it from random background molecules, the relevant synchronising waveform being obtained from a lamp and a photodetector. Because flight times of molecules between source and sample and between sample and detector must be known in order correctly to interpret data, it is essential to use a Knudsen cell as source—given an accurate measurement of the cell temperature, molecular velocities can then be readily calculated. At least two such cells are required in order to study the influence of, say, a gallium beam on the behaviour of arsenic molecules, when studying gallium arsenide surfaces. Experiments on alloys such as AlGaAs or InGaP clearly require three or more. An Auger electron spectrometer was available for monitoring the cleanliness of the sample surface, and the Auger electron gun could also be used as a source for the measurement of RHEED spectra. The sample was mounted on a goniometer so that it could be oriented towards the appropriate measuring device, and its temperature was measured by means of a

MBMS system constructed at Mullards

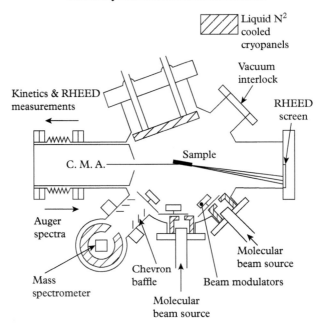

Figure 4.1 *The apparatus used at the Philips Laboratory in Redhill to study surface chemistry by modulated beam mass spectrometry. The Auger electron gun was used as a source for high energy electrons for RHEED measurements. Either the molecular sources or the mass spectrometer inlet were modulated by rotating choppers, and the mass spectrometer signal was analysed by Fourier transform methods. (From Foxon and Joyce 1981, courtesy North Holland.)*

thermocouple inserted in a small hole drilled through the substrate, reliable knowledge of sample temperature being essential for correct interpretation of data. We should also note the presence of several liquid-nitrogen-cooled cryopanels, designed to capture as many as possible of the randomly escaping molecules which might otherwise drown the mass spectrometer response. Finally, it is important to appreciate that the modulated beam intensity is very much less than that which might significantly perturb the surface—a typical value would be about 3×10^{12} molecules/cm^2/s, compared with the value of 6×10^{14} molecules/cm^2/s required to deposit a single monolayer in 1 s—though, as noted above, in many experiments, a second unmodulated beam of much greater intensity may be used to control surface properties.

In order to pursue further the question of data analysis, it is probably convenient at this point in our discussion to examine in somewhat greater detail the nature of a typical experiment and to consider the various parameters which it is designed to measure. Suppose we direct a modulated beam of molecules at a substrate and measure the nature and timing of the species reaching the mass spectrometer—what should we expect to see in the spectrometer signal? In the simplest case, these molecules will reach the sample after a transit time t_1, which depends on their thermal velocities, reside on the sample surface for a finite period (the surface lifetime τ) before desorbing, when some fraction (typically $10^{-2} - 10^{-3}$) of the desorbed species will arrive at the detector after a further transit time t_2, which is again determined by their (now possibly different) thermal velocities. Clearly, the amplitude of the detected signal will be much reduced, compared with that of the incident beam but the absolute amplitude is of little significance, determined, as it is, by imponderable geometrical considerations. Of more significance is the time delay and corresponding phase shift of the received signal with respect to the synchronising waveform from the chopper. It is helpful to estimate the respective contributions made by the various time delays and compare them with the period of the modulation, typically about 20 ms. Thermal velocities can be derived from equation (B2.6) in Chapter 2. Thus, if we assume a cell temperature of 800 K and a molecular weight of 300 for As_4, we obtain $c_1 \sim 200$ ms^{-1} and a flight time t_1 over a distance of 10 cm of approximately 500 μs. We may reasonably anticipate a comparable value for t_2. Typical surface lifetimes for As_4 are of order 100 μs to 10 ms (Foxon and Joyce 1975). However, we should note an important difference between the effect of transit times and lifetime on the spectrometer signal—whereas transit times simply shift the phase of the signal, surface lifetime affects the *shape* of the detected waveform.

During the time the modulator blades are open, the surface density N_s increases at a rate dN_s/dt, given by

$$dN_s/dt = N'_{in} - N_s/\tau \qquad (4.3)$$

where N'_{in} is the rate of supply from the Knudsen cell. Therefore, N_s builds up according to

$$N_s = N_{eq}[1 - \exp\{-t/\tau\}] \qquad (4.4)$$

Equilibrium is established after several lifetimes at a level of $N_{eq} = (N'_{in} \times \tau)$. At the end of the supply pulse N_s then decreases according to:

$$N_s = N_{eq} \exp\{-t/\tau\} \qquad (4.5)$$

The resulting time dependence of N_s is shown in Figure 4.2(b). However, the signal recorded by the mass spectrometer is proportional to the rate at which desorbing molecules reach it from the sample surface which, in turn, is proportional to the desorption rate N_s/τ—that is, the spectrometer signal amplitude depends inversely on the lifetime τ and is directly proportional to the density of molecules on the sample surface. We must also remember that it is delayed by a further transit time t_2. Figure 4.2(c) therefore shows the waveform we might expect to see at the spectrometer output on the assumptions made in this example. It suggests that we might hope to obtain a value for the surface lifetime from the *shape* of the output waveform.

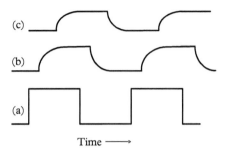

(c)

(b)

(a)

Time ⟶

Figure 4.2 *Typical waveforms encountered in modulated beam mass spectrometry; (a) shows the modulated source waveform, while (b) shows the time dependence of the species desorbing from the substrate and (c) the input to the mass spectrometer.*

Would that life were so straightforward! This neat and tidy analysis has ignored a rather important piece of physics—the thermal velocities used to estimate transit times are not characterised by unique values but must each show a significant spread, determined by the appropriate Maxwell–Boltzmann distribution. We are therefore faced with a considerable spread in transit times, both before and after contact with the sample surface, and the detected waveform is correspondingly distorted at both edges. Clearly, we can no longer derive a value for the surface lifetime simply by examining the shape of these edges—the analysis required is now significantly more complex and demands that we incorporate the appropriate distribution functions calculated from knowledge of the respective cell and sample temperatures. Only by deconvolving the various contributions can one learn anything of reliable significance from the experimental data and this led the Philips group to apply the method of fast Fourier transforms to simplify data analysis (Foxon et al. 1974). The point of all this is that it is much easier to perform the necessary deconvolution in the frequency domain, rather than in the time domain appropriate to Figure 4.2. Let us look at this for a moment.

As every aspiring engineer (and the occasional chemist?) knows, the square wave represented by Figure 4.2(a) can be represented as a sum of a constant term plus a series of harmonics; that is sine waves with frequencies given by $f_m = mf_1$ and amplitudes $A_m = A_1/m$, where f_1 is the fundamental frequency of the wave, A_1 its amplitude and $m = 1, 3, 5, 7$, etc., characterising each harmonic. Note that the amplitudes of the various harmonics decrease as m^{-1}, implying that the higher harmonics play a decreasing role in defining the wave; also, for the case of a wave which is antisymmetrical about the time $t = 0$, all even harmonics have zero amplitude. This is no place for any attempt to write a treatise on Fourier transform theory—we should simply be aware that there are always two alternative (and totally equivalent) ways of defining any waveform, one in terms of its time dependence and one in terms of its frequency makeup, and that precise mathematical techniques exist for converting between the two descriptions. Consider, for example, the waveform representing the spectrometer signal in Figure 4.2(c). Clearly, the symmetry of the original square wave has been lost and this complicates the corresponding Fourier series. In particular, it introduces two new features: first, the series includes both even and odd harmonics; second, it includes both sine and cosine terms. Alternatively, because of the well-known trigonometric relation

$$\sin(2\pi f_m t + \phi_m) = \sin 2\pi f_m t \cdot \cos \phi_m + \cos 2\pi f_m t \cdot \sin \phi_m \qquad (4.6)$$

this latter feature can be represented in terms of a phase shift ϕ_m, which will in general be different for each harmonic. In other words, the analysis of an experimental waveform, rather than being expressed as a search for characteristic *shapes* in the time domain, may alternatively be seen as a search for characteristic harmonics and phases in the frequency domain—what is more, the mathematical procedures required to effect the transformation can be programmed into a modest computer, thus relieving the experimentalist of any unwelcome burden. Nevertheless, the reader can be forgiven for wondering why the analysis should proceed via the frequency domain rather than the visually more comprehensible time domain, the reason being that it is much easier to separate the different experimental contributions to the signal. We shall leave it at that. Those wishing for deeper understanding are referred to the original paper by Foxon et al. (1974).

So much for signal analysis—though the use of Fourier transform techniques certainly enhanced experimental capability, the principal concern of this section is to outline the progress made in understanding crystal growth. Unsurprisingly, we begin with GaAs. Gallium is supplied as the monomer by evaporation from a liquid source and, at relatively low substrate temperatures, gallium has unity sticking coefficient on GaAs—though, as Arthur showed, it becomes significantly less at temperatures above about 750 K. Arsenic, on the other hand, can be supplied as either As_2, by evaporation from GaAs or from a high-temperature cracker, or as As_4, from solid arsenic.

Foxon and Joyce (1975) reported on the interaction between Ga and As_4 on GaAs {100} surfaces. They distinguished two temperature regimes, below and above 450 K. The behaviour in the low temperature region corresponded to non-dissociative chemisorption of As_4 molecules onto gallium atoms on the GaAs surface by way of a weakly bound precursor state (see Arthur 1974 for a similar concept in the adsorption of As_2 onto {111} GaAs). (Notice that, because this chemisorption process is *non-dissociative*, there is no growth of GaAs in this regime—it is essential for the molecule to break down into atoms if it is to be permanently incorporated.) In the absence of surface gallium, only physisorption into this precursor state occurs, from which desorption takes place with an activation energy of $E_D = 0.38$ eV (typical of a weakly bound state). The lifetime of As_4 was measured to be in the range 10^{-4} to 3×10^{-3} seconds over the temperature range 300 K to 400 K, and the sticking coefficient was effectively zero. It is interesting, however, that, even though the binding energy of this precursor state is relatively small, measurement of the surface-to-detector transit times showed that the As_4 molecules had thermally accommodated to the GaAs surface (i.e. their thermal velocities corresponded to the GaAs temperature rather than that of the arsenic cell from which they originated). When gallium atoms were present on the surface, the As_4 sticking coefficient was found to lie between 0.5 and 1.0, approaching unity as the temperature was lowered, and the surface lifetime was about a factor of ten longer, though showing the same activation energy as before. This seems to imply that desorption still occurs via this same precursor state and presents us with something of a dilemma—how is it possible to measure a non-zero sticking coefficient together with a finite surface lifetime? Surely, a finite lifetime implies a zero sticking coefficient, while non-zero sticking implies an infinite lifetime! The answer involves associating the lifetime with the precursor state, while molecules chemisorbed by binding to surface Ga atoms may stick with near-unity probability. Thus, molecules captured into the precursor state may either desorb with activation energy E_D or be trapped by Ga atoms. The sticking coefficient is then a measure of the branching ratio between these two 'fates'. Bearing in mind that the sticking coefficient is the ratio of molecules which stick to the total number of molecules which arrive, we can write

$$S_{As4} = k_2/(k_1 + k_2) \qquad (4.7)$$

where k_1 is the rate of desorption from the precursor state, and k_2 the rate of capture into chemisorbed states.

Chemisorption appears to be controlled by surface diffusion—As_4 molecules wander about over the surface until they find a suitable gallium atom—and we may reasonably anticipate this diffusion process, like the desorption process, to be thermally activated. Writing the activation energy for diffusion as E_λ (where $E_\lambda < E_D$), we see that

$$S_{As4}/(1 - S_{As4}) = k_2/k_1 \sim \exp\{(E_D - E_\lambda)/kT\} \qquad (4.8)$$

and this ratio therefore decreases as the temperature rises, consistent with the fact that S_{As4} also decreases with increasing temperature. From a logarithmic plot of the ratio against reciprocal temperature over the range 300 K–450 K, Foxon and Joyce obtained a value of $E_\lambda = 0.24$ eV, which appears eminently reasonable.

In the higher temperature regime, two significant results were obtained: first, the sticking coefficient of As_4 was found to increase with increasing gallium supply but under no circumstance did it take on a value greater than one-half: second, at low As_4 flux levels, the desorption rate of As_4 varied as the square of the As_4 supply rate, while at higher rates, it changed to a linear relationship. Finally, at temperatures above 600 K, the GaAs surface began to decompose by the thermal desorption of As_2 molecules, a process which effectively increased the supply of surface gallium. These observations therefore led them (Foxon and Joyce 1981) to propose a model for the growth of GaAs from Ga and As_4 which is outlined in Figure 4.3. As_4 is first adsorbed into the same weakly bound precursor state; it then migrates over the surface before being chemisorbed onto gallium sites on the GaAs surface. Here a pair of As_4 molecules interacts, four of the constituent atoms being incorporated into the surface, while the remaining four combine to form an As_4 molecule which desorbs. This can be seen to account for the second-order reaction kinetics and for the fact that the As_4 sticking coefficient $S_{As4} \leq 0.5$.

Returning briefly to the question of the accommodation coefficient of molecules desorbing from a GaAs surface, Arthur and Brown (1975) made this a subject for detailed

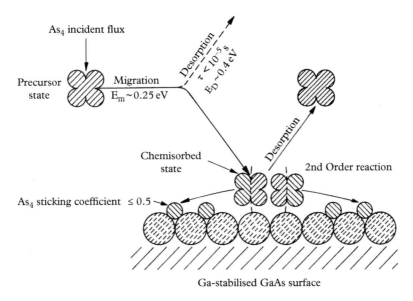

As$_4$ incident flux

Precursor state

Migration $E_m \sim 0.25$ eV

Desorption $\tau < 10^{-5}$ s $E_D \sim 0.4$ eV

Chemisorbed state

Desorption

2nd Order reaction

As$_4$ sticking coefficient ≤ 0.5

Ga-stabilised GaAs surface

Figure 4.3 *Growth model proposed by Foxon and Joyce (1981) for As$_4$ and Ga beams on {100} GaAs surfaces. A second order dissociative reaction between a pair of chemisorbed As$_4$ molecules on Ga sites results in four As atoms being incorporated, while the remaining four desorb from the surface. (From Foxon and Joyce 1981, courtesy North Holland.)*

investigation by measuring the spread in arrival times (at the detector) for Ga, As$_2$ and As$_4$ both scattering and evaporating from a (111)B GaAs surface over a temperature range from 300 K to 975 K. The most striking feature of their results was the fact that Ga atoms showed a distribution of arrival times which was well fitted by a Maxwell–Boltzmann velocity distribution, with a temperature corresponding to that of the GaAs substrate, whereas the corresponding data for As$_2$ required a somewhat lower temperature, and the As$_4$ temperature was slightly lower still. The most likely explanation seems to be that, in the case of the arsenic molecules, a portion of the total thermal energy goes into exciting internal vibrational modes, thus reducing their translational velocities. With this interesting caveat, their data appear to confirm that, in the majority of circumstances, molecules scattering off a GaAs surface do so with an accommodation coefficient close to unity. It also confirmed that, at temperatures of interest for GaAs growth, As$_2$ is the only arsenic species evaporating from GaAs.

In a later paper Foxon and Joyce (1977) also reported a study on the reaction of As$_2$ (from a Knudsen cell containing GaAs) with Ga on the same {100} GaAs surfaces. Their first interesting observation concerned the fact that, at temperatures between 300 K and 600 K, an As$_2$ beam impinging on the GaAs surface gave rise to desorption of As$_4$ molecules as a result of an association reaction between pairs of As$_2$ molecules. At 300 K and below, the lack of As$_4$ desorption is probably due to the non-dissociative chemisorption process discussed above, with near-unity sticking coefficient for As$_4$.

At temperatures above 600 K, the similar lack of As_4 desorption suggests that the association reaction no longer occurs. Turning attention to the behaviour of As_2 revealed that the As_2 sticking coefficient was 0 at 600 K in the absence of a Ga flux but increased with increasing supply of Ga, approaching unity when the Ga flux was approximately twice that of As_2. This should be contrasted with the corresponding behaviour of As_4, where the limiting value of sticking coefficient was 0.5. Also, in the absence of a Ga flux, the As_2 sticking coefficient increased with rising temperature, again approaching unity at a temperature of approximately 950 K. This was explained on the basis of increasing desorption of As_2 from the GaAs surface as the temperature was raised, leading to a situation where the loss of As_2 was just compensated by incoming molecules from the As_2 beam, the total flux of desorbed molecules remaining constant over the range from 600 K to 900 K. The model put forward to explain GaAs growth from gallium and As_2 (Foxon and Joyce 1981) is shown in Figure 4.4, As_2 molecules being captured into a physisorbed precursor state from which they may either be desorbed (with a very short lifetime) or incorporated into the GaAs surface by dissociative chemisorption. It is worth commenting that, on the basis of these models for growth from either As_4 or As_2,

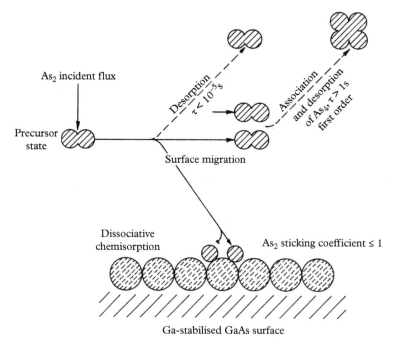

Figure 4.4 *Growth model proposed by Foxon and Joyce (1981) for the growth of GaAs from As_2 and Ga beams on {100} GaAs surfaces. As_2 molecules in a physisorbed precursor state may either desorb or undergo dissociative chemisorptions on the Ga-stabilised surface. (From Foxon and Joyce 1981, courtesy North Holland.)*

arsenic atoms are incorporated in pairs, though the details are significantly different, the sticking coefficients on a gallium-stabilised surface differing between 0.5 (As_4) and 1.0 (As_2). This incorporation of arsenic dimers will be seen to be of particular relevance when we discuss surface structure in the following sections.

Though GaAs was, perhaps, of greatest interest within the panoply of III-V semiconducting compounds, several others showed early promise of filling practical needs. GaP was studied as a material for red and green LEDs during the 1960s and early 1970s, InSb as a potential infrared detector material during the 1950s and 1960s, and InP as an exciting microwave material during the 1970s. Meanwhile, ternary alloys such as GaAsP, AlGaAs and InGaP were all investigated with a view to producing efficient visible LEDs and the quaternary InGaAsP was to make a name for itself in the development of infrared lasers for fibre-optic communications during the 1980s. The AlGaAsSb system was also of interest for the growth of long wavelength lasers and some interesting superlattices. Of these materials, GaP and GaAsP were grown by MBE during its formative years (Arthur and LePore (1969)) and InP, InGaP, AlGaAs, GaSbAs and InGaAsP were all grown by MBE with varying degrees of seriousness during the following decades. The point of interest here is that measurements of vapour pressures, sticking coefficients and growth mechanisms provided essential data for the optimisation of material quality and therefore demand our attention in the development of this section. In particular, we must address the problem of compositional control when growing alloy films—what, for example, determines the precise value of y in the compound $GaAs_yP_{1-y}$ when grown by MBE and how does this compare with the control of x in $Al_xGa_{1-x}As$?

One of the earliest studies of a III-V compound other than GaAs was undertaken by Robin Farrow at RSRE Malvern (Farrow 1974), when the microwave properties of InP were under serious scrutiny. He studied the conditions for MBE growth of InP using modulated beam mass spectrometry and showed that stoichiometric InP films could be grown from In and P_2 beams at temperatures of 573 K. Qualitatively, InP behaves similarly to GaAs but with significant quantitative differences. In particular, there is a region of congruent evaporation of In and P_2 from the InP surface up to a temperature of approximately 640 K (compared with $T_c = 930$ K for GaAs). Because of the very much lower temperatures involved, the maximum congruent evaporation flux was only 10^{-5} monolayers per second (compared with approximately 10^{-1} monolayers per second for GaAs), making heat-cleaning of InP far less efficient than that of GaAs. The sticking coefficient of In was found to be close to unity, while that of P_2 depended strongly on the presence of an In beam, varying from 0 in the absence of In to something approaching unity when the In beam was sufficiently intense. The similarity to the behaviour of Ga and As_2 behaviour on GaAs was comforting and suggested that many of the III-V compounds might be expected to follow suit—indeed only the nitrides appear to be rebellious.

One of the most important practical features of the III-V compounds is the possibility of forming ternary and quaternary alloys, thereby allowing a controlled choice of band gap and of electrical and optical properties, combined with that of lattice parameter which plays a central role in the growth of heterostructures and low-dimensional structures. It is therefore of some considerable importance that we understand the adsorption, desorption and segregation processes which affect the MBE growth of these alloys. The ternaries fall into two categories—those containing two Group III elements and those containing two Group V elements—which differ significantly in their surface chemistries. We shall deal with them separately. Of particular importance in the former category are AlGaAs, GaInAs and GaInP and of these AlGaAs was grown successfully by Cho in the early 1970s for DH laser applications. The In-containing compounds, however, proved less easy to control and led Foxon and Joyce (1978) to apply mass spectrometric methods to gain a better understanding of their surface properties. Films were grown on {100} GaAs substrates from beams of Ga, In, As_4 and P_4. Preliminary measurements on GaAs and InAs surfaces showed that, at temperatures above 600 K, As_2 desorbed much more readily from InAs than from GaAs. Thus, growth of alloy films at such temperatures takes place with an enhanced surface population of In and Ga and, though in principle this might be countered by increasing the As_4 (or P_4) flux, this proves impractical because of the impossibly large Group V flux needed. Annealing experiments on GaInAs surfaces revealed the fact that, under conditions when As_2 was suffering rapid desorption, Group III elements diffused to the surface from the bulk but with a considerably greater effect for In, thus producing an In-rich surface. A second significant observation was the relatively rapid loss of In from a GaInAs alloy surface at temperatures above 700 K. The activation energy for In desorption was measured as 1.5 eV, compared with the value of 2.50 eV for the enthalpy of sublimation of indium, and suggested that the rate-limiting step might be surface diffusion of In to a preferred desorption site. Clearly these effects must lead to serious difficulties in controlling the composition and uniformity of alloy films. The situation was summarised by their conclusion that 'the principal limitation to the growth of III-III-V alloy films is simply the thermal stability of the lesser stable of the two III-V compounds of which the alloy may be considered to be composed.' In practical terms, it is advisable to grow at a temperature below that at which the Group V molecules begin to suffer rapid desorption. In a final comment, they remarked that it was fortunate that AlGaAs was chosen as the first alloy to be grown, as it happens that AlAs shows greater thermal stability than GaAs and therefore it should be possible to grow films by MBE with good uniformity and control.

Problems associated with the growth of III-V-V alloy films are somewhat different. Arthur and Lepore (1969) first noted that, for $GaAs_yP_{1-y}$ grown from Ga, As_2, and P_2, the film contained approximately four times more arsenic than phosphorus, relative to the appropriate flux ratios. Similar results were found by Chang et al. (1977) for $GaSb_yAs_{1-y}$ films grown from Ga, Sb_4 and As_4, where the incorporation probability for antimony is much higher than that of arsenic. It was also apparent that the ratio between the two Group V species was very far from that to be expected on thermodynamic grounds—the respective incorporation ratios were clearly kinetically controlled. In the interest of understanding such behaviour, Foxon et al. (1980) therefore studied the interaction kinetics of As_4 and P_4 for both $GaAs_yP_{1-y}$ and $InAs_yP_{1-y}$ films by measuring the sticking coefficients and surface lifetimes of As_4 and P_4 using MBMS. Their results confirmed the much larger incorporation probability for As_4 compared with P_4

and showed that, if the As$_4$ flux was limited in relation to the Ga flux, together with an excess P$_4$ flux, there was a linear relation between the parameter y and the flux ratio J_{As4}/J_{Ga}. They also showed that this behaviour was not influenced by the nature of the Group III species. This suggested a relatively straightforward method of growing alloy films with reproducible composition—it was simply necessary to ensure growth with an excess flux of the Group V element characterised by the smaller incorporation probability, while controlling the ratio of the other Group V element with respect to that of the Group III element. Since the nature of the Group III element is unimportant, such procedures also apply to quaternary alloys such as $In_xGa_{1-x}As_yP_{1-y}$. However, we should note that these conclusions apply strictly to growth at moderate temperatures—at elevated temperatures, desorption of the Group V species introduces significant complications (Woodbridge et al. 1982). In passing, we might also note that later MBMS work by Evans et al. (1990) on $GaSb_yAs_{1-y}$ and $AlSb_yAs_{1-y}$ showed a similar dependence of the antimony incorporation rate with the Group III flux.

Following this initial burst of activity in which the understanding of MBE growth made exciting strides, there was a period when mass spectrometric studies went into decline—for the very good reason that MBE was now established as a practical method of producing films for a wide range of applications, both in the exploration of new materials and in the development of new devices. To many people it must have seemed that there was no further need for basic growth studies—all that mattered was the development of bigger and better MBE machines for device production. However, others saw things differently and in the 1990s it was recognised that not only was there still much to be learned about the details of crystal growth but that the MBMS technique had potential to illuminate some of that detail. The group at Sandia National Laboratories in Albuquerque took up the reins with aplomb and demonstrated that it was possible not only to understand steady-state growth conditions (for InAs, GaAs and AlAs) but actually to follow their time dependence during the deposition of a single bilayer of material. They mounted a mass spectrometer in a commercial MBE machine with additional liquid nitrogen cooling and were able to measure the arsenic flux leaving the sample as a function of time, with various gallium fluxes incident on the sample. In their first paper (Tsao et al. 1988) they confirmed that the steady-state sticking coefficient for As$_4$ had a maximum value of 0.5, as previously observed by Foxon and Joyce (1975), though under non-steady-state conditions it could take a slightly higher value—typically about 0.55. In a second paper (Brennan et al. 1992) they were able to correlate this observation with RHEED patterns, showing that the larger value of S_{As4} coincided with the transition from the Ga-rich (4×2) pattern through an intermediate (3×1) pattern to the arsenic-rich (2×4) pattern. They also demonstrated (Tsao et al. 1991) that the arsenic incorporation rate showed an oscillatory behaviour similar to the RHEED oscillations (now widely used as a measure of growth rate) originally reported by the Philips group. They found that the arsenic reaction rate went through a minimum at a point one-quarter of the way and a maximum at three-quarters of the way through the bilayer cycle and related this to changes in the As:Ga ratio during the deposition of a single bilayer. Very similar observations were later reported for the growth of AlSb films by Kaspi and Loehr (1997)—because antimony is much less volatile than arsenic and the background interference is correspondingly much less significant, these Sb desorption oscillations were very much better resolved than in the case of Tsao et al.'s results for As desorption.

Clearly, the use of MBMS to study the growth of III-V materials has yielded a deal of valuable information on growth mechanisms and surface chemistry but a full understanding requires us to relate this to the results of surface structure measurements and we now proceed to look at the information available from RHEED and STM studies.

4.3 RHEED (Reflection High Energy Electron Diffraction) Studies

A major advantage of the use of MBE as a growth method for thin films is its compatibility with in situ monitoring techniques and there can be little doubt that reflection high energy electron diffraction is one of, if not the, most significant. Its ability to obtain detailed structural information on the growing surface and to provide an accurate measure of growth rate has led to RHEED's becoming a standard item in almost all MBE machines. It was originated by Al Cho in his pioneering development of MBE for practical film growth and, as explained in the excellent revue article by Cho and Arthur (1975), it grew out of the earlier application of low energy electron diffraction (LEED) to the study of crystal surfaces during the 1960s. However, the use of an electron beam at normal incidence, as used in LEED, was not compatible with MBE growth so, if one wished to obtain structural information during growth, it was necessary to work at glancing angle, the diffracted beams being detected on a fluorescent screen placed directly opposite. To obtain information about surface atoms, LEED studies made use of electron energies of order 50 to 150 eV. RHEED, on the other hand, required correspondingly greater energies, in the region of 3–50 keV (Cho used 40 keV in much of his work). This being so, it was often convenient to use the electron gun associated with the Auger electron spectrometer as a source of electrons. Diffraction patterns might be in the shape of bright spots or streaks from which it was possible to infer something of the atomic arrangement at the sample surface. However, the distinction between spots and streaks turns out to be a particularly important one, to which we shall give due consideration in what follows. We should also emphasise the fact that, though LEED could yield a complete two-dimensional surface structure from a single diffraction pattern, this was not so for RHEED—in this case, it was necessary to record patterns with the incoming electron beam in several different azimuths—the sample therefore had to be capable of rotation about an axis normal to its surface. First, though, let us look a little more closely at Cho's pioneering work.

Cho's early work was concerned with the growth of GaAs, GaP and AlGaAs. At first, he chose to work with {111} substrates, mainly GaAs but in one instance CaF_2, later changing to {001} GaAs when his interests became concentrated on growing laser structures—on the very good practical grounds that this favoured the use of {110} cleavage planes as mirrors for the optical cavity. As it happens, both {111} and {001} are polar surfaces—that is they can be made up of either mainly Ga or mainly As atoms—a matter of some interest for the understanding of epitaxial growth mechanisms. Initially, he used a single GaAs effusion cell, choosing a cell temperature of about 1170 °C, for which the As_2 flux is an order of magnitude greater than that of Ga. Later he added a second cell containing either Ga or As in order to vary the ratio of beam intensities, independently of their absolute values.

An interesting, if somewhat bizarre, approach to epitaxy is described in one of Cho's first papers (Cho 1969) which does illustrate rather well the virtues of RHEED as an in situ growth monitor. GaAs substrates, polished using syton, were first examined, and these showed very poor diffraction patterns; but, following vacuum annealing at 580 °C, they showed well-defined RHEED streaks. Films of GaAs or GaP were deposited at room temperature, resulting in amorphous films, as indicated by the general blurring of the RHEED display screen. Subsequent annealing at various temperatures (up to 580 °C for GaAs and 650 °C for GaP) revealed a sequence of intermediate structures from polycrystalline (RHEED rings), through features indicative of twinning (twin spots) to single crystal features (streaks). In the case of GaAs films, the streak pattern was similar to that from the annealed substrate, though with less background scattering, while the GaP streak patterns were identical, apart from being roughly 5% more widely spaced (consistent with the smaller lattice constant of GaP—5.45 Å, compared with 5.654 Å for GaAs). Somewhat later, Cho (1970a) reported similar streak patterns from GaP films grown on {111} CaF_2 surfaces.

The observation of RHEED streaks as an indication of single crystal film growth was an obviously important step forward but perhaps of even greater import was Cho's report of different patterns representing specific surface reconstructions and their dependence on growth conditions (Cho 1970b). Earlier LEED studies had identified surface reconstructions with a $\sqrt{19}$ periodicity on {111} oriented bulk samples of silicon, germanium, InSb and GaSb—Cho observed a similar pattern in RHEED from {111} GaAs films and showed that this was characteristic of a high-temperature regime. Reducing the temperature resulted in a change to a {111}–2 structure (a structure with two-fold symmetry) and, what was more, the transition temperature was a strong function of growth conditions. The $\sqrt{19}$ structure was shown to be characteristic of a gallium-rich surface, the two-fold structure of an arsenic-rich surface. Clearly, the use of electron diffraction to study surfaces during growth had potential to advance the understanding of crystal growth as well as serving as a useful monitor of its progress. The year 1971 saw Cho's interest shift to the {001} surfaces of GaAs, for which he discovered (Cho 1971a, b) very similar behaviour to that evinced by their {111} counterparts. LEED measurements by Jona of IBM had demonstrated the existence of (100)–6 and (100)–C(2 × 8) reconstructions on bulk GaAs samples, and Cho was to find evidence of these in his RHEED spectra. In fact, he reported the existence of dominant C(2 × 8) and C(8 × 2) patterns (rotated by 90° about the [001] axis from one another) and, once again, he found that they were associated with Ga-rich (the latter) and As-rich (the former) surfaces, respectively. They could be made to interchange by varying the As_2/Ga flux ratio. Also mirroring the {111} surface structures, these interchanged at a transition temperature which was a function of beam intensities. In a nice little subsidiary experiment he showed that, when the As beam was removed from an As-stabilised surface, the time taken for the $C(2 \times 8)$ pattern to change to a $C(8 \times 2)$ varied exponentially with inverse temperature and that the associated activation energy $E_D = 3.9$ eV corresponded to the desorption energy of As_2 from a GaAs surface—loss of arsenic from the surface being responsible for the change from an As-stabilised to a Ga-stabilised surface condition. Finally, he noted that, if a Ga-stabilised (001)–$C(8 \times 2)$ surface was cooled in the absence of a Ga beam, a (001)–6 structure resulted.

These two papers are of even greater significance for the development of MBE growth techniques in respect of Cho's comparative study of RHEED spectra with electron microscope topology. In the second paper he presented a set of micrographs of Pt–C replicas of GaAs surfaces (at 38 400 magnification), together with their companion RHEED patterns, which we reproduce in Figure 4.5. These images feature again in Cho and Arthur's review of 1975 and have achieved a degree of fame in the MBE world which demands that they be taken with more than an average degree of seriousness. The first pair of images refers to a GaAs (001) substrate which had been polished with a Br–methanol etch, in what became a standard preparation method, and then annealed in vacuum at 855 K for 5 min. The RHEED pattern is clearly made up of spots rather than streaks, and the corresponding topograph reveals surface roughness with features of order of 1000 Å in dimension. Cho's explanation (which has been widely accepted) is illustrated in Figure 4.6—the glancing incidence geometry allows the electron beam to pass directly through asperities on the GaAs surface so that the diffraction pattern is one of transmission rather than reflection, and the resulting diffraction pattern is characteristic of bulk GaAs rather than of a surface. The second pair of photographs shows a much improved surface resulting from the epitaxial deposition of 150 Å of GaAs. The 'spotty' RHEED pattern is now showing a degree of 'streakiness', while the carbon replica micrograph shows a corresponding improvement in surface quality. The third pair represents the end result of depositing a micron of GaAs, sufficient to 'bury' the original surface irregularities. The topograph is now more or less featureless, while the RHEED pattern is composed of the almost uniform streaks that have come to be accepted as typical of a 'flat' surface. Such patterns represent the everyday experience of a wide spectrum of MBE growers and have acquired legendary status within the community. Indeed, there has grown up a reverential belief in the statement that 'RHEED streaks are characteristic of a perfectly flat surface'. As we show in Box 4.1, this is very far from the truth and, in any case, a moment's reflection on the resolution available in Cho's micrographs should alert us to the obvious conclusion that he was not, by a very long chalk, talking about 'atomically flat' surfaces. 'Flat' on a scale of 1000 Å is quite another matter! It brings us, therefore, to the point where we must give serious attention to the meaning of RHEED patterns and their interpretation in terms of surface reconstruction.

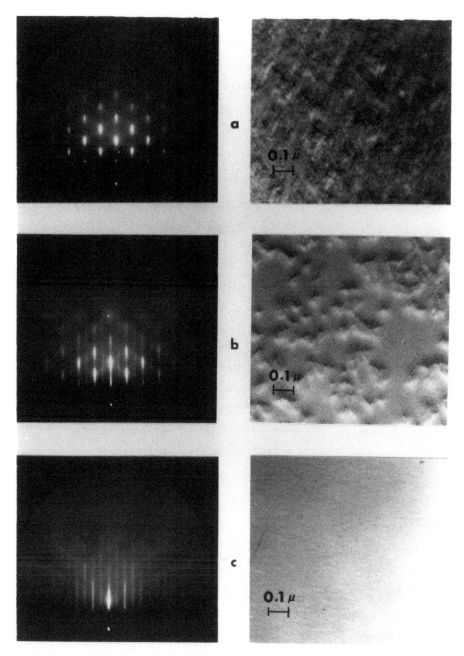

Figure 4.5 *RHEED patterns (40 keV,–1–10 azimuth) and corresponding electron micrographs (38, 400×) of Pt-C replicas of a GaAs surface (a) Br₂–methanol polish-etched and heated in vacuum to 580 °C for 5 min, (b) 150 Å film of GaAs deposited on (a), (c) 1 μm GaAs deposited on (a). (From Cho 1971b.) Reprinted with permission from Cho, A Y (1971b) J Vac Sci Technol 8, S31. Copyright 1971, American Vacuum Society.*

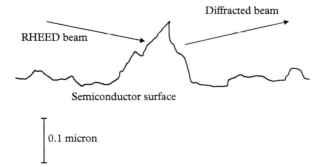

RHEED beam

Diffracted beam

Semiconductor surface

0.1 micron

Figure 4.6 *Schematic representation of a RHEED beam passing through an asperity on a rough semiconductor surface. The resulting diffraction pattern is composed of spots rather than streaks.*

Box 4.1 RHEED—basic theory

In this box we attempt to explain the origins of the streak patterns seen in a typical RHEED spectrum, it being far from obvious exactly how they occur. To be specific, we shall assume that the crystal surface under investigation lies in a horizontal plane, while the incoming electron beam lies in a vertical plane and makes a small angle θ with that surface. Scattered beams impinge on a fluorescent screen some 20 cm behind the sample, bright spots or streaks resulting from constructive interference of beams scattered from a sequence of atoms in the crystal surface plane (we neglect any contribution from atoms lying below the surface layer).

Electron diffraction depends, of course, on the fact that a beam of electrons has associated with it a wavelength, the de Broglie wavelength, proposed by de Broglie in 1924. His suggestion was that a beam of particles with momentum $\mathbf{p} = mv$ possessed undulatory properties characterised by a wavelength λ where

$$\lambda = h/p \tag{B4.1}$$

and h is Planck's constant.

It follows that λ is related to the energy of the particles, which, for electrons in free space, we take to be the kinetic energy $E = p^2/2m$ (we neglect relativistic effects). It follows that

$$\lambda = h/\{2mE\}^{1/2} \tag{B4.2}$$

Numerically, this can be written as:

$$\lambda = 12.24/\{E\}^{1/2} \tag{B4.3}$$

where E is measured in eV and the wavelength in Ångstroms. Thus, for 10 keV electrons, such as used in RHEED measurements, λ is approximately 0.12 Å.

The wave vector $\mathbf{k}$ which characterises the wave motion is related to the momentum $\mathbf{p}$ by

$$k = 2\pi/\lambda = 2\pi p/h \tag{B4.4}$$

The magnitude of **k** for 10 keV electrons is therefore $2\pi/0.12 \sim 50\text{Å}^{-1}$ and, for future reference, we can compare this with a typical crystal reciprocal lattice vector $2\pi/a_0 \sim 2\pi/3 \sim 2\text{Å}^{-1}$.

The RHEED streaks observed are usually separated by about 5 mm, so it is clear that the scattering angles are small—typically of the order of 1–2 degrees. To be specific, the incident beam makes an angle of one to two degrees in the vertical plane, while the diffracted beams make angles of a few degrees in directions between the horizontal and vertical planes.

To help with the understanding of such diffraction patterns, we begin by considering a particularly simple example where the relevant surface consists of a square array of atoms and where the incident beam lies in a vertical plane coincident with a line of these atoms. It is easy to see that constructive interference of beams diffracted from adjacent atoms (and also in the same vertical plane) will occur for diffraction angles equal to the incident beam angle θ and that this condition is *independent* of the lattice spacing. This beam, which is particularly bright, is referred to as the 'specular' beam. It is also easy to see that constructive interference will occur for other (greater) diffraction angles (also in the vertical plane) but that these angles *will* depend on the separation of the atoms a_0. To work out the very many possible diffraction angles in other more general directions clearly involves some complicated three-dimensional geometry and this is best done using a construction known as the Ewald Sphere. Just as the problem of mass spectrometer waveforms (discussed in the previous section) is best dealt with in the frequency domain, rather than the time domain, the diffraction problem is best dealt with in the reciprocal lattice domain and we shall now attempt to make clear, using our simple, idealised model, how this is done.

We need to appreciate that two mathematical conditions must be satisfied in order to define a diffraction spot. First, we are concerned with *elastic* scattering, which implies that the magnitude of the scattered wave vector is equal to that of the incident wave—only the direction of the **k** vector changes. This implies that, as shown in Figure 4.7, the point of the scattered wave vector must lie on a sphere of radius k defined by equation (B4.4), where

$$k = k_{in} = k_{out} \tag{B4.5}$$

The second condition is less obvious. It states that constructive interference occurs when the points of the vectors $\mathbf{k_{in}}$ and $\mathbf{k_{out}}$ both lie on points of the reciprocal lattice, and we show how this applies to the case of the specular beam in Figure 4.8. Again we assume an idealised square lattice of surface atoms which in reciprocal lattice space is represented by an array of infinite parallel lattice rods spaced by $2\pi/a_0$. (Note that, in the absence of broadening mechanisms, which we shall discuss later, these rods are characterised by an infinitely small cross-sectional radius.) Figure 4.8(b) shows that the end points of the two vectors $\mathbf{k_{in}}$ and $\mathbf{k_{out}}$ both lie on the 00 lattice rod, emphasising the point that the specular beam forms a spot which is independent of the lattice spacing. The angle θ is determined by the geometry of the measuring system and is therefore fixed at a value of the order of one to two degrees (Figure 4.8 is not drawn to scale!). Clearly, this gives rise to a single *spot* in the resulting pattern. Other diffraction spots are defined by the points at which other lattice rods intersect the Ewald sphere and these can be seen to represent electrons scattered partly vertically and partly sideways. Referring to Figure 4.8(a), we see that rods 00, 01, 02, 03, 04 and 0–1, 0–2, 0–3, 0–4 cut the sphere in a series of points lying on a symmetrical arc which peaks at the 00 position. In so far as the lattice rods can be represented as vertical *lines*, the resulting diffraction pattern is composed of a series of *points* lying on this arc. There is no reason to expect any streaking!

continued

Box 4.1 *continued*

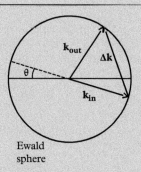

Ewald
sphere

Figure 4.7 *Sphere
construction for RHEED
scattering. The wave
vector k_{in} represents the
incoming electron beam at
glancing incidence angle θ,
k_{out} representing the
scattered wave whose point
lies on the sphere of radius
k_{in}. The vector Δk
corresponds to a reciprocal
lattice vector.*

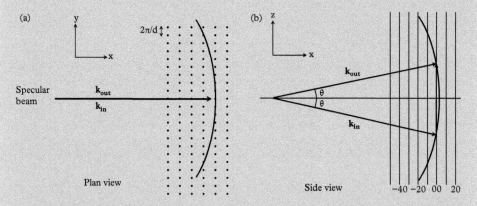

Figure 4.8 *Ewald sphere representation of scattering for the specular RHEED beam as seen in
plan view (a) and side view (b). The reciprocal lattice, for the idealised case of scattering from a
single plane of surface atoms of a cubic lattice, consists of an infinite array of lines parallel to the
surface normal, separated by a reciprocal lattice distance $2\pi/a_0$.*

Why, then, do the majority of experimental RHEED patterns consist of streaks? The complete answer is complex but Figure 4.9 makes clear that any mechanism which broadens the reciprocal lattice lines into rods with finite radii will lead to a streak pattern. Consider the spread angles for the scattered wave vector $\mathbf{k}_{out}$ for the special case of the specular beam. In the horizontal plane this amounts to

$$d\alpha = (d\Delta k)/k \tag{B4.6}$$

whereas in the vertical plane it is

$$d\theta = (d\Delta k)/k\sin\theta \tag{B4.7}$$

so the vertical spread in the specular spot is larger than the horizontal spread by a factor $\operatorname{cosec}\theta$. For an incoming angle of $\theta = 2$ degrees, this means that the vertical extension of the 'spot' will be some 30 times greater than the horizontal, quite sufficient to explain streak patterns. The detailed mechanism for rod broadening may be lattice vibrations, crystal disorder or other more esoteric effects. The point we wish to emphasise is that streaking occurs as a result of some such departure from perfection—it is not to be thought of as characteristic of an atomically flat crystal surface.

Finally, we note that, if the degree of broadening is small, the observed RHEED pattern will take the form of a series of short streaks lying on the semicircular arc referred to above, as shown in Figure 4.10. As the lattice broadening increases, these streaks will lengthen until the arc structure ceases to be apparent and the spectrum consists simply of a set of long parallel streaks, as observed in many practical situations. Clearly, such streaks do not imply the existence of a perfect surface—anything but!

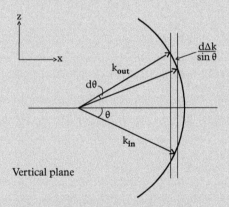

Figure 4.9 *Ewald sphere diagram demonstrating the effect of broadening the reciprocal lattice lines into 'rods', showing that the ratio of RHEED spot broadening in the vertical direction to that in the horizontal direction is a factor of cosec θ.*

continued

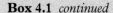

Box 4.1 *continued*

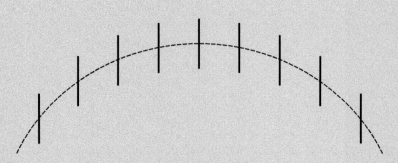

Figure 4.10 *Schematic diagram of specular RHEED pattern under circumstance of only moderate broadening of the reciprocal lattice rods.*

We should like to acknowledge the invaluable contribution of Peter Dobson to the presentation of this box.

Three aspects of the subject need comment and explanation. First, we examine the basic theory of electron scattering from an ideally flat surface and its extension to less than ideal circumstances; we then consider the question of surface reconstruction and the nature of the corresponding RHEED patterns; finally, we look at the practical use of RHEED in the everyday business of MBE film growth. This latter topic, of course, includes the important matter of oscillations in the RHEED intensity for measuring growth rate and we shall consider it in some detail. Let it be said at the outset that our aim here is not to present a comprehensive account of the subject but merely to try and illuminate its chief characteristics in as helpful a manner as possible. Readers wishing to gain deeper insight might like to consult the excellent reviews and books available, such as those by Dobson (1988), Braun (1999) and Ichimiya and Cohen (2004).

The basic theory of RHEED is outlined in Box 4.1, where we consider the two conditions imposed by (1) elastic scattering of incoming electrons and (2) constructive interference between electron waves reflected from a two-dimensional, square array of surface atoms. Satisfying these simultaneously implies that the scattered electron wave vector must terminate on a reciprocal lattice rod, where it cuts the Ewald sphere (the centre of which lies at the origin of the incoming wave vector k_{in}). As can be seen from Figure 4.8, the so-called specular beam forms a diffraction spot at an angle θ above the horizontal equal to the grazing angle of incidence. Other spots correspond to points where the vertically aligned rods intersect the sphere and, if we concentrate on the rods with coordinates $\pm 10, 20, 30$, etc., we see that they give rise to a set of spots lying on a semicircular arc. This set is referred to as arising from the zeroth-order Laue zone. A second, similar set of spots arising from the $\pm 1 - 1, 2 - 1, 3 - 1$,—rods corresponds to the first-order Laue zone, and so on. The overall spectrum therefore consists of spots lying on a series of concentric semicircles. Thus, for an ideal planar crystal lattice in which the reciprocal lattice rods are *lines* of zero lateral dimension, the anticipated RHEED pattern consists of spots—not streaks. While it is certainly true that the vast majority of published patterns consist of streaks, there is clear and unequivocal experimental evidence for this predicted spot pattern in the shape of the RHEED spectrum from a silicon (111) surface shown in Figure 4.11 (Nakahara and Ichimiya 1991). This spectrum was obtained at room temperature from a carefully prepared substrate, which had been annealed at 900 °C for ten minutes, and shows a characteristic (7 × 7) reconstruction—similar patterns were also observed from Si films grown on such substrates at temperatures in the range 500 °C–600 °C. Here, without doubt is a series of spots arranged on a semicircular framework, at least five Laue zones being clearly shown. This is the classic RHEED pattern from an atomically flat surface and poses the obvious question: what is so special about this sample that sets it apart from so many others? The answer probably centres on the simple fact that this is silicon, and that clean silicon {111} surfaces have a tendency to form large flat terraces which represent reasonable approximations to the ideal of the infinite atomically flat surface. However, to appreciate the full significance of this statement requires us to examine more closely the broadening mechanisms which result in streaky patterns in the majority of other samples.

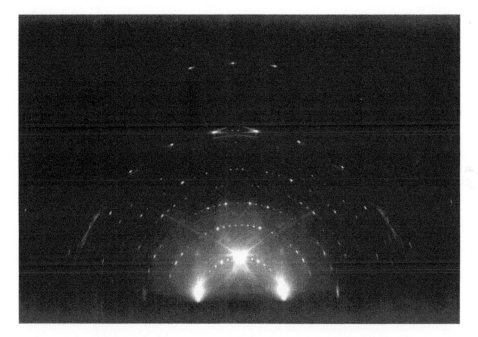

Figure 4.11 *RHEED pattern obtained (at room temperature) from a carefully annealed silicon (111) substrate surface. It represents a (7 × 7) reconstruction and demonstrates the diffraction* spot *pattern predicted for the case where the reciprocal lattice rods are* not *broadened. At least five Laue zones can be seen and the semicircular arrangement of the spots is clearly apparent. (From Ichimiya and Cohen 2004, courtesy of Cambridge University Press.)*

As explained in Box 4.1, streaks arise as a result of any mechanism which broadens the reciprocal lattice lines into rods with finite radii. In the case of only minor broadening, the resulting pattern consists of relatively short streaks which derive in an obvious manner from each of the spots, as shown in Figure 4.10. Increased broadening gives rise to a set of longer, parallel streaks which is the form frequently observed from real crystal surfaces, and, as we show in the box, the ratio of streak length to width is typically of the order of 30 times. As we also comment, there are many possible ways in which such broadening may come about. We must now examine some of them but, first, it is important to recognise the instrumental limitations to measurement which are imposed by imperfections in the electron beam. These come in two guises, first due to the finite spread in electron energy and second due to the spread in beam angle. The first effect defines a coherence length parallel to the beam direction:

$$l_{//} = 2\pi/d\Delta k$$
$$= (2/m)^{1/2}hE^{1/2}/dE \tag{4.9}$$

And, if we measure E in eV, this can be written conveniently as $l_{//} = 24.5\ E^{1/2}/dE$ Å. Thus, if $E = 10\ keV$ and $dE = 0.1\ eV$ (typical of the thermal spread in energies from a hot filament), we obtain a value of $l_{//} = 2.4\ \mu m$. What this implies is a measurement that is sensitive to surface imperfections on a scale *less* than about 2.5 μm. The finite size of the electron gun filament gives rise to a spread in beam angle dθ of about 1 mrad, which equates to a coherence length of

$$l_{//} = 2\pi/d\Delta k$$
$$= 2\pi/k\sin\theta \cdot d\theta \tag{4.10}$$

And, for $E = 10\ keV$, $\theta = 2°$, this gives $l_{//} = 0.35\mu m$, almost ten times smaller than the corresponding effect of energy spread. In other words, it is this latter effect which usually determines the maximum distance over which coherent interference can be detected. Note that the length of the RHEED streak is related directly to the coherence length—if we call this length L, it is clear that

$$L = Dd\theta \tag{4.11}$$

where D is the camera length (the distance to the luminescent screen). It then follows that we can write L as

$$L = 2\pi D/l_{//}k\sin\theta \tag{4.12}$$

And, putting in typical values for $D = 200\ mm$, $k = 50\ Å^{-1}$ and $\theta = 2°$, we arrive at a rough rule of thumb that $L \sim 10^3/l_{//}mm$ (where $l_{//}$ is measured in Å). Thus, a coherence length of $l_{//} = 1\ \mu m(10^4Å)$ corresponds to a streak length of about 0.1 mm, negligibly small.

Now let us look at some of the mechanisms which broaden the reciprocal lattice rods. Perhaps the one that springs first to mind is that of lattice thermal vibrations. Whereas the model of RHEED used in Box 4.1 makes the implicit assumption that the surface atoms are rigidly attached to their nominal lattice locations, it is clear that thermal vibrations result in blurring of these positions, with corresponding broadening of the reciprocal lattice rods. An obvious consequence is that one must anticipate RHEED streaks which show significant variation with sample temperature but, somewhat surprisingly, there appears to be no experimental evidence for such behaviour. We have to suppose that, in practice, other sources of streaking are dominant. In a word, these come down to departures from the ideal atomically flat surface. Any real surface is likely to include steps, either random or ordered. In the case of a vicinal surface (misorientated by a small angle from an exact principal plane), a terraced structure is to be expected, with terrace widths determined by the size of the angle, and these atomically flat terraces determine an effective coherence length for the diffraction process. Considering the case of the silicon sample of Figure 4.11, it is typical of high-quality silicon {111} surfaces that they form regular terraces with lengths as large as 1 μm, which implies that the corresponding RHEED patterns are likely to be limited by the properties of the electron beam, rather than by surface structure. (It is worth bearing in mind that such terrace lengths correspond to a vicinal angle of something like 10^{-3} degrees, which implies some considerable skill in sample preparation!) However, the majority of samples fall well below this standard, and more typical terrace lengths might be of the order of 10–100 Å, corresponding to streak lengths of roughly 1–10 cm. Of course, most surfaces will be characterised by random steps rather than well organised terraces and, in this case, the coherence length is approximated by the sum of the mean island size and that of the holes between them—streak lengths of order 10 cm are probably to be anticipated in most cases—effectively 'infinite' on the scale of a typical pattern. Another important broadening mechanism is that of atomic disorder. This may take many different forms but is often associated with surface reconstruction, where the associated atomic rearrangement occurs in a series of domains, and the coherence length is then determined by the extent of each domain. Once again, one may expect 'infinite' streaks in the diffraction pattern.

Needless to say, a full understanding of RHEED spectra demands considerably greater detail than provided by this elementary summary but our principal aim has been to provide a helpful basis for the reader to appreciate the applications of RHEED commonly met with in MBE growth, rather than any detailed explanation of the 'fine structure' of RHEED patterns. We can only hope that the above account will be seen to serve this lesser purpose. Let us now return to consider the everyday usage of RHEED as an adjunct to MBE growth. Almost without exception, RHEED patterns observed during MBE growth represent some form of surface reconstruction and, as the details of such reconstruction depend on growth conditions, in the way of temperature and beam fluxes, it is natural for RHEED patterns to serve as a useful monitor of those conditions. We have already referred to the example of growth on {001} GaAs and the use of appropriate RHEED patterns as indicators of growth temperature, which, in itself, implies that the relation between such patterns and growth conditions is extremely reproducible. In fact, quite a spectrum of reconstructions has been observed, as we shall see in a moment, but first it might be helpful to look in more detail at this whole question of reconstruction and its relationship with RHEED spectra.

Perhaps the simplest way to approach what is an extremely complex subject is to concentrate our attention on just one well-known example, the GaAs (001)–(2 × 4) surface reconstruction, which is characteristic of an arsenic-stabilised surface. Let us first be clear what such a designation means: with the electron beam aligned parallel to the [110] axis, the associated RHEED pattern shows a two-fold symmetry, while parallel to the [–110] axis it shows four-fold symmetry. This, in turn, implies that the spacing between RHEED streaks is (respectively) half and one-quarter of that associated with the unreconstructed (cubic) surface. It is essentially this spacing which provides the evidence for the relevant model of surface reconstruction. The 'cubic' streaks can be identified from their greater intensities, while the 'reconstructed' lines appear between them with appropriate separation. So much for identification—can we now understand how such surface symmetry comes about? The basic driving mechanism is one of so-called 'dimerisation' of surface arsenic atoms. If we look down on the unreconstructed (001) surface shown in Figure 4.12(a), which represents the case of 100% As coverage, we see an arrangement of atoms in the top layer with what we might call 'square', or (1 × 1) symmetry, the unit cell being indicated by the dashed square. Each As atom in the top layer is bonded to two Ga atoms below it in the second layer of atoms, while having

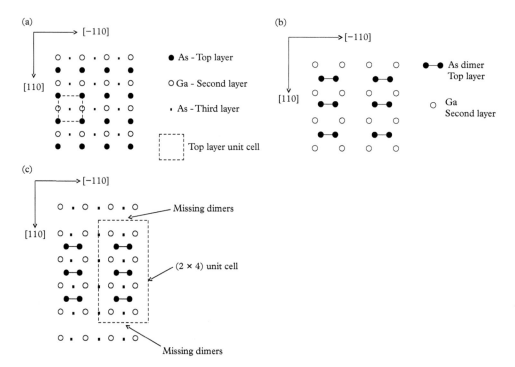

Figure 4.12 *Arrangement of atoms on the (001) GaAs surface, (a) in the ideal case of an unreconstructed surface, (b) showing the 'dimerisation' process in which pairs of As atoms are drawn together along the [–110] direction and (c) showing the (2 × 4) reconstruction resulting from missing dimers.*

two dangling bonds pointing upwards in the (110) plane. It turns out to be energetically favourable for these two bonds to draw As atoms together in pairs, as shown in Figure 4.12(b), thus forming arsenic dimers and lowering the symmetry in consequence. However, this is not the whole story—the (2×4) reconstruction corresponds to a situation with 75% As coverage in which 1 row of dimers (out of every 4) is missing. This structure is shown in Figure 4.12(c), where the unit cell can be seen to be 2 units long in the [−110] direction and 4 units long in the [110] direction. (Incidentally, this is not the only possible structure with (2×4) symmetry but total energy calculations by Chadi (1987) and STM studies by Pashley et al. (1988a) suggest that it is the most favourable one.)

As we have already noted, the particular reconstruction pattern observed during GaAs growth changes according to whether the surface is arsenic or gallium stabilised. Indeed, while the (001)–(2×4) structure discussed above is characteristic of an arsenic-stabilised surface, the change-over to gallium stabilisation generally results in a transition to a (4×2) pattern—in other words, the RHEED patterns corresponding to the electron beam's being parallel to the [−110] and [110] axes are interchanged—and this has been widely used as a method of monitoring surface conditions. However, it very soon became clear that these two reconstructions were not by any means all that could be observed and considerable careful effort was expended in many laboratories to clarify the details. An excellent summary of the so-called surface phase diagram has been provided by Daweritz and Hey (1990) and is reproduced in Figure 4.13. Measurements were made on vicinal samples, misoriented by 2° from the (001) surface towards the (1–11) As plane, and no less than 11 distinct RHEED patterns were recorded. Transitions between different reconstructions were all reversible but only the transition to the Ga-rich (4×2) structure was abrupt—in other cases there was evidence of ill-defined intermediate structures or coexistence of different domains, this being particularly so in the low temperature regime. Transition temperatures were claimed to be accurate to $\pm 5^0$. (Note that beam flux in Figure 4.13 is expressed in terms of beam-equivalent-pressure, or BEP, a convenient concept which we discuss in Section 4.7).

The perceptive reader will, at this point, be concerned at an apparent discrepancy between Figure 4.13 and our earlier discussion of the $C(8 \times 2)$ and $C(2 \times 8)$ reconstructions observed by Cho. As we remarked then, these latter patterns were first reported in studies of LEED rather than RHEED, and Cho took this as a basis for interpreting certain RHEED patterns which subsequently came to be referred to as (4×2) and (2×4). The truth of the matter is that distinguishing between these interpretations by RHEED is extremely difficult—for details, see Neave and Joyce (1978)—so it gradually became standard practice to adopt this latter terminology. In fact, Joyce et al. (1984) point out that (2×4) and $C(2 \times 8)$ may coexist on a GaAs (001) surface as a result of disorder. However, in the vast majority of applications to film growth, the fine detail is of little consequence—all that is needed is the ability to recognise a particular pattern as a reference point in establishing appropriate growth conditions. We shall leave the matter there and proceed to the final topic of this section, the exciting (and sometimes controversial) question of RHEED oscillations.

The discovery of RHEED intensity oscillations and the recognition that the period of these oscillations corresponds to the growth of a single monolayer of material has an

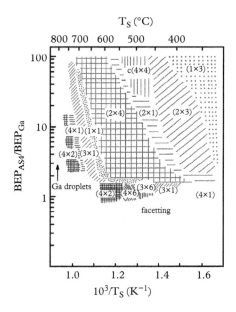

Figure 4.13 *Surface phase diagram for the GaAs (001) surface. The occurrence of the various reconstructions is shown as a function of both substrate temperature and As₄/Ga flux ratio (note that this ratio is expressed in terms of the beam-equivalent pressure (BEP), which is discussed in Section 4.7). (From Daweritz and Hey 1990.) Reprinted from Daweritz, L and Hey (1990) Surface Science 236, 15, with permission from Elsevier.*

interesting history. Such oscillations were probably first noticed towards the end of the 1970s at the Philips Research Laboratories in Redhill by Colin Wood (Wood 1981) but were attributed to vibrational input from the nearby London-to-Brighton railway line and were therefore dismissed as mere artefacts! It was several years later that they were studied in greater depth by Harris et al. (1981a) during a study of tin doping of GaAs MBE-grown films, when it was clearly demonstrated that there was a precise linear relation between the oscillation frequency and the gallium flux. Furthermore, the period of the oscillation was seen to agree with the time taken to deposit a single monolayer of GaAs at each value of Ga flux. The contribution of British Rail to the furthering of scientific progress could finally be discounted. Nevertheless, there was to be further controversy before the true nature of the oscillations could be fully appreciated. Because they were studying the incorporation of tin in GaAs when the oscillations were observed, Harris et al. concluded that tin was essential to their occurrence and developed an explanatory model based on the fact that tin shows a strong tendency to accumulate at the surface and, in doing so, changed the surface reconstruction of the growing film. Being a relatively large atom, they argued that it produced a large strain field when incorporated in a Ga site and, as growth continued, the strain energy increased to the point that the majority of tin atoms were forced out onto the surface, where their lattice-distorting influence was correspondingly minimised—though, at the same time their influence in modifying the RHEED pattern was correspondingly maximised. The suggestion that this expulsion occurred, not smoothly, but in regular 'bursts' as each tin atom was 'buried' under a monolayer of GaAs was seen as the explanation for an oscillatory change in surface reconstruction.

The next paragraph in the RHEED oscillation saga was penned from the other side of the Atlantic. Colin Wood had left the Philips laboratory and taken an appointment at Cornell, where he continued to study the MBE growth of GaAs and, in particular, the unconventional behaviour of RHEED patterns. In his letter to 'Surface Science' (Wood 1981) he made several important comments on the recently published letter by Harris et al. (1981a): first, the oscillations could be observed while growing films doped with a range of donors or acceptors whose atomic radii varied from being greater than, through being equal to, to being less than that of the gallium atom; second, oscillations could also be observed on nominally undoped films; third, the oscillations actually took the form of intensity variations which moved along bulk diffraction streaks, rather than a coherent brightening and darkening of the whole pattern. (On this latter point, we should note that Harris et al. had recorded their data via a photodiode coupled to the RHEED screen through an optical fibre, thus 'seeing' these movements as simple variations in intensity.) It began to look as though the oscillatory behaviour was not only independent of the presence of dopant atoms but was also subtly more complex than had originally been implied. Wood, himself, offered the hypothesis that it originated in the variation of the density of surface steps as growth proceeded. He emphasised that oscillations could only be seen following a brief interruption of growth (by shuttering the Ga beam) and that they then decayed in amplitude over a sequence of deposited monolayers. The effect of the interrupt was to optimise surface flatness, gallium atoms migrating to step edges, thus minimising the number of randomly occurring islands and consequently maximising the size of atomically flat terraces. When growth was re-established, new islands formed, increasing step-edge density and reducing terrace size, a process which waxed and waned periodically with each monolayer deposited. The slow decay in amplitude was presumably associated with a gradual approach to a situation where the random nucleation of new domains was balanced by the coalescence of existing domains.

The response from the European side of the Atlantic came immediately (Harris et al. 1981b), pointing out that, as we saw above, the effect of surface disorder invoked by Wood can only be responsible for changes in the sharpness and length of the RHEED streaks and not their intensity. In fact, the movement of intensity along RHEED streaks (which Harris et al. had also seen) implied some 'complex superposition of the scattered intensity distributions from the surface atoms' and cannot *simply* be explained in terms of variation in surface disorder. Clearly there was need for further work to clarify the essential mechanisms—on such a philosophy the protagonists were firmly agreed. The next step towards better understanding, however, came from Van Hove et al. (1983) at the University of Minnesota. Taking up Wood's suggestion, they developed a detailed model of the effects of surface disorder in the shape of a step distribution, varying periodically with the deposition of a series of monolayers. They argued that, for certain ratios of $\Delta k_{perp}:\Delta k_{paral}$ (where scattering from top and bottom of a step did *not* satisfy the Bragg condition), the reciprocal lattice rods would be strongly broadened by the presence of steps and that, in consequence, the RHEED streaks would be considerably lengthened. Crucially, they also pointed out that, because the total number of scattered electrons remains unchanged, this extension of the RHEED streaks implied a corresponding decrease in their brightness. Thus, a periodic variation in surface step density must lead to the observed oscillation in RHEED intensity. They proceeded to describe experimental evidence for the GaAs (001) surface that the length of the specular beam streak did, indeed, vary during the deposition of a single monolayer, in sympathy with the corresponding change in its intensity. Maximum length and minimum intensity occurred at the point when approximately half a monolayer of GaAs had been deposited, minimum length and maximum intensity corresponding to deposition of a complete monolayer. Here was clear confirmation that Wood's original hypothesis concerning the explanation of his mysterious intensity oscillations was close to the truth, and, as Van Hove et al. also pointed out, their data favoured a model in which the growth of GaAs involved competition between step-edge propagation and second layer nucleation. Growth by the former mechanism alone would imply no change in step density (and no RHEED oscillations), while dominance of the latter mechanism was inconsistent with layer-by-layer growth. Such ideas were later to find application to measurement of surface diffusion coefficients in efforts to gain still better understanding of growth dynamics. At this juncture, it is worth mentioning that oscillations were also predicted from Monte Carlo simulations of growth by Ghaisas and Madhukar (1985).

So far, so good, but two features remained to be explained—why did the oscillations decay towards zero amplitude over some modest number of periods and why did bright regions move periodically along RHEED streaks? The answer to the first question was clearly hinted at by Wood and can be understood in terms of the 'competition' model championed by Van Hove et al. Suppose that, following the shuttering of the gallium beam, the surface becomes perfectly smooth. Deposition of the first layer of GaAs will initially lead to an increase in island density, then by a subsequent decrease, as the spaces between islands are filled in, until the surface once again becomes flat—or, rather, because of the competition between edge propagation and nucleation, until it becomes *nearly* flat. In practice, a second layer will nucleate before the first is complete and this tendency will gradually lead to a smaller variation in island density between the 'nearly flat' turning points. The third layer will be rougher than the second, the fourth rougher still, until eventually an equilibrium will be established—the amplitudes of the corresponding RHEED oscillations decreasing in sympathy. The early stages of this process are illustrated in Figure 4.14 (following Joyce et al. 1986), which prompts an interesting alternative way of looking at the scattering process, in terms of an optical scattering model. The wavelength associated with the electron beam is typically about 0.1 Å and the size of atomic steps of the order of 3 Å, which might be compared to diffuse optical scattering (wavelength ~ 0.5 μm) from surface asperities of the order of 10 μm in dimension. The result is that electrons are scattered in all directions, thus reducing the intensity

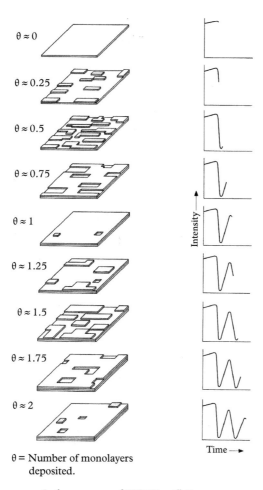

θ ≈ 0

θ ≈ 0.25

θ ≈ 0.5

θ ≈ 0.75

θ ≈ 1

θ ≈ 1.25

θ ≈ 1.5

θ ≈ 1.75

θ ≈ 2

θ = Number of monolayers
deposited.

Surface coverage and RHEED oscillations.

Figure 4.14 *Model of film growth proposed by
Joyce et al. (1986), showing the evolution of the
film in terms of the number (θ) of monolayers
deposited, starting from a perfectly flat surface.
The model shows the relationship between surface
roughness and RHEED intensity and illustrates
how growth on the third layer before completion of
the second layer leads to gradually increasing
roughness and corresponding reduction in RHEED
oscillation amplitude.*

of the appropriate RHEED streak. Note that this mechanism for reducing the intensity is quite different from that proposed by Van Hove et al. and serves to alert us to the fundamental complexity of the scattering process. Indeed, the so-called kinematic model of scattering used by Van Hove et al. (and, incidentally, by us in Box 4.1), which assumes a single scattering mechanism, is far from adequate to explain all the ramifications of observed RHEED oscillations, such as, for example, the occasional observation of a second harmonic component in the waveform. Indeed, the second question posed above can only be answered by recognising that electron scattering involves multiple processes—scattering by atoms located in below-surface layers, excitation of surface plasmons and phonons, interaction between electron waves scattered by different mechanisms, etc. For more detail, the reader may consult the papers by Joyce et al. (1986), Dobson et al. (1987) and Zhang et al. (1987), in which the need for such complications is made very clear. We shall say no more at this juncture, except to emphasise that the use of RHEED oscillations to measure growth rate with sub-monolayer resolution has well-established validity, no matter that theoretical understanding may still lack some degree of totality. For practical use, it is simply necessary to choose appropriate diffraction conditions, as pointed out in the above papers.

While we appreciate that this has been a long section, it should be apparent that there is excellent reason for it—RHEED is, by far, the most widely used technique for monitoring MBE growth of thin films. In conclusion, therefore, it may be worthwhile summarising its various applications, some of which will be discussed in greater detail in later chapters. In the first place, RHEED was used to check the smoothness of the substrate on a roughness scale of 1000 Å by observing a streak pattern (though, as we have already emphasised, an *atomically* flat surface is characterised by spots, rather than streaks). Second, RHEED is widely used to establish the surface conditions required for growth by checking the appropriate reconstruction pattern (such as the arsenic-stable (001) (2 × 4) pattern when growing on a GaAs substrate). Third, RHEED oscillations provide an invaluable aid to measuring growth rates on a monolayer scale, a facility which has been applied to the measurement of re-evaporation rates (Foxon 1986), alloy composition (Dobson et al. 1986), quantum well thickness (Dawson et al. 1985) and surface diffusion rates (Neave et al. 1985). In view of its obvious importance, we feel it imperative that the reader should be aware of the inherent complexities that we have attempted to explain in this section.

4.4   Scanning Tunnelling Microscopy

Given that the coherence length employed in RHEED is of the order of 1μm, there can clearly be no possibility of achieving spectral resolution on an atomic scale. Even the best of scanning electron microscopes struggles to achieve resolution better than 1μm (except in the case of very thin films), so it therefore came as exciting news to the surface science community when in 1982 Binnig et al. from the IBM Zurich Research Laboratory reported the first scanning tunnelling microscope (STM) studies of single atom steps on an otherwise flat surface. The basis of their method was to scan a very fine metal point over the surface, while measuring the current which tunnelled through the vacuum space between tip and surface. The tip-surface separation s was controlled by a piezoelectric drive, and the mode of operation involved scanning the tip, while measuring the voltage on the piezo-drive required to maintain a constant tunnel current. Because this current depends exponentially on s, the drive voltage is also an extremely sensitive function of s, thus offering a capability for detecting single atom steps. Their success represented a triumph of mechanical engineering on the scale of the infinitesimal! It was later to be developed into a technique capable of resolving single atoms in a flat surface and, therefore, providing the ideal method of studying surface reconstruction. In a later review of the development of the STM, Binnig and Rohrer (1987) showed the first atomic scale STM image of the much-prized Si (111) (7×7) reconstructed surface, which they had obtained as early as 1982.

The key to achieving such atomic resolution lies in two areas of expertise: those of anti-vibrational mounting (about which we shall say no more than that it involves cleverly designed springs!) and of fine tip manufacture. The original tips were made from 1 mm diameter tungsten rods, ground to a point, having a radius on the order of 0.1–1.0 μm but relying on the accidental formation of a small number of sharp 'minitips', the longest of which served as active probe (because the tunnel current varies rapidly with s, the longest minitip carries almost all the current). Though this approach undoubtedly worked, it still left something to be desired in respect of reliability and reproducibility— there was clearly scope for improvement and this, after much trial and error with other techniques, came from the development of electrochemical etching. Pashley et al. (1988b), who studied the (001) MBE-grown GaAs surface, refer to tungsten tips sharpened by electrochemical etching, followed by vacuum annealing at 700 °C to remove any oxide. Krause et al. (2005) similarly describe etched tips made from single crystal tungsten and heated by an electron beam. Fotino (1993) reviews tip preparation methods and describes the preparation of tungsten tips with apex radii of 1 nm or less by a two-step process in which 0.5 mm diameter tungsten wire is etched in 2N NaOH, first with the tip pointing downwards into the solution, then, second, in reverse orientation. The much improved sharpness obtained in the second stage originates from the difference in the flow of bubbles over the etched conical surface. The small radius of curvature, combined with a narrow cone angle, makes these tips ideal for atomic resolution STM application.

Once the validity of the STM approach to surface studies was accepted by the surface science community, it was rapidly adopted in a wide range of investigations (Binnig and Rohrer 1987) but, from the MBE point of view, the first significant development took place at the North American Philips laboratory at Briarcliffe Manor in New York state (Pashley et al. 1988a, b). GaAs films were grown on (001) substrates at the IBM Research Centre, Yorktown Heights, and transferred to Briarcliffe for STM studies. This was a matter of some significance, not only technically but also 'politically'—that two rival commercial laboratories could combine to pursue an important fundamental study is remarkable in itself but the fact that samples could be taken from the MBE machine and transferred to another laboratory for investigation raises interesting questions of surface probity. The intention was to study the (001) (2×4) and C(2×8) reconstructions representative of arsenic-stable conditions so, though it was easy enough to grow an

MBE layer under appropriate conditions, it was also vital to ensure that the resulting surface remained unchanged and uncontaminated during transfer and this was achieved by the deposition of a layer of amorphous arsenic to protect it. This was grown near room temperature, after the film had cooled from 600 °C, the temperature at which the GaAs film was deposited. Once the sample was safely ensconced in the STM UHV system, the arsenic was then removed by heating to 370 °C for several minutes, and the probity of the surface was checked by AES and LEED measurements. It was found to be free of oxygen or carbon contamination and to show the anticipated C(2 × 8) diffraction pattern. In other words, all was well and STM measurements could be undertaken with a degree of confidence.

The films were doped n-type with silicon donors at a level of $2 \times 10^{24}\,\mathrm{m^{-3}}$ in order that the tunnelling current could flow to an external contact but of more immediate interest was the question of how this current might flow between tip and sample. It turned out that stable tunnelling could only be obtained when the tip was made positive with respect to the sample, consistent with electron flow out of filled electronic states at the GaAs surface and this, in turn, was consistent with theoretical prediction that such states should be associated with As atoms on the surface. This therefore provided further evidence that the surface under investigation was, indeed, arsenic stabilised. But what of the STM images obtained? The resolution was not quite good enough to pinpoint individual atoms but more than adequate to reveal the arrangement of arsenic dimers and, crucially, the unit cell containing three dimers and one missing dimer, as shown in Figure 4.12(c). Thus, the STM results confirm the validity of the missing dimer model of the (2 × 4) reconstruction but also provide evidence for the coexistence of the C(2 × 8) reconstruction, resulting from surface disorder (calculations of total energy by Chadi (1987) suggest that these two configurations differ in energy by only 0.12 eV, the C(2 × 8) being lower). Domains of (2 × 4) and C(2 × 8) symmetry occur side by side, the difference between them depending on the nature of the boundaries, running along the [–110] direction. These two arrangements of the basic (2 × 4) unit cell are illustrated in Figure 4.15. In (2 × 4) domains, rows of dimers are all aligned along the [110] direction whereas, in domains having C(2 × 8) symmetry, a jog exists, as shown in Figure 4.15(b), resulting in a unit cell with double the length along [110]. As we have already intimated, this dual surface reconstruction was predicted by the proponents of RHEED studies but experimental evidence was far from totally convincing—Pashley et al.'s STM images certainly raised the level of belief to a significantly higher level. They were also able to demonstrate that the (001) surface of their samples contained small islands and holes with step heights of 2.8 Å, corresponding to a single layer of GaAs (i.e. one layer of Ga plus one of As) and no other configuration, a result clearly of some interest for the understanding of growth mechanisms (see e.g. the model shown in Figure 4.14).

What was of even greater interest was the nature of the islands revealed by the STM images. They were seen to be considerably extended in the 2× direction [–110] (i.e. along the missing dimer rows), while being only 1 or 2 units wide in the 4× direction [110]. In other words, relatively long step edges occur along the [–110] direction but only very short edges along [110], of vital importance to ideas of the step-edge growth

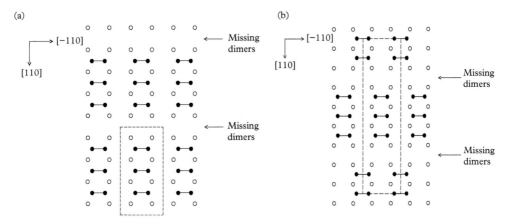

Figure 4.15 *Atom arrangement on the (001) GaAs surface making comparison between the (2 × 4) reconstruction (a) and the (2 × 8) reconstruction (b). Note the 'jog' in the alignment of the atoms along the [110] direction in case (b).*

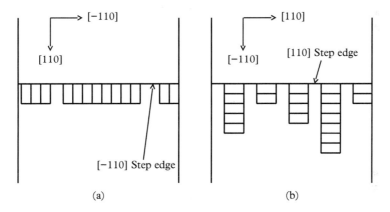

Figure 4.16 *Schematic diagram to show the difference in step-edge growth along a [−110] step-edge (a), as compared with a [110] step edge (b). The rectangles represent the surface unit cells shown in Fig 4.15(a).*

mechanism. It is of particular interest to consider the effect on GaAs samples grown on vicinal substrates (where the (001) surface is tilted by a small angle in either the [110] or [−110] directions). In the former case, the naturally occurring edges are aligned parallel to the vicinal step edges, while in the latter case the opposite is true (see Figure 4.16) and this has a marked effect on the nature of the step edges actually observed in STM images—see Figure 4.17 (Pashley et al. 1991). In the first case, step edges are found to be relatively straight, while in the second case, they are extremely jagged, implying a considerable difference in the surface density of sites to which incoming gallium atoms

(a)

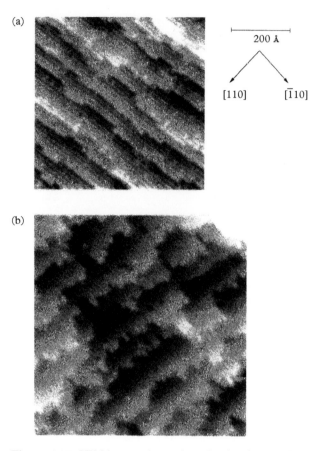

200 Å

[110]   [$\bar{1}$10]

(b)

Figure 4.17 *STM images of step edges, showing the smooth growth on a [–110] edge, compared with the very rough [110] edge. (From Pashley et al. 1991.) Reprinted with permission from Pashley, M D, Haberern, K W and Gaines, J M (1969) Appl Phys Let 58, 406. Copyright 1969, American Physics Institute.*

may attach themselves. Earlier RHEED studies had suggested the likelihood of some considerable difference between the two orientations but these STM images certainly gave a very much clearer picture of the true state of affairs.

The growing sophistication of our understanding of GaAs surface structures can be well illustrated by a brief look at the physical explanation of these unexpected developments, in terms of a very simple model, the 'electron counting model' which was applied by Pashley (1989) to explain the (2 × 4) structure of the (001) GaAs surface. From a knowledge of the energies of the sp^3 hybridised orbitals of Ga and As atoms and their relationship to the GaAs conduction and valence band edges, it is possible to establish conditions that give rise to an electrically neutral semiconducting surface. Because the

As bond energy lies near the GaAs valence band edge while the Ga bond energy lies close to the conduction band edge, it is clear that the total energy will be minimised by an arrangement in which the As dangling bonds are all filled while their Ga counterparts are empty. Looking then at an individual As dimer, each dimer requires six electrons in the top layer of atoms, two in each dangling bond and two in the dimer bond. Eight electrons are also required to bond the dimer to the lower layer Ga atoms. Thus, 14D electrons are required for the number D of dimers in the unit cell. How many electrons are available? This number can be derived from a knowledge of the atom valencies, $V_{As} = 5$ and $V_{Ga} = 3$. The number of electrons available from the upper layer of atoms is $2V_{As}D$ and the number from the lower layer is $2V_{Ga}N/2$ (where N is the periodicity of the unit cell in the direction normal to the dimer rows). Note, here, that half the electrons are needed to bond the second layer atoms to the third layer, hence the factor of ½. Equating these two expressions leads to the following relationship between the number of dimers and the periodicity N:

$$4D = 3N \qquad (4.13)$$

and the smallest unit cell which satisfies this condition is the (2 × 4), with three dimers and one missing. The electron counting model maintains that this condition gives rise to a stable (i.e. the lowest energy) state and we see why the basic (2 × 4) building block is found to dominate the (001) surface of GaAs. (Incidentally, Pashley also applied it to the ZnSe (001) surface, with a seriously different outcome.) Extending the concept to the electronic structure of islands on the GaAs (001) surface, Pashley was also able to explain the fact that, whereas such islands may be infinitely long in the 2× direction [–110], any extension along the perpendicular 4× direction is energetically unfavourable. Such asymmetry between the two types of step edge is clearly important for any model of step-edge growth. While it is energetically unfavourable for growth to occur on the [–110] edge, there is no such constraint to growth along a [110] edge.

As we see, then, the combination of MBE growth, the near-atomic resolution of STM images and a relatively simple piece of theory produced an extremely valuable advance in the understanding of this particular surface. Needless to say, it represented only one special case from a veritable plethora of possibilities but it certainly represented a very important breakthrough in our understanding of surface structure and its application to the study of crystal growth. MBE may often be judged in terms of its ability to produce well-controlled thin films for exotic new electronic and optical devices but we would be seriously failing in our aim to produce a balanced account of its contribution to the well-being of mankind if we were to overlook such advances as this in basic understanding.

As we have already made clear, the (2 × 4) reconstruction of the GaAs (001) surface is only one of many possible reconstructions (Daweritz and Hey 1990) and it would be surprising if the STM technique had not been applied to studying atomic structures of many others. However, we should note an important prerequisite—the (2 × 4) structure is characteristic of an arsenic-stabilised surface, and the samples measured by Pashley et al. were transferred from growth chamber to STM under a protective layer of arsenic. In order to study the whole range of such structures *not* characterised by arsenic-stable conditions, it was first necessary to combine the two chambers within a single vacuum system. The first such facility appears to have been built by Biegelsen et al. (1990a) at the Xerox laboratory in Palo Alto and these authors fired off a series of papers reporting STM studies of numerous GaAs surfaces, together with some interesting probing of the early stages of the heteroepitaxial growth of GaAs on silicon substrates.

First, we shall look (very briefly) at their analysis of a range of different reconstructions on the (001) surface, resulting from MBE growth, followed by UHV annealing (Biegelsen et al. 1990b). In descending order of the As/Ga surface concentration ratio, these were $C(4 \times 4)$, (2×4), (1×6), (4×6) and (4×2) (though it is difficult to reconcile these attributions with the phase diagram shown in Figure 4.13). We might begin by noting that the analysis of the (2×4) reconstruction follows very closely that outlined above, though the authors claim their samples showed considerably less disorder than those studied by Pashley et al.—presumably an artefact of the protective arsenic layer used in the earlier study. The $C(4 \times 4)$ surface was prepared by growing a 300 nm layer at 640 °C under As-stabilised conditions, then cooling the sample to 300 °C in an arsenic atmosphere (how this can be reconciled with the phase diagram shown in Figure 4.13 is, once again, difficult to understand!), then performing the STM measurements at room temperature. The images obtained show a series of rectangular 'bright' regions each containing three dimer-like features oriented at right angles to those associated with the (2×4) surface shown in Figure 4.14. The gallium-stabilised (4×2) (or $C(8 \times 2)$) structure was shown to be characterised by a unit cell containing two Ga–Ga dimers, together with two missing dimers, the orientation of the dimers being at right angles to that of the As–As dimers of the (2×4) reconstruction. Atomic models consistent with electron counting principles were proposed to explain all their STM images. In a further paper, Biegelsen et al. (1990c) reported similar studies of the GaAs (–1–1–1) surface. For further details we refer the reader to the original papers. We should also recognise that this was never likely to be the end of a complex story. There were to be numerous further studies of the surface structure of GaAs and attempts to relate it to growth conditions but this would involve us in wading too deeply into the quagmires of surface science—we can do no more than offer a useful summarising reference in the form of Pashley et al. (2005).

Two other early applications of STM imaging to MBE growth were concerned with the growth of GaAs on silicon (001) substrates (Biegelsen et al. 1990d, e) and with an MBE-grown AlGaAs/GaAs multiple quantum well structure (Gomez-Rodriguez et al. 1990). This latter work is of some interest in so far as it addresses the question of how STM contrast arises when examining a surface made up of chemically different materials (differences in work function, band bending, band gap?) but is of negligible interest from our present viewpoint. It involved cleaving through the MQW structure to expose the required surface, which was also heavily oxidised because the measurement was made in a rather modest vacuum. Biegelsen et al.'s work on heteroepitaxy, on the other hand, represented an interesting use of STM to study an important problem of crystal growth—if GaAs could be grown with device quality (whatever that might demand in whatever application!) on silicon substrates, it would certainly offer a significant saving on the cost of substrates. From our present perspective, it makes sense merely to outline their approach, which consisted in studying first the clean silicon substrate, then the result of depositing an initial layer of arsenic and finally that of adding a first layer of GaAs. The evidence suggested an initial growth process in which a level 'sea' of pseudo-morphic GaAs (i.e. a layer with the lattice constant of silicon, rather than that of GaAs) is formed, followed by island growth of relatively unstrained GaAs. They demonstrated that STM was able to throw considerable light on the detailed surface structures of these various stages but we shall say no more here other than to comment on its obvious potential to help greater understanding. To arrive at any sort of complete picture of such a complex process was going to require much hard work.

4.5 Doping

Once the ability to grow single crystal films of device quality had been established, the next obvious step towards practical application of these films lay in the direction of controlled doping, both n- and p-type (not forgetting the occasional requirement for semi-insulating material). At the low end of the doping spectrum, it was, of course, essential to minimise the density of accidentally incorporated impurities, while at the high end one was faced with problems of dopant solubility and auto-compensation. In the case of GaAs, there was already much prior understanding based on experience with bulk crystals and with thin films grown by liquid and by vapour phase epitaxy so it was immediately apparent where to start looking for suitable dopants to use in MBE growth. Indeed, as early as about 1975, the behaviour of no less than eight different doping elements, Si, Ge, Sn, Mn, Te, Zn, C and Mg, had been explored and documented (Cho and Arthur 1975), while by 1980, with the addition of Be, S, Cr and Fe, all the likely candidates were well understood. The earlier studies of doping in LPE and VPE growth were, of course, helpful but we should remember that MBE generally takes place at significantly lower temperatures and under arsenic-rich growth conditions so it was not altogether surprising to find a number of important differences in dopant behaviour. We shall comment on each as we trace their chronological emergence.

Let us first look at n-type dopants. In 1971 Cho and co-workers reported on the Group IV elements Si, Ge and Sn, all of which could be used to dope GaAs n-type, even though they are potentially amphoteric, being n-type when on the gallium site and p-type on the arsenic site. Electron concentrations as high as $5 \times 10^{24}\,\mathrm{m}^{-3}$ were obtained with Si and Ge, while Sn doping achieved even higher levels, up to $1.2 \times 10^{25}\,\mathrm{m}^{-3}$. Several factors may affect which site is preferred but we might reasonably anticipate a degree of compensation as a result of dopants electing to sit on both sites. A glance at the table of covalent radii presented in Table 4.1 suggests that silicon could be accommodated on either site, while germanium might prefer the gallium site, and tin would have difficulty in fitting into either site but would surely favour gallium. In other words, we should not be surprised, on this basis alone, to find that they could all function as n-type dopants. However, surface conditions during growth may also play an important role—under arsenic-rich conditions we must expect a preponderance of Ga vacancies to be present, thus, again, favouring n-type behaviour, though gallium-rich conditions obviously favour p-type. The effect of surface conditions is dramatically demonstrated

Table 4.1 *Covalent radii of atoms appropriate to the doping of various semiconductor materials*

Element	Covalent radius (Å)
Al	1.26
Ga	1.26
In	1.44
N	0.70
P	1.10
As	1.18
Sb	1.36
C	0.77
Si	1.17
Ge	1.22
Sn	1.40
Be	1.06
Mg	1.40
Zn	1.31
S	1.04
Se	1.14
Te	1.32

by the work of Wang et al. (1985) from IBM, as they studied silicon doping on a range of different GaAs sample orientations. For growth on (111)A, (211)A and (311)A surfaces, they found that silicon behaved as an acceptor, whereas on the corresponding B faces it acted as a donor, a result which they explained in terms of surface bonding. Can we also anticipate the effect of the much lower temperatures appropriate to MBE growth, compared to LPE and VPE? We shall discuss the thermodynamic approach to understanding doping behaviour in more detail in the next section but, for the moment, it is sufficient to note that, the presence of gallium vacancies being essential to n-type doping with Group IV elements, the equilibrium density of such defects can be expected to be significantly lower at MBE growth temperatures. This may seem to mitigate against high doping levels with these elements but there is another subtlety which bears serious mention, that being the presence of an acceptor complex $[V_{Ga}Si_{Ga}^-]$ (i.e. a gallium vacancy with silicon on nearest neighbour gallium site). This complex appears to be responsible for a degree of auto-compensation present in bulk and epitaxial samples and which limits maximum doping levels to roughly mid 10^{24} m^{-3}. In the case of MBE growth, one may anticipate a significant reduction in the equilibrium density of these defects and it turns out that n-type doping with silicon can achieve levels as high as mid 10^{25} m^{-3} (see the discussion by Hurle 1999). In terms of convenience, we might recall that the evaporation of silicon requires high temperatures—in fact, when growing silicon films, it is necessary to use electron beam evaporation—but a conventional MBE cell can be used to achieve fluxes appropriate for doping (though the high temperature required can result in contamination by spuriously generated impurities). Many years later, Briones et al. (1985) also reported the use of silane for successful n-type doping of GaAs. Nevertheless, much early work was performed using either germanium or tin, which require significantly lower cell temperatures. Germanium has the possible advantage of true amphoteric behaviour—Cho demonstrated the possibility of producing p–n junctions by simply changing the arsenic-to-gallium flux ratio—while tin is capable of achieving higher doping levels. On the other hand, tin suffers from the disadvantage of strong surface segregation, which results in much broadened doping profiles. (For a detailed discussion of this behaviour, see Wood and Joyce 1978.)

So much for Group IV donors—we next consider the Group VI elements S, Se and Te, arsenic site donors which were much used in LPE and VPE. Direct doping with sulphur from a thermal source is difficult on account of its high vapour pressure at typical MBE growth temperatures, which demands particularly large flux levels to counter the high rate of desorption from the growing surface. Two distinct disadvantages of such an approach are evident—first, the high sulphur vapour pressure would lead to undesirable contamination of the vacuum system; second, even during bakeout of the MBE chamber, there would be significant emission of sulphur into the system. However, various schemes have been adopted to overcome this deficiency. Wood (1978), for example, used PbS as a so-called 'captive source' to dope GaAs by way of a 'surface exchange' process. PbS evaporates congruently and only decomposes when it arrives on the GaAs surface, where the sulphur is trapped by association with gallium atoms and incorporated naturally into arsenic sites. There was no evidence for lead being incorporated into the grown film—it apparently segregates to the surface and evaporates. He also obtained similar results with PbSe. A related technique involved the use of hydrogen sulphide (H$_2$S) to dope GaSb by Gotoh et al. (1981), a method taken up later for doping GaAs but with only limited success. Also of interest was the approach adopted by Davies et al. (1981) at the British Telecom Research Laboratories, who developed an electrolytic cell to generate a beam of S$_2$ molecules (whereas a thermal source produces largely S$_8$) and achieved doping levels in the range 10^{21} m^{-3} to 10^{24} m^{-3}. The rapid response of this cell enabled them to obtain sharply varying doping profiles simply by modulating the cell EMF. They later used this approach to achieve sulphur doping of InP at levels of 5×10^{22} to 6×10^{25} m^{-3} (Iliadis et al. 1986). Doping with tellurium was investigated by Arthur (1974). Though Te also has a high vapour pressure, it appeared to interact chemically with gallium on the GaAs surface, thus allowing it to be readily incorporated into the growing film. However, though it was possible to achieve doping levels in the region of 10^{25} m^{-3}, tellurium showed a strong tendency for surface segregation, which made accurate control extremely difficult. In an attempt to overcome this problem, Collins

(1979) employed SnTe as a source of donors, both elements producing shallow donor levels in GaAs. He obtained values of n ~ $10^{21} - 10^{24}$ m^{-3}, with acceptable electron mobilities, though at the higher concentrations there was again evidence of Te segregation at the GaAs surface. This approach was later adopted by Chen and Cho (1991) as a method of doping GaSb. As an alternative source of Te, the group at the Max Planck Institute in Stuttgart (De-Sheng et al. 1982) extended Wood's earlier use of PbS and PbSe by employing PbTe. They achieved doping levels between 2×10^{22} m^{-3} and 2×10^{25} m^{-3} and found no evidence of Te surface segregation, provided the source consisted of a true Pb-saturated material.

It thus appears that useful n-type doping may be achieved in many different ways but, in practice, most workers have opted to use either silicon or tin. As we commented earlier, tin is easy to handle and allows high doping levels to be obtained but suffers from the disadvantage of surface segregation. This requires that the surface concentration of tin be such as to be in equilibrium with that in the bulk, so that the incorporation rate is just balanced by the arrival rate and this implies a rather slow response to changes in tin flux. It has the distinctly undesirable effect of making it impossible to generate sharp changes in doping level. Silicon doesn't suffer at all from this problem and can provide GaAs doping levels as high as 5×10^{25} m^{-3} at a growth temperature of 410 °C. Also, unlike germanium, it shows no tendency for amphoteric behaviour. Its principal drawback is the need for high cell temperatures in order to generate the necessary fluxes but this appears to have become much less of a problem with the development of pyrolytic boron nitride (PBN) insulating components, to replace the alumina material used in early cell design. The result was a significant reduction in the incidence of unwanted impurities.

Of p-type dopants in GaAs, magnesium, zinc and cadmium were all successfully used in bulk crystals and in LPE and VPE film growth. Germanium also acted as an acceptor in LPE GaAs on account of the gallium-rich growth conditions. However, zinc was, by a long chalk, the most commonly used acceptor and it was natural for the MBE pioneers to explore its possible use in their rapidly developing new technology. Arthur (1973) was first to point out, however, that zinc was not at all suitable as a dopant in MBE because of its very high vapour pressure. In fact, the Murray Hill researchers made an early choice of magnesium as their preferred p-type dopant (Cho and Panish 1972), even though the sticking coefficient of Mg on GaAs at 560 °C was as low as 10^{-5}. However, it increased quite dramatically when a few percent of aluminium was included in the growing film. The explanation for this was unclear but may have been related to competition between Al and Mg for any oxygen present at the surface. Another drawback to the use of magnesium was its rapid diffusion in GaAs at MBE growth temperatures, which rather spoiled the MBE virtue of providing sharp doping profiles. Nevertheless, Cho and Panish reported hole densities in GaAs of up to 10^{25} m^{-3} and, for a number of years, magnesium remained the only working acceptor in the world's MBE facilities. It was in 1975 that several alternatives came into focus, when Marc Ilegems, of the rival Murray Hill MBE group, reported that both carbon and manganese acted as p-type dopants in GaAs (Ilegems et al. 1975). 'Accidental' doping levels as high as 10^{22} m^{-3} were found in several instances and, when Mn was deliberately incorporated, hole concentrations of 10^{24} m^{-3} were achieved. This latter figure represented an upper limit on account of the

Mn acceptor level being 113 meV above the valence band edge, significantly deeper than the value of roughly 30 meV appropriate to more 'conventional' acceptors. Nevertheless, it is worth noting that the first CW GaAs laser grown by MBE made use of manganese as the p-type dopant (Cho et al. 1976). Carbon, on the other hand remained unacceptable on the grounds that there was no suitable source available. Meanwhile, an interesting alternative was suggested by Takahashi from the Tokyo Institute of Technology (Naganuma and Takahashi 1975). By ionising the zinc atoms emerging from a conventional MBE cell, their sticking coefficient was very much enhanced and doping levels up to 3×10^{25} m^{-3} were obtained. As the authors pointed out, this approach could equally well be applied to other acceptor sources but there appears little evidence of its being taken seriously. The likelihood of damage to the growing crystal resulting from this (admittedly low energy) ion implantation process, allied to the additional complexity introduced by the ioniser, appears to have frightened potential followers of this particular initiative.

Perhaps the really important breakthrough in p-type doping came two years later, when Ilegems reported his investigation of beryllium as a dopant for GaAs and AlGaAs (Ilegems 1977). Layers were grown under As-rich conditions at temperatures of 580 °C–600 °C. Be could be evaporated from a conventional source at temperatures between 600 °C and 800 °C, and hole densities from roughly 10^{22} to 5×10^{25} m^{-3} were obtained, with hole mobilities close to those measured on GaAs samples grown by other techniques. In addition, he showed that the diffusion coefficient of Be in GaAs was low enough that diffusion would not constitute 'a problem which would limit the usefulness of Be doping for device applications', an important advantage over the then current favourite p-type dopant, magnesium. Beryllium was rapidly adopted by the majority of MBE growers as the preferred acceptor element and remained so for many years—it was also used successfully for p-type doping of InP (Kawamura et al. 1983) and GaInAs (Cheng et al. 1981). However, we should note that it later came to be seriously rivalled by carbon, once appropriate controlled sources were discovered. The first deliberate carbon doping studies were reported by Weyers et al. (1986), who were in process of using metal-organic MBE to study the chemistry of MOVPE growth. They used a mixture of trimethyl and triethyl gallium and it turned out that carbon was incorporated from the TMG but not from the TEG. Thus, by varying the ratio of TMG to TEG, they were able to control carbon doping over the range 10^{20} to 10^{25} m^{-3}. Of course, this approach was hardly appropriate to conventional MBE but it certainly proved the point that carbon could act as an effective p-type dopant, if a suitable source could be found. The first of these took the simple form of an electrically heated graphite filament (Malik et al. 1988) which was capable of producing well-controlled hole densities in the range 10^{23} to 10^{26} m^{-3}. Not only was the upper bound of these figures significantly greater than those obtained with other acceptors but the authors also demonstrated significantly reduced diffusion at a temperature of 800 °C as compared with Be doping. Here, at last, was the ultimate p-type dopant, which didn't even need an effusion cell for its incorporation—we use it as standard in our own Nottingham Lab.

Finally, we remember our earlier caution not to overlook the occasional need for semi-insulating material, which demands a deep level impurity. In fact, two have been employed to render GaAs highly insulating. Morkoc and Cho (1979) produced such material as a 'substrate' for field-effect transistors by doping with chromium, while Covington et al. (1980) used iron. We shall say no more for the moment but move on to our final section.

4.6   Theoretical Aspects

We round off this survey of the initial establishment of MBE as a practical growth technique with a brief summary of some theoretical studies which provided important aids to the understanding of film growth and doping behaviour. They came in various forms: first, we consider the application of thermodynamics to MBE growth, as exemplified by the work of Heckingbottom et al. at the British Telecom Laboratories, then look at a group of papers from Anupam Madhukar and collaborators from the University of Southern California, who developed Monte Carlo models of the growth of GaAs and applied them to the study of interface quality and quantum well behaviour in AlGaAs/GaAs structures. This is followed by modelling of growth by Vvedenski at Imperial College, London, Farrell et al. from Bell Communications Research and Cohen et al. from the University of Minnesota and by tight binding calculations of the energetics of surface reconstruction on the GaAs (001) surface (Chadi of Xerox). It is beyond the limited ambition of an historical outline to delve into very much detail of such esoteric undertakings—we shall do no more than attempt to set the work in context and explain its relevance to the experimental background that we have covered in previous sections.

There was a certain amount of dispute concerning the application of thermodynamics to MBE growth, based on the fact that MBE, unlike LPE and VPE, proceeds under conditions very far from thermal equilibrium. Indeed, first thoughts suggest that it is entirely kinetically limited. For example, the fact that Zn cannot be used as an acceptor in MBE GaAs is explained in terms of its low sticking coefficient (a kinetic limitation?) whereas it works very well in the equilibrium-controlled LPE and VPE. In similar vein, Arthur's equilibrium vapour pressure data (Cho and Arthur 1975) show that $P_{Ga} = P_{As} \sim 10^{-6}$ Torr at $T_{substrate} \sim 690\,°C$, whereas much MBE growth of GaAs takes place near $580\,°C$—clearly very far from equilibrium. It was therefore of some interest that the British Telecom group should publish a number of papers during the early 1980s in which they took a second look at this assumption (summarised by Heckingbottom et al. 1983). In particular, they considered the incorporation of a wide range of dopants into GaAs, and their overall conclusions were that, with only one or two exceptions, thermodynamics represented a useful tool with which to predict their incorporation. They made the important distinction between what they called 'facile' and 'hindered' processes, where, in the latter case, 'kinetic barriers' hinder the progress of a reaction towards its ultimate (i.e. thermodynamic) end point. As an example of such a process, we note the well-established fact that it is possible to form a sharp interface between GaAs and AlAs films, while the obvious thermodynamic end point for such a structure requires complete interdiffusion of the Ga and Al species. Thus, the existence of a sharp interface implies a strongly kinetically hindered process, the rate of interdiffusion being extremely low at normal growth temperatures. As an example of a surface process, we might consider the case of GaAs growth from Ga and As_4 beams, where, as shown by Foxon and Joyce (1975), the sticking coefficient of As_4 is never greater than 0.5, even on a Ga-rich surface, while it becomes vanishingly small under As-rich conditions. Thermodynamics suggests that, on a Ga-rich surface, all the As_4 should be condensed, up to the completion of a single monolayer of GaAs—the fact that half of it is desorbed, as a consequence of the surface chemistry, represents a rate-limiting process which hinders (only slightly in this case) the thermodynamics.

Let us now look at an example of dopant incorporation. The thermodynamic model of silicon doping in GaAs proceeds as follows: silicon is incorporated in GaAs to provide n-type doping according to the reaction

$$Si(g) + V_{Ga} = Si_{Ga}^+ + e^- \tag{4.14}$$

where $Si(g)$ is a silicon atom in the gaseous state, V_{Ga} is a gallium vacancy in GaAs, Si_{Ga}^+ is a positive silicon ion on a gallium lattice site and e is a free electron in the GaAs conduction band. Straightforward application of thermodynamic reasoning suggests that the density of donors incorporated should increase as the square root of the silicon-beam pressure, whereas experimental data show a linear dependence. (The explanation of this anomaly has been the subject of much discussion but is best understood in terms of Hurle's (1999) model of compensation by native defects.) The pressure of silicon gas is adjusted to give the required doping level. Assuming unity sticking coefficient and a gallium beam pressure of 10^{-6} Torr, a silicon pressure of 10^{-10} Torr can be expected to produce a free-electron density of about 2×10^{24} m^{-3}. Thermodynamic data indicate that, for a typical substrate temperature of about 600 °C, at all pressures greater than about 10^{-13} Torr, silicon should condense (implying unity sticking coefficient). Thus, assuming the solubility limit of Si in GaAs is not exceeded, the doping level will be a linear function of the silicon-beam pressure. This is what is found in practice up to levels of about 10^{25} m^{-3} but at higher concentrations a degree of compensation sets in and the free-electron concentration goes through a maximum as a function of silicon doping level. This deviation from the straight and narrow behaviour is again explained by Hurle (1999) in his comprehensive treatment of point defects and doping in GaAs. It depends upon two factors: the compensating effect of the [V_{Ga} Si_{Ga}^-] acceptor and the competition between arsenic and silicon fluxes at high silicon-beam pressures. The important general point seems to be that the whole range of silicon doping behaviour can be modelled in simple thermodynamic terms, a feature which is replicated for very nearly all the dopant species considered by the British Telecom group. In summary, it is probably fair to say that these thermodynamic calculations have rarely been predictive but may certainly be seen as useful aids to understanding, particularly with regard to the importance of native defects in modifying dopant incorporation.

Turning now to a very different theoretical approach to MBE growth, we take a brief look at the use of Monte Carlo calculations to simulate the growth of GaAs on the

(001) crystal plane. The basic theory was described by Singh and Madhukar (1983). They considered growth from beams of Ga and As_2, the latter of which required careful modelling of the arsenic incorporation mechanism. As proposed by (Foxon and Joyce 1981), the As_2 molecule is assumed to be physisorbed into a lightly bound precursor state, allowing it to diffuse between surface sites until it is dissociatively chemisorbed at a suitable gallium site—for example, a pair of Ga atoms to which both As atoms may be bonded. The Monte Carlo model takes account of the appropriate surface geometry (i.e. detailed atomic arrangement), surface migration and evaporation and allows calculation of the developing surface structure during the deposition of a number of monolayers of GaAs. The main conclusion was that the nature of the resulting surface was critically dependent on the degree of surface migration and, bearing in mind that diffusion is a thermally activated process, on growth temperature. In particular, the authors demonstrated that, at temperatures below about 950 K, growth was essentially two-dimensional (layer-by-layer growth), while above this, it became three-dimensional, with consequent surface roughness. Clearly, this had important consequences for the quality of the interface between GaAs and AlGaAs which, as quantum well and superlattice structures were under intense investigation during the 1980s, was of major practical importance. According to the model, the explanation lay with the increased mobility of Ga atoms at the higher growth temperatures, which increased the probability of a pair of Ga atoms being available to bind the two As atoms from the As_2 molecule.

In a later paper (Ghaisas and Madhukar 1985), the Monte Carlo model was extended to explore the consequences of varying the parameter describing the As_2 dissociative reaction rate. In the case of a very slow reaction rate, the growth process is limited by the reaction rate (a reaction-limited incorporation, RLI growth mechanism) and the model predicts a small (~5%) degree of oscillation in the growth rate. In the case of a fast reaction rate, the growth is limited by the configuration of surface atoms (a configuration-limited reactive incorporation, CLRI) and the calculated growth rate is very nearly constant throughout the growth period. In both cases the model predicts damped RHEED oscillations in general agreement with experimental observation. However, in an accompanying paper, Singh and Bajaj (1985) suggest that, under normal growth conditions, it is the cation migration rate that determines the surface and interface quality, rather than the As_2 reaction rate. Their calculations were based on combining Monte Carlo and statistical fluctuation theory but it is not clear exactly where the difference between these two conflicting accounts arose. The progress of science throws up such mysteries not infrequently—in our historical approach we are concerned merely to outline the kind of work that was under way, rather than to dot every 'i' and cross every 't'. An interesting extension of these models which emphasises the importance of dissociative reaction of As_2 molecules at steps in the Ga sublattice was presented by Thomsen and Madhukar (1987). They calculated that a relatively small number of such reactions could improve surface smoothness quite significantly. They also predicted the existence of an optimum V/III ratio for surface smoothness which was consistent with (then) recent photoluminescence measurements on AlGaAs/GaAs quantum well structures. It is clear that these Monte Carlo simulations had much to offer in improving our understanding of MBE crystal growth, though we might also note that they took no account of surface reconstruction, perhaps because the detailed arrangement of surface atoms only became clear with the advent of STM studies at the end of the 1980s. Another possible criticism concerns the undoubted complexity of the models employed—later Monte Carlo simulations by Clarke and Vedensky (1987) suggested that similar predictions concerning RHEED oscillations could be obtained from a much simpler model in which only the cations were considered. Their simplified model considered only monatomic species, randomly deposited on the surface but also took account of surface migration. They predicted damped RHEED oscillations which, on vicinal surfaces, disappeared when the substrate temperature was high enough to produce step-edge growth—that is when the surface diffusion length became comparable with terrace length. The advantage of this approach was that it minimised the number of adjustable parameters (whose values may not be known reliably!) though the rival protagonists would reasonably argue that crystal growth is a complex process and can therefore be represented accurately only by a complex model. As mere historians, we shall not venture to cross swords with either party!

A very different approach to growth modelling is represented by the work of Farrell et al. (1987) of Bell Communications Research. They concentrated attention on the specific atomic structures relevant to each stage of the growth cycle, starting from the (100) (2 × 4) reconstruction shown in Figure 4.15(a), characterised by three arsenic dimers aligned along the [–110] direction, together with a single dimer vacancy, and this starting point they justified on the basis of Chadi's free energy calculations to which we have already referred (Chadi 1987). Their model divided the growth of a single monolayer of GaAs into four stages and looked in detail at the situation pertaining to one-, two-, three- and four-quarters, respectively, of each monolayer deposited. Stage 1 proposed the incorporation of gallium dimers, with [110] orientation, between a pair of arsenic dimers and allowed for relaxation of the original As dimers. Stage 2 involved the addition of further Ga dimers, together with As dimers to fill the original dimer vacancies and further As dimers atop the newly incorporated Ga dimers. This, therefore, started the second half of the cycle, which was completed by further Ga dimers, then, finally, a Ga dimer and two As dimers, to arrive back at the original surface arrangement. (A glance at the original paper will make all this clear.) The point to grasp is that each stage was constructed carefully in the light of appropriate chemical bonding and free energy considerations. Interestingly, and perhaps understandably, considering the authors' industrial association, one of their motives for undertaking the work was the hope that better understanding of the growth process might facilitate improvements in MBE techniques in the direction of lowering growth temperatures. They point out that, if the deposition temperature could be reduced by a factor of two, it would then be possible to grow GaAs films on top of previously processed structures without degrading them. They make two suggestions, resulting from their growth model, which might lead to such a development: first, they note that an overpressure of arsenic is required only at certain stages of the cycle and that pulsing the As_2 beam might therefore be advantageous;

second, the necessary breaking of certain chemical bonds might be facilitated by the use of monochromatic light, rather than relying on thermal energy. Such thinking lies behind many of the modified MBE techniques that we shall be considering in our next chapter.

As we have emphasised earlier, reflection high energy electron diffraction proved to be an invaluable aid to the understanding of both surface reconstruction and the monitoring of MBE growth. In Section 4.3 we provided an elementary account of RHEED theory and of the phenomenon of RHEED oscillations. Perhaps the best qualitative illustration of the origin of these oscillations can be seen in Figure 4.14, which demonstrates how surface roughness oscillates in sympathy with the cycle of layer deposition, showing a gradual approach to a steady state after the deposition of a certain number of layers. RHEED intensities simply reflect this oscillating degree of roughness and can therefore be represented (qualitatively, at least) as a damped sine wave. As we also hinted, there is much subtle detail in the experimental data, implying a need for some deeper theoretical probing in order to establish a quantitative understanding. The Monte Carlo simulations already described include RHEED oscillations in their description of growth but make no clear connection with details of surface atomic structure. Such connection is made in much simpler terms by the so-called 'birth–death' models of Cohen et al. (1989), who set out to examine the effects of surface migration and the presence of multiple atomic layers for growth on both low-index and vicinal surfaces. In their model the coverage and surface roughness are calculated, together with RHEED intensity, for various growth conditions, including (1) perfect layer-by-layer growth, (2) non-diffusive growth on a low-index surface and (3) diffusive growth on a low-index surface. In cases (1) and (3), RHEED oscillations were predicted but with varying waveforms. Perfect layer-by-layer growth produced waveforms with cusps at their maxima, while diffusive growth gave shapes varying between cusped and sinusoidal according to the probability of inter-layer hopping. Non-diffusive growth, on the other hand was characterised by a rapid decay of RHEED intensity over the first monolayer—in other words, a rapid, non-oscillatory increase in roughness. Encouragingly, many of the predicted features are mirrored by experimental observations.

In summary, we can simply offer the obvious comment that the derivation of any detailed, atom-based account of such a complex process as crystal growth must inevitably be fraught with difficulty. That these various theoretical approaches start from and finish with different viewpoints is hardly surprising; that, collectively, they certainly advanced understanding is a matter for congratulation; and, finally, we might reasonably claim that MBE, of all crystal growth methods, probably offers the best possible hope for detailed understanding on the ground that it alone allows reliable measurement of growth parameters.

4.7   MBE Operating Procedures

This chapter is concerned with the early history of MBE film growth and the relevant surface science, the various aspects each being discussed in isolation—it therefore seems appropriate to provide a brief summary of 'real-life' growth procedures which depend upon these features, in an effort to stitch them all together into an integrated whole. We base our account on the procedures adopted in the Nottingham GEN-III system used to grow AlGaInAs structures and various other materials such as Group III nitrides and dilute magnetic semiconductors but these procedures are very similar to those employed over a wide spectrum of growth activities throughout the MBE world.

On taking delivery of a new MBE machine, it is standard practice to bake overnight while pumping until, after cooling, the system pressure reaches about 5×10^{-11} Torr but an interesting question arises with regard to the best practice to follow when it is subsequently necessary to let the system up to air (to recharge cells or rectify a vacuum leak). In the early days of MBE growth, it was usual to re-bake each time, and this appears to be standard practice in many laboratories even today. However, in Nottingham we have come to the conclusion that the risk of a vacuum leak developing as a result of thermal stress tilts the balance in favour of a procedure which omits any form of bakeout. The system is let up to an atmosphere of dry nitrogen and kept under a slight overpressure to prevent the ingress of water vapour or oxygen during the turnround, then pumped down again without further ado. In normal use, the system would be cooled overnight by flowing gaseous nitrogen through source and principal cryopanels, while the arsenic valve is closed and the Group III sources are kept at temperatures of 300 °C (In and Ga) and 800 °C (Al) to ensure the metals remain in the liquid state. Aluminium, in particular, if allowed to freeze, readily cracks the crucible and hence ruins the cell. The pressure, at this stage, will normally be mid 10^{-9} Torr and the quadrupole mass spectrometer will show it to be made up of hydrogen (mass 2), and residual arsenic (mass 75—a cracking fragment from As_4) but with small peaks at mass 28 (N_2 or CO) and 91 (AsO—a cracking fragment from As_3O_4). This small residual arsenic oxide peak is a by-product of water vapour (driven off by heating of the substrate), which oxidises the inevitable arsenic coating on the chamber walls. In a system not using nitrogen in the growth process, one can check for small vacuum leaks by comparing the signal at mass 14 (a cracking fragment from N_2) with that at mass 15 (CH_3). Provided the latter is significantly greater than the former, all is well—nitrogen can only appear from outside the vacuum system, methane only from within. When growing nitride semiconductors, such a test loses its validity and the operator must employ more subtle methods, based on long experience!

On flowing liquid nitrogen through the cryopanels, the pressure should drop to a little above 10^{-10} Torr and, if left long enough, would reach the gauge limit. However, in practice, the metal sources are now run up to their operating temperatures and their respective 'beam-equivalent-pressures' (BEPs) are measured by moving the ion gauge so as to intercept the beams. This terminology reflects the fact that the ion gauge measures pressure (strictly, density), rather than flux but it should be understood that the resulting values of BEP are merely convenient (but arbitrary) numbers which, though having little fundamental significance, may be relied upon in making comparison between 'today's' and 'yesterday's' growth conditions. At the same time, the power input to and temperature of each cell are recorded and similarly related to previous values. All being well, the growth sequence can begin, using, initially, a small (5×5 mm^2) RHEED sample to set the appropriate beam fluxes.

The arsenic valve in the cracker is opened to allow setting of the V–III BEP ratio. Assuming growth on a (001) surface, stoichiometry can be established by looking for the RHEED transition from a (2×4) pattern to a (3×1) at a temperature of about 500 °C (see Figure 4.13). In our system this occurs at a BEP ratio of 5:1 or 7:1 for As_2, and 10:1 or 14:1 for As_4, the uncertainty depending on exactly how the arsenic pressure is measured. On first opening the arsenic shutter, one observes a rapid initial rise due to the direct beam, followed by a much slower increase as the background arsenic pressure builds up. For AlGaAs growth, which takes place under arsenic-rich conditions, RHEED oscillations are measured for both gallium and aluminium, providing arrival-rate data for both species. On the other hand, if an alloy contains indium, it is necessary to replace the GaAs RHEED substrate with a small sample of InAs to measure the indium flux (taking care to correct for the difference of lattice parameter between GaAs and InAs). For growth of these alloys at modest temperatures, this data allows accurate estimates of growth rate and alloy composition but it is often necessary to grow at higher temperatures to obtain material of adequate quality and, in this case, Ga may evaporate (typically of the order of 0.2 MLs^{-1} ($\sim 1.3 \times 10^{14}$ atoms/cm^2/s) at T = 680–700 °C), thus reducing the growth rate and increasing the proportion of aluminium in the film.

The RHEED samples are now removed and stored in the buffer chamber while the 'real' substrate, which has previously been heated to remove adsorbed gas, is introduced into the growth chamber. Sample rotation is started, typically, at a rate of one rotation for every three monolayers, and surface oxide removed by heating slowly to 620 °C –630 °C. Above about 400 °C, the arsenic shutter is opened and heating continued until a (2×4) reconstruction is seen over the whole wafer, when the temperature is set to the required growth temperature and growth is commenced. Substrate temperatures are measured with a high degree of accuracy in our machine by using the 'k-space BandiT' (band gap measuring) system but one must be aware of possible inaccuracies in making comparison with similar results from other laboratories, where temperatures are measured with a thermocouple or pyrometer. At the end of the growth sequence, the sample is cooled to 400 °C and the arsenic shutter closed (unless a coating of amorphous arsenic is required for surface passivation).

In growing films containing phosphorus or antimony, pressures and temperatures would differ from those quoted above but the overall procedure would certainly be similar. However, when growing nitrides, RHEED oscillation monitoring is much more difficult due to the lack of lattice-matched substrates and to the fact that, in plasma-assisted MBE, growth proceeds under excess Group III conditions, so layer thickness is determined by the nitrogen arrival rate. Growth rates must therefore be determined ex situ using optical interference methods (which, in turn, require an accurate knowledge of the appropriate refractive indices), combined with BEP measurements, X-ray diffraction and photoluminescence.

In the case of production systems, which tend to involve very much longer growth runs, the usual procedure is to rely on test structures to optimise fluxes in terms of device performance, there being far less reliance on the sophistication of RHEED and BEP measurements. The essential point here is that, once conditions are right, nothing will be changed until the end of the run—the idea is to repeat the same sequence of layers as accurately as possible, in contrast to the research environment, where change is at the very heart of the business.

....

REFERENCES

Arthur, J R (1967) Vapor pressures and phase equilibria in the Ga–As system. J Phys Chem 28, 2257.

Arthur, J R (1968) Interaction of Ga and As2 molecular beams with GaAs surfaces. J Appl Phys 39, 4032.

Arthur, J R (1973) Adsorption of Zn on GaAs. Surf Sci 38, 394.

Arthur, J R (1974) Surface stoichiometry and structure of GaAs. Surf Sci 43, 449.

Arthur, J R and Brown, T R (1975) Velocity distributions of As_2 and As_4 scattered from GaAs. J Vac Sci Technol 12, 200.

Arthur, J R and LePore, J J (1969) GaAs, GaP, and $GaAs_xP_{1-x}$ epitaxial films grown by molecular beam deposition. J Vac Sci Technol 6, 545.

Biegelsen, D K, Bringans, R D, Northrup, J E and Swartz, L -E (1990b) Surface reconstructions of GaAs(100) observed by scanning tunneling microscopy. Phys Rev B 41, 5701.

Biegelsen, D K, Bringans, R D, Northrup, J E and Swartz, L -E (1990c) Reconstructions of GaAs (1 1 1) surfaces observed by scanning tunneling microscopy. Phys Rev Lett 65, 452.

Biegelsen, D K, Bringans, R D, Northrup, J E and Swartz, L -E (1990d) Early stages of growth of GaAs on Si observed by scanning tunneling microscopy. Appl Phys Let 57, 2419.

Biegelsen, D K, Bringans, R D, Northrup, J E and Swartz, L E (1990e) Materials Research Society symposium proceedings 198, 359.

Biegelsen, D K, Swartz, L -E and Bringans, R D (1990a) GaAs epitaxy and heteroepitaxy—a scanning tunnelling microscopy study. J Vac Sci Technol A 8, 280.

Binnig, G and Rohrer, H (1987) Scanning tunneling microscopy—from birth to adolescence. Rev Mod Phys 59, 615.

Binnig, G, Rohrer, H, Gerber, Ch and Weibel, E (1982) Surface studies by scanning tunneling microscopy. Phys Rev Lett 49, 57.

Braun, W (1999) 'Applied RHEED: Reflection High-Energy Electron Diffraction During Crystal Growth', Springer Tracts in Modern Physics 154, Springer-Verlag, Berlin.

Brennan, T M, Tsao, J Y and Hammons, B E (1992) Reactive sticking of As_4 during molecular beam homoepitaxy of GaAs, AlAs, and InAs. J Vac Sci Technol A 10, 33.

Briones, F, Golmayo, D, Gonzalez, L and de Miguel, J L (1985) SiH4 doping of MBE GaAs and Al_xGa_{1-x}As. J Vac Sci Technol B 3, 568.

Chadi, D J (1987) Atomic structure of GaAs(100)-(2 × 1) and (2 × 4) reconstructed surfaces. J Vac Sci Technol A 5, 834.

Chang, Chin-An, Ludeke, R, Chang, L L and Esaki, L (1977) Molecular-beam epitaxy (MBE) of $In_{1-x}Ga_x$As and $GaSb_{1-y}As$y. Appl Phys Lett 31, 759.

Chen, J F and Cho A Y (1991) J Appl Phys 70, 277.

Cheng, K Y, Cho, A L and Bonner, W A (1981) Characterization of Te-doped GaSb grown by molecular beam epitaxy using SnTe. J Appl Phys 52, 4672.

Cho, A Y (1969) Epitaxy by periodic annealing. Surf Sci 17, 494.

Cho, A Y (1970a) Epitaxial growth of gallium phosphide on cleaved and polished (111) calcium fluoride. J Appl Phys 41, 782.

Cho, A Y (1970b) Morphology of epitaxial growth of GaAs by a molecular beam method: the observation of surface structures. J Appl Phys 41, 2780.

Cho, A Y (1971a) GaAs epitaxy by a molecular beam method: observations of surface structure on the (001) face. J Appl Phys 42, 2074.

Cho, A Y (1971b) Film deposition by molecular-beam techniques. J Vac Sci Technol 8, S31.

Cho, A Y and Arthur J R (1975) Molecular beam epitaxy. Prog Sol St Chem 10, 157.

Cho, A Y, Dixon, R W, Casey, H C Jr and Hartman, R L (1976) Continuous room-temperature operation of GaAs-Al$_x$Ga$_{1-x}$As double-heterostructure lasers prepared by molecular-beam epitaxy. Appl Phys Lett 28, 501.

Cho, A Y and Panish M B (1972) Magnesium-doped GaAs and Al$_x$Ga$_{1-x}$As by molecular beam epitaxy. J Appl Phys 43, 5118.

Clarke, S and Vvedenski, D D (1987) Origin of reflection high-energy electron-diffraction intensity oscillations during molecular-beam epitaxy: a computational modeling approach. Phys Rev Lett 58, 2235.

Cohen, P I, Petrich, G S, Pukite, P R, Whaley, G J and Arrott, A S (1989) Birth-death models of epitaxy: I. Diffraction oscillations from low index surfaces. Surf Sci 216, 222.

Collins, D M (1979) The use of SnTe as the source of donor impurities in GaAs grown by molecular beam epitaxy. Appl Phys Lett 35, 67.

Covington, D W, Comas, J and Yu, P W (1980) Iron doping in gallium arsenide by molecular beam epitaxy. Appl Phys Lett 37, 1094.

Davies, G J, Andrews, D A and Heckingbottom, R (1981) Electrochemical sulfur doping of GaAs grown by molecular beam epitaxy. J Appl Phys 52, 7214.

Daweritz, L and Hey R (1990) Reconstruction and defect structure of vicinal GaAs(001) and Al$_x$Ga$_{1-x}$As(001) surfaces during MBE growth. Surf Sci 236, 15.

Dawson P, Duggan, G, Ralph, H I and Woodbridge, K (1985) Positions of the sub-band minima in GaAs–(AlGa)As quantum well heterostructures. Superlattice Microst 1, 231.

De-Sheng, J, Makita, Y, Ploog, K and Queisser, H J (1982) Electrical properties and photo-toluminescence of Te-doped GaAs grown by molecular beam epitaxy. J Appl Phys 53, 999.

Dobson, P J (1988) 'An Introduction to Reflection High Energy' in 'Surface and Interface Characterisation by Electron Optical Methods', NATO ASI Series B, Physics, Vol 191 (eds A Howie and U Valdre), Plenum Press, New York, p 159.

Dobson, P J, Foxon, C T and Neave, J H (1986) Patent Application US19830552653 19831117, GB198220033778 19821126.

Dobson, P J, Joyce, B A, Neave J H and Zhang, J (1987) Current understanding and applications of the RHEED intensity oscillation technique. J Cryst Growth 81, 1.

Evans, K R, Stutz, C E, Yu, P W and Wie, C R (1990) Mass-spectrometric determination of antimony incorporation during III–V molecular beam epitaxy. J Vac Sci Technol B 8, 271.

Farrell, H H, Harbison, J P and Peterson, L D (1987) Molecular-beam epitaxy growth mechanisms on GaAs(100) surfaces. J Vac Sci Technol B 5, 1482.

Farrow, R F C (1974) Growth of indium phosphide films from In and P$_2$ beams in ultra-high vacuum. J Phys D: Appl Phys 7, L121.

Fotino, M (1993) Tip sharpening by normal and reverse electrochemical etching. Rev Sci Instrum 64, 159.

Foxon, C T (1986) Kinetic processes in molecular beam epitaxy growth of III–V materials. J Vac Sci Technol B 4, 867.

Foxon, C T, Boudry, M R and Joyce, B A (1974) Evaluation of surface kinetic data by the transform analysis of modulated molecular beam measurements. Surf Sci 44, 69.

Foxon, C T, Harvey, J A and Joyce, B A (1973) The evaporation of GaAs under equilibrium and non-equilibrium conditions using a modulated beam technique. J Phys Chem Solids 34, 1693.

Foxon, C T and Joyce, B A (1975) Interaction kinetics of As_4 and Ga on {100} GaAs surfaces using a modulated molecular beam technique. Surf Sci 50, 434.

Foxon, C T and Joyce, B A (1977) Interaction kinetics of As_2 and Ga on {100} GaAs surfaces. Surf Sci 64, 293.

Foxon, C T and Joyce, B A (1978) Surface processes controlling the growth of $Ga_xIn_{1-x}As$ and $Ga_xIn_{1-x}P$ alloy films by MBE. J Cryst Growth 44, 75.

Foxon, C T and Joyce, B A (1981) 'Fundamental Aspects of Molecular Beam Epitaxy' in 'Current Topics in Materials Science', Vol 7 (ed E Kaldis), North Holland, Amsterdam, p 1.

Foxon, C T, Joyce, B A and Norris, M T (1980) Composition effects in the growth of $Ga(In)As_yP_{1-y}$ alloys by MBE. J Cryst Growth 49, 132.

Ghaisas, S V and Madhukar, A (1985) Monte-Carlo simulations of MBE growth of III–V semi-conductors: the growth kinetics, mechanism, and consequences for the dynamics of RHEED intensity. J Vac Sci Technol B 3, 540.

Gomez-Rodriguez, J M, Baro, A M, Silveira, J P, Vazquez, M, Gonzalez, Y and Briones, F (1990) Observation of AlGaAs/GaAs multiquantum well structure by scanning tunneling microscopy. Appl Phys Lett 56, 36.

Gotoh, H, Sasamoto, K, Kuroda, S, Yamamoto, T, Tamamura, K, Fukushima, M and Kimata, M (1981) S-doping of MBE-GaSb with H_2S gas. Jap J Appl Phys 20, L893.

Harris, J J, Joyce, B A and Dobson, P J (1981a) Oscillations in the surface structure of Sn-doped GaAs during growth by MBE. Surf Sci Lett 103 L90.

Harris, J J, Joyce, B A and Dobson, P J (1981b) RED intensity oscillations during MBE of GaAs: comment. Surf Sci 108, L444.

Heckingbottom, R, Davies, G J and Prior, K A (1983) Growth and doping of gallium arsenide using molecular beam epitaxy (MBE): thermodynamic and kinetic aspects. Surf Sci 132, 375.

Hurle, D T J (1999) A comprehensive thermodynamic analysis of native point defect and dopant solubilities in gallium arsenide. J Appl Phys 85, 6957.

Ichimiya, A and Cohen, P I (2004) 'Reflection High Energy Electron Diffraction', Cambridge University Press, Cambridge.

Ilegems, M (1977) Beryllium doping and diffusion in molecular-beam epitaxy of GaAs and $Al_xGa_{1-x}As$. J Appl Phys 48, 1278.

Ilegems, M, Dingle, R and Rupp, L W Jr (1975) Optical and electrical properties of Mn-doped GaAs grown by molecular-beam epitaxy. J Appl Phys 46, 3059.

Iliadis, A, Prior, K A, Stanley, C R, Martin, T and Davis, G J (1986) Influence of growth conditions on undoped and sulfur-doped InP grown by molecular-beam epitaxy. J Appl Phys 60, 213.

Joyce, B A, Dobson, P J, Neave, J H, Woodbridge, K, Zhang, J, Larsen, P K and Bolger, B (1986) RHEED studies of heterojunction and quantum well formation during MBE growth—from multiple scattering to band offsets. Surf Sci 168, 423.

Joyce, B A, Neave, J H, Dobson, P J and Larsen, P K (1984) RHEED studies of heterojunction and quantum well formation during MBE growth—from multiple scattering to band offsets. Phys Rev B 29, 814.

Kaspi, R and Loehr, J P (1997) Bilayer growth period oscillation of the Sb_2 reactivity during molecular beam epitaxy of AlSb (001). Appl Phys Lett 71, 3537.

Kawamura, Y, Asahi, H and Nagai, H (1983) Electrical and optical properties of Be-doped InP grown by molecular beam epitaxy. J Appl Phys 54, 841.

Krause, M, Stollenwerk, A, Awo-Affouda, C, Maclean, B and LaBella, V P (2005) Combined molecular beam epitaxy low temperature scanning tunneling microscopy system: enabling atomic scale characterization of semiconductor surfaces and interfaces. J Vac Sci Technol B 23, 1684.

Malik, R J, Nottenburg, R N, Schubert, E F, Walker, J F and Ryan, R W (1988) Carbon doping in molecular beam epitaxy of GaAs from a heated graphite filament. Appl Phys Lett 53, 2661.

Morkoç, H and Cho, A Y (1979) High-purity GaAs and Cr-doped GaAs epitaxial layers by MBE. J Appl Phys 50, 6413.

Naganuma, M and Takahashi, K (1975) Ionized Zn doping of GaAs molecular beam epitaxial films. Appl Phys Lett 27, 342.

Nakahara, H and Ichimiya, A (1991) Structural study of Si growth on a Si(111)7 × 7 surface. Surf Sci 241, 124.

Neave, J H, Dobson, P L, Zhang, J and Joyce, B A (1985) Reflection high-energy electron diffraction oscillations from vicinal surfaces—a new approach to surface diffusion measurements. Appl Phys Lett 47, 100.

Neave, J H and Joyce B A (1978) Structure and stoichiometry of {100} GaAs surfaces during molecular beam epitaxy. J Cryst Growth 44, 387.

Pashley, M D (1989) Electron counting model and its application to island structures on molecular-beam epitaxy grown GaAs(001) and ZnSe(001). Phys Rev B 40, 10481.

Pashley, M D, Haberen, K W, Friday, W, Woodall, J M and Kirchner, P D (1988a) Structure of GaAs(001) (2 × 4) – c(2 × 8) determined by scanning tunneling microscopy. Phys Rev Lett 60, 2176.

Pashley, M D, Haberern, K W and Gaines, J M (1991) Scanning tunneling microscopy comparison of GaAs(001) vicinal surfaces grown by molecular beam epitaxy. Appl Phys Lett 58, 406.

Pashley, M D, Haberen, K W and Woodall, J M (1988b) The (001) surface of molecular-beam epitaxially grown GaAs studied by scanning tunneling microscopy. J Vac Sci Technol B 6, 1468.

Pashley, D W, Neave, J H and Joyce, B A (2005) Long-range disorder effects on the GaAs(0 0 1) β2(2 × 4) surface. Surf Sci 582, 189.

Singh, J S and Bajaj, K K (1985) Theoretical studies of the intrinsic quality of GaAs/AlGaAs interfaces grown by MBE: role of kinetic processes. J Vac Sci Technol B 3, 520.

Singh, J S and Madhukar, A (1983) Far from equilibrium vapour phase growth of lattice matched III–V compound semiconductor interfaces: some basic concepts and monte-carlo computer simulations. J Vac Sci Technol B 1, 305.

Thomsen, M and Madhukar, A (1987) Computer simulations of the role of group V molecular reactions at steps during molecular beam epitaxial growth of III-V semiconductors. J Cryst Growth 80, 275.

Tsao, J Y, Brennan, T M and Hammons, B E (1988) Reflection mass spectrometry of As incorporation during GaAs molecular beam epitaxy. Appl Phys Let 53, 288.

Tsao, J Y, Brennan, T M and Hammons, B E (1991) Oscillatory As₄ surface reaction rates during molecular beam epitaxy of AlAs, GaAs and InAs. J Cryst Growth 111, 125.

Van Hove, J M, Lent, C S, Pukite, P R and Cohen, P I (1983) Damped oscillations in reflection high energy electron diffraction during GaAs MBE. J Vac Sci Technol B 1, 741.

Wang, W I, Mendez, E E, Kuan, T S and Esaki, L (1985) Crystal orientation dependence of silicon doping in molecular beam epitaxial AlGaAs/GaAs heterostructures. Appl Phys Lett 47, 826.

Weyers, M, Pütz, N, Heineke, H, Heyen, M, Lüth, H and Balk, P (1986) Intentional ρ-type doping by carbon in metalorganic MBE of GaAs. J Electron Matls 15, 57.

Wood, C E C (1978) "Surface exchange" doping of MBE GaAs from S and Se "captive sources". Appl Phys Lett 33, 770.

Wood, C E C (1981) RED intensity oscillations during MBE of GaAs. Surf Sci 108, 441.

Wood, C E C and Joyce, B A (1978) Tin-doping effects in GaAs films grown by molecular beam epitaxy. J Appl Phys 49, 4854.

Woodbridge, K, Gowers, J P and Joyce B A (1982) Structural properties and composition control of $GaAs_yP_{1-y}$ grown by MBE on VPE $GaAs_{0.63}P_{0.37}$ substrates. J Cryst Growth 60, 21.

Zhang, J, Neave, J H, Dobson, P J, Joyce, B A, (1987) Effects of diffraction conditions and processes on RHEED intensity oscillations during the MBE growth of GaAs. Appl Phys A 42, 317.

5

Modified Growth Techniques

5.1 Introduction

Our discussion so far has been based on the assumption that film growth is dependent on the continuous supply of appropriate elemental molecular beams impinging upon a suitable clean substrate, the typical example being that of GaAs growth from beams of Ga atoms and As_2 or As_4 molecules. However, during the associated studies of growth mechanisms, it was found helpful to employ discontinuous (i.e. pulsed) beams, which led to gradually improving understanding of the time dependence of the growth process at the atomic level, the discovery and detailed understanding of RHEED oscillations obviously playing a central role. In turn, this led to the idea that the quality of the growing film might be optimised by supplying appropriate atoms or molecules in carefully controlled pulses, thus allowing for rearrangement of surface atoms to occur during the interim. At the same time, experience in epitaxial growth from the vapour phase suggested that there might be some advantage in supplying the necessary atoms in the form of more complex molecules, such as arsine (AsH_3) instead of arsenic. An obvious advantage to this was the fact that the gas arsine could be injected from a source external to the vacuum system, thus avoiding the interruption of a growth run in order to replenish the arsenic cell. Indeed, we might recall that the original MBE work by Bruce Joyce at Plessey was concerned with growing silicon films from silane and its derivatives (see Section 2.3). Nor should we overlook the fact that, by the beginning of the 1980s, MOVPE growth had been employing metal-organic compounds for a long time, with obvious success, which meant that suitable high-purity sources were readily available (see Section 1.2). It would be surprising, therefore, if ambitious young scientists had not experimented with such modifications, if only because 'they were there'. On the other hand, there were good reasons to experiment, based on the demands of the ever-expanding electronics and optoelectronics industries (which, we must never fail to remember, were providing the driving force and much of the funding in support of such innovative thinking).

As we noted briefly in Chapter 1 (Section 1.4), the rapid development of fibre-optic communications during the late 1970s led to a demand for lasers operating in the wavelength region between 1.3 μm and 1.6 μm and it soon became clear that the InGaAsP material system was well suited to satisfying it. There was also a continuing interest in the growth of high-quality GaP for light-emitting devices and, somewhat later, in the alloy AlGaInP specifically for high-efficiency red and yellow LEDs. All these compounds contained phosphorus, which was far from easy to handle in conventional MBE technology, so there was a clear requirement for improved techniques to produce suitable material. MOVPE took the lead here but, as we shall see, gas source MBE could make a worthwhile contribution. At much the same time (i.e. the 1980s) the explosion of interest in low-dimensional structures presented material scientists with a challenge to which, in the event, they would rise with aplomb (see Chapter 6). The early work on superlattices which we described in Chapter 2 (Section 2.6) had to be refined and amplified to satisfy an ever-widening demand for complex structures grown with atomic precision and with interfaces between different materials which were flat to monolayer accuracy. Such exciting developments obliged MBE growers (who, in this instance certainly did take the lead) to polish and refine their skills to the nth degree. Control of the deposition process at the monolayer level was to challenge both skill and understanding as it had never been challenged before and modified growth techniques were an essential component of the MBE armoury brought to bear.

This short chapter will attempt to summarise the various 'alternative' MBE techniques which materialised during the 1980s, providing an outline of the motivation and principal objectives and achievements. Here we shall of necessity concentrate on the structural aspects of growth, thus rounding off the first half of the book, which aims to cover techniques and growth processes. In contrast, the remainder of the book will deal with specific applications of MBE, emphasising the relevant physics and engineering aspects of each topic, while illustrating the important role of MBE in furthering these practical developments. We divide the chapter into two parts, the first dealing with what one might call 'mechanical' variants of MBE, that is, 'modulated beam techniques' (Section 5.2) and 'use of surfactants' (Section 5.3), while the second will be concerned with 'chemical' modifications, under the heading 'Gas Source MBE' (Section 5.4). Both categories of MBE variants emerged in the early 1980s, the first in response to the demands of a major international initiative in LDS (low-dimensional structures), the second, as we have already hinted, in the equally important development of the Group III phosphides. But, before plunging into detail, it would be well to emphasise the important reservation made above. In several cases a proper understanding of the motivation for some specific initiative in MBE growth implies an understanding of subtle aspects of, for example, the physics of quantum wells and their optical properties, so a full explanation must await the presentation of such physical subtleties in our next chapter. This chapter will concentrate solely on the structural aspects which, we can, of course understand well enough on the basis of our current knowledge.

5.2   Modulated Beam Techniques

The starting point for any account of modulated beam MBE must be that of RHEED oscillations and their interpretation in terms of step-edge density. As a marker, we might refer to the paper by Neave et al. (1983), which sets out the state-of-the-art understanding of RHEED oscillations at the beginning of the 1980s. Let us try to list the essential features, taking as our example the case of the GaAs (001)–(2 × 4) reconstruction, characteristic of typical GaAs growth conditions. First, we should emphasise that RHEED oscillations are a quite general phenomenon, not peculiar to any specific material or dopant—the original impression that they might be specific to tin-doped GaAs was well and truly outmoded by the beginning of the 1980s. It was also well established that the period of oscillation corresponded precisely to the time taken to deposit a single monolayer of material—in the case of GaAs, one monolayer of Ga plus one of As (in the case of silicon or germanium, a single layer of atoms). The oscillations begin immediately after initiating growth (by turning on the Ga beam—assuming the As beam intensity to be held constant) and they are almost always fairly strongly damped, with a time constant of the order of ten periods (see Figure 5.1). (In fact, the damping is not always strictly exponential but we may ignore the fine details in pursuit of our present purpose.) Whenever the Ga beam is shut off, even for times as short as one second, then restarted, the oscillations recommence at an intensity much greater than that measured just prior to switch-off. Oscillatory behaviour follows the same pattern for the alloy AlGaAs and, interestingly, if an Al beam is switched on after growing GaAs for some time,

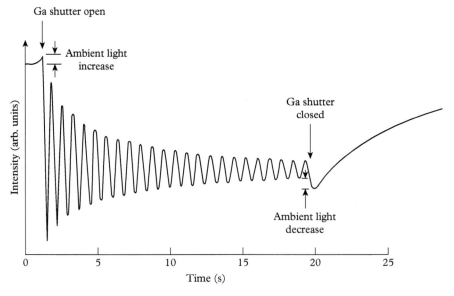

Figure 5.1 *Intensity oscillations of the specular RHEED beam from a GaAs (001) 2 × 4 reconstructed surface, [110] azimuth, showing the typical decay in amplitude as the surface progressively roughens. Note the recovery following the termination of growth but with the arsenic beam still switched on. The period corresponds to the growth of a single monolayer of GaAs (Ga + As). The inflections at beginning and end of growth result from changes in light level at the fluorescent screen when shutters are opened and closed. (From Neave et al. 1983.) Reprinted with permission from Springer.*

oscillations reappear with much enhanced amplitude. Similar effects can be seen for all reconstructions and for all diffraction features, though the specular 00 beam shows the greatest oscillation amplitude and is therefore chosen in most cases for experimental convenience. However, there are also differences in regard to the orientation of the RHEED beam with respect to the crystallographic axes, the amplitude being greater when aligned parallel to the [110] (2-fold) axis, rather than the [-110] (4-fold) axis. The amplitude of oscillation is temperature dependent, being maximum in the temperature range 540 °C–590 °C, which corresponds to that chosen for much practical MBE growth. (In passing, we should note that different RHEED features show different phases, the initial effect for the specular and half-integral streaks being a decrease in amplitude, while the integral diffraction streaks show an initial increase. This is yet another feature that we shall ignore in discussing modulated beam MBE methods!) On cessation of growth, the RHEED intensity (for the 00 specular feature) shows a gradual increase, with an initial rapid response, followed by one much slower. The fast time constant is seen to be temperature dependent, with an activation energy (over the temperature range 500 °C–600 °C) $E_A = 2.3 \pm 0.2\,\text{eV}$, which suggests that it involves bond-breaking (the cohesive energy of GaAs being approximately 1.7 eV) as well as surface diffusion. Results for the slower time constant are much less coherent, while the damping time constant is (perhaps surprisingly?) not temperature dependent at all. So much for the experimental data—what do they all mean?

It was generally accepted that, as we saw in Chapter 4, growth takes place in terms of a competition between step-edge propagation and three-dimensional nucleation (see, in particular, Van Hove et al. 1983). If we assume it to commence from a perfectly flat surface (maximum reflectivity), the initial nucleation results in surface roughening, which tends to a maximum at half-monolayer coverage (minimum reflectivity), followed by a recovery as the resulting holes are filled in. Damping occurs by virtue of the fact that a second layer starts to grow before the first layer is completed (see Figure 4.14) and this overall roughening effect proceeds until a steady state is reached, characterised by a constant density of step-edges, at which point the oscillations have died away completely. The observation that shutting off the Ga beam leads to recovery of the oscillation amplitude (Figure 5.1) implies that, during a period of non-growth, the surface becomes smoother, a process which involves atoms in an outer layer breaking their bonds and diffusing to a step edge, where they are captured again. Such a process clearly involves a reduction in overall step density and results in an increase in surface reflectivity for electrons in the RHEED beam. It also explains the thermally activated time constant associated with the recovery process. (The final, slow recovery is believed to be associated with surface disorder.) The dependence on electron beam orientation can be understood in terms of the STM measurements of surface reconstruction, shown in Figure 4.16, the increased tendency for growth along the [-110] direction resulting in a greater step density seen by the beam oriented along [110], as compared with the [-110].

As outlined in the introduction to this chapter, it was the appearance of low-dimensional structures on the Group III-V scene towards the end of the 1970s which first called for atomically smooth interfaces between, in the first instance, AlGaAs and GaAs layers, and, later, other pairs of compounds such as InGaAs and InP; and this, in turn, demanded of the crystal grower an ability to prepare films with atomically flat surfaces. It occurred to various MBE groups during the 1980s that this might be facilitated by using 'growth interrupts' to smooth surfaces, as evidenced by the then current understanding of RHEED oscillations—in particular the recovery in intensity following shutting off the Ga beam, when growing GaAs. If the observed recovery of RHEED intensity to a value appropriate to that associated with an initially flat surface could occur in a time of order one to two minutes, it should be possible, for example, to obtain a flat interface by stopping GaAs growth momentarily before initiating the growth of an AlGaAs layer. The first explicit use of such procedures was reported by Sakaki et al. (1985) at the University of Tokyo, followed almost immediately by another Japanese group at the Optoelectronics Joint Research Laboratory in Kawasaki (Fukunaga et al. 1985). Both groups were interested in photoluminescence resulting from exciton recombination in AlGaAs/GaAs quantum wells and monitored the quality of their structures in terms of the measured line positions and (in particular) line widths. Comparison of results from quantum wells grown with and without interrupts clearly demonstrated the value of interrupting growth to achieve smoother interfaces. Specifically, the use of interrupts, typically of 1–2 min-duration, resulted in atomically flat regions in excess of the exciton diameter (~25 nm). (Much shorter interrupts, like 5 s, were shown to be entirely insufficient.)

This preliminary data was published in two independent bursts of enthusiasm, in the form of letters; but the Tokyo group then took a little time to reflect on the matter before publishing longer accounts (see e.g. Tanaka and Sakaki 1987) dealing with quantum well luminescence in greater detail. Mature reflection brought to the fore the distinction between 'top' and 'bottom' interfaces (often referred to as 'normal' and 'inverted' interfaces), the important difference being the presence of aluminium in the barrier layers. Aluminium atoms are more strongly bound to arsenic than are gallium atoms, implying less surface mobility and consequently less smoothing during the interrupt periods. In the case of AlAs/GaAs wells, it is clear that smoothing of the bottom interface must be seriously inhibited, resulting in an atomically rough interface on a scale much smaller than the exciton diameter. Unexpectedly, perhaps, this is seen by the exciton as a pseudo-flat interface, implying a luminescence spectrum largely determined by the top interface. In the case of AlGaAs/GaAs wells, one must consider the aluminium content of the barriers—Tanaka and Sakaki distinguished between a 'high' Al content, when $x > 0.5$, and a 'low' content, when $x < 0.3$—thus allowing for the smoothing effect of the more mobile Ga atoms. The overall situation is summarised (following their paper) in Figure 5.2, which illustrates five distinct pairings of different interfaces. However, a full understanding of the PL line-width involves a knowledge of the number of different levels in a direction normal to the surface, as shown by Zhang et al. (1990).

We shall leave further discussion of this work until the next chapter, when we shall be better equipped to understand the background but it is important to recognise its significance for the world of crystal growth—MBE could be seen to offer a degree of control scarcely dreamt of by other practitioners. It also stimulated a degree of thinking about the nature of an 'atomically flat' surface. We might emphasise two important features. In the first instance, no real surface can be perfectly flat—the only meaningful question

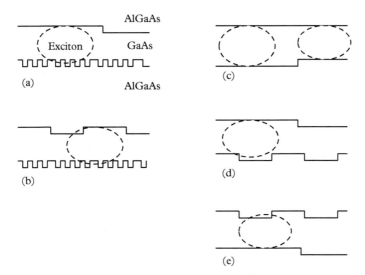

Figure 5.2 *Atomic scale models of various types of interfaces between GaAs quantum wells and AlGaAs barriers grown with and without growth interrupts, compared with the size of the exciton orbit (dashed lines). Note that the quality of the interfaces depends not only on the use of interrupts but also on the Al content of the barriers, because Al atoms are less mobile on the surface than those of Ga. (Following Tanaka and Sakaki 1987).*

to be asked is 'Over what distance does it remain flat?' and this must then be compared with the dimension of the measurement probe (for instance, the diameter of an exciton or the size of the X-ray beam used to obtain an XRD spectrum). It is also worth bearing in mind that a misorientation of the substrate by just 1 degree implies a maximum terrace dimension of roughly 15 nm. To obtain surfaces flat to dimensions greater than the 25 nm of a typical exciton demands accuracy of substrate preparation to considerably better than half a degree. Second, the ultimate quality of an interface between two lattice-matched materials is inevitably limited by interdiffusion of the relevant species (in this case aluminium and gallium) and in this regard the relatively low temperatures appropriate to MBE growth clearly represent a significant advantage.

So much for the use of growth interrupts but this does not exhaust the possible uses of the RHEED oscillation phenomenon in controlling growth surfaces. One should not, for example, be surprised to encounter its use as an aid to the growth of short-period superlattices, where the accurate and reproducible generation of large numbers of identical mono- or bilayer structures obviously depends on the crystal grower's ability to work with atomically flat interfaces. As we saw in Chapter 2, the interest in superlattices in the 1970s served as an important stimulus to the development of MBE growth of III-V compounds, though early attempts relied simply on controlling shutter opening and closing times rather than on any intrinsic measurement of layer thicknesses.

The appreciation that layer thickness and RHEED oscillation frequency were fundamentally linked therefore represented an important advance in the accurate control of superlattice periodicity, it merely (!) being necessary to activate source shutters at the appropriate phase of the oscillation cycle. Given that the oscillation maxima correspond to optimum flatness, this was clearly the moment to choose in switching between sources, and of even greater importance was the fact that it provided a precise measure of film *thickness*. Thus, no matter that switching a shutter had the inevitable result of perturbing cell temperature and therefore beam flux, relying on RHEED oscillations neatly eliminated such perturbations and provided an altogether tidier approach to superlattice preparation. What is more, it was a method singularly available to the MBE grower. However, life is rarely ideal and the crystal grower had to accept *some* restriction on the capabilities even of MBE—in order to record the necessary RHEED intensities the substrate had to remain stationary, the customary rotation essential to uniform film deposition having to be eschewed. In other words, the precise atomic accuracy of film thickness was available only in the location where the RHEED beam impinged on the substrate.

Once again, it was a Japanese initiative that sparked the practical development of these ideas and, again, remarkably enough, two separate groups published their results almost simultaneously. First, by a short head, were Sano et al. (1984), of Kwansei-Gakuin University, who reported the growth of both mono- and bilayer AlAs/GaAs superlattices; they were followed, a month later, by Sakamoto et al. (1984), jointly of the Electrotechnical Laboratory and the Japan Aviation Electronics Industry Co Ltd, who also grew $(AlAs)_2(GaAs)_2$ superlattices. Sano et al. made use of the specular beam from the (001)–(2×4) reconstructed surface to monitor the progress of layer growth, their measured RHEED intensities for the growth of a $(GaAs)_1(AlAs)_1$ superlattice being shown as a function of time in Figure 5.3. On opening the Ga shutter, the RHEED intensity first decreased, then increased again, the Ga shutter being closed at the point of maximum intensity (i.e. after growth of a monolayer of GaAs). The next step (of approximately 3 s)

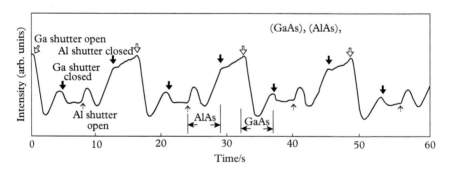

Figure 5.3 *RHEED intensity oscillations of the specular beam from the (001) (2 × 4) reconstructed surface during growth of a single monolayer (GaAs)(AlAs) superlattice. Note the 3 s 'smoothing' period between each monolayer of growth. (From Sano et al. 1984.) Copyright 1984, The Japan Society of Applied Physics.*

involved the application of a flux of As only, with the intention of further smoothing the GaAs layer, followed by an AlAs monolayer and a second smoothing period. This whole sequence was then repeated some 600 times to produce the superlattice, the amplitudes of the various RHEED peaks repeating in uniform fashion throughout the growth (as we have already indicated, RHEED intensity recovers each time the appropriate molecular beam is switched on, there being no damping in this mode of observation). Sano et al. used X-ray diffraction to monitor the effectiveness of their growth method, observing sharp diffraction features appropriate to the relevant superlattice periodicities. They also found 'fairly good' agreement between measured and calculated intensities for these various features. Clearly, the use of RHEED oscillations allowed good control of these short-period superlattices, though one is inevitably left wondering exactly how they compared with earlier structures, grown by less sophisticated means.

Sakamoto et al. preferred to study their RHEED oscillations with the electron beam aligned along a (100) azimuth, rather than the more usual (110) direction, a procedure which allowed them to count over 400 oscillations during growth of an AlGaAs film and thereby providing much improved accuracy in the measured oscillation frequency (f). They estimated the aluminium fraction x in two distinct ways, from measurement of $f(AlAs)$, $f(GaAs)$ and $f(Al_xGa_{1-x}As)$, respectively. Thus,

$$x = f(AlAs)/[f(AlAs) + f(GaAs)] \qquad (5.1)$$

and

$$x = [f(AlGaAs) - f(GaAs)]/f(AlGaAs) \qquad (5.2)$$

obtaining good agreement between the two methods over the whole range of x-values between 0 and 1. They then proceeded to grow a 699-period bilayer AlAs/GaAs superlattice using 'phase-locked epitaxy' (PLE) in precisely the same manner as employed by Sano et al. As a method of monitoring the quality of the resulting structure, they compared Raman spectra from this superlattice and a 50% alloy film grown under similar conditions. As had earlier been demonstrated by Merlin et al. (1980), the superlattice showed splitting of the resonance lines, resulting from layering and strain, whereas the spectrum from the alloy film showed no such effect.

Before moving on, a brief comment is probably in order concerning the difference in damping of the RHEED oscillations reported by Sakamoto et al. for the (100) azimuth, compared with that usually observed for the (110) azimuth. Their explanation is based on observation of various diffraction features, by scanning an optical fibre over the RHEED pattern. In the (110) direction they showed that the intensity from the specular beam position was actually made up of two contributions: one from the specular beam itself and another from an integral order diffraction streak, which turned out to have nearly opposite phase. The rapid damping was the result, therefore, of mutual interference between these two contributions, whereas with the electron beam aligned along a (100) azimuth, the phase difference between them was very much smaller and damping was correspondingly less. Indeed, in a later paper (Sakamoto et al. 1985) the same group counted as many as 700 distinct oscillations during the growth of an $Al_{0.41}Ga_{0.59}As$ film.

(In passing, we may note that a small change in the angle of incidence of the electron beam is also sufficient to remove the coincidence between the two contributions and thus avoid their mutual interference.)

The challenge of growing accurately reproducible superlattice structures with molecular thicknesses is clearly a considerable one but there are other multilayer structures of importance within the sphere of optoelectronics. One such is the Bragg reflector, a vital component of many semiconductor lasers, particularly the surface-emitting laser, which plays a major role in fibre-optic communication systems. A conventional semiconductor laser might employ an optical cavity typically of some 500 μm in length, consistent with high round-trip gain, and, in consequence demanding only modest reflectivity from its end mirrors. By contrast, the surface-emitting version is limited in cavity length to typical semiconductor layer thicknesses—say 1 μm—and this makes much more serious demands on mirror reflectivity—say R ~ 0.9, which can be obtained with thin metallic films. However, in many instances, practical devices came to rely on the gain provided by a single quantum well of thickness 10 nm and this called for values of R ~ 0.999—an altogether different order of difficulty, obtainable only with multilayer Bragg stacks of some 30 AlAs/GaAs pairs. Each individual layer had to be a quarter wavelength in thickness, typically about 80 nm, or some 300 monolayers. The crystal growth problem was hardly one of controlling thickness to a fraction of a monolayer but advantages of 'producibility' (the ease and reproducibility of practical device production) were well appreciated by MBE growers involved in developing relevant methods. In particular, the use of PLE, with its ease of computer-control, offered obvious advantages. Typical of the PLE approach to VCSEL growth was that described by Walker et al. (1990 and 1991). They were concerned to establish a totally reliable method of producing an InGaAs QW laser emitting at wavelength close to 900 nm and using AlGaAs/GaAs Bragg reflectors. Because of the difficulty in maintaining surface flatness when growing high-aluminium alloys, they decided to simulate the alloy with an appropriate short-period superlattice—in one example the $Al_{0.67}Ga_{0.33}As$ alloy was actually grown as an $(AlAs)_6(GaAs)_3$ superlattice—and the aluminium mole fraction was graded in the region of the AlGaAs/GaAs interface so as to reduce the series resistance associated with an abrupt step in the appropriate band edge. They also took advantage of the opportunity to put dopant atoms only in the GaAs regions where they could be expected to give low resistivity material. While this is not the place to become involved in any further complexities of VCSEL design, it is worth recognising that the use of PLE MBE growth was seen as well worth the effort required to achieve a high degree of structural control, resulting in much superior reproducibility of device performance over a wafer and between wafers. The fact that the substrate could not be rotated was a distinct disadvantage, leading to an approximately 5% per centimetre variation in layer thickness across the wafer but, in a later letter, Walker et al. (1993) described a simple method of overcoming even this. Following the growth of each pair of Bragg layers, growth was stopped for a matter of a few seconds while the substrate was rotated by 180°, then the sequence repeated until all the layers were complete. It resulted in alternate layers being slightly thinner or thicker than average but the long range periodicity was reliably maintained and the lasing wavelength remained constant across a wafer within about 1.5%.

A particularly interesting approach to modulated beam MBE was devised by Horikoshi et al. (1986) of NTT in Tokyo. It was based on the idea that layer-by-layer growth depended on the surface migration of the Group III elements and the possibility of controlling it by modulating the Group V flux. Many years earlier Arthur (1966) had observed that gallium atoms were very much more mobile on a GaAs surface in the absence of an arsenic flux than under an arsenic-rich atmosphere, the likely explanation being that in the latter case gallium and arsenic atoms combined to form less mobile GaAs molecules. Horikoshi's innovative idea, which he referred to as 'migration-enhanced epitaxy' (MEE), made use of this effect to change the balance between step-edge growth and island nucleation. He argued that, if the arsenic flux was turned off during the supply of gallium, Ga atoms would migrate rapidly to steps and growth would proceed largely by edge propagation. Only when a complete coverage of gallium had been achieved would the arsenic beam be turned on (and the Ga beam off), to convert all those gallium atoms into molecules; then the two stages could be repeated ad infinitum. Two features were of the essence: the gallium flux had to be chosen such that a single monolayer of gallium was deposited during the 'Ga-on' pulse and the timing of the pulses was monitored by synchronising them with RHEED oscillations. Noteworthy was the fact that, in contrast to the case of conventional MBE, these oscillations showed no sign of damping when measured under MEE conditions and they also persisted down to temperatures as low as 300 °C, further evidence that the enhanced Ga surface mobility favoured step-edge growth over surface nucleation. In a later paper (Yamaguchi and Horikoshi 1989), the same group pursued these ideas by studying MEE growth on vicinal surfaces (misorientated by 2° from the (001) plane) where, again, they were able to demonstrate growth by step-edge propagation, indicating a gallium atom surface diffusion length of at least 80 Å at 560 °C. Under normal MBE conditions, when the arsenic beam was kept permanently 'on' the diffusion length was clearly much shorter, giving rise to modulation of the RHEED oscillations, consistent with competition between step-edge and nucleation processes.

So much for the principle—what about its practical virtues? Of particular significance was the possibility of growing high-quality GaAs films at much lower temperatures than had been possible previously, thus facilitating the growth of films on top of sensitive optoelectronic device structures and the growth of films with extremely sharp doping profiles. This is consistent with the known activation energy for Ga surface migration (1.3 eV) which suggests that at low growth temperatures Ga atoms would be totally immobile in the presence of an arsenic flux, while still being adequately mobile in MEE conditions. (The same is even more true of aluminium, which has a still greater binding energy with arsenic and therefore an even larger activation energy for surface migration.) In the event this group succeeded in growing GaAs, showing excellent photoluminescence properties, at temperatures as low as 200 °C (Horikoshi et al. 1986) and high-quality quantum well structures at 300 °C (Horikoshi et al. 1987). Single quantum well $Al_{0.5}Ga_{0.5}As$/GaAs structures with well widths of 1.7, 3.3 and 5.0 nm showed low-temperature luminescence line widths closely similar to and intensities some ten times greater than those obtained from conventional MBE structures. As an example of a practical application, we might also note the use of MEE to grow a single quantum well

AlGaAs/GaAs laser diode at 350 °C (Asai et al. 1988), though, in the interest of complete honesty, we should note that only the quantum well region was grown by MEE—the rest of the structure was grown by conventional MBE in the interest of saving the source shutters from overwork! Indeed, one might reflect on the fact that growing one micron of GaAs by MEE involves something like 8000 movements of each shutter—life rarely yields favours at no cost. On the other hand, the ability to grow high-quality films at temperatures as low as 300 °C was surely worth a certain amount of inconvenience! There was no record of even moderately satisfactory GaAs films being grown by conventional MBE below 400 °C.

While it is apparent that the development of these various modulation techniques had a distinctly Japanese emphasis, we should not overlook the European contribution from Fernando Briones and his collaborators in Madrid. They, too, took up the challenge of developing a low-temperature growth capability by controlling both Group III and Group V beams in synchronisation with RHEED oscillations. In 1987 they demonstrated the use of phase-locked epitaxy to grow, at 400 °C, AlGaAs/GaAs quantum well structures in which the confining layers were short-period superlattices (Briones et al. 1987). For example, the quantum well consisted of 20 monolayers of GaAs, while the 0.1 μm thick confining layers were made from $(AlAs)_5(GaAs)_7$ superlattices. Growth rates were very practical ones, 1 ML/s for GaAs and 0.6 ML/s for AlAs. Two distinct methods were employed: in the first case, only the As_4 beam was pulsed in sympathy with RHEED oscillations, the Al and Ga beams being run continuously, as and when required; in the second modification all 3 beams were modulated, the superlattices being constructed from sequences of 7 pulses of Ga together with 7 synchronised pulses of As_4, followed by 5 pulses of Al together with 5 of As_4 and the quantum well from 20 synchronised Ga and As_4 pulses. (RHEED oscillations and shutter operations are shown in Figure 5.4 for the second scheme.) The quality of the resulting structures was monitored by low-temperature photoluminescence, strong and narrow ($\Delta E = 7$ meV) PL lines being obtained from the QWs in both cases, perfectly in line with similar measurements on QW structures grown at 'conventional' temperatures. Perhaps the most interesting and practically important discovery made by the Briones group was their observation that it was not necessary to maintain a constant computer-controlled synchronisation between RHEED oscillations and growth conditions throughout the whole growth sequence. Once steady-state growth conditions had been established (after as little as ten periods), the required sequences could be stored in the computer and used to control the growth process automatically, without further reference to RHEED. Apparently, the phase relationship between RHEED oscillations and shutter operation was not sensitive either to long-term drifts in beam fluxes or to small flux gradients across the substrate area, so the RHEED measurement could be switched off once steady-state conditions had been established, and the substrate could be rotated in the standard manner to smooth out non-uniformities over the slice. Together with the practical growth rates, this feature offered a considerable appeal to anyone wishing to grow complex structures at relatively low substrate temperatures.

In a later paper (Briones et al. 1989), the Madrid group expanded on their earlier work and coined the names 'atomic layer MBE' (ALMBE) for the technique in

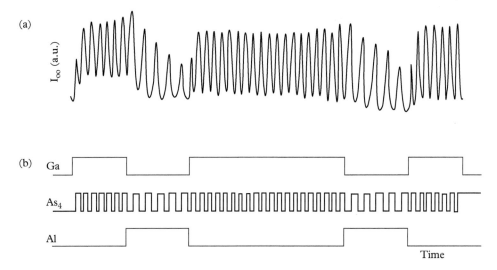

Figure 5.4 *RHEED oscillations of the specular beam and shutter operation of the Ga, Al and As beams during growth of a 20 ML GaAs quantum well confined by an $(AlAs)_5(GaAs)_7$ superlattice. The growth temperature was 400 °C. In (a) only the As beam is pulsed, while in (b) all the beams are pulsed in synchronism with the RHEED oscillations are shown In (a), while in (b) is shown the synchronised pulsing of all three beams. (From Briones et al. 1987.) Copyright 1987, The Japan Society of Applied Physics.*

which both Group V and Group III beams were pulsed, and 'nucleation-enhance MBE' (NEMBE) for the technique in which only the Group V beams were pulsed. Of the various growth methods explored by them, they considered NEMBE to be the most practical. In the interest of clarity, let us recapitulate—gallium was supplied continuously to the surface, while arsenic was supplied in controlled pulses, timed to coincide with the peak of the RHEED oscillation. These pulses were not generated by a standard shutter but by a specially designed quartz valve which minimised mechanical wear. The object was to supply a single monolayer of arsenic in the pulse, which was then converted to a monolayer of GaAs by the impinging gallium, growth being by nucleation, rather than by step-edge propagation. The process works at low temperatures by virtue of the fact that it does *not* depend on (thermally activated) surface migration. They used it to grow a range of different structures at 350 °C, including heavily silicon-doped GaAs, GaAs quantum wells, AlAs/GaAs superlattices and modulation-doped superlattices, in which the silicon dopant atoms were supplied only to the two central monolayers of GaAs. They also demonstrated the capability for growing highly mismatched layers, such as InAs on GaAs substrates in which the InAs acquired an excellent surface morphology after only a few monolayers. Finally, they explored its application to the growth of alloys containing two Group V elements, such as As and P. In standard MBE, the competition between As and P leads to structures containing much larger proportions of arsenic than suggested by the relative beam intensities, making it very difficult to control the As:P ratio. Briones et al. overcame this difficulty by supplying the Group V elements as single monolayers, thus obviating the problem of competition between the two species. Thus a $GaAs_{0.67}P_{0.33}$ alloy could be simulated in the shape of a $(GaAs)_6(GaP)_3$ superlattice with a high degree of compositional control. Finally, as we have previously intimated, the advantages of such low temperature growth are various: interdiffusion of (for example) Al and Ga at interfaces is minimised, doping profiles with improved sharpness can be obtained, surfaces can be prepared free of oval defects, thermal strain effects in mismatched structures are reduced, while growth of new device structures on top of preprocessed circuits can be contemplated without damaging the underlying features.

At this point the reader may well feel the need for a brief summary of the apparently random collection of new growth techniques (not to mention the corresponding new acronyms) presented in this, perhaps confusing, section. We must surely, therefore, oblige. It seems possible to distinguish five different modifications of the basic MBE process, involving the use of (1) growth interrupts, (2) phase-locked epitaxy (PLE), (3) migration-enhanced epitaxy (MEE), (4) nucleation-enhanced MBE (NEMBE) and (5) atomic layer MBE (ALMBE). How are we to rationalise them all? First, it might be well to re-emphasise the fact that they all emerged in response to the burgeoning demands of the low-dimensional structures, or LDS initiative which was born towards the end of the 1970s, grew to maturity during the 1980s and which forms the subject of our next chapter. The essential requirement was for semiconductor heterostructures demanding control of materials on the atomic scale. It was necessary to form interfaces between different compounds which were flat on the atomic scale and to place dopant atoms in position with similar accuracy. There can be little doubt that MBE was remarkably well suited to satisfying such demands, based, as it was, on one form or other of monolayer-by-monolayer growth. What these various modifications purported to do was to provide superior control, tightening up the process so as to maintain accurate monolayer accuracy over an unlimited sequence of monolayers. They also provided useful information concerning the fine detail of growth mechanisms, and the fact that several of them also offered high-quality material at significantly lower growth temperatures was an important bonus. Let us now attempt to point out the essential features of each one.

The idea behind the use of growth interrupts was one of allowing an initially rough layer to smooth by surface migration of atoms on the 'second layer', causing them to fall back into the 'first layer'. It was clearly temperature dependent and therefore relevant only to growth at 'conventional' temperatures in the region of 550 °C–600 °C. Phase-locked epitaxy made use of the fact that the period of RHEED oscillations provided a precise measure of the deposition of a single monolayer which could be used to control the growth of mono- and bilayer AlAs/GaAs superlattices. Arsenic was supplied constantly, while the aluminium and gallium beams were switched at the peak of the RHEED intensity waveform. Migration-enhanced epitaxy depended on the fact that Ga atoms diffuse much more rapidly on a GaAs surface in the absence of arsenic, so gallium was supplied while the As_4 beam was turned off; then arsenic was supplied to convert the Ga layer into GaAs. It required accurate monitoring of the gallium supply so as to deposit a precise monolayer, thus avoiding formation of gallium droplets on the surface. Because of the enhanced migration rate, MEE allowed high-quality GaAs growth at much reduced temperatures—typically 350 °C–400 °C. While MEE was designed to encourage step-edge growth, nucleation-enhanced epitaxy, on the other hand, aimed to foster nucleation of islands. A controlled pulse of arsenic formed an As monolayer on the surface which was then converted into GaAs by the Ga beam. Gallium droplets were prevented by the arrival of the next accurately timed layer of arsenic. 'Second layer' nucleation was prevented by the fact that Ga atoms stick much more strongly to As than to Ga. Note that, in this case, it worked at low temperatures because it did *not* depend at all on surface migration. Atomic layer MBE differed only slightly from NEMBE. Again, it depended on a pulse of arsenic to form a surface monolayer but, in this case, the gallium beam was pulsed as well, rather than being supplied continuously. Figure 5.5 provides a summary of the shutter sequences appropriate to these various methods.

(At this point, we feel obliged to add an important rider to the above account in the shape of a comment by our colleague Bruce Joyce, who points out that RHEED oscillations are influenced not only by morphological changes on the growth surface but also by changes in surface reconstruction. There is no clear evidence which of these factors is dominant in any particular case so, though there can be no doubt that these modified growth techniques 'work' in practice, we should be a little cautious in accepting the veracity of the modelling used to explain their function.)

Finally, we should make mention of yet another variant of MBE which has come to be known as 'droplet epitaxy'. It was apparently 'invented' at the National Research

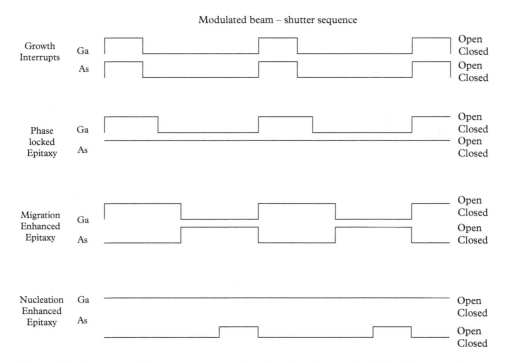

Figure 5.5 *Summary of shutter sequences used in the various forms of modulated beam growth.*

Institute for Metals in Tsukuba (Chikyow and Koguchi 1990) and later applied by the same group to growing GaAs quantum dots (Koguchi et al. 1993) and, very recently, quantum rings (Zhou et al. 2013). It was developed as a means of growing quantum dots in a lattice-matched material system, rather than relying on strain, as in the InGaAs/GaAs system which we discuss in Chapter 6 (Section 6.5). The idea was to use a gallium beam to deposit Ga droplets on a specially treated GaAs surface, then to convert them to GaAs microcrystals with an arsenic flux.

5.3   Use of Surfactants

A very different approach to the modification of surface structure was also developed during the 1980s, making use of large atoms like lead as surfactants—that is species which modify the growth process without actually being incorporated in the resulting film. We have already commented on the behaviour of tin, used as a dopant in MBE-grown GaAs, in so far as it showed a strong tendency to segregate on the GaAs surface, and, in so doing, modified the surface structure sufficiently that RHEED oscillations could be recognised for the first time (Harris et al. 1981). The use of sulphur to modify the properties of the (001) GaAs surface had also been reported by Massies et al. (1980) in their study of Schottky barriers—in particular the observation of its effect on surface reconstruction. Lead, being an even larger atom, can be expected to behave similarly and evidence was soon acquired to confirm such notions. What was certainly not anticipated was the possibility that it might significantly modify the *bulk* electrical properties of a film, in spite of its being incorporated only to a very small extent in the grown film. Probably the first observation of such an effect came with the early development of the alloy GaInP as a material for visible wavelength LEDs and lasers. $Ga_{0.51}In_{0.49}P$ is lattice-matched to a GaAs substrate, with a direct band gap of 'approximately' 1.9 eV (see Orton 2004, p. 291, for an explanation of the use of the word 'approximately') and emits light at a wavelength of about 670 nm in the red part of the spectrum. MBE proved a practical method for growing suitable films but there were various difficulties in achieving the desired material quality. One such was the observation by the Philips group (Blood et al. 1982b) that it was very difficult to achieve adequate p-type doping with Be, the (then) preferred dopant for MBE growth of III-V compounds. Apparently, some form of compensation was occurring as a result, not of impurity incorporation, as might have been expected but of structural imperfection in the films, a problem which was overcome by the addition of a beam of lead atoms to the MBE configuration. The fraction of lead incorporated was estimated as being only about 2×10^{-4} of that supplied but there was a marked effect on the nature of surface reconstruction, the lead flux resulting in a change from the usual (001) (2×1) pattern to a much more sharply defined (001) (1×2) pattern. It was then found possible to obtain controlled p-type doping at levels down to 10^{22} m^{-3}, altogether unavailable in earlier attempts. That the addition of a lead flux could, indeed, reduce the density of deep level traps in MBE-grown GaAs was shown very clearly by Blood et al. (1982a) and by Akatsu et al. (1987), using DLTS (deep level transient spectroscopy).

A particularly interesting application of this same technique was described by the French group at CNRS, Valbonne (Grandjean et al. 1992; Massies et al. 1992). Along with many others, they were concerned to exploit strained layers of InGaAs grown on GaAs substrates for both optoelectronic and microwave devices. However, the lattice mismatch between InAs and GaAs being close to 7%, attempts to grow InAs on GaAs usually resulted in heavily dislocated epitaxial films. Typically, no more than two mono- layers of InAs could be grown pseudomorphically before strain energy built up to the point where dislocations were initiated and three-dimensional growth took over. Pro- portionately greater thickness became possible as the fraction of GaAs in the alloy was increased but, even so, available thicknesses were inadequate for most device require- ments. The concept of applying surfactants to ameliorate such a problem had been demonstrated for the similarly mismatched Si/Ge system (see Chapter 7) so it was rea- sonable to seek solace in such methods in this case too. Based on their earlier work on the use of sulphur to modify the surface properties of GaAs, they decided to try tellur- ium as surfactant and found it to work surprisingly well. Having grown a buffer layer of GaAs, they subjected the surface to a Te beam, which produced a $(001)–(6 \times 1)$ re- construction, with about one monolayer of Te on the surface. Growth of InAs on this surface proceeded up to thicknesses of 6 monolayers before the onset of dislocations, and as many as 60 monolayers were grown without severe degradation of the RHEED pattern. In other words, the growth mode remained two-dimensional for at least 60 ML, whereas for the normal (2×4) GaAs surface, three-dimensional growth set in after the growth of only 2 monolayers of InAs. For the case of the alloy $In_{0.25}Ga_{0.75}As$, no less than 170 ML ($\sim$100 nm) were deposited on this same Te-modified surface with no sign of RHEED degradation and, bearing in mind that this particular alloy was appropriate to the preparation of 980 nm lasers for pumping Er-doped fibre amplifiers, this result was clearly of considerable practical significance.

There followed a fairly intense discussion as to the precise mechanism responsible for these observations. Snyder and Orr (1993) challenged the use of the word 'surfactant' on the grounds that a surfactant was defined by the fact that it reduced the surface free energy and that this would promote island growth rather than two-dimensional growth; but Grandjean and Massies (1993) defended its use, claiming that this would only be true under thermodynamic equilibrium which was not the case here. Perhaps of greater significance was their later work (Massies and Grandjean 1993), making the important distinction between what they termed 'non-reacting' and 'reacting' surfactants, Sn and Pb being examples of the first category, Te of the second. While it was important that the surfactant should segregate on the surface, this was not enough to define its mode of operation. Some segregants certainly did promote two-dimensional growth but others did not—in the case of InGaAs/GaAs epitaxy, Te did so, while Sn did not. The strict distinction between the two types was made in terms of their effect on the migration length of adatoms. Thus, a non-reactive surfactant was weakly bonded to the semicon- ductor and occupied an interstitial site. It operated by decreasing the semiconductor inter-atomic bond strength, thereby reducing the energy barrier for surface migration and thus increasing the migration length. A reactive surfactant, on the other hand, occupied a substitutional site and encouraged the bonding of adatoms, thus reducing their migration lengths. Massies and Grandjean (1993) showed that Te decreased and Pb increased diffusion lengths of Ga atoms on the GaAs surface during epitaxy, thus confirming the distinction between the two types of surfactant. A helpful summary is provided by Tournie et al. (1995)—we shall leave the inquisitive reader to look therein for further enlightenment.

5.4   Gas Source MBE

The introduction of gaseous sources into MBE growth technology represented a considerably more radical innovation than anything involved in the 'mechanical' modifications outlined in the two previous sections. Replacing solid sources with their gaseous counterparts called for significant change to the MBE vacuum system in that it involved the provision of gas lines feeding through the vacuum shell, each incorporating a tiny leak and an isolation valve. It is convenient to distinguish three separate variants, depending on whether only the Group V sources were replaced with appropriate hydrides (originally entitled gas source MBE or GSMBE by Panish, otherwise known as hydride source MBE, or HSMBE), only the Group III sources were replaced with metal-organic compounds (metal-organic MBE, or MOMBE) or both sources were replaced (chemical beam epitaxy or CBE, but sometimes referred to as GSMBE or MOMBE). We shall deal with them in this order. Bearing in mind the need to distinguish these new techniques from the original 'conventional' MBE we shall refer to the latter as solid source MBE (SSMBE) though we should note that the term 'elemental source MBE' (ESMBE) is also used. Sadly, there appears to be little agreement on terminology so we are obliged to state our own preference, to which we hope to adhere throughout the present account. This is for the use of the term GSMBE as a generic description of any method employing gas sources, HSMBE for techniques which replace just the Group V source with hydrides, MOMBE when only the Group III source is replaced by organometallics and CBE for the combined Group V plus Group III replacements. We believe that there is some fairly widespread agreement on this particular terminology, though clearly it does not amount to totality. We should perhaps also point out that it was unusual to replace dopant sources with gaseous versions because the standard solid sources lasted very much longer than those used to supply constituents of the film itself. Finally, before proceeding to outline the various aspects of GSMBE, it should be emphasised that the subject has been extremely well reviewed in the book by Panish and Temkin (1993)—anyone requiring greater detail will certainly find it here—though defined by yet another terminology!

The first move to use gaseous sources was made by Mort Panish at Bell Labs, Murray Hill (Panish 1980). He chose to grow GaAs and InP using arsine (AsH_3) and phosphine (PH_3) as sources of the Group V elements and advanced a number of arguments to justify this approach. Solid sources suffered from the fact that they became depleted far more rapidly than Group III sources, thus leading to time-varying beam fluxes and the frequent need to open the vacuum system in order to replenish the source, a drawback exaggerated by a trend to employ higher substrate temperatures in the interest of obtaining better material quality (the higher the temperature, the greater the loss of the Group Vs from the sample surface and the lower the efficiency with which they were used). What was more, solid sources produced As_4 and P_4 rather than the dimer species which were thought to result in better material quality (this being before the availability of high-temperature cracker sources). Perhaps more importantly, the use of an elemental phosphorus source led to complications due to its being made up of variable propor-tions of different phosphorus allotropes, having different vapour pressures, making flux control difficult. Another advantage of using gas sources was the possibility of supply-ing both arsenic and phosphorus through the same leak valve, thus generating identical spatial distributions at the sample surface.

Clearly, the design of the gas source was crucial, and Panish's leak source is shown in Figure 5.6. Arsine or phosphine was supplied at a pressure of order 1 atm via a suitable gas-handling system and heated to 800 °C–900 °C in order to crack the hydride according to

$$4AsH_3 = As_4 + 6H_2 \tag{5.3}$$

The furnace projected beyond the leak so as to complete the conversion to the dimer species on the low-pressure side of the leak, and this effusion section was also intended to thermalise the beam so as to generate a cosine distribution of molecules at the sample. (This it singularly failed to do but there was obviously scope for redesign to improve this particular aspect.) Because the pumping system was faced with removing a considerable

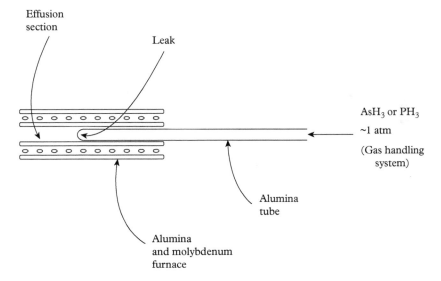

Figure 5.6 *The leak source used by Panish to crack arsine and introduce a beam of*
As$_2$ molecules into his MBE system. This is an example of a high-pressure source,
arsine being cracked on the high-pressure side of the leak. (From Panish 1980.)
Reproduced by permission of The Electrochemical Society.

volume of hydrogen, he decided to replace the original ion pump with an oil-diffusion pump, providing a minimum pressure of 5×10^{-9} Torr, the leak being designed so that the working pressure was approximately 5×10^{-5} Torr. This was a system designed to demonstrate principals, rather than to grow high-purity material. Nevertheless, Panish was able to demonstrate growth of reasonably good GaAs films at temperatures as high as 700 °C, and InP films on InP substrates at temperatures up to 600 °C. Growth rates were in the region of the 1 μm h^{-1} typical of solid source MBE.

In contrast, the next venture into gas sources (Calawa 1981) was destined to produce some of the highest purity GaAs available at the time. Calawa designed his own version of an arsine cracker (in which cracking took place on the low-pressure side of the leak) and examined the arsenic species produced with a quadrupole mass spectrometer. The results showed a complex mix of AsH$_3$, AsH, H$_2$, As$_1$, As$_2$ and As$_4$ but at the optimum temperature of 620 °C the dominant arsenic species was As$_1$, thus introducing yet another new twist to the MBE saga. Nominally undoped samples of GaAs, grown at 1μm h^{-1}, at a substrate temperature of 580 °C, showed free-electron concentrations close to 10^{20} m^{-3}, with peak mobilities as high as 13.3 m^2 V^{-1} s^{-1}, superior (at the time) to anything grown by VPE or LPE. Interestingly, undoped samples grown in the same machine, using a solid arsenic source, were p-type. Could it be that the presence of As$_1$ in the cracking fraction of arsine was responsible for the high quality of these 'gas source' GaAs samples? Calawa demonstrated that higher cracker temperatures, which generated lower fractions of As$_1$, resulted in lower mobilities (probably as a result of silicon contamination from the quartz liner in the cracker) but this still left scope for speculation. Whatever the answer to this specific question, future developments saw both VPE- and LPE-grown samples reaching significantly greater peak mobilities, while gas source MBE has taken many different routes to fame.

It was clear from the first exploratory attempts that HSMBE was intended to further the growth of the quaternary alloy InGaAsP which involved the accurate control of the As:P and In:Ga ratios. From a device-oriented viewpoint, this alloy offered the alluring promise of material lattice-matched to InP with direct band gaps ranging from 0.75 eV to 1.35 eV, spanning the important range of wavelengths between 1.3 μm and 1.55 μm (hυ = 0.95 eV and hυ = 0.80 eV) which correspond to minimum dispersion and minimum loss in quartz optical fibres. If we write the alloy as Ga$_x$In$_{1-x}$As$_y$P$_{1-y}$, the lattice-matching condition can be simply written as

$$y/x = 2.21 \tag{5.4}$$

and the values of x and y appropriate to wavelengths of 1.3 μm and 1.55 μm are respectively x = 0.28, y = 0.61 and x = 0.41, y = 0.90. The possibility for designing double heterostructure lasers operating at these wavelengths was immediately apparent—the practicality of actually making them depended, in MBE terms, on adequate control of four beam fluxes. Readers with appropriately good memories will recall that we discussed the problems of alloy growth in Chapter 4 (Section 4.2), where it was clear that even ternary alloys could prove difficult—the accurate control of the quaternary must inevitably present a greater challenge.

Control of the In:Ga ratio is essentially a question of growing at a temperature low enough that the desorption of In (the more volatile of the two elements) from the surface of the film is effectively negligible. Thus, Panish et al. (1985) describe successful growth of InP and GaInAsP lasers by HSMBE at temperatures in the vicinity of 500 °C, rather than the 600 °C typical of GaAs growth. Under such conditions, one can expect the In:Ga ratio in the film to be determined simply by the ratio of beam intensities. On the other hand, this is far from true for the P:As ratio. Foxon et al. (1980) had shown, for example, for the case of GaAsP grown from As_4 and P_4 that P was incorporated with a probability some 40 times less than that of As. Added to that was the uncertain composition of any solid P source, already referred to above. The chances of success with conventional MBE appeared to be miniscule. Furthermore, the very poor take-up of phosphorus must result in both a heavy demand on the vacuum pumps and the deposition of large amounts of this highly flammable element within the vacuum system. More than one MBE group has experienced the embarrassment of an internal conflagration—it is not a possibility to be taken lightly. The alternative of growing from thermally cracked arsine and phosphine provides a well-controlled supply of As_2 and P_2 to the growing surface which results in a far more favourable uptake of phosphorus, the ratio of P:As in the film being typically about $\frac{1}{4}$ (rather than the 1/40 associated with the use of the corresponding tetramers). Panish et al. (1985) observed that, once the relevant beam intensities had been calibrated, the P:As ratio remained highly reproducible during a lengthy sequence of depositions. They went on to demonstrate the growth of a wide range of InGaAsP laser structures, including double heterostructures, single and multiple quantum wells and a graded index device in which the various ratios were changed in tiny steps to simulate a smoothly varying composition. The resulting laser devices showed excellent performance, with threshold current densities as good as or better than any reported in the literature at that time. Two years later Panish published a further paper (Panish 1987) describing a number of GaInAs(P)/InP quantum well and superlattice structures grown by HSMBE and which well-illustrated the excellent capabilities of the technique to produce high-quality multilayers with monolayer accuracy. Clearly this was a growth method which could rival any other when phosphorus-containing material was concerned.

It remains for us to add a few comments concerning instrumental aspects of HSMBE. As already intimated, both high- and low-pressure crackers have been developed with success, the former depending on the large number of collisions inherent in the high molecular density to enhance reaction rates, the latter relying on the use of a tantalum heater coil or an assembly of tantalum chips to catalyse the cracking process. At typical operating temperatures of about 800 °C–1000 °C, over 99% of the hydrides are cracked to yield As_2 or P_2 as the dominant products, though there is evidence in some cases for relatively large proportions of the monomer As_1, which appears to result in better quality GaAs. It is convenient to supply both arsine and phosphine to the same cracker, thus providing a uniform As:P ratio over the substrate surface. The low-pressure cracker used by Calawa (1981) employed a quartz liner within a tantalum tube and, at temperatures much above 600 °C, this led to contamination of the film with silicon, as we have already noted. Later models made use of BN, rather than quartz, to minimise such effects. High-purity arsine and phosphine were readily available as a result of their being used in MOVPE systems and, by the same token, appropriate techniques for their safe handling were also well established. Finally, we must emphasise the fact that vacuum pumps have to deal with large volumes of hydrogen—for example, Panish et al. (1985) relied on cryopumps capable of pumping H_2 at a rate of 1000 litres s^{-1}.

Turning attention now to the use of organometallic compounds, we should recall that these were introduced by Mansevit in 1968 in his pioneering study of MOVPE growth. He showed that GaAs could be grown by reacting gallium trimethyl ($Ga(CH_3)_3$), or TMG, for short, with arsine and that it could be doped n- or p-type by the addition of H_2Se or diethyl zinc (DEZ). These organic compounds were liquid at room temperature but could readily be transported by bubbling pure hydrogen through them. As intimated in Chapter 1 (Section 1.2), MOVPE rapidly developed into a technique capable of growing the ultra-thin films demanded by low-dimensional structures and we shall meet it again in our next chapter in the guise of a very serious rival to MBE. Ironically, the first use of organometallic compounds in MOMBE and CBE was as a means of studying reaction processes appropriate to MOVPE. The first references date from 1981 (Veuhoff et al. 1981; Vogjdani et al. 1982), closely following Panish's introduction of arsine as a replacement for solid arsenic. These authors demonstrated the possibility of growing GaAs films from gaseous sources for both Group III and Group V elements (though Veuhoff et al. apparently achieved success somewhat by accident, not realising that it was necessary to pre-crack arsine before introducing it into the vacuum system!). Vogjdani et al., at the French Philips research laboratory, achieved surprisingly high growth rates (up to 7 $\mu m\ h^{-1}$) using TMG and As_4 but their GaAs films were heavily p-type. The use of arsine (by way of a cracker) to replace As_4 reduced both growth rate and p-type doping level but, more importantly, also allowed them to demonstrate selective epitaxy through oxide or nitride masks, there being no growth on the masks. Thus, the combination of hydrides plus organometallics could already be seen as a very promising form of MBE growth. However, clinging tenaciously to the plan outlined above, we shall first discuss the various attempts to grow with a combination of organometallics for the Group III element and solid sources for the Group V.

The practical growth of GaAs, using As_4 from a solid source and TMG or TEG (triethyl gallium) seems to have been essentially a Japanese affair. The first report came from Takahashi's group at the Tokyo Institute of technology (Tokumitsu et al. 1984, 1985), Takahashi having been involved in MBE growth from a very early date (see Section 2.7). They were closely followed by rival groups at Fujitsu (Kondo et al. 1986) and at the Optoelectronics Joint Research Laboratory (Kimura et al. 1987). Perhaps the first point to make is that TMG and TEG were introduced into the vacuum system by way of a leak valve, their vapour pressures being high enough that it was not necessary to use a flow of hydrogen to transport them. Second, we should note that the organometallics are cracked on the substrate surface, rather than in the gas stream, as in MOVPE. In the Takahashi work, GaAs films were grown at substrate temperatures between 550 °C and 630 °C, with growth rates being in the range from 0.5 $\mu m\ h^{-1}$ to 1.0 $\mu m\ h^{-1}$. When under excess arsenic conditions and at temperatures above 550 °C, the growth rate, using either TMG or TEG, was proportional to the organometallic flux and was insensitive to growth temperature, indicating that it was not limited by the rate of thermal decomposition. Interestingly, while growth occurred in windows through silicon oxide masks, there was no growth at all on top of the oxide, suggesting this as an ideal technique for selective epitaxy. There was, however, a serious drawback to the method in so far as large amounts of carbon were incorporated in the films, resulting in heavy p-type doping. When using TMG p-type doping levels were close to 10^{26} m^{-3}. While use of TEG reduced this by two orders, this was still far too high for many applications and attempts were made to reduce it still further by introducing a beam of ionised hydrogen—presumably in the hope that hydrogen would combine with the carbon and prevent its incorporation in the growing film. Though there was some further reduction in free-carrier density, carrier mobilities were generally low and it seemed clear that, unless one had a need for very heavily p-type material, this was not a technique with any great promise for device fabrication.

How wrong could one be! The work of the Fujitsu group published about one year later demonstrated that GaAs films grown from TEG and As$_4$ (without hydrogen) could certainly compete on level terms with material grown by solid source MBE. The p-type carrier densities were as low as 10^{21} m^{-3}, with peak hole mobilities close to 0.6 m^2 V^{-1} s^{-1} and a compensation ratio of about 0.16. The 'secret' appeared to be the use of larger As$_4$/TEG ratios and this was confirmed soon afterwards by the group at the Optoelectronics Joint Research Laboratory who showed hole densities decreasing systematically with increasing As$_4$:TEG ratio. Using much larger ratios than commonly used in SSMBE, their lowest hole density was measured as 3×10^{20} m^{-3} (though no value for hole mobility was quoted). They also showed low-temperature photoluminescence data consistent with the dominant acceptor being carbon. Mass spectra of species leaving the substrate were extremely complex but suggested that the excess arsenic was combining with carbon to prevent its incorporation in the GaAs films. In similar vein, yet another Japanese group (at the NTT lab in Atsugi) reported the use of triethyl indium (TEI) and solid phosphorus to grow films of InP on InP substrates (Kawaguchi et al. 1984). The films were n-type but were fairly closely compensated.

Finally, let us pick up on the 'ultimate' technique of CBE, involving substitution of gas sources for both Group III and Group V elements. As we have already mentioned, Veuhoff et al. (1981) at the Aachen Technical University appear to have been first in combining organometallics and hydrides to grow GaAs films by MBE but their real interest seems to have been in MOVPE so they did not pursue the subject very far. A parallel venture by the NTT group made use of TEI and phosphine to grow InP and achieved considerably better electrical properties than those obtained with a solid source of phosphorus. Their lowest electron density was 4×10^{21} m^{-3} with electron mobility of 2.4 m^2 V^{-1} s^{-1} at 77 K—compensation ratios lying in the range of 0.29 to 0.31. Thus, we can see evidence for successful use of gas sources for both constituent elements but by far the strongest drive to develop CBE as a practical growth method came from Wan Tsang at Bell Labs. Tsang started his work at Murray Hill but very soon moved to the Holmdell Laboratory which was concerned considerably more with development activities, a move which well illustrates the importance accorded to CBE by Bell at a time when fibre-optic communications were under intensive development.

Tsang introduced a further innovation by using organometallic compounds for the Group V elements, as well as for the Group III, arsenic and phosphorus beams being formed by thermal decomposition (at 950 °C–1200 °C) of trimethyl arsine (TMA) or triethyl phosphine (TEP) to produce beams of As_2 or P_2. His reason was mainly one of greater safety, the Group V alkyls not only oxidising rapidly in air but also having the useful characteristic of being 'extremely malodorous' and therefore readily recognised, should they escape. TMA and TEP were both introduced via the same cracking tube, thus ensuring uniform As:P ratio in compounds containing both Group V elements. In his first paper Tsang (1984) described the growth of both GaAs and InP, demonstrating that growth rates were independent of substrate temperature (above a certain turnover temperature) and proportional to the Group III flux. Surface morphologies were good and undoped samples showed free-carrier densities of $n \sim 10^{23}$ m^{-3} for InP and $p \sim 5 \times 10^{22}$ m^{-3} for GaAs. Photoluminescence spectra showed single band-edge emission lines, consistent with at least moderately good quality material, even though the Group V alkyls were less pure than the hydrides used in earlier work.

Having thus established his new technique, Tsang then went on to demonstrate its versatility in a series of letters (Tsang 1985a, b, c). Selective area epitaxy through silicon masks was reported for both GaAs on GaAs and $In_{0.53}Ga_{0.47}As$ lattice-matched to InP. He also performed an interesting experiment to prove the molecular beam nature of the deposition process, supporting a fine wire mesh in front of the substrate which threw a 'shadow' on the grown film, thus proving that he was growing by MBE, rather than by MOVPE. Next, he showed that CBE had the capability of growing either GaAs or $In_{0.53}Ga_{0.47}As$ free of so-called 'oval defects'. This was a problem which had bedevilled MBE growth for many years, an apparently random density of surface defects which probably emanated from the use of hot Ga sources. Either the result of Ga_2O formed within the gallium crucible or of gallium spitting from the crucible, these defects promised to defeat all attempts to make optoelectronic integrated circuits on III-V substrates. It followed that, if such explanations were correct, the use of a gas source for gallium in which elemental Ga was only formed on the heated substrate should offer an effective solution, and Tsang showed unequivocally that such was, indeed, the case. Finally, he described the growth of high-quality InGaAs on InP substrates using a high-purity TMA source. He first proved it possible to grow highly uniform layers without the need for substrate rotation, partly as a result of supplying both Ga and In from the same source tube and partly by careful design of the source itself. The electrical properties of the layers showed significant variation with the batch of TMA used, as shown in Figure 5.7 but were comparable to the best results obtained by MOVPE. Photoluminescence results showed some improvement over similar samples grown by solid source MBE. Here, beyond doubt, was a growth technique with excellent prospects for the further development of optoelectronic devices and circuits.

As one example of this, we might look briefly at another Tsang paper (Tsang and Miller 1986) which describes the application of CBE to the growth of AlGaAs/GaAs quantum wells and DH lasers. The apparatus employed is shown in Figure 5.8. While organometallics were used to supply the Group III elements, Tsang and Miller reverted to the use of a low-pressure arsine cracker as a source of arsenic, presumably in the

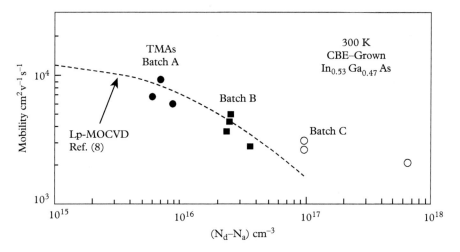

Figure 5.7 *Measured electron Hall mobilities for $In_{0.53}Ga_{0.47}As$ films grown by CBE, lattice-matched to InP substrates and using three different batches of TMA. The dashed curve represents the results from material grown by low pressure MOVPE. (From Tsang 1985c.) Reproduced with permission from Tsang, W T (1985c) Jnl Appl Phys 58, 1415. Copyright 1985, AIP Publishing LLC.*

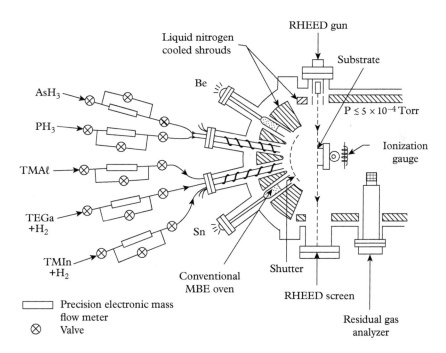

Figure 5.8 *The gas-handling system and growth chamber used by Tsang for CBE growth of III-V alloys. (From Tsang and Miller 1986.) Reproduced with permission from Elsevier.*

interest of purity. Growth temperature was chosen to be ~650 °C, for which case the growth rate was independent of temperature and proportional to the Group III flux. They list the advantages of CBE as: (1) the use of *room-temperature* Group III sources simplifies multiwafer scale-up; (2) sources are effectively semi-infinite; (3) very accurate flux control is effected by using precision mass flow meters; (4) both Group III beams are supplied through a common inlet, ensuring constant Al:Ga ratio over the substrate; (5) oval defects are eliminated, even at high growth rates; and (6) growth rates of the order of 3–6 μm h^{-1} are possible, considerably greater than typical of SSMBE. It was clear that the authors were looking forward to commercial application of the technique to optoelectronic devices. The excellent control of material quality made possible was first demonstrated by measuring photoluminescence and excitation spectra from a series of GaAs/Al$_{0.5}$Ga$_{0.5}$As quantum wells, with well widths of the order of 100–150 Å. The narrow line widths observed suggested that interfaces were sharp to within a single monolayer, and reproducibility demonstrated by the relative sharpness of the spectrum from a multiquantum-well sample containing 18 notionally identical wells. Detailed analysis also suggested that the material was relatively free from unwanted impurities—though no electrical data were given. Finally, results were reported on the emission properties of broad-area DH lasers, showing excellent uniformity of threshold current, with current densities of the order of 500 A/cm^2. A plot of threshold current density as a function of active layer thickness compared very closely with similar results from lasers grown by SSMBE and, finally, current densities were as low as the best values measured on MOVPE- and LPE-grown structures. The versatility of the CBE method was further illustrated by its application to the growth of quantum wells and heterostructures in the GaInAs/InP system (Tsang 1987).

Discussion of GaAs lasers grown by CBE provides an appropriate stimulus for us to mention other device applications in which gas source MBE played a role. We make no attempt to go into detail—for that the reader is recommended to consult Panish and Temkin (1993)—but it is worthwhile noting the range of applications which had accrued by the end of the 1980s. In no particular order, we may reference the HBT (heterojunction bipolar transistor), the long wavelength laser, the corresponding long wavelength photodetector and the visible laser. Implicit in the 'philosophy' of the HBT is a heterojunction between emitter and base designed to favour minority carrier injection into the base and prevent the reverse current from base to emitter. Thus an effective lattice-matched structure makes use of an n-type InP emitter together with a p-type InGaAs base and an n-type InGaAs collector, the larger band gap of the InP serving to minimise the reverse base-to-emitter current. As we have seen, this particular combination of materials could conveniently be grown by HSMBE, while it must also be apparent that they could equally well be grown by CBE. Such, indeed, is the case—both techniques were successfully applied to the growth of HBT structures during the 1980s. We have already discussed the principles behind the use of InP/InGaAsP as the basis for communication lasers operating at wavelengths of 1.3 μm and 1.55 μm. We need only add that quantum well lasers were also made by CBE in 1991 with threshold current densities as low as 170 A/cm^2, much lower than anything seen from alternative techniques. The material system InP/InGaAs was also used for making avalanche photodiodes, providing fast detectors for optical communications. The alloy $In_{0.53}Ga_{0.47}As$, with a band gap of 0.75 eV, is an effective absorber of photons with wavelengths of 1.3 and 1.55 μm, while the avalanche region of the diode is conveniently made from InP and such photodiodes, grown by CBE date from 1987. Finally, we should note that visible injection lasers, based on the alloy AlGaInP were also grown by HSMBE in 1990.

Attempting to summarise the status of MBE at the end of the 1980s, we can see that it had blossomed from being an esoteric technique for scientific exploration, with emphasis on its ability to unravel the secrets of crystal growth, to a highly promising method of growing complex structures with 'real' applications. What was more, it had broadened into a *set* of techniques, all based on the use of molecular beams but supplying these beams in a variety of different ways. But, without doubt, the importance of MBE could be seen to lie in its ability to grow semiconductor structures with monolayer accuracy and with correspondingly sharp interfaces between different materials. Indeed, this ability had already been perfectly demonstrated in the growth of so-called low-dimensional structures, which will form the subject of our next chapter. It was this venture that brought MBE to maturity—and without which it might well have remained simply a convenient academic method of studying crystal growth.

....

REFERENCES

Akatsu, Y, Ohno, H, Hasegawa, H and Hashizume, T (1987) Effect of a coincident Pb flux during mbe growth on the electrical properties of GaAs and AlGaAs layers. J Cryst Growth 81, 319.

Arthur, J R (1966) Gallium arsenide surface structure and reaction kinetics: field emission microscopy. J Appl Phys 37, 3057.

Asai, M, Sato, F, Imamoto, H, Imanaka, K and Shimura, M (1988) Low-temperature (350 °C) growth of AlGaAs/GaAs laser diode by migration enhanced epitaxy. J Appl Phys 64, 432.

Blood, P, Harris, J J, Joyce, B A and Neave, J (1982a) Deep states and surface processes in GaAs grown by molecular beam epitaxy. J Phys Colloques 43, C5-351.

Blood, P, Roberts, J S and Stagg, J P (1982b) GaInP grown by molecular beam epitaxy doped with Be and Sn. J Appl Phys 53, 3145.

Briones, F, Gonzalez, L, Recio, M and Vazquez, M (1987) Low-temperature growth of AlAs/GaAs heterostructures by modulated molecular beam epitaxy. Jap J Appl Phys 26, L1125.

Briones, F, Gonzalez, L and Ruiz, A (1989) Atomic layer molecular beam epitaxy (Almbe) of III–V compounds: growth modes and applications. Appl Phys A 49, 729.

Calawa, A R (1981) On the use of AsH$_3$ in the molecular beam epitaxial growth of GaAs. Appl Phys Lett 38, 701.

Chikyow, T and Koguchi, N (1990) MBE growth method for pyramid-shaped GaAs micro crystals on ZnSe (001) surface using Ga droplets. Jap J Appl Phys 29, L2093.

Foxon, C T, Joyce, B A and Norris, M T (1980) Composition effects in the growth of Ga(In)As$_y$P$_{1-y}$ alloys by MBE. J Cryst Growth 49, 132.

Fukunaga, T, Kobayashi, K L I and Nakashima, H (1985) Photoluminescence from AlGaAs-GaAs single quantum wells with growth interrupted heterointerfaces grown by molecular beam epitaxy. Jap J Appl Phys 24, L510.

Grandjean, N and Massies, J (1993) Grandjean and Massies reply. Phys Rev Lett 70, 1031.

Grandjean, N, Massies, J and Etgens, V H (1992) Delayed relaxation by surfactant action in highly strained III-V semiconductor epitaxial layers. Phys Rev Lett 69, 796.

Harris, J J, Joyce, B A and Dobson, PJ (1981) Oscillations in the surface structure of Sn-doped GaAs during growth by MBE. Surf Sci Lett 103, L90.

Horikoshi, Y, Kawashima, M and Yamaguchi, H (1986) Low-temperature growth of GaAs and AlAs-GaAs quantum-well layers by modified molecular beam epitaxy. Jap J Appl Phys 25, L868.

Horikoshi, Y, Kawashima, M and Yamaguchi, H (1987) Photoluminescence characteristics of AlGaAs-GaAs single quantum wells grown by migration-enhanced epitaxy at 300 °C substrate temperature. Appl Phys Lett 50, 1686.

Kawaguchi, Y, Asahi, H and Nagai, H (1984) MBE growth of high-quality InP using tri-ethylindium as an indium source. Jap J Appl Phys 23, L737.

Kimura, K, Horiguchi, S, Kamon, K, Mashita, M, Mihara, M and Ishii, M (1987) Molecular beam epitaxial growth of GaAs using triethylgallium and As_4. Jap J Appl Phys 26, 419.

Koguchi, N, Ishige, K and Takahashi, S (1993) New selective molecular-beam epitaxial growth method for direct formation of GaAs quantum dots. J Vac Sci Technol B 11, 787.

Kondo, K, Ishikawa, H, Sasa, S, Sugyama, Y and Hiyamizu, S (1986) MBE growth of high-quality GaAs using triethylgallium as a gallium source. Jap J Appl Phys 25, L52.

Massies, J, Dezaly, F and Linh, N T (1980) Effects of H2S adsorption on surface properties of GaAs {100} grown in situ by MBE. J Vac Sci Technol 17, 1134.

Massies, J and Grandjean, N (1993) Surfactant effect on the surface diffusion length in epitaxial growth. Phys Rev B 48, 8502.

Massies, J, Grandjean, N and Etgens, V H (1992) Surfactant mediated epitaxial growth of $In_xGa_{1-x}As$ on GaAs (001). Appl Phys Lett 61, 99.

Merlin, R, Colvard, C, Klein, M V, Morkoc, H, Cho, A L and Gossard, A C (1980) Raman scattering in superlattices: anisotropy of polar phonons. Appl Phys Lett 36, 43.

Neave, J H, Joyce, B A, Dobson, P J and Norton, N (1983) Dynamics of film growth of GaAs by MBE from RHEED observations. Appl Phys A 31, 1.

Orton, J W (2004) 'The Story of Semiconductors', Oxford University Press, Oxford.

Panish, M B (1980) Molecular beam epitaxy of GaAs and InP with gas sources for As and P. J Electrochem Soc 127, 2729.

Panish, M B (1987) Gas source molecular beam epitaxy of GaInAs(P): gas sources, single quantum wells, superlattice pin's and bipolar transistors. J Cryst Growth 81, 249.

Panish, M B and Temkin, H (1993) 'Gas Source Molecular Beam Epitaxy', Springer-Verlag, Berlin.

Panish, M B, Temkin, H and Sumski, S (1985) Gas source molecular beam epitaxy of GaInAs(P): gas sources, single quantum wells, superlattice pin's and bipolar transistors. J Vac Sci Technol B 3, 657.

Sakaki, H, Tanaka, M and Yoshino, J (1985) One atomic layer heterointerface fluctuations in GaAs-AlAs quantum well structures and their suppression by insertion of smoothing period in molecular beam epitaxy. Jap J Appl Phys 24, L417.

Sakamoto, T, Funabashi, H, Ohta, K, Nakagawa, T, Kawai, N J and Kojima, T (1984) Phase-locked epitaxy using RHEED intensity oscillation. Jap J Appl Phys 23, L657.

Sakamoto, T, Funabashi, H, Ohta, K, Nakagawa, T, Kawai, N J, Kojima, T and Bando, Y (1985) Well defined superlattice structures made by phase-locked epitary using RHEED intensity oscillations. Superlattice Microst 1, 347.

Sano, N, Kato, H, Nakayama, M, Chika, S and Terauchi, H (1984) Mono-and bi-layer superlattices of GaAs and AlAs. Jap J Appl Phys 23, L640.

Snyder, C W and Orr, B G (1993) Comment on "Delayed relaxation by surfactant action in highly strained III-V semiconductor epitaxial layers". Phys Rev Lett 70, 1030.

Tanaka, M and Sakaki, H (1987) Atomistic models of interface structures of GaAs-$Al_xGa_{1-x}As(x = 0.2 - 1)$ quantum wells grown by interrupted and uninterrupted MBE. J Cryst Growth 81, 153.

Tokumitsu, E, Kudou, Y, Konagai, M and Takahashi, K (1984) Molecular beam epitaxial growth of GaAs using trimethylgallium as a Ga source. J Appl Phys 55, 3163.

Tokumitsu, E, Kudou, Y, Konagai, M and Takahashi, K (1985) Metalorganic molecular-beam epitaxial growth and characterization of GaAs using trimethyl-and triethyl-gallium sources. Jap J Appl Phys 24, 1189.

Tournie, E, Grandjean, N, Trampert, A, Massies, J and Ploog, K H (1995) Surfactant-mediated molecular-beam epitaxy of III–V strained-layer heterostructures. J Cryst Growth 150, 460.

Tsang, W T (1984) Chemical beam epitaxy of InP and GaAs. Appl Phys Lett 45, 1234.

Tsang, W T (1985a) Selective area growth of GaAs and $In_{0.53}Ga_{0.47}As$ epilayer structures by chemical beam epitaxy using silicon shadow masks: a demonstration of the beam nature. Appl Phys Lett 46, 742.

Tsang, W T (1985b) Elimination of oval defects in epilayers by using chemical beam epitaxy. Appl Phys Lett 46, 1086.

Tsang, W T (1985c) Chemical beam epitaxy of InGaAs. J Appl Phys 58, 1415.

Tsang, W T (1987) Chemical beam epitaxy of $Ga_{0.47}In_{0.53}As$/InP quantum wells and heterostructure devices. J Cryst Growth 81, 261.

Tsang, W T and Miller, R C (1986) GaAs/AlGaAs quantum wells and double-heterostructure lasers grown by chemical beam epitaxy. J Cryst Growth 77, 55.

Van Hove, J M, Lent, C S, Pukite, P R and Cohen, P I (1983) Damped oscillations in reflection high energy electron diffraction during GaAs MBE. J Vac Sci Technol B 1, 741.

Veuhoff, E, Pletschen, W, Balk, P and Lüth, H (1981) Metalorganic CVD of GaAs in a molecular beam system. J Cryst Growth 55, 30.

Vodjdani, N, Lamarchand, A and Paradan, M (1982) Parametric studies of GaAs growth by metalorganic molecular beam epitaxy. J Phys Colloques 43, C5-339.

Walker, J D, Kuchta, D M and Smith, J S (1991) Appl Phys Lett 59, 2079.

Walker, J D, Kuchta, D M and Smith, J S (1993) Wafer-scale uniformity of vertical-cavity lasers grown by modified phase-locked epitaxy technique. Electron Lett 29, 239.

Walker, J D, Malloy, K, Wang, S and Smith, J S (1990) Vertical-cavity surface-emitting laser diodes fabricated by phase-locked epitaxy. Appl Phys Lett 56, 2493.

Yamaguchi, H and Horikoshi, Y (1989) Step-flow growth on vicinal GaAs surfaces by migration-enhanced epitaxy. Jap J Appl Phys 28, L1456.

Zhang, J, Dawson, P, Neave, J H, Hugill, K J, Galbraith, I, Fawcett, P N and Joyce, B A (1990) Influence of growth interruption on inverted interface quality in single AlAs-GaAs quantum wells grown by molecular beam epitaxy. J Appl Phys 68, 5595.

Zhou, Z Y, Zheng, C X, Tang, W X, Tersoff, J and Jesson, D E (2013) Origin of quantum ring formation during droplet epitaxy. Phys Rev Lett 111, 036102.

6

Low-Dimensional Structures

6.1 MBE Grows Up

Looking at the history of semiconductor research, one must, surely, recognise the tremendous boost provided by the invention of the transistor at Bell Labs, Murray Hill, in 1947 and of the integrated circuit at Texas Instruments and Fairchild in the following decade. Though initiated with germanium, this centred, of course, very firmly on the development of silicon technology, tending to leave all other semiconductor materials trailing forlornly in its wake. Nevertheless, the rise of the low-dimensional structures (LDS) initiative had an almost identical effect on research into III-V materials—though on a very much smaller scale. Using the size of the commercial market as a measuring rod, the ratio between the two is something like a factor of 30 but it would be a short-sighted commentator who could ignore the importance of GaAs- and InP-based devices, many of which depend for their efficient functioning on the use of low-dimensional structures. And, if the development of LDS can be seen as central to the long-term future of the Group III-V compounds, it is equally important that we recognise the essential contribution of MBE to the growth of low-dimensional structures. This is clearly reflected in the rapid rise in the (worldwide) number of active MBE groups, following the first reports of the optical properties of quantum wells in 1974 and of the exciting transport behaviour of the two-dimensional electron gas in 1979. Figure 6.1(a) provides an outline of the growth in MBE facilities based on the number of groups presenting papers at four International Conferences on MBE, which took place in 1978, 1982, 1986 and 1990. The rapid rise after 1982 reflects the impact that LDS had on investment in MBE facilities, with consequent increase in a wide range of associated research. From its birth in the early 1970s MBE grew into a major international activity, with well over a hundred separate research groups, most of which were concerned with the growth of III-V compounds. Unsurprisingly, this was reflected in the number of research papers published annually. If we look at the number of papers recorded on the 'Web of Science' website and whose title or abstract contains the words 'molecular beam epitaxy' or 'molecular beam epitaxial', we find a similar abrupt rise round about the year 1990. In 1985 the number was just over 200, and in 1995 it was

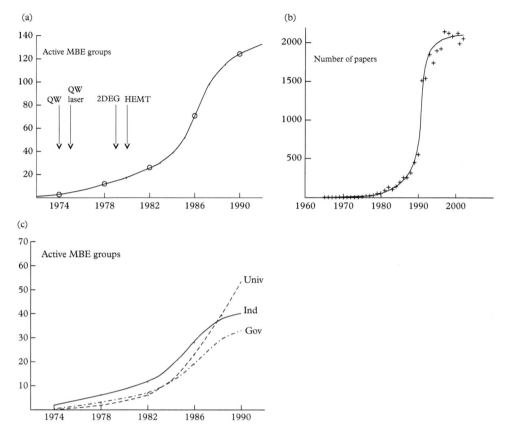

Figure 6.1 *(a) shows the way in which the total number of active MBE groups increased during the two decades from 1970 to 1990, based on the number of different groups submitting papers to the International MBE Conference, held in 1978, 1982, 1986 and 1990. These numbers show a rapid increase following the first reports on quantum well and 2DEG properties. In (b) we show the corresponding increase in published papers. In part (c) the numbers of research groups are broken down between industrial, government and university laboratories, showing the early dominance of the industrial effort; only after 1986 was it exceeded by university-based activity.*

a little under 2000(!), at which level it then settled down (see Figure 6.1(b)). Such a dramatic increase may not be unique in the annals of science but it is very far from commonplace.

An interesting feature of the growth statistics presented in Figure 6.1(c) is the balance between industrial, government and academic funding for the various activities. In the early days this was dominated by industry, a situation which continued until well into the 1980s. It was only towards the end of that decade that the number of university-based research groups came to exceed those in the industrial sector. Such an observation appears to fly in the face of the conventional view that new ideas flow from academia into industry, where they can be commercially exploited. However, it probably exemplifies the behaviour of many materials technologies—when industry perceives them as being central to device development, the necessary funding can usually be found in house. Then, only when industry has declared a serious interest, does funding for, what is, after all, a rather expensive growth technology, become more widely available. (At a very rough estimate, the total investment up to 1990 probably exceeded $100 m, which was something like 5% of the, then, market for III-V devices, showing just how seriously this burgeoning III-V technology was being taken. Note that, in the majority of cases, growers would purchase a commercial MBE machine, thus stimulating a separate, but closely associated market.) We suggest (without proof!) that the development of many other crystal growth techniques probably followed a similar pattern. In terms of the distribution between nations, our data show a broad equality between the USA, Japan and Europe, as revealed in Table 6.1. Though the United States do appear to be taking a significant lead towards the end of the 1980s, one has a distinct impression that the Europeans, having missed out rather badly on the surge in silicon, were determined not to repeat their mistake in the case of the III-Vs. They were clearly 'hanging in' in the MBE stakes.

A particularly interesting aspect of the 'politics' of MBE funding concerns the re-action of the UK government. While criticism could certainly be (and was!) generated by its singularly poor response to earlier international developments in transistor and integrated circuit science, there could be no complaint about funding for LDS. In this instance, nearly £40 m was invested specifically in LDS research over the seven years between 1984 and 1991, under a programme known as the 'Low Dimensional Structures and Devices Initiative'. Box 6.1 provides a short account of the programme, with a very brief list of its main highlights. So far as we can discover, no comparable funding initiative was mounted in any other country, and the UK community should surely be grateful that, in this instance, it was given a considerable kick-start in LDS science by the Science and Engineering Research Council (SERC), as it was then known. One overall consequence for the UK was the development of expertise in the growth and study of compound semiconductors which far outstripped that of silicon or germanium.

Table 6.1 *The geographical distribution of the number of active MBE research groups.*

	1978	1982	1986	1990
USA	4	6	23	51
Japan	2	10	20	29
Europe	5	8	25	37

Box 6.1 The UK LDS initiative and MBE development

The UK LDS initiative was a directed programme of research funded by the Science and Engineering Research Council (SERC) and ran from 1984 to 1991. Prior to this, MBE activity in the UK was centred on industrial and government laboratories, such as Mullard/Philips, British Telecom and RSRE. The funding of nearly £40 m from the initiative enabled the balance to shift to the universities, including, as it did, the purchase of 1 MOCVD and 7 MBE machines within the university community.

The initiative was instigated by the physics committee of SERC, the main funding body in the UK supporting research in academia. It commissioned a report by the late Professor John Beeby, which he presented to the Science Board and which was approved in 1983. At the same time there was separate funding for low-dimensional devices from the Engineering Board of the Information and Technical Advisory Board (ITAB), which also supported a central growth facility at the University of Sheffield, the idea being that Sheffield would grow samples to order for other university researchers who had been successful in winning funds for innovative ideas in LDS.

The programme was placed under the control of a coordinator, initially in the person of Professor Beeby himself, who was largely responsible for its initial success. Later, Dr John Walling was given part-time leave of absence from his post at the Philips Laboratory to take up the appointment, and, in the final phase of the programme, one of the present authors (TF) was appointed to oversee the running of the various MBE machines.

The work supported included growth of three material systems: III-Vs, II-VIs and Si/SiGe. An independent report published in 1992 drew attention to a number of highlights: growth of the highest purity GaAs by Dr Colin Stanley at Glasgow, growth of high-mobility two-dimensional hole gases at Cambridge University, growth of II-VI dilute magnetic semiconductors by Mr Bernard Lunn at Hull University, growth of world state-of-the-art Si/SiGe 2D hole gases at Warwick University and work on ballistic transport in 1D electron systems by Prof. Michael Pepper at Cambridge—also, the growth of quantum pillars, the detection of the quantum Hall effect using optical means by Dr Robin Nicholas and Dr John Ryan in Oxford, the observation of superlattice band folding by Dr Robin Nicholas, several beautiful experiments on the emission and absorption of phonons by Prof. Lawrie Challis at Nottingham, the study of interfaces using ballistic electron emission microscopy by Prof. Robin Williams at Cardiff, the study of resonant tunnelling by Prof. Lawrence Eaves at Nottingham and the study of strained-layer structures by Prof. Alfred Adams of Surrey University.

There can be no doubt that, as a direct result of this initiative, the amount and quality of UK research on LDS topics increased dramatically, as did the provision of material structures grown by MBE—though this, in turn, also assisted indirectly in the development of MOCVD. At the same time, a significant side effect was the considerable strengthening of the commercial base of the principal UK manufacturer of MBE machines, VG Semicon.

Having thus emphasised the vital importance of low-dimensional structures in the world of III-V semiconductor development and the continuing future of MBE technology, it would be well to define its essence. What was it that so excited the world's scientists that so many of them switched to studying it at remarkably short notice? In a word, it concerned interesting modification to the properties of electrons in semiconductors when faced with abrupt changes in energy gap, as for example between GaAs and AlAs. In fact, we have already come across an example in the shape of a structural superlattice (Section 2.6), which is characterised by an effectively different band structure from those of its constituents. As we pointed out then, the key parameter is one of size—if the thickness of a semiconductor layer is reduced until it becomes comparable to an electron wavelength, we must expect significant modification to that electron's behaviour (just as we anticipate the behaviour of photons to change when constrained by wavelength-sized objects). To illustrate the point, it is convenient to consider two different aspects of such behaviour: those of 'electron tunnelling' and of so-called quantum confinement. Let us examine them in this order.

Figure 6.2 represents an electron approaching a potential barrier E_0 from left to right. In the case of a classical particle, whose behaviour is ruled by classical mechanics, there can be no possibility of its appearing on the right-hand side of the barrier unless it possesses sufficient energy to go over the barrier—the probability of its 'seeping' through the barrier is identically zero. As is well known, this is not so in the case of quantum mechanics, the electron wave function being finite on the right-hand side even when its energy is well below that of the barrier. The criterion to be satisfied in order that the probability of transmission through the barrier is of practical significance concerns the height E_0 and thickness W of the barrier. Assuming a *finite* barrier, the electron can tunnel through it when its thickness is less than a certain value which, as we show in Box 6.2 for typical AlGaAs/GaAs barriers, corresponds (approximately) to the electron de Broglie wavelength λ. Assuming the electron has thermal energy, this, at room temperature, implies that λ ~ 25 nm, so we should anticipate significant tunnelling through the barrier when its thickness becomes less than about 25 nm. The fact that the transmission probability falls away exponentially when W is greater than this (see Figure 6.3) indicates that such quantum effects turn on rather abruptly, justifying the drawing of a sharp dividing line between low-dimensional structures and their ordinary three-dimensional counterparts.

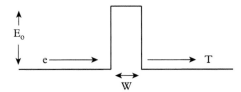

Figure 6.2 *Schematic diagram of an electron approaching a rectangular barrier of height E_0 and thickness W from the left; T represents the probability of the electron tunnelling through the barrier and appearing on the right-hand side.*

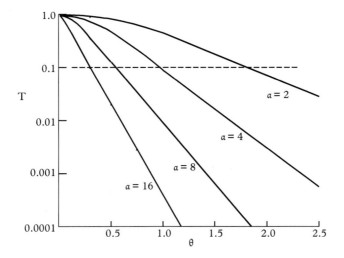

Figure 6.3 *Plots of the tunnelling probability T through the rectangular barrier of Figure 6.2 as a function of the parameter $\theta = W/\lambda$ (where λ is the electron wavelength). The various curves correspond to different values of $\alpha = E_0/E$, the ratio of barrier height to electron energy.*

Box 6.2 **Electron tunnelling through a rectangular barrier**

Electron tunnelling through a rectangular barrier is a classical problem in quantum mechanics textbooks. For example, Schiff (1955) derives an expression for the tunnelling probability T as

$$T = [1 + (E_0{}^2 \sinh^2 \beta W)/4E_0(E_0 - E)]^{-1} \qquad (B6.1)$$

where $\beta = \{2m(E_0 - E)/h^2\}^{1/2}$, m is the electron mass, E its energy and the other parameters are defined in Figure 6.2.

For convenience, we can restate this in terms of the dimensionless ratios $\alpha = E_0/E$ and $\theta = W/\lambda$, where λ, the de Broglie electron wavelength, is given by

$$\lambda = h/\{2mE\}^{1/2} \qquad (B6.2)$$

Thus, we have

$$T = \left[1 + \left\{\alpha^2/4(\alpha - 1)\right\} \sinh^2 \left\{\theta(\alpha - 1)^{1/2}\right\}\right]^{-1} \qquad (B6.3)$$

Figure 6.3 demonstrates how the transmission coefficient varies with the ratio θ, for several values of α. Notice that, when the argument of the sinh term is much greater than unity, one can approximate it as $\sinh^2 x \sim \exp(2x)/4$ and then T becomes

$$T \sim \left\{16(\alpha - 1)/\alpha^2\right\} \exp\left\{-2\theta(\alpha - 1)^{1/2}\right\} \qquad (B6.4)$$

continued

Box 6.2 *continued*

which explains the linear sections of these curves, that is, the transmission probability falls away exponentially with increasing barrier thickness. The point we aim to make from Figure 6.3 is that, if we take a value of $T \sim 0.1$ as representing a 'significant' transmission coefficient, it is clear that this occurs when the barrier thickness is roughly equal to the electron wavelength (i.e. θ lies between 0.5 and 2).

Finally, we can use equation (B6.2) to calculate appropriate values for the electron wavelength in terms of its energy E. Assuming that E represents its thermal energy, that is, that $E \sim kT \sim 0.025\,eV$ at room temperature, and taking a representative effective mass of $m_e = 0.1\,m$, we arrive at a value of $\lambda \sim 25\,nm$. This, therefore, gives us a working estimate for the thickness to which the barrier must be reduced before it becomes transparent to an electron with thermal energy.

Note that typical values of barrier height associated with AlGaAs/GaAs heterostructures lie in the range 0.1 eV to 0.4 eV, corresponding to the values of $\alpha \sim 4 - 16$ used in Figure 6.3.

In the case of tunnelling, the electron wave function is largely excluded from the barrier region whereas, in the case of a quantum well, a region of low energy sandwiched between two barrier regions, the wave function is strongly confined within the well. In fact, for infinite barriers, the wave function penetrates the barrier not at all. We shall discuss the details of the rectangular quantum well in Section 6.3 but it is helpful, at this point, to note just one essential feature. Electron or hole confinement in a direction normal to the plane of the well (usually designated the z-direction) has the result of modifying the density of states within the appropriate energy band—i.e. the conduction or valence band. While carriers are still free to move in the xy plane within a band of states having an approximately parabolic density-of-states function, quantum confinement results in an overall density of states consisting of a series of steps separated in energy by amounts which depend specifically on the width of the well. The narrower the well, the greater are these energy differences. So, just as we could ask what was the criterion for tunnelling through a narrow barrier, we can similarly ask for a criterion which defines the significance of quantum confinement. Once again this depends on thermal energy—if, for example we perform an optical absorption measurement which takes a valence band electron into the bottom of the lowest confined conduction band, and then we ask whether this particular transition can be distinguished from one involving the next higher band, the criterion must be that the energy separation between these two bands is greater than the electron's thermal energy. Thus, at room temperature, if confinement energies are less than about 25 meV, experiment will be incapable of detecting quantum confinement effects and, in a typical well of perhaps 200 meV depth, this happens when the well is greater than about 25 nm in width—a criterion close to that which determines quantum tunnelling through a barrier of similar height. Thus, in a simplistic manner we arrive at the conclusion that low-dimensional structures require us to control semiconductor film thickness in the range below 25 nm, or, in molecular terms, from 1 to 100 monolayers. In the 1970s, this was, of course, the sole realm of

MBE, and the current chapter is concerned to describe just how well MBE rose to this exciting challenge. Needless to say, the initial experimental demonstrations were made using the highly convenient AlAs/GaAs material system, and most of what we shall have to say here will concern this near-perfectly lattice-matched combination. Other combinations will be covered in later chapters when we expand our horizons to take in the wide range of alternatives.

6.2 Structural Superlattices

As we saw in Section 2.6, structural superlattices (thus designated to distinguish them from doping superlattices) made an appearance on the LDS horizon at a remarkably early date. The paper by Esaki and Tsu (1970) and which outlined the anticipated properties of a superlattice appeared some years before the authors were in a position to attempt the growth of a practical example. Apparently, the idea of applying MBE to the growth of suitable structures came from Al Cho at Bell Labs but the IBM group could claim one important innovation when they built a computer-controlled MBE machine to facilitate the preparation of the required multilayer structures (Chang et al. 1973a). Beam fluxes were measured with a quadrupole mass spectrometer and the information fed to the computer, which then controlled the timing of shutter operations so as to generate the desired structure. Typical of such was a 100-period $Al_{0.5}Ga_{0.5}As$–GaAs superlattice with well thicknesses of 6 nm and barriers of 1 nm, though, numerous other combinations emerged during the following years.

The basic idea behind the superlattice was beautifully simple. The regularly repeated potential step between well and barrier provided a second (one-dimensional) crystal lattice potential superimposed on that of the crystalline material, and the electron wave function which described its motion in the direction normal to the superlattice planes could similarly be written as a Bloch function but with the periodicity of the superlattice, rather than that of the crystal. Thus, the resulting E–k (energy–momentum) diagram took the same general form as that of the original crystal but with a much reduced zone width of π/d (rather than π/a_0), where d was the superlattice periodicity. The net effect was to divide up the crystal zone into a series of 'mini-zones' such as are shown in Figure 6.4. If d was, say, ten times greater than a_0, then there would be ten such mini-zones within the original Brillouin zone of the crystal and, as is obvious from Figure 6.4, each mini-zone would extend over a much smaller energy span than that of the crystal zone. Whereas typical electron energies were very much smaller than the crystal zone energy, this was no longer true of the mini-zones, and electrons in the lowest mini-zone could readily occupy energy states over its total span. While the crystalline E–k diagram is conveniently approximated (near its minimum) by a parabola, the mini-zone must be described in its entirety and, again as seen in Figure 6.4, this allows electrons to reach that part of the curve above the point of inflection, where quite unusual things can happen to them!

As we show in Box 6.3, electrons in energy states above the mini-zone inflection point will experience a *negative* acceleration under an applied electric field, an increase in field resulting in a decrease in velocity, suggestive of a negative resistance. However, to derive

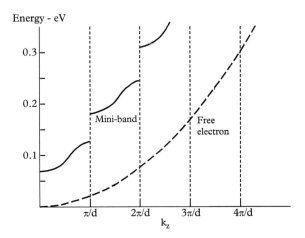

Figure 6.4 *Schematic plot of the one-dimensional E-k diagram for a typical AlGaAs/GaAs superlattice, showing the formation of mini-bands. Whereas electrons in bulk GaAs are confined to the bottom of the Γ conduction band, those in the lowest mini-band may readily achieve energies to take them above the inflection point, resulting in negative differential resistance.*

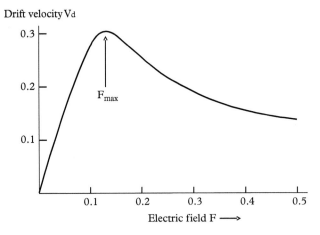

Figure 6.5 *Electron velocity-field curve as calculated by Esaki and Tsu for the case of a typical AlGaAs/GaAs superlattice (arbitrary units). The peak field F_{max} satisfies the condition $eF_{max}\tau d/(h/2\pi) \sim 1$.*

meaningful numbers from the calculation requires that electron scattering (characterised by a scattering time τ) be taken into account. Thus, Esaki and Tsu showed that, when the *average* drift velocity was evaluated, the velocity–field curve took the form shown in Figure 6.5, with a peak at a field F_{max} where $eF_{max}\tau d/(h/2\pi) \sim 1$. It followed that the desired negative differential resistance could only be seen experimentally provided the electron mean-free-path was at least three times the superlattice period d, with corresponding scattering time τ greater than 0.2 ps. This suggests that the electron mobility should exceed about 0.3 m^2 V^{-1} s^{-1}, a not unreasonable demand for electrons in good quality GaAs, even at room temperature. What, then, of the practicality?

Box 6.3 Negative resistance in a structural superlattice

The negative differential resistance in a superlattice arises from the fact that the superlattice potential splits the crystal Brillouin zone into mini-zones, as shown in Figure 6.4. Electrons in the lowest mini-zone may readily acquire sufficient energy to reach the upper part of the zone, where their effective mass is negative. In this box we provide a simplified mathematical account of this phenomenon.

Suppose that we apply an electric field F in a direction normal to the superlattice planes, thereby applying a force eF to an electron in the mini-zone and accelerating it up the E-k curve. We should like to know how the acceleration dv_z/dt varies as the electron moves up the curve and this is conveniently done by finding a relationship between the acceleration and the curvature of the E-k diagram. To do this, we first recall that the electron momentum p can be written as

$$p = mv_z = (h/2\pi)k_z \tag{B6.5}$$

and its energy E as

$$E = mv_z^2/2 = \left(h^2/8\pi^2 m\right)k_z^2 \tag{B6.6}$$

We then write the equations of motion as

$$eF = mdv_z/dt = (h/2\pi)dk_z/dt \tag{B6.7}$$

and

$$v_z = (h/2\pi)k_z/m = (2\pi/h)dE_z/dk_z \tag{B6.8}$$

from which it is straightforward to obtain a relationship between the acceleration of the electron and the curvature of the mini-zone E-k diagram. Thus, differentiating equation (B6.8), we have

$$dv_z/dt = (2\pi/h)\left(d^2E_z/dk_z^2\right)dk_z/dt$$
$$= (2\pi/h)^2\left(d^2E_z/dk_z^2\right)eF \tag{B6.9}$$

This shows that, above the point of inflection of the E–k diagram, the acceleration is negative which implies negative differential resistance. (Another way of saying all this is to note that, above the point of inflection, the effective mass of the electron for motion along the z-direction is negative.)

continued

Box 6.3 *continued*

To describe the mini-zone transport properties, we must calculate the electron drift velocity v_d as a function of electric field F and, according to Esaki and Tsu (1970), this may be done as follows:

$$v_d = \int \exp\left(-\frac{t}{\tau}\right) dv$$

$$= (2\pi/h)^2 \, eF \int \left(d^2E_z/dk_z^2\right) \exp(-t/\tau)dt$$

(B6.10)

where τ is the electron scattering time, that is, the mean time between scattering events. Esaki and Tsu showed that a plot of v_d vs electric field had the form shown in Figure 6.5, with a maximum at a field F_{max} which satisfied the following condition:

$$eF_{max}\tau d/(h/2\pi) \sim 1$$

(B6.11)

For a superlattice period d = 10 nm and a modest value of $F_{max} = 5 \times 10^5 \, V\,m^{-1}$, this leads to a scattering time of $\tau = 0.13$ ps, the significance of which may be made clear by relating it to the electron mean-free-path λ and mobility μ. Thus,

$$\mu = e\tau/m$$

(B6.12)

and

$$\lambda = \tau v_{th}$$

(B6.13)

where v_{th} is the average thermal velocity

$$v_{th} = [3kT/m]^{1/2}$$

(B6.14)

At room temperature $v_{th} \sim 4 \times 10^5 \, ms^{-1}$ and a value of $\tau = 1.3 \times 10^{-13}$ s corresponds to values of $\mu \sim 0.25 \, m^2 \, V^{-1} \, s^{-1}$ and $\lambda \sim 50$ nm. This implies that the electron mean-free-path is some five times the superlattice period d and demands an electron mobility appropriate to reasonably good quality GaAs. Such arguments clearly gave rise to a degree of confidence that the appropriate negative resistance might be obtained in real-life superlattices, once a suitable growth technology became available.

Once the necessary MBE equipment was in place, success came remarkably quickly, in the shape of the pronounced negative resistance shown in Figure 6.6 (Chang et al. 1973b). While not quite as large as the effect predicted theoretically, there can be no mistaking the qualitative agreement with the curve shown in Figure 6.5, vindicating their dedication to this innovative cause. This result, one of the best they were to obtain, represented the I-V characteristic of a GaAs/Al$_{0.5}$Ga$_{0.5}$As superlattice with GaAs layers 6 nm thick and AlGaAs layers 1 nm thick, sandwiched between two GaAs contact layers, doped at 5×10^{23} m^{-3} and 200 nm thick. The device area was a 6 μm diameter circle, defined photolithographically, to minimise the probability of involving a defective region of the sample. As can be seen, the ratio of the negative to positive slopes is about 0.1, while theoretical prediction suggested a value of roughly 0.2, a degree of agreement

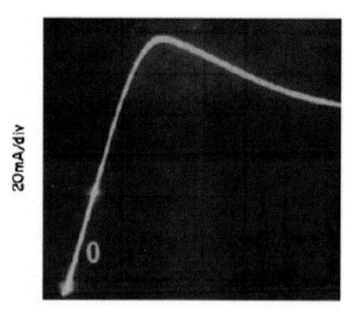

Figure 6.6 *Experimental current–voltage characteristic of a 7 nm period AlGaAs/GaAs superlattice measured with 100 ns voltage pulses with a 10 kHz repetition rate. The observed negative resistance results from the negative electron effective mass associated with conduction in the first one-dimensional superlattice mini-band. (From Chang et al. 1973b.) Reproduced with permission from Chang, L L, Esaki, L, Howard, W E and Ludeke, R (1973a) J Vac Sci Technol 10, 11. Copyright 1973, AIP Publishing LLC.*

which could very reasonably be regarded as representing success, given all the difficulties involved in both calculation and experimental realisation.

To control the quality of their GaAs and AlGaAs films, the IBM group employed the same SEM technique as used by Cho at Bell labs to check surface quality, and RHEED measurements to confirm surface reconstruction, under appropriate growth conditions. They measured Hall mobility vs doping level on a set of GaAs films, obtaining results only marginally less good than published data on LPE samples. They measured cathodoluminescence at low temperatures, finding line widths and luminescence intensities also comparable with LPE samples. They used Raman scattering to show the phonon spectra appropriate to both GaAs and AlGaAs in a superlattice structure. They were unable to obtain TEM micrographs of their structures because the necessary techniques were not then available to them but X-ray scattering measurements and a combination of argon ion sputtering and Auger electron spectroscopy were both employed to confirm the structural veracity of their superlattices (Esaki and Chang 1976). The advantage of using computer control of deposition sequences was clearly vindicated in the accuracy and regularity of the resulting structures.

Esaki's main research interest had always been the phenomenon of negative differential resistance, however it was to be generated, and there were alternatives to the superlattice mini-band method. First came that of resonant tunnelling via a double-barrier quantum system (Tsu and Esaki 1973), which we shall discuss in a moment, and somewhat later the idea put forwards by Döhler et al. (1975), also involving a super-lattice but dependent on a different mechanism. Very briefly, it concerns conduction through a superlattice in which the barrier layers are somewhat thicker than those in the previous example, typically 4 nm, rather than 1 nm. One consequence is the narrow-ing of the mini-bands, with greater localisation of the electrons in the wells. So long as adjacent mini-bands are energetically in touch with one another (i.e. are actually over-lapping), conduction can occur via an interwell band conduction mechanism, but, as the applied voltage is increased and the voltage drop across each mini-band exceeds the width of the bands, they lose touch with one another and the current inevitably drops—in other words, a negative differential resistance occurs. Experimental verification of such behaviour was obtained from measurement of photocurrents at low temperatures (Esaki and Chang 1976) but we shall look no further into the details. Suffice it to say that MBE's ability to deliver well-controlled repetitive structures with layers less than a few nanometres thick had clearly opened a new field of endeavour with many subtle ramifications.

The ultimate goal of the superlattice devotee was, of course, the structure in which each alternating layer consisted of a single monolayer, a challenge to MBE's precocity first met at Bell labs by Art Gossard in 1976 (Gossard et al. 1976). As many as 10 000 al-ternate layers of $GaAs_n$ and $AlAs_m$ were deposited by MBE on (100) GaAs substrates. Transmission electron microscopy was used to study the crystallinity and periodicity of the structures, the anticipated periodicity being confirmed in spite of the presence of some disorder. Of particular interest was the question of how the effective band gap of such a superlattice might compare with that of the corresponding random alloy, and measured optical absorption and photoluminescence at low temperature showed evi-dence for their being very similar. In particular, the absorption edge of an $n = m = 1$ superlattice appeared consistent with an indirect gap, as expected for a 50% alloy, while a sample with $n = 8$ $m = 1$ showed a sharp exciton peak, consistent with a direct en-ergy gap. As a simple demonstration of the use of such short-period superlattices, they measured the absorption spectrum of a 20 nm GaAs quantum well contained within barriers of $(GaAs)_2 (AlAs)_1$ material. The sharp exciton absorption lines observed made clear that this superlattice material could confine both electrons and holes just as the corresponding 33% random alloy might have done.

Of particular interest for MBE practitioners were the measurements made on short-period superlattices by TEM imaging, TEM diffraction and X-ray diffraction, reviewed by Gossard (1982). TEM images of samples with 10 nm layers showed well-ordered patterns, while diffraction measurements suggested that interfaces were sharp to within 1 nm or less (i.e. a few monolayers). Annealing at 900 °C for 2 h produced interdiffusion profiles characterised by interdiffusion of order 5 nm. Thinner layered samples showed clear, coherent images down to a few monolayers but a $(GaAs)_2(AlAs)_2$ superlattice revealed a number of defects suggestive of a lack of ideal planarity. This was hardly surprising, perhaps, when one considered the difficulty in preparing substrates with perfect crystalline orientation and the already well-established possibility of island growth during MBE deposition. Diffraction studies showed that growth temperature played a significant role in determining ordered layer growth but it was not possible to determine the precise amplitude of compositional modulation. This was better provided by X-ray diffraction studies which revealed a cosine wave profile from an approximately monolayer superlattice. In other words, the interpenetration of Ga and Al was of order one monolayer. Interestingly, in some cases, interface flatness on the (001) surface was seen to be anisotropic, being better in a (110) direction than for a (–110) direction, in accord with the known reconstruction which we discussed earlier. As Gossard pointed out, such measurements on short-period superlattices offered a valuable tool for the study of MBE crystal growth mechanisms.

While not employing a superlattice, the phenomenon of resonant tunnelling through a double-barrier structure has such close similarity to conduction through a superlattice that it makes good sense to discuss it here. Again, the early running was taken up by the IBM group, the basic theory having been worked out by Tsu and Esaki (1973) and the first experimental observation of the effect having been presented by Chang et al. (1973c). The concept is illustrated graphically in Figure 6.7. The pair of narrow barriers contains a confined energy state in much the same manner as in the quantum wells we discuss in Section 6.4, though, because of the partial transparency of the barriers, electrons can tunnel out of this level and only remain within the well for a finite time τ. The Heisenberg uncertainty principal therefore requires that the confined level be broadened to an extent given by $\Delta E \sim (h/2\pi\tau)$, as shown schematically in the figure. Either side of the barrier structure is a contact layer doped to such an extent that the Fermi level lies within the conduction band, and a 'sea' of electrons is available to tunnel through the barriers under the influence of an applied bias. At small values of bias the tunnelling probability is small, being roughly the product of the probabilities for tunnelling through the individual barriers, but when the bias is sufficient to bring the confined state into line with the contact Fermi level (as shown in (b), the current increases sharply due to so-called 'resonant tunnelling'. As shown in (c), further increase in bias shifts the confined state below the electron sea, removes this resonance and leads to a sudden drop in current—once again giving rise to a negative resistance which can be used as the basis of a high-frequency oscillator (an important parameter being the ratio of the peak current to that at the minimum, the so-called peak-to-valley ratio). In this regard, the device is similar to the Gunn oscillator that we discussed in Chapter 1 (Section 1.2), though, of course, the mechanism differs considerably, and an important distinction can be made in respect of the ultimate limitation to the possible frequency. Whereas the Gunn oscillator frequency is limited by intraconduction band transfer rates, the resonant tunnelling diode depends on tunnelling probability and is potentially capable of operating at very much greater frequencies. It is interesting to examine, briefly, the history of its development.

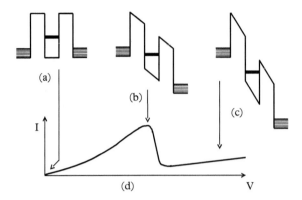

Figure 6.7 *Conduction-band energy level diagram illustrating the phenomenon of resonant tunnelling through a double-barrier structure. In (a) the applied bias is zero, as is the current. In (b) the applied bias has shifted the confined energy level so that it coincides with the 'Fermi sea' of electrons in the heavily doped contact layer— current flows as shown in (d). In (c) a further increase in bias has shifted the confined level below the sea of electrons, and current has dropped considerably, thus demonstrating a negative differential resistance.*

Needless to say, most of the early work was concerned with the AlAs/GaAs material system but the observed peak-to-valley current ratio proved disappointing, practically useful values only being available at low temperatures, a serious disadvantage for any device aimed at the electronics market. Nevertheless Solner et al. (1983) were able to demonstrate the fundamental high-frequency capability of the RTD (resonant tunnelling diode) by measuring oscillation frequencies as high as 2.5 THz. The inevitable search for room temperature operation led to the exploration of the InP-based In-GaAs/AlAs and the InSb-based InAs/AlSb systems, which showed considerably greater peak-to-valley ratios, and, in the latter case, significantly greater peak current density. Notwithstanding, the majority of workers today prefer the former, utilising, as it does, much of the technology developed for communication lasers. A typical structure is illustrated in Table 6.2, from which it can be seen just how complicated a very simple device can become. A number of comments are in order.

The combination of AlAs barriers with InGaAs wells results in increased barrier height compared to earlier AlGaAs/GaAs structures and effects a significant reduction in the valley current. The use of moderate doping levels for emitter and collector layers produces a narrower distribution of the electron sea and sharpens the resonance with the energy level within the well but it then requires a more heavily doped contact layer to reduce the inevitable series resistance. The undoped InGaAs spacer layers were originally introduced with the intention of reducing diffusion of dopant atoms into the central structure but it soon came to be recognised that they also reduced device

Table 6.2 *Typical RTD structure based on InP/InGaAs/AlAs.*

1	$In_{0.53}Ga_{0.47}As$	$n = 5 \times 10^{24} \text{ m}^{-3}$	100 nm	Contact layer
2	$In_{0.53}Ga_{0.47}As$	$n = 10^{23} \text{ m}^{-3}$	100 nm	Collector layer
3	$In_{0.53}Ga_{0.47}As$	Undoped	2 nm	Spacer layer
4	AlAs	Undoped	2 nm	Barrier layer
5	$In_{0.53}Ga_{0.47}As$	Undoped	6 nm	Quantum well
6	AlAs	Undoped	2 nm	Barrier layer
7	$In_{0.53}Ga_{0.47}As$	Undoped	2 nm	Spacer layer
8	$In_{0.53}Ga_{0.47}As$	$n = 10^{23} \text{ m}^{-3}$	100 nm	Emitter layer
9	$In_{0.53}Ga_{0.47}As$	$n = 5 \times 10^{24} \text{ m}^{-3}$	300 nm	Contact layer
10	InP	SI		Substrate

capacitance, important from the point of view of high-frequency operation. Finally, the use of a semi-insulating InP substrate requires that the device be defined photolithographically in the shape of a mesa to facilitate metallisation of the lower contact layer. In fact, device areas must be defined in the range of roughly 1–5 μm^2 in order to minimise capacitance, a requirement which inevitably reduces the amount of power available from a single device. Typically, modern devices generate power levels of the order of 10 μW at frequencies of the order of 0.5 THz, though it is possible to combine a number of devices so that they oscillate coherently. We shall attempt no further exploration of the necessary circuit design—interesting details can be found in the review article by Asada et al. (2008). However, we should take up one vital matter concerning the MBE growth of these device structures, the question of layer uniformity.

As discussed in the review article by Liu and Sollner (1994), the frequency limit on RTDs is set by the lifetime of electrons in the resonant confined state, and this, in turn, is determined by the thickness of the barriers. Lifetimes as short as 1 ps, appropriate to terahertz operation, require barriers of order 1 nm thickness, which puts severe constraints on the uniformity of growth structures. The use of such thin barriers (of the order of just a few monolayers), in order to allow adequate tunnelling probability, inevitably leads to a degree of sensitivity to minor variations in layer thickness. Even though practical devices utilise small areas, we should recognise that a lateral dimension of just 1 μm still includes something like 2000 unit cells, and the probability of finding monolayer variations in thickness over a 1 μm^2 area is obviously far from tiny. What effect should one expect from monolayer variations in thickness? With regard to the barriers, narrow regions give rise to an increase in non-resonant tunnelling and therefore an increased valley current. Variations in well thickness broaden the confined energy level and result in a less sharp resonance, not only reducing the peak current but also the slope of the negative resistance region. Note, too, that narrow barriers also broaden the resonance by shortening the lifetime of carriers within the well. Even worse, is the fact that thin barrier regions result in current leakage which effectively short-circuits the required device characteristics.

It is clear, then, that one would expect a significantly smaller peak-to-valley ratio and corresponding degradation in performance as a result of imperfections in layer smoothness and it is therefore imperative that every effort be made to grow structures with smooth interfaces. As we saw in Chapter 5, Section 5.2, one appropriate technique makes use of growth interrupts to effect smoothing of, say, a GaAs layer before depositing AlAs upon it, and an interesting application has recently been described by Zhang et al. (2011) of the Chinese Academy of Science, Beijing. They were concerned with diodes based on AlAs/InGaAs/InP and studied the effect of growth interruptions of between 10 and 60 seconds at the AlAs–InGaAs interfaces. TEM images of the barrier layers showed clear evidence of improved smoothness, with an optimum interrupt time of 40 s. As the authors point out, two factors are involved; the well-established layer smoothing but also the effect of strain (AlAs and $In_{0.53}Ga_{0.47}As$ are mismatched by nearly 4%) and the interplay between them seems likely to account for the optimum interrupt time.

6.3 Doping Superlattices

While the concept of a structural superlattice is probably simpler to appreciate, we should recognise that any form of repetitive modification to a semiconductor can equally well be used to modulate its properties. The doping superlattice, first proposed by Döhler (1972), is one such, having the advantage over its structural counterpart of not being dependent on lattice-matching of different materials. Rather than using a sequence of layers having different band gaps, the idea was to alternate nanoscale regions of n-type and p-type doping. A typical example might consist of a sequence of 20 GaAs layers, some 30–40 nm thick, doped alternately with silicon (n-type) and beryllium (p-type), though this might often be modified by the inclusion of undoped (intrinsic) layers between them. This led to the use of the term 'n-i-p-i superlattice'. The main interest in such structures lay in the possibility of controlling the effective band gap and the recombination lifetime of electrons and holes, with consequent practical application in photodetection, though, as we shall see, various other applications materialised, once the necessary growth technology became available.

In the well-established theory of p–n junction behaviour, the depletion region which separates n- and p-type material contains ionised donor atoms (positively charged) adjacent to the n-side and ionised acceptors (negatively charged) on the p-side. These fixed charges are responsible for a strong electric field which sweeps electrons and holes into the n- and p-materials, respectively. Similar behaviour is obviously to be expected in a n-i-p-i structure, being repeated at each junction and resulting in a modulation of the conduction and valence bands of the constituent semiconductor, as shown in Figure 6.8. If we consider the special case where the total positive charge is equal to the total negative charge, that is,

$$N_D d_n = N_A d_p \qquad (6.1)$$

where N_D and N_A are the densities of donors and acceptors, while d_n and d_p are the appropriate thicknesses of the doping regions, it follows that all the electrons freed from donors will be captured by acceptors and, conversely, for holes. There will therefore

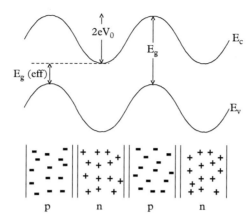

Figure 6.8 *Plots of conduction and valence band edges compared with the positions of doping planes in a n-i-p-i superlattice. Ionised donor and acceptor atoms produce electric fields which modulate the band edges and result in a significant reduction in effective band gap, compared with that of the host semiconductor.*

be no free carriers within the structure at equilibrium. However, one important feature apparent in Figure 6.8 is the large change in effective band gap which results from the periodic doping, E_g(eff) being considerably less than the semiconductor band gap E_g. Thus,

$$E_g(\text{eff}) = E_g - 2eV_0 \tag{6.2}$$

where this modulation of the conduction and valence bands can be calculated by the usual double integration of the Poisson equation. The result is

$$V_0 = \rho d_n / 8\varepsilon\varepsilon_0 \tag{6.3}$$

where $\rho = eN_D d_n$ is the charge per unit area per layer, and ε is the relative dielectric constant of the semiconductor. Thus, if $N_D = 5 \times 10^{24} \text{ m}^{-3}$ and $d_n = 10$ nm, we obtain a value of $2eV_0 = 0.19$ eV (though $2eV_0$ can even be made as large as E_g by appropriate choice of doping level and superlattice period, thereby producing a n-i-p-i semimetal).

  This reduction in band gap implies that optical absorption can occur for photon energies well below the usual absorption edge. Clearly, the appropriate transition involves a spatial jump between adjacent doping regions but, since the superlattice period is small, there is adequate overlap of appropriate electron wave functions for the transition to be allowed. On the other hand, recombination of electrons and holes occurs with much reduced probability, compared to that in the host crystal, leading to a much enhanced photoconductive response. (Bear in mind that, though transport normal to the doping planes is strongly inhibited by the potential barriers apparent in Figure 6.8, conduction in the plane of the layers is freely allowed.) Two exciting possibilities follow from these simple ideas: it is obviously possible to 'tune' the effective band gap by varying the doping levels and, rather less obviously, it is possible to effect a similar change either by applying an external voltage normal to the doping planes or by absorption of near-band-edge photons. Döhler was responsible for a considerable burst of anticipation when he published his theoretical findings in 1972 but it was to take almost a decade before MBE techniques could be sufficiently refined to demonstrate their practical realisation. Let us briefly examine how it all came about.

When Döhler published his original articles, he was working at the Max Planck Institute in Stuttgart, where there was, at the time, no possibility of fabricating suitable practical structures. In 1974 he took time off as a visiting scientist at the IBM Yorktown Heights Laboratory, where he worked with Esaki on the theory of structural superlattices and where he became aware of the capabilities of MBE to grow suitable layers (the IBM group reported its first practical realisation of superlattices in 1973). It was also during 1973 that an aspiring young chemist, Klaus Ploog, accepted an offer to set up an MBE facility in Stuttgart, with the intention of growing ultra-thin multilayer structures and, on his return from America, Döhler joined him in this worthy enterprise. Their first joint activity was, not surprisingly, aimed at growing doping superlattices in GaAs and this, of course, required the ability to dope the material both n- and p-type at densities of $10^{24}\,\mathrm{m}^{-3}$ or more. It was also important that the dopants should stay where they were put during growth—significant diffusion at the growth temperature could not be countenanced. The choice of donors lay between Si, Ge and Sn but Sn could be ruled out for this application by its tendency to segregate at the growth surface, thus preventing the achievement of a sharp doping profile. In contrast, both Si and Ge acted as satisfactory donors under the commonly used As-stabilised growth conditions. The choice of acceptors lay between Mn, Mg, Zn and Ge. The Mn acceptor level was too deep for comfort, Mg had a tendency to diffuse rather rapidly, Zn could only be incorporated in the form of positive ions (an inconvenience, to say the least) and Ge acted as an acceptor only if growth took place under Ga-stabilised conditions. It was Hobson's choice, and Döhler and Ploog (1979) chose Ge for both donor and acceptor, on the basis that they could vary the As flux to control its doping behaviour. Thus, in 1976, having set up the necessary growth system, they were ready to tackle the problems of growing n-i-p-i structures. The first publication (Ploog et al. 1979) reported promising results for the use of Ge as a donor atom but there were obvious difficulties with Ga-stabilised growth and a further difficulty in that Ge suffered from a 'memory effect'—it lingered in the MBE system and appeared in all subsequent films even when not deliberately incorporated—and this meant that light n-type doping was very difficult to achieve. Salvation came from the work of Marc Ilegems at Bell Labs in demonstrating the virtues of Be as a well-behaved acceptor (Ilegems 1977) and a change to Si as the preferred donor. By 1980, all was well and the first successful doping superlattices were available for further study (Ploog et al. 1981a).

The lengthy struggle to reach this stage certainly proved worthwhile—within a matter of a single year, not only was high-quality GaAs grown but the basic properties of n-i-p-i structures were clearly demonstrated. Optical absorption at photon energies well below the GaAs band gap, band-gap tuning both with an applied voltage and with intense radiation levels, and photoluminescence from confined energy states within the conduction-band wells were all unequivocally measured and successfully modelled. Good quantitative agreement between theory and experiment was obtained for the first time. Döhler's long crusade to establish the theory was finally vindicated, Ploog's pioneering efforts to set up a viable growth activity were crowned with success and MBE could look forwards to an even brighter future.

It was first necessary to produce bulk layers of suitably doped GaAs and to demonstrate its quality. The growth temperature was chosen to be 530 °C in order to minimise impurity diffusion. Silicon-doped films were grown with $(N_D - N_A) = 4 \times 10^{21} \, \text{m}^{-3} - 6 \times 10^{24} \, \text{m}^{-3}$, and beryllium-doped films with $(N_A - N_D) = 4 \times 10^{21} \, \text{m}^{-3} - 4 \times 10^{25} \, \text{m}^{-3}$. Free-carrier mobilities were measured which compared very favourably with similar material grown by LPE or VPE, and low-temperature photoluminescence showed sharp exciton lines as further confirmation of material quality, p-type layers being of better quality than any previously grown by MBE as a result of the use of ultra-pure beryllium. Various n-i-p-i superlattices were then grown and their optical absorption spectra compared. Two samples were of particular interest, being doped equally at $n = p = 10^{24} \, \text{m}^{-3}$ but with different doping layer thicknesses of $d_n = d_p = 20 \, \text{nm}$ and 40 nm, respectively. According to equations (6.2) and (6.3), both samples should be characterised by effective band gaps significantly less than that of GaAs; indeed, both showed optical absorption below E_g but the second sample (having $d_n = 40 \, \text{nm}$) showed 'remarkably' greater absorption than its counterpart (again, as predicted by equation (6.3), V_0 being proportional to $d_n{}^2$). The flood gates were opened and streams of further data could now pour through.

In a second paper, following close on the heels of the first, the same team (Ploog et al. 1981b) reported on the modulation of the free-carrier concentration in both the n-type and p-type regions of a doping superlattice by the application of an external potential difference. Samples consisted of 20 alternately n-doped and p-doped layers to which electrical contacts were made by the alloying of small Sn or Sn/Zn balls. These made contact to n- and p-type layers, respectively, and allowed separate measurement of the modulation of electron and hole conductivities (at 4.2 K). Three different types of sample were grown, characterised by (a) $N_D d_n > N_A d_p$, (b) $N_A d_p > N_D d_n$ and (c) $N_D d_n = N_A d_p$, representing n-type, p-type and compensated samples, respectively but the relatively large values of d_n and d_p (of the order of 200 nm) resulted in the structure behaving like a semimetal at zero bias. An external voltage, applied between the n- and p-type contacts, had the effect of splitting apart the electron and hole quasi-Fermi levels (thus effectively tuning the energy gap) and forcing them further into the respective energy bands, thereby increasing free-carrier densities. In practice, this resulted in modulation of both electron and hole areal densities by as much as $10^{16} \, \text{m}^{-2}$ in each layer. It is important to recognise that the *simultaneous* modulation of both carrier types is made possible by the extremely long recombination lifetimes associated with these structures (due to the spatial separation of conduction-band minima and valence-band maxima).

In a third paper the Max Planck group (with a little help from the Technical University of Munich) demonstrated the possibility of tuning the energy gap by means of an incident photon flux (Döhler et al. 1981). The MBE samples used consisted of 20 periods, with $N_D = N_A = 10^{24}$ m^{-3} and $d_n = d_p = 40$ nm. The experiments involved photoluminescence and Raman spectroscopy at 4.2 K. PL was excited by a Kr$^+$ gas laser with varying intensities of up to 10^8 Wm^{-2}. At maximum intensity, the measured emission lay close to the GaAs band edge at 1.52 eV, while gradual reduction in intensity produced a downward shift in emission energy by 200 meV, a measure of the decrease in band gap resulting from a corresponding decrease in free-carrier density. Another aspect of the observed photoluminescence concerned the precise nature of the recombination process. As shown schematically in Figure 6.9, it involved electrons in confined states in the conduction-band quantum well recombining with holes trapped on acceptors, and we need to understand three aspects. First, because the well is roughly parabolic in shape, the confined states are approximately equally spaced—the precise separation of about 20 meV having been measured directly by Raman scattering. Second, because the beryllium acceptor level is much deeper than the corresponding donor state, acceptors near the valence band maximum are not ionised at 4.2 K. Third, the overlap of electron wave function with acceptor states is greater for the more energetic confined levels and this leads to an emission line which is sharper on the high energy side, as observed experimentally. Detailed calculation of the occupancy of these confined states (i.e. the position of the electron Fermi level) as a function of excitation intensity resulted in a theoretical line shape and emission energy in excellent agreement with measured values, confirming not only the recombination model but also the overall understanding of the electronic properties of these n-i-p-i structures. Nor should we forget that such detailed comparison depended critically on the fact that the growth process was under near-perfect control and the growth temperature was low enough to minimise impurity diffusion—in short, that the structures were exactly what they were supposed to be!

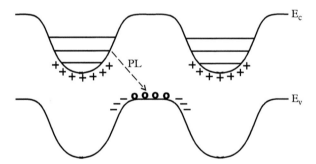

Figure 6.9 *Schematic plot of conduction and valence bands at 4.2 K for a superlattice having $N_D = N_A = 10^{24}$ m^{-3} and $d_n = d_p = 40$ nm. Photoluminescence results from recombination between electrons in conduction-band confined states and holes trapped on acceptor atoms near the valence band maxima.*

So far in our discussion of doping superlattices, we have been concerned with what were essentially n-p-n-p structures—the 'i' bits of the n-i-p-i were mere matters of speculation. However, a final paper ('final' only from the viewpoint of this discussion) from the group in Stuttgart (Schubert et al. 1985) put this very much to rights by describing experiments on a 'sawtooth semiconductor superlattice', in which the i-layers dominated. In fact, the doping was confined not merely to within a dimension of order 30 nm but to single atomic planes, that is, of the order of 0.3 nm. The technique, variously described as 'delta doping', planar doping', 'sheet doping' and 'atomic plane doping', was first established by Colin Wood and collaborators at Cornell University (Wood et al. 1980) who used it to synthesise unusual doping profiles in MBE-grown GaAs by combining several such planes of the n-type dopant germanium. Confining the dopant atoms to a single atomic plane was achieved by shuttering off the Ga beam while opening the Ge shutter. Thus, the growth of GaAs was momentarily halted while the dopant was deposited under As-stabilised conditions (which ensured that the Ge substituted on Ga sites), and then normal growth was continued by reversing the procedure. Schubert et al. referred to it as an 'impurity growth mode'. Wood et al. were unable to achieve p-type delta doping with Ge but the Stuttgart group had no difficulty in obtaining both types by alternating Si and Be doping planes—in fact, their sawtooth structure was just that—a repetitive sequence of equally doped Si and Be planes separated by 10 nm. It was of special interest in being quite remarkably different from the earlier structures in some of its optoelectronic properties.

Because the space between each doping plane was intrinsic, the profiles of conduction and valence band edges showed strictly linear variation, as illustrated in Figure 6.10, hence the 'sawtooth' designation. And, because of the rather short periodicity of 20 nm, the modulation of the band edges was fairly modest. For this case, it is easy to show that the amplitude of the zigzag is given by

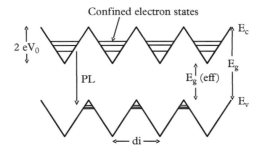

Figure 6.10 *'Zigzag' doping superlattice produced by the use of Si and Be planar doping. Typical doping levels might be $2 \times 10^{16}\ m^{-2}$, and interlayer spacing, 10 nm. Luminescence occurs by recombination of electrons and holes in CB- and VB-confined states.*

$$2V_0 = \rho d_i / 2\varepsilon\varepsilon_0 \qquad\qquad (6.4)$$

where ρ is again the charge in each doping plane and d_i is the distance between Si and Be planes. Thus, for a planar-doping density of 10^{16} m^{-2} and $d_i = 10$ nm, we find $2V_0 = 70$ mV. Also, because of the zigzag format, we must expect to find confined electron states within the conduction-band profile and corresponding hole states in the valence band. Finally, because the periodicity is short, there is good overlap of electron and hole wave functions, as shown schematically in Figure 6.10. In other words, we must expect strongly allowed interband optical transitions of similar character to the quantum well structures we discuss in Section 6.4. Note that this implies significantly shorter recombination lifetimes, compared to those appropriate to the earlier n-i-p-i structures. One important consequence is that it is no longer possible to tune the band gap with an incident photon flux—the rapid recombination prevents any build-up of free carriers, which, in the earlier structures, screened the fixed depletion layer charges and resulted in an effective change of $E_g(\text{eff})$.

So much for expectation—what of experiment? Samples were grown by MBE at a temperature of 550 °C, planar-doped at levels of 2×10^{16} m^{-2}, with periodicity of 20 nm and characterised by measurements of photoluminescence at both 2 K and 77 K. Three comments about the PL spectra are in order: first, the peak of the emission occurred just 70 meV below the GaAs band edge (at both temperatures), second, the emission energy was independent of excitation intensity over more than four orders of magnitude and, third, the width of the emission line lay in the range 40–50 meV and was nearly independent of temperature and excitation intensity. There was no sign of luminescence at the GaAs band edge, an observation which confirmed the model of recombination between confined electron and heavy-hole states, and detailed modelling showed good agreement with the 70 meV shift of emission energy. The lack of sensitivity to excitation intensity is consistent with the anticipated short radiative recombination lifetime, an estimate of which, from the measured onset of band filling, gave a value of 3.3 ns, comparable with values typical of bulk GaAs. The large, almost temperature-independent line width seemed to imply significant statistical fluctuations in either the planar-doping level or doping-plane separation, resulting from variations in growth parameters.

Clearly, these particular samples behaved very differently from the earlier n-i-p-i samples but it was also clear that the reason for this was the strong overlap of electron and hole wave functions resulting from the short periodicity. Later work (Ploog 1987) showed that, when the periodicity of these sawtooth superlattices was increased, the recombination lifetime also increased and band-gap tuning became possible, just as found in the original n-i-p-i structures. Here was another excellent example of the understanding to be achieved when theoretical acumen, experimental skill and MBE growth precision could be brought together in one laboratory. The n-i-p-i story is surely one for its authors to feel proud of but we should, perhaps, add one caveat—the unexpectedly large PL line width referred to above suggests that control of MBE growth parameters may sometimes be more difficult than its practitioners might imagine.

6.4   Quantum Wells

We chose to discuss superlattices first in this chapter largely because the original proposals for their utilisation appeared in 1970 (structural superlattices) and 1972 (doping superlattices) but there can be little doubt that an even greater stimulus to the development of low-dimensional structures came with the experimental demonstration of the optical properties of quantum wells in 1974. As with so much else in the history of MBE, this advance was made in the Bell Laboratories in Murray Hill and published under the title 'Quantum states of confined carriers in very thin $Al_xGa_{1-x}As$-GaAs-$Al_xGa_{1-x}As$ heterostructures' (Dingle et al. 1974; see also the review, Dingle 1975). These 'very thin' GaAs layers were, in fact, multilayers (with as many as 50 repeats) separated by relatively thick barriers (> 25 nm) with x = 0.2 and, as the electron wave functions penetrated no more than 2.5 nm into the barriers, they were effectively isolated wells—here was no suggestion of superlattice behaviour. The multilayer structure was introduced simply in order to increase the optical absorption which was used to characterise the confined energy states within the wells. Absorption measurements at 2 K, made on samples from which the GaAs substrates had been removed using a selective etch, showed a series of exciton peaks corresponding to transitions between confined hole states in the valence band well, and confined electron states in the conduction-band well (see Figure 6.11). Seven samples with differing well widths, between 7 nm and 48 nm, provided information on the well-width dependence of these excitonic features and allowed detailed comparison with a theoretical model. The positions of all the observed features were calculated with a remarkable degree of accuracy, confirming beyond reasonable doubt the nature of the absorption lines and, at least to first order, the correctness of the calculated confinement energies. Here was clear evidence of completely new semiconductor behaviour, dependent solely on the quantum properties of very thin layers. It was a remarkable demonstration of the ability of MBE to generate well-controlled ultra-thin films and it opened a whole new spectrum of scientific and device possibilities. Many subtleties were involved and we need to examine some of the features in greater detail.

We shall begin by looking at the theoretical modelling. In order to determine the energy of each exciton peak it is necessary to know a whole range of parameter values, such as the band gap of the GaAs, the aluminium content of the AlGaAs, the so-called band offset ratio which defines the fraction of the energy difference that appears in the conduction band (and thereby also in the valence band), the well width (indeed the well *widths* in the case of a multiwell structure), the exciton binding energy and the electron and hole effective masses both in the well material and in the barriers. Obviously, some of these factors are directly determined by the degree of control of the MBE process which is used to build the structures, some are reasonably well-established semiconductor parameters (others not so well established), while others still lie in the realms of theoretical subtlety. We need to look carefully at them and try to estimate the degree of accuracy and reliability associated with each before we can feel entirely confident that we understand the data in all its complexity. Needless to say, it is an enormous field and we can do no more than offer a superficial outline. Our chief interest must be, of course, those aspects which relate directly to MBE growth but we shall begin with the question

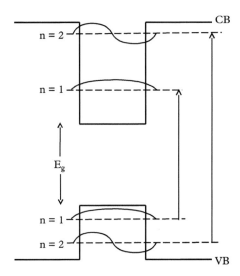

Figure 6.11 *Schematic illustration of optical absorption in a single square quantum well. Electrons are excited from confined states in the valence band into similar states in the conduction band. Allowed transitions are characterised by Δn = 0. Note that the diagram is very far from being to scale—E_g is typically about 1.5 eV, while confinement energies are of the order of 100 meV. Note too that the valence band in GaAs should be represented by light- and heavy-hole states, which are split apart by the reduced symmetry of the quantum well. Thus, each of the transitions shown in the figure would be split into two components, designated (lh_n–e_n) and (hh_n–e_n).*

of how to calculate the confined energies—the central feature of the low-dimensional aspect of the story.

A convenient starting point is the standard text-book calculation of confined energy states in an infinite quantum well, summarised in Box 6.4 (taken from Chapter 6 of Orton 2004). The essential result is that the energy of electrons or holes may be written as

$$E = E_n + \left(h^2/8\pi^2 m_w\right)\left(k_x^2 + k_y^2\right) \tag{6.5}$$

with

$$E_n = \left(h^2/8m_w\right)\left(n^2/L_z^2\right) \tag{6.6}$$

Box 6.4 Quantum confined energy levels

The optical properties of semiconductor quantum wells are defined largely in terms of the confinement energies of electrons in the CB well and of holes in the VB well. It is convenient to treat the calculation of these energies in two parts—first considering the case of an infinite well, then dealing with the modifications appropriate to a finite well. We begin, then, with the straightforward problem of an electron in an infinite well, for which we reproduce the account given in Orton (2004) (Box 6.1).

The confinement of an electron in a potential well forms a parallel with a classic elementary problem in wave mechanics which finds a niche in most quantum mechanics textbooks—that of a particle in a finite box. The quantum well problem is actually somewhat simpler on account of its one-dimensional nature. It consists of solving the Schrödinger equation for a free electron which is constrained within coordinates $z = -L_z/2$ to $z = +L_z/2$—in fact, the physics is really contained in the boundary conditions. If we assume that the well is infinitely deep, there is zero probability of finding the electron inside the barrier so, in order to match the wave functions in the well and in the barrier, it is necessary to specify that $\psi(z) = 0$ at $z = \pm L_z/2$. To give the problem maximum simplicity, we define the zero of energy to lie at the bottom of the well, so that, within the well, $V(z) = 0$ everywhere. The Schrödinger equation then takes the form

$$- (h^2/8\pi^2 m_w) d^2 \psi_n/dz^2 = E_n \psi_n \tag{B6.15}$$

where we treat the electron as essentially free but take account of the fact that it resides in a semiconductor crystal by giving it a mass m_w rather than the free electron mass m_0. It is easy to see that the solution has the form

$$\psi = A\cos kz \quad \text{or} \quad A\sin kz \tag{B6.16}$$

where $[\psi(z)]^2$ represents the probability of finding the electron at the point z.

Consider the cos kz solution. In order to satisfy the boundary conditions, we require that $\cos(kL_z/2) = 0$, or that $kL_z/2 = \pm\pi/2, \pm 3\pi/2, \pm 5\pi/2$, etc. In other words, $k = n\pi/L_z$, where $n = 1, 3, 5$, etc. These are the solutions with even symmetry:

$$\psi_n = A\cos(n\pi z/L_z) \quad \text{with } n = 1, 3, 5, \text{ etc.} \tag{B6.17}$$

Similarly, the odd solutions are

$$\psi_n = A\sin(n\pi z/L_z) \quad \text{with } n = 2, 4, 6, \text{ etc.} \tag{B6.18}$$

The constant A is not completely arbitrary because the probability for the electron to be somewhere within the well must be unity, so we can write

$$\int \psi^2 dz = 1 \tag{B6.19}$$

from which it follows that $A = \{2/L_z\}^{1/2}$, though this is not a result we shall make much use of.

Much more importantly, we can obtain the allowed energies by substituting these wave functions back into equation (B6.15) and find

$$E_n = h^2 k^2/8\pi^2 m_w = n^2 h^2/8 m_w L_z{}^2 \tag{B6.20}$$

continued

Box 6.4 *continued*

Note that these energies are measured upwards from the bottom of the well—they are referred to as 'confinement energies'—and form a sequence with increasing spacing as the energy increases (because of the proportionality to n^2). Because the well is infinitely deep, there is an infinite number of confined states.

This account shows some resemblance to the solution of the simple harmonic oscillator, which corresponds to the case of a parabolic well, rather than the square well assumed here. In that case, of course (see any text on quantum mechanics), the energies are proportional to n (rather than n^2) and are therefore equally spaced but again there is an infinite number of allowed states.

It may be helpful to think of these solutions as representing an electron bouncing backwards and forwards between the impermeable barriers at $z = \pm L_z/2$. By virtue of the motion, it possesses a momentum p which varies with z and reverses at $z = \pm L_z/2$. The average value of p must be zero because the electron spends as much time travelling towards positive z as towards negative z. However, the average of p^2 is clearly non-zero. In fact,

$$< p_n^2 >_{av} = 2m_w E_n = n^2 h^2 / 4L_z^2 \tag{B6.21}$$

This result is interesting in that it enables us to find a value for the de Broglie wavelength of the electron λ_n:

$$\lambda_n = h/p_n = 2L_z/n \tag{B6.22}$$

In other words, $L_z = n\lambda_n/2$, showing that the de Broglie wave of the electron corresponds precisely to the wave function found from the Schrödinger equation.

A similar concept of wave function matching emerges from the old quantum theory of the hydrogen atom. The quantum condition is introduced into this model by way of the Planck condition on the angular momentum p_ϕ.

$$\int p_\phi d\phi = nh \tag{B6.23}$$

And, as angular momentum is a constant of the motion, $\int p_\varnothing\, d\phi = p_\phi \int d\phi = 2\pi p_\phi$. Therefore,

$$p_\phi = nh/2\pi \tag{B6.24}$$

Using the relation $p_\phi = ap$, where a is the radius of the orbit, allows us to write the expression for the de Broglie wavelength as

$$\lambda_n = h/p = ha/p_\phi = 2\pi a/n \tag{B6.25}$$

Thus, the Planck condition implies that the circumference of the orbit must be an integer number of wavelengths, $2\pi a = n\lambda_n$.

So much for the infinite well. How must the above account be modified for the case of a finite well of depth eV_w? The essential difference is that the electron wave function no longer goes to zero at the boundaries $z = \pm L_z/2$ but penetrates a short distance into the barriers, where it decays exponentially with distance from the interface. Thus, for states with even symmetry, we can write the wave function in the form

$$\psi_n = B \exp(\kappa_n z) \quad z < -L_z/2$$

$$\psi_n = A \cos(k_n z) \quad -L_z/2 < z < L_z/2 \qquad (B6.26)$$

$$\psi_n = B \exp(-\kappa_n z) \quad z > L_z/2$$

with a corresponding sin $(k_n z)$ set for odd symmetry. The condition for matching at the interface requires that the amplitudes are equal, so $A \cos(k_n L_z/2) = B \exp(-\kappa_n L_z/2)$, and that the particle current $(1/m)(d\psi/dz)$ is continuous, so $(k_n/m_w) A \sin(k_n L_z/2) = (\kappa_n/m_b) B \exp(-\kappa_n L_z/2)$. Combining these two conditions yields

$$\tan(k_n L_z/2) = m_w \kappa_n / m_b k_n \qquad (B6.27)$$

and feeding the wave functions (B6.26) back into the Schrödinger equation then yields two further equations:

$$E_n = h^2 k_n^2 / 8\pi^2 m_w \qquad (B6.28)$$

and

$$eV_w - E_n = h^2 \kappa_n^2 / 8\pi^2 m_b \qquad (B6.29)$$

Finally, eliminating k_n and κ_n from these results in a single equation which can be solved numerically to evaluate the confinement energy E_n:

$$eV_w = E_n \left\{ 1 + (m_b/m_w) \tan^2 \left[(\pi/2)(E_n/E_1^*)^{1/2} \right] \right\} \qquad (B6.30)$$

where E_1^* is the first confined state energy in an infinite well of the same width as the one under consideration and can be calculated by using equation (B6.20). For states with odd symmetry the $\tan^2[]$ term is replaced by $\cot^2[]$. Note that setting $n = 1$ implies an angle within the range $0 - \pi/2$, $n = 2$ within the range $\pi/2 - \pi$, $n = 3$, $\pi - 3\pi/2$ and so on.

where m_w is the electron or hole effective mass in the well, L_z is the well width and $n = 1, 2, 3, \dots$. Equation (6.5) describes the band behaviour in the plane of the well, k_x and k_y representing the components of electron or hole momentum in the xy plane, but shows that the appropriate band minima lie at energies E_n above the bottom of the well (for electrons) or below the top of the well (for holes). As shown in Box 6.4, for the case of the infinite well, the wave functions in the z-direction are sine or cosine waves which go to zero at the well–barrier interfaces. In other words, these functions are confined totally within the well and do not penetrate the barrier material. There is an infinite number of confined levels, with energies increasing as n^2. However, for a finite well, neither of these statements is true—the number of levels decreases as L_z decreases (though there is always at least one), while the wave functions penetrate a finite distance into the barrier material, as shown in Figure 6.11. One further complication must be taken into account—the GaAs valence band consists of three components: a light-hole band, a heavy-hole band and a spin-orbit split-off band. In the cubic symmetry of the bulk material the light- and heavy-hole bands are degenerate at $k = 0$, while the split-off band lies approximately 0.35 eV below. In the axial symmetry of the quantum well, the light and heavy holes are split apart, with the heavy-hole states lying above their light-hole counterparts (i.e. the *confinement energies* of the heavy-hole states are smaller than the corresponding light-hole states).

To calculate the various confinement energies in *finite* wells we follow, for example, Weisbuch and Vinter (1991) or Nelson (2001). By matching both wave function and particle current at the well-barrier interface (see Box 6.4) it is straightforward to show that the confinement energy E_n is found by numerical solution of the following equations:

$$eV_w = E_n \left\{ 1 + (m_b/m_w) \tan^2 \left[(\pi/2)(E_n/E_1{}^*)^{1/2} \right] \right\} \quad (n = 1, 3, 5, \text{etc.}) \qquad (6.7)$$

$$eV_w = E_n \left\{ 1 + (m_b/m_w) \cot^2 \left[(\pi/2)(E_n/E_1{}^*)^{1/2} \right] \right\} \quad (n = 2, 4, 6, \text{etc.}) \qquad (6.8)$$

where eV_w is the well depth, m_b and m_w are the masses of the appropriate carrier in barrier and well, respectively, and $E_1{}^*$ is the confinement energy of the lowest state in an infinite well. Thus,

$$E_1{}^* = \left(h^2/8m_w L_z{}^2 \right) \qquad (6.9)$$

Finally, we also note that the number N of states within any specific well is given by the formula

$$N = 1 + \text{Int} \left[\left(8m_w eV_w L_z{}^2/h^2 \right)^{1/2} \right]$$
$$= 1 + \text{Int} \left[(eV_w/E_1{}^*)^{1/2} \right] \qquad (6.10)$$

where 'Int' means 'the integer part of'. For example, if we take $m_w = 0.08m_0$, $eV_w = 38$ meV and $L_z = 10$ nm, we find $N = 1 + \text{Int}[0.90] = 1$—there is only one light-hole state in such a well. However, if we increase L_z to 14 nm, we obtain $N = 1 + \text{Int}[1.26] = 2$—two light-hole states are now possible. Alternatively, if we increase the well depth to 50 meV, two light-hole states are again possible. In general, the wider the well, the greater the number of confined energy states—in the case of very large well widths, they become so close together that they merge into a continuum, as they obviously must.

Let us now return to the problem of fitting experimental absorption energies with an appropriate model. It is first necessary to take account of the selection rule $\Delta n = 0$, which implies that the energy of the *exciton* feature associated with the nth hole and electron states (see Figure 6.11) is given by

$$h\upsilon_n = E_g + (E_{ne} + E_{nh}) - E_{ex} \qquad (6.11)$$

where E_g is the band gap of the well material, E_{ne} is the confinement energy of the electron state, E_{nh} that of the hole state and E_{ex} is the (quasi-two-dimensional) exciton binding energy. Thus, for GaAs at 2 K, $E_g = 1.520$ eV, $(E_{ne} + E_{nh})$ amounts to something like 20–200 meV, and E_{ex} to roughly 10 meV, so we can expect $h\upsilon_n \sim 1.53$–1.70 eV, as indeed was observed by Dingle et al. (1974). For a detailed analysis, we need to know the well width L_z and both the conduction and valence band-well depths eV_w, so we can calculate the confinement energies, as well as the way in which E_{ex} depends on well width. Neither is entirely straightforward. For the case of $Al_xGa_{1-x}As/GaAs$ structures, the CB well depth is given by

$$eV_w = 1.247x \times Q_c \qquad (6.12)$$

where Q_c is the fraction of the band-gap difference which appears in the conduction band, while the corresponding valence band-well depth is

$$eV_w = 1.247x \times Q_v \qquad (6.13)$$

where $Q_v = 1 - Q_c$. In 1974 no one had any firm idea what value to assign to Q_c, so Dingle et al. treated it as a variable to be derived from their experimental data.

In bulk GaAs the exciton binding energy was already well established, having a value of approximately 4.2 meV, so the exciton peak lay just 4.2 meV below the band edge, that is, at a photon energy of 1.515 eV. However, in the case of the quantum well, there was considerably greater uncertainty in the value to be anticipated for the quasi-two-dimensional exciton. In the limit of ideal two-dimensionality, the binding energy should be just four times that in three dimensions, that is, about 17 meV but in a finite quantum well one must expect a somewhat smaller value which varies with well width. At large well widths E_{ex} will approach the bulk GaAs value while, as L_z tends to zero, electrons and holes must be associated with the bulk AlGaAs barrier material, rather than the GaAs well material, so the value of E_{ex} obviously goes through a maximum. Calculations (Greene et al. 1984) have suggested a peak value of about 9 meV at $L_z \sim 5$ nm but experimental evidence (Dawson et al. 1986b; Moore et al. 1986) favours a peak value closer to 15 meV. This work at the Philips Research Laboratory in Redhill provided an excellent example of the importance of high-quality MBE growth—quantum well samples with well widths in the range 7.5 to 11.2 nm showed spectra with line widths narrow enough to reveal transitions associated with the 2s state of the exciton and therefore allowed the binding energy to be directly measured. It was the first time that precise experimental evidence of 2D exciton structure had been obtained. In Figure 6.12 we have therefore taken the liberty to scale up the theoretical curve to match this experimental data. There are inevitably slight differences between light- and heavy-hole excitons but this curve can probably be relied upon to an accuracy of about 2 meV. The point we wish to emphasise here is simply that all this leads to a degree of uncertainty when attempting to fit experimental data to theoretical models. None the less we are (at last!) in a position to look at Dingle et al.'s analysis.

First, it is helpful to examine some of their measured spectra. Figure 6.13 shows 2 K absorption spectra from 5 quantum well samples, together with a sample of bulk GaAs. Several features are of interest; first that the QW exciton lines are much broader than the bulk GaAs exciton, second, the n = 1 exciton lines shift to higher energy as the well width is reduced, third, they also show a splitting which increases as the well width is reduced

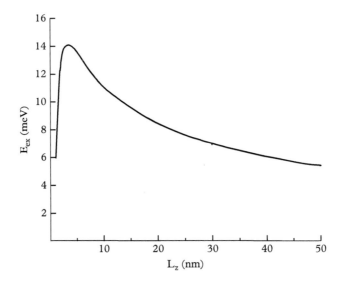

Figure 6.12 *Estimate of the quasi-two-dimensional exciton binding energy in an AlGaAs/GaAs quantum well as a function of the well width L_z. This empirical curve represents a scaled-up version of the results calculated by Greene et al. (1984) to match experimental data from Dawson et al. (1986). Small differences (of the order of 1 meV) between light- and heavy-hole excitons are ignored!*

and, fourth, there is clear evidence for n = 2, 3 and 4 excitons which again shift to higher energy with decreasing L_z. The large line widths are probably due to variations in well width between different wells in these multiwell structures, the shift to higher energy with decreasing well width is a straightforward consequence of the fact that confinement energies increase with decreasing width, and the existence of n = 2, 3 and 4 absorption lines is to be anticipated in wide well samples. However, the splitting of the n = 1 lines bears more careful examination. It was explained by Dingle et al. as resulting from the splitting of light- and heavy-hole states—in other words, the fact of their having different effective masses results in two separate confined hole states for each value of n (see equations (6.7), (6.8) and (6.9)). What was more, they noted an obvious lack of any similar splitting in the n = 2, 3 and 4 exciton absorption lines and they interpreted this to mean that the valence band well contained only one light-hole state—according to equation (6.10). This, in turn, implied that the well was shallow and led them to the conclusion that the valence band offset was unexpectedly small—that is, Q_v = 0.12, consistent with eV_w = 30meV. (All their samples were grown with $Al_{0.2}Ga_{0.8}As$ barriers, so x = 0.2 in equation (6.13)). So, once this value was determined, the rest of the spectral features could be modelled, using the known values of effective mass given in Table 6.3. Note that they assumed the same values of effective mass in well and barrier, for which

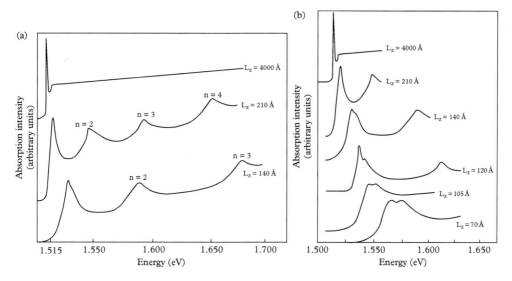

Figure 6.13 *Absorption spectra measured at 2 K on a series of Al$_{0.2}$Ga$_{0.8}$As/GaAs square quantum wells with various well widths, showing, in (a), clear evidence of $\Delta n = 0$ exciton transitions for $n = 1, 2, 3$ and 4 confined states and, in (b), the splitting of the $n = 1$ transitions due to the different confinement energies of the light- and heavy-hole states. (From Dingle 1975, courtesy Springer).*

Table 6.3 *Effective masses used in calculating confinement energies in AlGaAs/GaAs QWs.*

	GaAs			AlGaAs		
	m_e/m_0	m_{lh}/m_0	m_{hh}/m_0	m_e/m_0	m_{lh}/m_0	m_{hh}/m_0
Dingle et al.	0.067	0.080	0.45	0.067	0.08	0.45
'Modern' values	0.067	0.094	0.34	0.092	0.11	0.46

case equations (6.7) and (6.8) can be written in the simpler form

$$eV_w = E_n \sec^2 \left[(\pi/2)(E_n/E_1{}^*)^{1/2} \right] \tag{6.7A}$$

and

$$eV_w = E_n \csc^2 \left[(\pi/2)(E_n/E_1{}^*)^{1/2} \right] \tag{6.8A}$$

It is effectively these equations which they used to analyse their data.

The modelling, as we said earlier, resulted in an excellent agreement with experimental data over a range of nine samples with well widths ranging from 7 nm to 40 nm and encompassing up to seven pairs of confined states. The value of Q_v derived from the model was 0.12 in the first instance (Dingle et al. 1974) but was modified to 0.15 in Dingle (1975) and this gave rise to much discussion over the following years as evidence gradually accumulated that a value of order 0.3–0.4 was more likely to be correct (see the critical review by Duggan 1985) and, finally, in 1987 direct, irrefutable measurements by Dawson et al. (1987) that the value should lie in the range 0.33–0.34. This was 13 years after the original estimate! Why did it take so long and how did the original estimate come to be so far in error? How was it possible to achieve such excellent agreement with experimental data using an erroneous value of one of the critical parameters?

That the original value of Q_v = 0.15 should long continue to represent accepted wisdom is well illustrated by a review paper written by Art Gossard in 1982 (Gossard 1982). In this he shows absorption spectra from three quantum well samples with L_z = 19.2, 11.6 and 5.0 nm, showing much narrower line widths and better resolved exciton peaks but, in discussing their theoretical modelling, he makes the statement that 'in general, the positions of the lines are well represented by energy levels calculated using bulk electron and hole masses and bulk band gaps, with 0.85 of the band-gap difference comprising the conduction band potential barrier and the remainder the valence band barrier.' However, just two years later came the first serious challenge to the well-established value of Q_v = 0.15, following a remarkable development of MBE growth technology, and it came from the same stable as produced the original estimate (Miller et al. 1984a). This was the growth, not of square, but of parabolic wells. It was achieved by computer-controlled deposition of alternate layers of GaAs and $Al_{0.3}Ga_{0.7}As$ during which the GaAs layer thickness was gradually increased, as the AlGaAs thickness was decreased, to the bottom of the well, which was pure GaAs, and then the procedure was reversed to generate the opposite side of the well. The finished structure contained 20 layers of AlGaAs and 21 layers of GaAs, with average thickness of approximately 1 nm. The thickness of the Nth layer of AlGaAs from the centre of the well was given by

$$L_N = [(N - 0.5)/10]^2 \times L_z/20 \tag{6.14}$$

where L_z is the total width across the top of the well. It is interesting to note that this approach to MBE growth of a specialised band-gap profile is akin to the use of planar doping to simulate various doping profiles which we referred to earlier in connection with the growth of n-i-p-i superlattices (Wood et al. 1980). The technique is concerned, of course, to find a way round the problem of generating a continuously varying Al flux and, while thinking about this, we might briefly note the use of pulsed beam MBE, in which the Al beam is modulated with a rotating chopper blade (Kawabe et al. 1983). This allows the same Al cell to be used to grow layers with two different Al contents, though it would be difficult to extend the idea to the growth of a continuously varying fraction, such as required to grow parabolic wells.

However, returning to our theme, three different parabolic well samples were reported with values of L_z = 32.5, 33.6 and 51.0 nm and their spectra measured by a technique known as photoluminescence excitation spectroscopy (PLE) which was, by this time being widely adopted, in preference to absorption. It involved measuring the intensity of the photoluminescence as a function of the wavelength of the excitation source, the obvious advantage being that it was no longer necessary to remove the substrate from the sample. The reason for going to this degree of growth complexity was two-fold—first, the confined energy states in a parabolic well are evenly spaced (just as they are in the case of a simple harmonic oscillator) and the allowed transitions between valence-band and conduction-band states are no longer limited to those for which $\Delta n = 0$; second, and of particular interest for our present discussion, the partitioning of the band offset appears directly in the expression for the confinement energies, unlike the situation in square wells. This obviously leads to greater clarification of its role and implies a better chance of obtaining a reliable value from measured spectra. The spectrum from the 51 nm sample is shown in Figure 6.14 and clearly reflects the even spacing of the confined states and shows the presence of several transitions characterised by $\Delta n = 2$ and which arise because the wave functions in this case differ in their spatial distributions from those appropriate to square wells and such transitions are no longer 'parity forbidden'. From our present point of view, it is sufficient to note that attempts to fit these spectra with the same parameter values previously used for the spectra from square wells

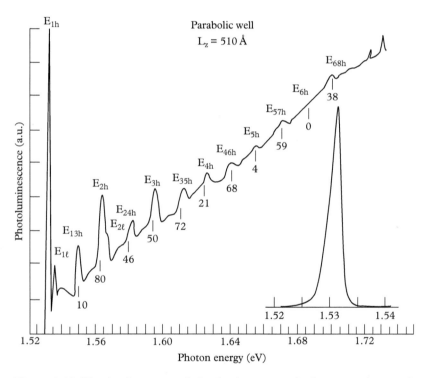

Figure 6.14 *Photoluminescence and photoluminescence excitation spectra (measured at 5 K) from a parabolic well, with $L_z = 51$ nm, in the GaAs/Al$_{0.3}$Ga$_{0.7}$As system. The linear dependence of the confined state energies on n is evident from the excitation spectrum (compare with Fig 6.13). Note the number of transitions for which $\Delta n = 2$. Relative intensities calculated on the assumption of equal CB and VB well depths are indicated immediately below each exciton line. (From Miller et al. 1984a, courtesy American Institute of Physics).*

proved impossible and led the authors to suggest an altogether different band offset ratio $Q_c/Q_v \sim 1$, that is, $Q_c \sim Q_v \sim 0.5$.

Interestingly, once the idea of new parameter values achieved respectability, it was immediately applied to the analysis of square wells, too. In a second paper, hard on the heels of the first, Miller et al. (1984b) presented an analysis of both square and parabolic wells using a modified set of effective masses, $m_e = 0.0665m_0$, $m_{hh} = 0.34m_0$ and $m_{lh} = 0.094m_0$, together with the values $Q_c = 0.57$, $Q_v = 0.43$. They also introduced the concept of barrier effective masses different from those in the well, effectively making use of equations (6.7) and (6.8) rather than (6.7A) and (6.8A) and considered possible effects (rather small) of the non-parabolicity of electron and light-hole bands. In examining the fit to a wide range of square well samples they concluded that, in some cases, it was necessary to modify the well widths by as much as 10% from those estimated from growth rates in order to obtain satisfactory agreement. Here was evidence that MBE growth might not always yield layers with precisely the thickness anticipated, a theme taken up by the Philips group in Redhill during the following year (Dawson et al. 1985). Their approach made use of a technique due to Geoff Duggan known as contour plotting, in which they took both the band offset ratio and the well width to be unknown, while seeking a coincidence between several different transitions (note that it was important to include so-called 'forbidden' transitions such as HH_3–E_1 which appeared as weak lines in many spectra). This hopefully provided reliable values for both these parameters and led them to suggest that $Q_c \sim 0.75$ (we might add here that later refinement of this method led to a value of $Q_c = 0.68 \pm 0.02$—see Orton et al. 1987). Finally, to complete the international temper of this 'revolution' in band offsets, a study by Meynadier et al. (1985) on structures containing a narrow well within a wider well came to the conclusion that Q_c must be close to 0.59. Clearly, once the flood gates were opened, a veritable torrent of evidence was unleashed into the previously placid backwater of square well science. However, we should note that, though a consensus had emerged to the effect that the original value of $Q_v = 0.15$ was surely wrong, there was yet a serious degree of doubt as to its precise value. What was required to remove such concerns was a direct spectroscopic determination which did not depend on fitting a questionable theory to a set of experimentally uncertain spectra and this was soon to emerge from a link forged between Bell Labs and PRL (Philips Research Laboratory, UK).

In 1985 Phil Dawson from Redhill spent a year working at Murray Hill and, during that secondment, studied an unusual type of quantum well structure in which the lowest electron states are associated with the X minima in the barrier, while the hole states remain at the Γ valence band maxima in the well (Dawson et al. 1986a). This so-called 'type II' structure results in relatively weak photoluminescence transitions characterised by long radiative lifetimes but which allowed direct measurement of the valence band offset, as we shall see in a moment. In order that the Γ minimum in the well should be at higher energy than the X minima in the barrier, they chose a structure with wells consisting of $Al_{0.37}Ga_{0.63}As$ enclosed in AlAs barriers (see Figure 6.15). Photoluminescence was measured at $T = 6\,K$, and the line at 1.919 eV was interpreted as a transition from a confined X-electron state in the AlAs barrier to a confined Γ heavy-hole state in the AlGaAs well. Thus, by allowing small corrections for the appropriate confinement

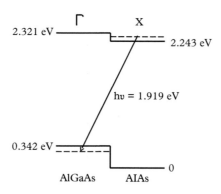

Figure 6.15 *Energy bands of the type II MQW structure measured by Dawson et al. (1986), showing the indirect luminescence transition from a confined X-electron state in the AlAs barrier to a heavy-hole confined state in the $Al_{0.37}Ga_{0.63}As$ well. The valence band offset of 0.342 eV could be obtained directly from the measured photon energy and the known parameters of the AlAs/GaAs system.*

energies, they were able to estimate the valence band offset as being 342 ± 4 meV, which represents a fraction 0.30 of the Γ band-gap difference. For the first time, the value of Q_v had been measured directly and unequivocally. All that remained was to perform similar measurements on samples with GaAs wells to confirm this result, but suitable samples would require well widths narrow enough to push the lowest (n = 1) confined electron state above the X state in the barrier. It was relatively easy to calculate that this required well widths less than about 3 nm and, on his return from America, Phil Dawson combined with his PRL colleagues to pursue the matter to a satisfactory conclusion (Dawson et al. 1987; Duggan and Ralph 1987; Dawson et al. 1988). Measurements on two GaAs/AlAs quantum well samples with well widths of 2.3 nm and 3.0 nm showed clear evidence of type II spectra which led them to conclude that 'this means Q_v lies in the range 0.33–0.34'. At last, the uncertainty was reduced to a mere 3%(!), most of which was concerned with uncertainties in the precise well widths and in the heavy-hole effective mass needed to calculate the heavy-hole confinement energy. It is only fair to say that the Philips group was not alone in reaching this conclusion—measurements in Spain/Germany (Madrid/Stuttgart), France (CNRS Bagneux) and the USA (Bell Labs Murray Hill and Holmdel) showed similar evidence and helped to build a solid international consensus in favour of a band offset ratio of $Q_c/Q_v = 0.67/0.33$. Thus was concluded one of the more bizarre controversies of recent semiconductor physics. Can we understand how it all came about?

It would obviously be inappropriate to launch into a lengthy critique of the modelling which led to the original estimates of $Q_v = 0.12$ or 0.15 but it is possible to make one or two (hopefully) helpful comments. First, if one uses equations (6.7) and (6.8) to calculate confinement energies,* it is not difficult to show that the calculated absorption spectra are surprisingly insensitive to the value of Q_v employed. Changing it from 0.15 to 0.30 results in shifts in photon energy of only a few meV, changes which are readily counterbalanced by small changes in well width or effective mass. The physical explanation lies in the fact that such changes in Q_v affect electron and hole confinement energies in opposite senses, so an increase in electron confinement energy is balanced by a decrease in hole energy. As Geoff Duggan showed, using his contour plotting of photon energy against well width and band offset ratio, it is far from easy to pin down an accurate value of Q_v even with a sophisticated approach to the calculation. It also gradually became clear that only samples with well widths much less than 10 nm—that is, significantly narrower than those used in the early studies—showed the necessary sensitivity to Q_v. There were, in fact, far too many uncertainties in the parameter values used to model the early spectra to allow a firm conclusion concerning the optimum value for Q_v—well widths and barrier compositions were not known with sufficient accuracy, accepted values for effective masses have been considerably modified (see Table 6.2) and the effect of band non-parabolicity has now been recognised. (Note, particularly, that the light-hole/heavy-hole splitting is sensitive to the effective mass values.) However, once the value 0.15 became lore, it rapidly became folklore (and perhaps that is the criticism before which the semiconductor community should collectively bow—we were not, ourselves, sufficiently critical?). We shall leave the matter there—with a sigh of relief that a reliable answer did finally emerge.

'Uncertainty' is an apposite word with which to begin our next topic, the veracity of the MBE-grown samples which had stimulated these exciting developments in low-dimensional structures. One feature should be relatively clear—the precise quality of the interfaces between well and barrier could be seen to play a major role in determining the quality of the physical phenomena observed, and, what was more, the narrower the wells, the greater would be its significance. A random fluctuation in well width, a gradual variation in well width over a sample area or a variation from well to well in a multi-quantum well system, for instance, would have far greater effect on wells consisting of a few monolayers than on those whose widths were measured in tens of nanometres. In the pioneering work from Bell Labs, the question of well rectangularity was clearly seen as important and the conclusion reached that 'as grown structures are considered to have interfaces that are sharp to the order of a few atomic layers.' However, the narrowest well investigated was still 7 nm wide, relatively large compared with the 0.28 nm appropriate to a single monolayer of (100) GaAs. Also, as we have already intimated, the exciton features in Figure 6.13 were considerably broader than that of the bulk GaAs sample used for comparison, which suggested some variability in well width. As a marker, we simply note that, for a 7 nm well, a variation of ± 1 ML would result in a broadening of

*We are much indebted to Dr Richard Campion for writing the computer program used.

about 10 meV, comparable to the measured line widths in Dingle's (1975) paper. But what exactly was the truth of the situation? What, precisely, were the well widths? What exactly was the structure of the AlGaAs/GaAs interfaces? And how could these essential parameters be reliably measured?

As we discussed in Chapter 3 (Box 3.1), the nature of the MBE process is obviously prone to generate a degree of non-uniformity across a wide substrate, even with substrate rotation, and that can be readily accommodated, but of more significance in this instance are the fluctuations on an atomic scale. Growth involves step edges, which imply monolayer steps at interfaces, and precise substrate orientation has an effect on terrace length. The difference in mobility of Ga and Al atoms on the growing surface is responsible for differences between the quality of the 'normal' (AlGaAs on GaAs) and 'inverted' (GaAs on AlGaAs) interfaces. Also, it may be anticipated that Al and Ga atoms would interpenetrate their respective layers to some extent, resulting in a potential step which is not strictly vertical. While diffusion is negligible at normal growth temperatures, the fact that growth takes place simultaneously on several different atomic levels implies that interfaces will be at least a few atomic layers wide. Such were the likely departures from ideality—how could they be characterised in practice? And how, for that matter, could such basic factors as well width and aluminium fraction be measured reliably? These latter could, of course, be determined from measured growth rates but one must surely question the accuracy of such estimates before accepting them as gospel. Three other methods immediately come to mind—those of transmission electron microscopy (TEM), X-ray diffraction (XRD) and, of course photoluminescence (PL and PLE). While there were numerous protagonists of all these techniques, we shall begin by quoting from a 'home-grown' publication (Orton et al. 1987) which surveyed the overall programme at PRL, Redhill, in the mid 1980s.

MBE growth rates can be measured simply by calibrating beam intensities with an ion gauge but are more reliably obtained from RHEED oscillations. However, we should not forget that this technique involves measurements on a static substrate, while sample growth usually takes place on a rotating substrate. Because of the small angle of incidence, it is also difficult to be sure exactly where on the substrate the RHEED beam impinges, a problem sometimes solved by using a very small, centrally mounted substrate for calibration, before changing to the larger substrate for practical growth. In any case, estimates of film thickness or alloy composition depend on the effusion cells functioning in stable fashion and it was difficult to be certain that this was always the case. Note, in particular, that growth of AlAs/GaAs structures involves switching the Ga cell on and off, with inevitable temperature fluctuations and consequent instability in beam flux. A similar remark holds true in estimating the Al content of barrier layers—though, if they are thick enough, this may be no more than a trivial effect.

TEM has the advantage of providing a dramatic image of sample structure, particularly impressive in the case of MQW or superlattice samples but has the disadvantage of requiring elaborate and time-consuming sample preparation. It is necessary to thin the sample by ion-beam etching to a thickness of a few tens of nanometres so that the electron beam is incident at right angles to the sample growth normal. The transmitted beam then forms an image (referred to as a dark-field image), deriving contrast from the different scattering factors associated with Ga and As atoms. A combination of microscope resolution and interface monolayer steps effectively broadens the interface so that estimates of well or barrier thickness are limited to a precision of about ±0.5 nm but the method is extremely useful for the study of well and barrier uniformity throughout a multilayer structure. High-resolution microscopy can be used to obtain better resolution but only where a full simulation is available, making even greater demand on time and effort. Such measurements can only be performed in a very few test cases to check important details.

XRD depends on interpretation of diffraction patterns, and two aspects are of importance in the characterisation of multilayer structures. The spacing of the principal diffraction spots provides a measure of the average lattice parameter $<a>$, while satellite spots whose spacing either side of a principal spot yields a value for the MQW or superlattice period L. Furthermore, analysis of their intensities and widths can be used to measure both L_z and L_B independently and to estimate the degree of interface grading and the uniformity of the period throughout the structure. Because of the small difference in lattice parameter between GaAs and AlAs, the MQW structure gives rise to a diffraction peak which is shifted from that of the GaAs substrate, and the value of $<a>$ derived from this shift can then be used, via Vegard's law, to derive a value for the Al content x of the alloy barrier regions:

$$<x> = [<a> - a_{GaAs}] / [a_{AlAs} - a_{GaAs}] \tag{6.15}$$

and

$$x = <x>(L/L_B) \tag{6.16}$$

Note, however, that more recent work (Wasilewski et al. 1997) has shown that Vegard's law is not strictly accurate for this most studied of alloys—to obtain an accurate value for the composition from X-ray measurements, a small quadratic term should be included. Thus, without correction, a 50% alloy is measured as 53–54%, while a 33% alloy is measured as 35–36%, and this must also signal a necessary degree of caution when working with other alloy systems.

Typically, interfaces are found to extend over four monolayers (Fewster et al. 1991), significant mixing occurring over two monolayers and weak dilution on the two adjacent atomic planes. Estimated accuracies for MQW parameters are ±0.1 nm for L and about 1% for L_z. Finally, we should emphasise that the XRD technique averages data over an area several millimetres square, very much larger than the atomic-scale dimensions with which we are principally concerned and in stark contrast to that possible using high-resolution TEM.

PL and PLE measurements were carried out at low temperatures (T < 10 K) in order to maximise spectral resolution. A typical PLE spectrum is shown for a 5.5 nm well sample in Figure 6.16, the PL being detected on the n = 1 electron heavy-hole emission line and the various exciton absorption lines being labelled. As can be seen, they consist of strong $\Delta n = 0$ lines, together with weak $\Delta n = 1, 2$ lines all of which were used in

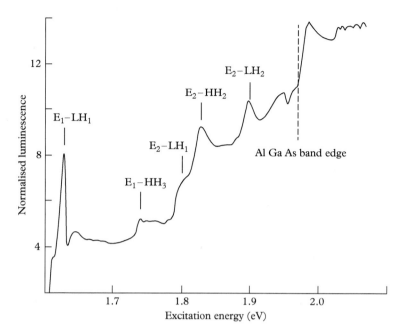

Figure 6.16 *Photoluminescence excitation (PLE) spectrum of an*
$Al_{0.36}Ga_{0.64}As/GaAs$ quantum well ($L_z = 5.6\,nm$) measured at $T = 8K$. The
detection energy was in the low energy wing of the E_1—HH_1 emission line. The
peaks identified are the various confined state exciton transitions—note the
presence of two 'forbidden' transitions. (From Orton et al. 1987, courtesy
Institute of Physics)

the contour plotting routine to obtain values for Q_c and L_z, the latter to be compared
with that estimated from the GaAs growth rate. Confinement energies were calculated
on the assumption of square well profiles and the good agreement with measured spectra suggests that interfaces were sharp to within about one monolayer, in agreement
with the X-ray data discussed above. Turning, now, to luminescence, two features are
of importance: the line width and the so-called Stokes shift of the emission line from its
corresponding absorption line. Because of the low temperature used, emission is only
seen from the lowest energy-confined states, E_1–HH_1 and E_1–LH_1. (Indeed, even the
light-hole line would not be expected—it appears that the effective carrier temperature
is somewhat higher than the lattice temperature.) We have already discussed the fact
that luminescence line widths can be used to infer non-uniformity in MQW structures,
either from well widths that vary within a single well or between different wells and the
Stokes shift can be used to distinguish between these alternatives. The Stokes shift arises
because excitons tend to thermalise within a well to the wider regions where their energy
is minimised and this can be taken as evidence for intrawell variation of L_z. Thus, if
the luminescence is measured over a range of temperatures, such thermalisation can frequently be observed. However, well-width variations between different wells do not show

any such effect, and the existence of two or more emission lines remains independently of temperature. Meanwhile, we have said it before but it is worth emphasising that the spatial resolution available with PL is set by the exciton diameter, of order 20 nm, and one final comment should be made—background doping, if high enough (perhaps of the order of 10^{23} m^{-3} or greater) can also result in significant line broadening but the effect is independent of well width, in contrast to the broadening due to well width variation.

As already noted, TEM measurements are extremely useful in checking uniformity of both well and barrier widths and, in one particular case, where the Ga cell was known to be unstable, it served to show surprisingly large (as much as 20%) variation over a 60 well sample. However, this was certainly atypical—most samples studied by TEM showed values of L_z to be constant within about one monolayer. At the same time, there were sometimes discrepancies between TEM, XRD and PLE estimates of L_z, also with estimates derived from growth rate. While, generally speaking, agreement was very satisfactory, it was clear that inexplicable variations (of the order of 10%) did occasionally occur, emphasising the importance of checking dimensions whenever possible, rather than relying on any single estimate. Similar remarks were found to be applicable to the measurement of Al content in barrier layers.

To bring this all-too-short account of quantum well structures to a conclusion, we shall now return to the detailed examination of photoluminescence and its interpretation in terms of interface quality (the brief account given in Chapter 5 (Section 5.2) set the scene but requires a closer look if we are to appreciate its significance to the full) and, finally, we shall briefly examine its relevance to the optical properties of short-period AlAs/GaAs superlattices.

Several references have already been made to the broadening of luminescence lines by random variations in quantum well widths, and emphasis laid on its increasing importance as the well width L_z is reduced. For example, when a 5 nm GaAs well is grown by conventional MBE, one would normally expect to measure (at low temperatures) a FWHM line width for the E_1–HH_1 transition of approximately 10 meV, increasing to 25 meV when $L_z = 2.5$ nm. This corresponds to a fluctuation in L_z of about 1 ML and that it should be the result of interface roughness has been confirmed by several studies of the effect of growth interrupts. This is well illustrated by the work of Sakaki and colleagues at the University of Tokyo (Tanaka et al. 1986; Tanaka and Sakaki 1987), as they grew a range of $Al_xGa_{1-x}As/GaAs$ quantum wells (with x = 0.2 to x = 1) both with and without growth interrupts. Their initial observation was based on the comparison of 4.8 nm well-width AlAs/GaAs structures where both interfaces were grown with (type A) and without (type D) growth interrupts. Type A samples showed line widths of 4.0 meV, compared to 15 meV in type D samples. What is more, type A samples showed evidence of two emission lines, separated by 15 meV, suggesting the presence of two distinct well regions characterised by values of L_z differing by a single monolayer (0.283 nm). They then recognised that the upper and lower interfaces might be expected to behave differently, on the basis that Al atoms are more strongly bound than are Ga atoms, so should diffuse less readily over the growth surface and therefore be less susceptible to the smoothing effect of growth interrupts. This led them to add two more sample types to their repertoire: type B, in which interrupts were used only on the top interface, and type C, only on the bottom. In a word, then, type B samples duplicated the spectra of type A, while type C replicated those of type D. As anticipated, it was clearly the top interface that was responsible for the line broadening in type C and D samples. Their next question was, what should we expect when the barrier layers are formed from alloys with only a small Al content—for example, x = 0.2? In this case, the bottom interface contains a large proportion of mobile Ga atoms, so should show a favourable response to the use of growth interrupts, in spite of the sluggish response of the Al atoms. One might therefore anticipate a significant improvement in samples of type C and this is, indeed, what they measured—the line width came down from 15 meV to about 10 meV. However, the surprise came for type B samples—when they were grown with x = 0.2, the line width went *up* from 5 meV to 13 meV. 'Improving' the bottom interface had actually caused an *increase* in width. How could they reconcile these apparently conflicting results?

The answer is contained in Figure 6.17, replicated from their papers. Again, we need to remember the importance of the size of the 'probe' used to interrogate the interface structure, in this case the exciton diameter of about 20 nm. Whether an interface appears rough or smooth depends on how large are its atomically flat sections in comparison with this probe size. Thus, Tanaka et al. envisaged three types of interfaces: type I, in which the flat terraces were larger than 20 nm, type II, in which they were much

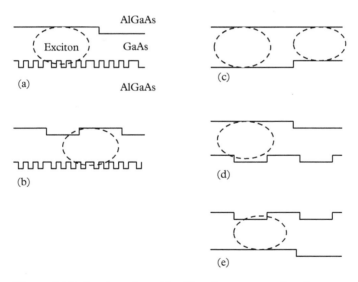

Figure 6.17 *Atomic-scale models of interface structure of various Al$_x$Ga$_{1-x}$As/GaAs quantum wells (x = 0.2 – 1.0) prepared with and without growth interrupts: (a) shows the model of type A and type B QWs with high Al content in the barriers (x > 0.5), whereas (b) shows that of type C and type D QWs with low Al content (x < 0.3); (d) and (c) show that of type B and type C QWs, respectively. See text for details. (Following Tanaka and Sakaki 1987).*

shorter than 20 nm and a third, type III, the line broadening 'culprit', where they were comparable to 20 nm. The first looked smooth because it *was* smooth, the second looked smooth because the exciton effectively averaged the barrier potential over its small scale fluctuations, while the third simply looked rough and led to line broadening. Notice too that this rough interface could occur at either the top or bottom of the well, according to the details of the growth. When the aluminium content of the barriers was equal to or greater than x = 0.6, the bottom interface was always of type II (see (a) and (b)) but when x ≤ 0.3, the bottom interface could be of type I when growth interrupts were used (see (c) and (e)) but was rough (type III) in conventional growth (see (d)). Life can, indeed, be surprisingly complicated!

Tanaka et al. assumed that only single steps occurred at each interface and were able to model PL line widths with satisfactory accuracy. However, as we saw above, XRD data suggest that typical interfaces actually cover something like four atomic planes, which implies that some step heights must be of two or more monolayers, a position adopted by Zhang et al. (1990) in their interpretation of interface properties. They grew single well AlAs/GaAs structures with well widths of 6 nm, using three different temperatures (630 °C, 650 °C and 680 °C) and used both interrupted and conventional growth methods. The samples were characterised by measurement of RHEED oscillations and 6 K PLE, while comparing the results with a Monte Carlo model of growth dynamics. They too observed a reduction in PLE line width following the use of interrupts at the bottom (AlAs/GaAs) interfaces but proposed slightly different criteria for this to happen. They showed that the condition for successful smoothing was that the RHEED intensity should increase during growth interruption and that the associated reduction in step density occurred only above a critical growth temperature (somewhere in the region of 600 °C). The major factor controlling line-width reduction was shown to be the number of molecular layers over which the growth front was distributed—in other words, the 'thickness' of the interface in units of monolayers. This factor is obviously of ever-increasing significance as the thickness of the constituent layers is reduced and this has important consequences for the optical properties of short-period superlattices. We shall now look briefly at this topic in terms of $(AlAs)_m(GaAs)_n$, where m and n are the numbers of monolayers in the respective layers and where m,n ≤ 10.

The use of multi-quantum well structures to investigate superlattice properties was first reported by Dingle et al. (1975), though this paper was concerned with relatively wide wells (~5–20 nm). By reducing the barrier thickness to 1.2–1.8 nm, they observed the effect of coupling between wells, the relatively sharp exciton features seen in isolated wells gradually blurring into superlattice bands with little structure. The following year Gossard et al. (1976) reported the growth of ultra-short-period superlattice structures with well and barrier widths down to single monolayer dimensions and obtained TEM images which confirmed their coherence, at least down to values of m ~ n ~ 5. They also measured (T = 2 K) an absorption edge on an m = n = 1 sample at an energy near 2eV (Cf E_g = 2.1eV for a 50% AlGaAs alloy). A decade later, the study of type II MQW samples, which we described earlier, led to speculation about the precise nature of the electron states in such short-period superlattices, a question taken up by the Philips group in papers published in 1987 and 1988 (Moore et al. 1987, 1988). The uncertainty hung on the result of calculations which suggested that, as the barriers became thinner, the Γ electron state in the GaAs well would be lowered in energy, bringing it down below the X state in the AlAs. In other words, the system would revert from being of type II to the more usual type I configuration, a prediction that was accurately confirmed by measurements of PL and PLE at low temperatures. Five MBE-grown structures were studied, having GaAs wells of constant thickness (measured by XRD as $L_z = 2.5 \pm 0.2$ nm), while the AlAs barrier thicknesses were reduced progressively from 3.9 nm to 0.57 nm. The results are summarised in Figure 6.18, showing the dramatic reduction in energy of the $E_1(\Gamma)–HH_1$ transition, while the type II $E_1X–HH_1$ transition moves to higher photon energy. The crossover from type II to type I occurred at an AlAs thickness of 1.3 nm (between 4 and 5 monolayers).

Following up on this result, low-temperature PL and PLE measurements were then applied to understanding the effective band structure of AlAs/GaAs superlattices with m = n = 2 to 8. XRD confirmed that the average Al fraction was close to 0.5 and that the periods lay within 0.5 MLs of their nominal values. The samples were grown without interrupts so it was anticipated that interfaces would be graded; but the time-consuming analysis required to confirm this was not carried out—it would be necessary to rely on the luminescence data to infer the extent of such departure from the ideal. Careful measurements of PL and PLE at low temperatures showed that, for the m = n = 4, 5,

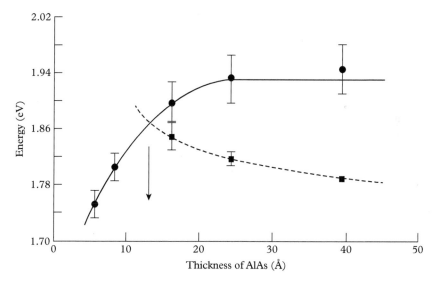

Figure 6.18 *Experimental data (T ~ 7K) showing the type II–type I crossover in the $(AlAs_m–GaAs_m)$ system as the number of atomic layers m is reduced. The solid line represents the $(E_1\Gamma–HH_1)$ exciton transition, as measured by PLE spectroscopy, the broken line the type II emission line. The crossover occurs at a layer thickness of ~ 1.3nm. (Following Moore et al. 1987).*

6 and 8 samples, the lowest conduction-band state was a folded X_z minimum, and the superlattices could be regarded as 'pseudo-direct' (in contrast to the 50% alloy, which is indirect). On the other hand, the m = n = 2 sample showed a luminescence spectrum closely similar to that of a 50% alloy, a conclusion not entirely unexpected when one remembers that AlAs/GaAs interfaces tend to spread over four atomic planes!

Needless to say, this account of the early history of the optical studies of quantum well and superlattice samples does no more than skate lightly over a very large and densely populated 'data rink'. Our intention has been simply to give a flavour of the exciting range of new physics thrown up by the ability of MBE growers to produce remarkably precise structures with resolution down to *almost* single atomic dimensions.

6.5   Quantum Wires and Quantum Dots

The quantum well achieved its success through its ability to confine electrons in just one dimension, thus altering their behaviour in this particular direction and, once this possibility had been well and truly demonstrated, it was a matter of only a short time before the materials scientist was exploring quantisation in both two and three dimensions. If all manner of exciting new physics could accrue from the restriction of electron motion in one dimension, perhaps even more might arise by repeating the trick in the other two? Well, perhaps—but it was to prove very much more difficult to achieve. There was no *obvious* way in which crystal growth could be adapted to such a task—it appeared that it would have to rely on quite different fabrication methods. But first we should look briefly at the motivation. What specific advantages might materialise if ways could be found to produce either wires or dots with nanoscale dimensions? Two quite distinct ideas were put forward by Hiroyuki Sakaki from the University of Tokyo: one concerned with electron transport, the second with semiconductor lasers. In the first paper (Sakaki 1980) he pointed out that, if electrons could be confined in a quantum wire, their mobility at low temperatures would be considerably enhanced because ionised impurity scattering would be strongly limited by the fact that an electron with initial momentum $+k_1$ could only be scattered into a final state with momentum $-k_1$. (In 2D or 3D no such restriction applies.) The second paper (Arakawa and Sakaki 1982) predicted a significant improvement in the temperature dependence of laser threshold current for quantum wire or quantum dot lasers, when compared to their 3D or 2D counterparts. The fact that laser threshold currents increased rapidly with increasing temperature was an established problem and it had already been demonstrated that quantum well (i.e. 2D) lasers suffered less from this drawback than normal DH lasers. Arakawa and Sakaki showed theoretically that considerable further improvement could be anticipated for 1D or 0D devices—indeed, the latter should show no temperature dependence whatsoever. Furthermore, the absolute value of threshold current was expected to improve significantly with the use of wires or dots—calculations published by Asada et al. (1986) suggested that this should scale roughly as 1000, 400, 150 and 50 Acm^{-2} for 3D, 2D, 1D and 0D structures, respectively. The extreme difficulty of making practical quantum wires with effective contacts meant that rather little came of the first idea, whereas optical properties of wires and dots were much easier to investigate experimentally; so, we shall concentrate on the second.

From the point of view of laser threshold current, the important difference between the various structures is one of the density-of-states functions $\rho(E)$ in the semiconductor conduction and valence bands. These are shown schematically in Figure 6.19. In 3D, $\rho(E)$ varies as $E^{1/2}$, in 2D as a step function, in 1D as $E^{-1/2}$ and in 0D it is delta function—in other words, the dot levels are sharp and atom like. The thermal behaviour of laser threshold results from the spreading of electrons (or holes) over a range of energies within the appropriate band and this clearly becomes less of a problem as the density-of-states function narrows, in the sequence from 3D to 0D. As Arakawa and Sakaki pointed out, there is no possibility for thermal spreading when the confined states have no width, so the quantum dot laser represented an ideal—if only there were practical means of achieving it. Arakawa and Sakaki demonstrated a somewhat *impractical* method of making wires by subjecting a conventional DH laser to a strong magnetic field (~24 T) applied in the z-direction, that is, perpendicularly to the plane of the p–n junction. This had the effect of confining electrons in the xy plane, while they were free to move along the z-direction. In other words, they looked rather like a bundle of wires! The effect was to confirm the predicted improvement in the temperature dependence of threshold

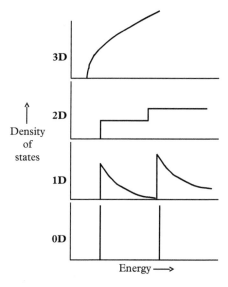

Figure 6.19 *Schematic diagram of the various density-of-states functions associated with 3D, 2D, 1D and 0D electron populations. Note that the DOS becomes more sharply defined through the sequence. In the case of zero dimension, the allowed states become atom like.*

current, thereby raising the level of desire for *practical* quantum wire and quantum dot structures. As we shall see, a number of moderately successful attempts were made to create QWRs but, in the end, QDs came out on top and we shall therefore focus our attention largely on them.

Just how quantum dots came into practical being makes an intriguing story but first we should make one important point concerning their confined energy levels. We made the same observation earlier in relation to quantum wells (see Section 6.1)—if one is concerned with observing well-resolved quantum states, it is essential that the energy separation of these states is large compared to the thermal energy kT—25 meV at room temperature, and 0.35 meV at 4 K. This implies a limitation to dot size, which must be much less than about 25 nm at room temperature or 250 nm at 4 K. This turned out to be an important criterion in respect of attempts to form wires or dots by lithographic definition, where, even for electron-beam lithography, resolution tended to be in the region of 500 nm. This serves to emphasise the distinction between so-called 'mesoscopic' systems (with dimensions of order 1 μm), which we shall discuss briefly in the next section, and true quantum structures, which we shall be concerned with here. However, in the important case of dots, there is a further qualification to be made—because bound electrons are confined in all directions, there is a *minimum* size for a dot if it is to be characterised by a confined electron state, given (for spherical dots of diameter D) by (see Bimberg et al. 1999, p. 6)

$$D_{min} = \pi h/2(2m_e \Delta E_c)^{1/2} \tag{6.17}$$

where m_e is the electron effective mass and ΔE_c is the conduction-band step between dot and barrier materials. A similar condition applies to hole states, the overall result implying that dots should be no *smaller* than about 4 nm.

Initial attempts to form quantum wire and quantum dot structures relied on the fact that MBE growth of GaAs and AlAs on high-index planes no longer proceeded by the layer-by-layer process appropriate to (001) surfaces but rather in a three-dimensional mode which arose as a result of growth rate varying quite strongly as a function of crystallographic direction. They also made use of the fact that electrons in quantum wells tend to congregate in wider regions of the well, where their energies are minimised—note that this implies the confined energy *differences* must be greater than kT, which, in turn, implies the use of narrow wells. Two approaches to achieving suitable structures were used; first came the use of V-grooves defined lithographically and etched in a (001) GaAs surface and then, second, growth on substrates cut with high-index surfaces. The use of pre-patterned substrates could be traced back to the early days of MBE, when Tsang and Cho (1977) had studied the growth of GaAs and AlGaAs/GaAs on channels etched along [110] and [–110] directions. They were not concerned to produce 1D structures but the principle of forming directional waveguides, for example, was clearly established. A decade later, Kapon et al. (1987) showed how quantum wires could be made by this method and in 1989 went on to report stimulated emission from wires made by etching a V-groove in a GaAs (001) surface and growing a 7 nm-thick quantum well structure in the groove (though it has to be regretted that, in this instance, they actually chose

MOVPE, rather than MBE to grow the structure!) (Kapon et al. 1989). Supportive evidence that the emission did, in fact, come from wires, rather than wells was provided by a measurement of an electron subband energy separation of approximately 10 meV, in agreement with accompanying calculations. A similar approach was followed by Rajikumar et al. (1993) to make quantum dots. They showed that pre-patterning a (111)B GaAs surface, followed by wet etching, could yield truncated triangular pyramids, and subsequent MBE growth of AlGaAs/GaAs QW structures resulted in QDs, as observed in TEM images. In much the same way, wires could also be formed on ridges, as demonstrated by Koshiba et al. (1994). They formed mesa stripes along the <110> direction on a GaAs (001) surface and subsequent growth led to the formation of ridges with gradually narrowing tops (see Figure 6.20). By following this with the growth of a GaAs QW, they were able to form narrow wires (as little as 10 nm wide) aligned along <110> and which showed strong photoluminescence. They distinguished luminescence lines, from QWs on the facets and from wires. The wire luminescence appeared at a photon energy 45 meV lower than that from a standard reference sample, showing that the wire was indeed thicker than a QW grown on a (001) substrate, whereas the QW line occurred 21 meV above the standard, showing that it was thinner. (An obvious geometrical factor comes into play, as shown in Figure 6.20 but, in addition, Ga atoms are more mobile on the (111)B face and tend to diffuse to the ridge.) Given that the volume of the wires was very much smaller than those of the wells, while its luminescence intensity was comparable, one may also conclude that the QWR efficiency was significantly higher—an important observation, as defect-free wires and dots were seen to be particularly advantageous for optoelectronic devices.

Another approach to the problem of growing wires and dots, which did not involve pre-patterned substrates, was reported by Nötzel and Ploog from Stuttgart. It had previously been noted that, when QW structures had been grown on non-(001) surfaces, their luminescence spectra were shifted to lower energies compared to those grown on (001), and various explanations had been proposed to explain this. Nötzel and Ploog (1992)

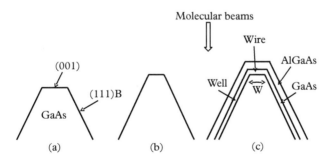

Figure 6.20 *Schematic diagram showing the growth of a quantum wire on the top of a GaAs (001) ridge. Growth is more rapid on the (001) surface so the GaAs quantum well is thicker on top of the ridge, resulting in reduced confinement energy.*

analysed RHEED patterns to show that the true explanation was the formation of facets with a regular periodicity, which minimised the surface energy. For example, growth of AlGaAs/GaAs QWs on the (311)A GaAs surface produced wires on ridges with (331) facets, oriented along the [−233] direction. In the case of other high-index planes, the formation of pyramid structures led to the growth of quantum dots. An excellent account of this work can be found in their review paper (Nötzel and Ploog 1993).

Another interesting, though complicated, method of forming quantum wires is illustrated in Figure 6.21. It depends on the process of 'cleaved edge overgrowth', in which two MBE-grown quantum wells are arranged to intersect at right angles to form a 'T' shape, the wire being represented by the junction of the wells (see e.g. Pfeiffer et al. 1990). It probably has one important advantage over other processes in so far as it allows much superior control over the wire dimensions. Recently (Yoshita et al. 2012), this technique has been used in the study of exciton absorption, well-resolved features representing the ground state, first excited state and higher excited states and thus providing a precise value of the exciton binding energy (12 meV) in wires of dimension 14 nm × 6 nm. We shall refer to this technique again when we discuss the growth of quantum dots.

However, at this point it is appropriate to look at a totally different method of forming quantum wires and which had been available since the 1960s, though it might well have slipped through our literature-searching net on the strength of its referring to them as 'nanowires' (or, sometimes, as nanocolumns or nanowhiskers). A convenient starting point is the paper by Wagner and Ellis (1964), in which they discussed the mechanism of

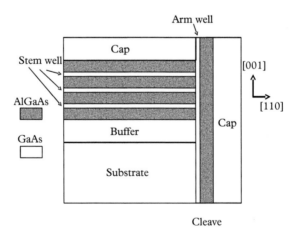

Figure 6.21 *Schematic diagram of the CEO method of growing quantum wires. The junction between the (001) plane 'stem wells' and the (110) plane 'arm wells' offers carrier confinement in wires aligned perpendicular to the plane of the diagram.*

wire formation in the VPE growth of silicon. In this case, wire growth was critically dependent on the presence on the substrate surface of a small particle of gold—a so-called nanoparticle (other metals may also be used but we shall concentrate on the use of gold, as most reports refer to it alone). According to Wagner and Ellis' model (known as the vapour–liquid–solid or VLS mechanism) the molten Au drop dissolves silicon from the substrate to form the Au–Si eutectic alloy. In wire growth, Si is deposited from this eutectic in much the same way as occurs in LPE, while the liquid drop takes up more Si from the vapour phase. The wire grows beneath the liquid drop and the drop remains perched on top, even up to wire lengths of several microns. Wagner and Ellis showed a micrograph of a 100 nm diameter wire but subsequent work on other semiconducting materials resulted in wires with somewhat smaller diameters, typically in the range 10–50 nm (though we should note that even this is barely into the range where quantum effects become of major significance).

The first report (that we are aware of) of the use of MBE to grow nanowires came from Korea (Lee et al. 2001). It described the growth of GaAs wires on (100) GaAs substrates on which had been deposited a pre-layer of silicon, either 100 nm or 200 nm thick. Apparently, this pre-layer served the same purpose as the gold dots used by Wagner and Ellis, forming a eutectic with GaAs, from which the wires grew by the same VLS mechanism. It would appear that the liquid drops were self-organised, perhaps as a result of the large lattice mismatch between GaAs and Si. Wires were constant in diameter ($d \sim 70$–80 nm) over lengths of up to 5 μm. In the case of the 100 nm Si layer, they grew accurately perpendicular to the substrate surface, and, in the case of the 200 nm layer, at angles of approximately 45°. High-resolution TEM measurements showed them to be single crystal, while energy-dispersive X-ray fluorescence (EDX) showed them (unsurprisingly) to be Si doped. The way was open for MBE growers worldwide to follow this admirable precedent, and many of them did.

The group at the Lund University in Sweden followed more conventional lines in using gold nanoparticles to stimulate the growth of both GaAs and InAs whiskers by CBE (Ohlsson et al. 2001; Björk et al. 2002; Ohlsson et al. 2002). In particular, they made use of size-selected Au nanoparticles, deposited in an aerosol machine, typical sizes being of order 20 nm, producing wire diameters in the range 40–80 nm. A fascinating 'extra' was their use of an atomic force microscope to push these Au particles into place along the substrate surface. Along with many others, they found that whiskers could be grown much more successfully on GaAs (111)B substrates, rather than (100). Interestingly, they were able to grow GaAs/InAs heterojunctions which, though strained, were dislocation free—thought to be a consequence of the narrow width of the structures. They also reported InAs wires which included several short InP sections, TEM measurements showing interfaces sharp on the atomic scale. At much the same time a group in Toronto studied wire growth by conventional MBE in the GaAs/AlGaAs system, again using (111)B substrates but aiming to achieve a narrower distribution of wire diameters (Wu et al. 2002, 2003, 2004). They therefore made use of 2–4 nm thick gold dots deposited through a nanochannel alumina mask. Dot diameters were of the order of 45 nm, and the resulting wires showed a diameter distribution of 37 ± 2.5 nm, compared to a control sample grown without masking of 33 ± 7.5 nm. They grew GaAs/AlGaAs multilayer wires with layer thicknesses of 20 nm and measured efficient photoluminescence at 4 K. They also observed that the percentage of Al in the AlGaAs (13%) differed considerably from that measured under identical growth conditions for a continuous film (24%), an observation which was consistent with different growth mechanisms for wires and films.

Up to this point it had been natural to assume that whisker growth did, indeed, follow the VLS process, and this seemed to be appropriate for samples grown by MOVPE; but there were growing doubts about its applicability to MBE, where the incoming constituent atoms were directionally controlled. The question was addressed in Sweden by Jensen et al. (2004) and at the Ioffe Institute in St Petersburg by Dubrovskii et al. (2005), both of whom suggested that whisker growth was actually limited by surface diffusion of the appropriate cation. Thus, in the growth of GaAs whiskers, Ga atoms deposited on the substrate surface should diffuse to the base of the whisker (under As-rich conditions) and then up the sides of the whisker (under As-deficient conditions—because of the directionality of the As beam) to the Au/GaAs growth interface. Such a mechanism was consistent with the observation of the Russian group that the whisker length was several times longer than the nominal thickness of GaAs deposited. Also, on this basis they calculated the relation between whisker length and diameter and obtained good agreement between theory and experiment. Jensen et al. grew InAs whiskers on InAs (111)B substrates by CBE and showed that 80% of whisker growth was by surface diffusion of In. They estimated the In diffusion length on the whisker sides to be greater than 10 μm. The topic was taken up again by the Lund group in 2007 (Persson et al. 2007). They grew InAs wires by CBE, using electron beam lithographically patterned Au dots whose spacing was varied from 0.25 μm to 4 μm, thus varying the collection areas appropriate to the whiskers. Again, they concluded that growth was limited by diffusion of the In to the growth interface, rather than by the conventional VLS mechanism. Finally, the Russian group (Tonkikh et al. 2006) showed that AlGaAs whiskers were characterised by growth rates which decreased as the Al content increased and interpreted this as due to the lower mobility of Al atoms on the substrate surface, compared with that of Ga.

The interest in nanowires has continued up to the present day on the basis of possible applications in 'nano-electronics' and 'nanophotonics'. Two recent papers which illustrate this have been published from the Ioffe Institute (Cirlin et al. 2009; Novikov et al. 2010), the former being concerned with establishing precise criteria for the growth of InAs, InP and GaAs wires on mismatched substrates, the latter with the luminescence properties of GaAs wires grown on (111)B GaAs substrates. Of great significance for material science is the fact that nanowires can be grown on severely mismatched substrates without incurring the disadvantage of dislocation formation and this is obviously seen as important for their device applications. (This is particularly relevant to the growth of photonic materials, such as GaAs, InAs or InP on silicon substrates.) It results from the ease with which strain can be accommodated in a thin wire but this implies a limit to the thickness of such coherent growth as this depends on the degree of mismatch between wire and substrate—the greater the mismatch, the smaller must be the wire diameter—and it was this aspect which Cirlin et al. investigated. They grew wires of InAs, InP and GaAs on (111) Si substrates and InAs wires on (111)B GaAs substrates and confirmed such a pattern of behaviour, the critical diameters being 24 nm, 39 nm, 110 nm and 44 nm for the corresponding pairings, values which were in good agreement both with earlier reports for samples grown by MOCVD and with theoretical predictions. In common with several other groups, Novikov et al. observed that GaAs nanowires generally contain both zinc blende and wurtzite phases, though, by careful selection of samples, they were able to distinguish the PL spectra appropriate to each phase. Their data suggest that the band gap of the WZ phase is approximately 40 meV lower than that of the ZB phase—somewhat surprising when we compare this with both phosphides and nitrides, where the difference has the opposite sign. The line widths of their PL spectra also suggest that material quality leaves much to be desired. There is clearly some way to go before wire quality can compare with that of two-dimensional films and, as we have already noted, the dimensions so far recorded are probably too great to offer significant quantum confinement effects. We shall therefore return to the subject of quantum dots.

By far the most exciting method of growing quantum dots was discovered very much by accident. In fact, the original motivation of those working in the field was to avoid or eliminate the very mechanism which gave rise to what has become known as 'self-organised quantum dots'. A huge literature on this topic has grown up over the past two-and-a-half decades so we can do no more than outline some of the important features—readers seeking greater detail might like to consult the book by Bimberg et al. (1999) or the reviews by Franchi et al. (2003) and by Joyce and Vvedensky (2004). Clear evidence for the formation of dots could be seen in the work of Lewis et al. (1984) of the California Institute of Technology and of Goldstein et al. (1985) of CNRS Bagneux, as both studied RHEED behaviour during the growth of InGaAs on GaAs substrates. The former were probably first to demonstrate a clear transition from layer-by-layer growth to three-dimensional growth at a sharply defined critical thickness of InGaAs, RHEED patterns being streaky during the initial growth, changing to spotty thereafter. Recognising that the lattice constant of InAs is 7% larger than that of GaAs, it is clear that continuous Frank - Van Der Merwe, layer-by-layer growth is highly unlikely and, as the InGaAs layer thickness increases, some form of strain-relief mechanism must come into play. The concept of a critical thickness for the growth of strained layers was far from new, of course, having been treated theoretically by (amongst others) Van Der Merwe (1962), Mathews and Blakeslee (1974), and, more recently, by People and Bean (1985) and by Dodson and Tsao (1987). However, these authors were concerned with instability leading to the formation of misfit dislocations, rather than with three-dimensional growth, while, in this instance, strain appeared to be relieved by the Stranski–Krastanow mechanism (see Bimberg et al. 1999, p. 38). An initial 'wetting layer' of InGaAs grew with an in-plane lattice parameter equal to that of the GaAs substrate, while, beyond the critical thickness, small islands grew with the relaxed InGaAs parameter. First reactions were probably to see all this as something of a nuisance, getting in the way of growing high-quality films of InGaAs, but gradually the realisation dawned that it might be put to good use.

Goldstein et al., studying the growth of InAs/GaAs superlattices by MBE, confirmed these observations. By growing with a reduced In flux (0.2 ML/s) in order to observe the growth mode of thin InAs films, they were able to distinguish two stages. For initial growth of up to ~2 ML, RHEED patterns showed a diffuse streakiness, probably indicative of a degree of disorder associated with strain, but for thickness greater than 2 ML this was replaced by a spotty pattern characteristic of three-dimensional growth. They showed by STEM that the resulting islands were roughly 40 nm × 20 nm in area and arranged in rows, some 60 nm apart. GaAs grown over these 3D islands gradually reverted to normal RHEED behaviour after some 20–30 ML. Subsequent InAs deposition produced further island growth and, significantly, the next layer of islands was aligned with the layer below, suggesting that the strain associated with lattice mismatch propagated through the whole structure and triggered the formation of each new set of islands. Photoluminescence spectra at 77 K from these superlattices showed two distinct lines, according to whether the InAs 'wells' were 2D ($h\upsilon = 1.42\,\text{eV}$) or 3D ($h\upsilon = 1.13\,\text{eV}$). The 2D line was sharp ($\Delta E = 10\,\text{meV}$), while the 3D line was much broader ($\Delta E = 50\text{–}100\,\text{meV}$), presumably the result of a random spread in size of the islands, a characteristic—and a problem—to be met with in future, when deliberate attempts were made to grow self-organised dots.

There emerged a serious question as to the nature of strain relief and the relation between the mechanisms of dislocation formation and 3D growth, together with the related question of whether the islands themselves might contain defects. An important contribution to the discussion came from Berger et al. (1988), who proposed that, in the S–K growth mode, as the strain energy increases, with increasing strained-layer thickness, the free-energy minimum surface is no longer flat but is three dimensional. In other words, they predicted that the wetting layer should show a marked ripple which would act as a precursor to the formation of islands and that this would imply a *gradual*, rather than an abrupt transition from 2D to 3D growth. They then employed RHEED to monitor the change of lattice parameter of InGaAs films grown on (001) GaAs surfaces as a function of layer thickness. This was, indeed, shown to be gradual, with a rate which increased with increasing growth temperature in the range 400 °C to 520 °C, suggesting a correlation with the surface mobility of the appropriate cations. They also showed that increasing strain (increasing In content in the alloy) resulted in more rapid change from 2D to 3D growth, as predicted by their theoretical model. The correctness of this interesting idea was subsequently confirmed by TEM measurements on a totally different material system—Ge-on-Si. Eaglesham and Cerullo (1990) grew Ge films on Si substrates by MBE and showed clear evidence for the existence of substrate curvature, prior to island formation. They also emphasised the fact that the islands were defect free up to thicknesses very much greater than the critical thickness appropriate to an ideal 2D growth system. A few months later, Guha et al. (1990) demonstrated similarly defect-free InGaAs islands in MBE growth on (001) GaAs substrates and showed that strain existed not only in the islands themselves but also in the substrate, as far as 15 nm below the interface. Here was valuable evidence towards the understanding of growth mechanisms as well as confirmation of the surprisingly high structural quality of the islands, which was of critical importance to those concerned with exploitation of quantum dots in optical devices.

Gradually, a more complete picture of InGaAs island growth was pieced together. In particular, Chang et al. (1989) showed that dislocations did, indeed play a role but only at the point where islands were large enough to coalesce—edge dislocations forming between them. Elman et al. (1989) studied the behaviour of the critical thickness t_c as a function of x in the alloy $In_xGa_{1-x}As$ and found it to vary in a non-linear fashion with In content. Below x = 0.25, t_c increased rapidly towards values greater than 100 nm, while for values of x between 0.25 and 0.5, it levelled off at about 4 ML (as mentioned above, the value then falls to about 2 ML for x = 1). Photoluminescence measurements on quantum wells confirmed that good quality InGaAs layers could be grown for composition x < 0.25, whereas, for x > 0.25, emission lines were broad, indicative of relaxed layers containing dislocations (the authors made no mention of possible 2D growth). Snyder et al. (1991, 1992) investigated the growth of $In_xGa_{1-x}As$ at 520 °C and 320 °C for values of x up to 0.5, using RHEED, STM and TEM and concluded that growth proceeded from a 2D rippled surface by the formation of 2D islands which could then grow into 3D islands, provided the temperature was high enough to allow surface adatom mobility. They also pointed out that the RHEED pattern during InGaAs growth changed from the well-defined (2 × 4) to a diffuse (1 × 1) pattern, indicating that the alloy grew by some mechanism different in detail from that well-established for GaAs on the (001) surface. Petroff and DenBaars (1994), in reviewing their recent work, refined the critical thickness for InAs growth to 1.7 ML, rather than the 2 ML previously reported. However, a major change in 'belief' concerning dot growth was to emerge from the group at Imperial College London (B. Joyce et al. 1997 and P. Joyce et al. 1998—no relation so far as we know!).

They made two crucial contributions to our understanding of InAs dot formation. First, they demonstrated that dots form *only* on the GaAs (001) surface—on (110) and (111)A surfaces, InAs layers were characterised by unchanging streaky RHEED patterns which indicated a steady layer-by-layer growth mechanism, at least up to 30 MLs (the maximum thickness investigated). In the case of the (001) surface, the initial (2 × 4) reconstruction gave way immediately to a C(4 × 4) (cf. the (1 × 1) pattern reported by Snyder et al.), then to a spotty pattern characteristic of 3D growth. Clearly, this distinction between different surfaces proved that strain, alone, could not account for dot formation—the implication had to be that the atomistic details of growth were an essential ingredient. Second, they measured (by STM and AFM) the total volume of material in dots as a function of the amount of InAs deposited (beyond 1.6 ML). At temperatures in the range 450 °C–500 °C they found the former to be far greater than the latter—in other words, the dots must contain both In and Ga and this was far from being a small effect—indeed, at 500 °C the Ga content was some four times greater than the In, meaning that the dot was an alloy of $In_{0.2}Ga_{0.8}As$, with a band gap of 1.2 eV, rather than the 0.35 eV of pure InAs. Clearly, the Ga came from either the substrate or from the wetting layer (by way of the substrate) and this implied a Ga bulk, rather than a surface, diffusion mechanism. At lower temperatures, in the range 350 °C–400 °C, the two amounts were approximately equal—the dots were made up entirely of InAs, consistent with a much smaller diffusion coefficient. They also emphasised that the wetting layer, itself, was an alloy of InAs and GaAs. Finally, they pointed out that the spotty RHEED spectra also had something to say about the shape of the dots. Whereas earlier work had led to a widely accepted belief that dots were pyramidal, the Joyces argued convincingly in favour of a lenticular shape. Growth studies apart, this was of considerable consequence for attempts to calculate the energies of confined states within the dot. To say the least, this work not only changed conceptions of growth but provided device engineers with valuable data on which to base their material 'design'. For readers bent on acquiring further depth of understanding, we can do no better than recommend that they study the review by Joyce and Vvedensky (2004).

Turning, then, to the question of practical usefulness, we might note a number of developments which materialised during the corresponding period. One important fact which emerged from the growth studies was that InGaAs dots were essentially defect free. In other words, there were no dislocations to act as non-radiative recombination centres and this should surely be good news for any optical device, the quantum dot laser in particular. In a similar vein, one must raise the question of InAs–GaAs interface quality and, though this was somewhat less clear, there seemed to be a consensus that it, too, was of high quality. What of the nature of luminescence from a dot? Theoretical concepts concerning the density of states suggested that luminescence lines should be extremely sharp, though measured spectra showed line widths of order 50–100 meV. This was attributed to the inevitable statistical spread of dot sizes but it was obviously desirable to confirm such thinking by finding a way of measuring the spectrum of an individual dot and this was first achieved by the group at CNRS Bagneux (Marzin et al. 1994). Their method was straightforward, involving the etching of small mesas (of the order of 0.5 μm) which contain only a limited number of dots and therefore allow the resolution of individual dot lines. Gratifyingly, the narrowest lines (at 10 K) had a width smaller than the resolution of their spectrometer (0.1 meV), while the spectrum from a large mesa (5 μm) showed a Gaussian line of width of 50 meV. The emission energy of 1.28 eV agreed reasonably with calculations of the confined energies of the dot ground state and an estimate of size distribution led to a satisfactory reproduction of the broad emission spectrum from a large number of dots. One other point of interest was the measurement of very short capture and relaxation times for electrons and holes onto the dots, a matter of importance for high-frequency modulation of quantum dot lasers. Grundmann et al. (1995) and Leon et al. (1995) later reported similar measurements on individual dots by cathodoluminesence. This technique has the advantage of allowing a highly focussed electron beam to excite only a small surface area, thus avoiding the need to etch very small mesas. (Note, however, that it also requires rapid capture of electrons and holes onto the dots if diffusive spreading is not to spoil the effect.) Gammon et al. (1996) employed small apertures through which to excite and detect luminescence, with a similar aim in view.

The distribution of dot sizes is obviously a parameter of crucial importance from the applications viewpoint, as is the density of dots, and a number of authors have published measurements. Let us try to summarise the available data: for typical growth temperatures of 520 °C, InGaAs dots grow with lateral dimensions of the order of 15–20 nm and heights of the order of 5 nm. Size distributions are Gaussian, with widths of approximately ±10%. Dots tend to grow in lines, associated with specific crystal directions, and are spaced approximately 50 nm apart and, as shown very clearly by the work of Xie et al. (1995), it is possible to grow a series of dot planes spaced as little as 10 nm apart (see Figure 6.22). Thus, we can contemplate an areal density of something like 3×10^{15} m^{-2} and a volume density of 3×10^{23} m^{-3} of defect-free dots, comparable with a midrange semiconductor doping level. However, having said that, we should be aware that these figures do depend on growth conditions—for example, recent work by Shamirzaev et al. (2008) on InAs/AlAs QDs demonstrates a clear dependence of dot size and density on growth temperature and on the length of growth interrupt employed. Increasing temperature increases dot diameter but reduces density, while length of interrupt also increases diameter while decreasing density. Perhaps of even greater significance is the fact that intermixing of Al into the InAs dots becomes significant at temperatures above about 480 °C and growth interrupts longer than 50 s. This degree of control is clearly of value when designing dots for any specific application.

Reference to the growth of InAs/AlAs dots stimulates acknowledgement that the InGaAs/GaAs system is certainly not the only one available for the fabrication of quantum dots, nor, of course, is it the only material system which MBE is capable of mastering. The InAs/AlAs system was first explored by collaborators at NRC, Ottawa and the University of California (Raymond et al. 1995) with a view to making dots which emit visible light (hυ ~ 1.9 eV), while InP dots in In$_{0.485}$Ga$_{0.515}$P, emitting at hυ ~ 1.7 eV, were studied by Kurtenbach et al. (1995). This particular InGaP alloy is lattice-matched to GaAs and has a direct band gap of about 1.9 eV. In another interesting experiment, Wang and Forchel (1998) used GaSb self-organised dots, grown on top of an AlGaAs/GaAs structure, to induce strain into a 6 nm GaAs quantum well so as to reduce the GaAs band gap locally and thereby produce GaAs dots. These generated photoluminescence at 1.55 eV, while the undisturbed quantum well emitted at 1.57 eV. As might be anticipated, the vertical separation of the InSb dots from the well

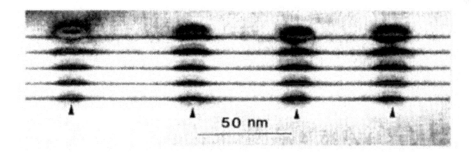

Figure 6.22 *Typical dark-field TEM picture of a set of InAs/GaAs self-organised quantum dots, separated by 36 ML GaAs spacer layers. Note the formation of vertical lines of dots, due to the influence of strain in triggering the growth of the next layer of dots. (From Xie et al. 1995, courtesy American Institute of Physics).*

was a crucial factor in determining their efficacy in producing GaAs dots, a value of about 15 nm being optimum. Also of interest was the fact that all these experiments made use of the (001) GaAs surface—we know of no self-assembled dots on any other GaAs surface. However, we do know of quantum dots being grown in numerous other materials—for example (see Mariette 2005), strain-induced dots have been reported in the II-VI systems CdTe/ZnTe and CdSe/ZnSe, in the Group IV Si/Ge system and in the Group III nitrides, GaN/AlN. Of some interest to the MBE grower is the observation that, in the case of the nitrides, plasma-assisted MBE results in Stranski–Krastanow dot formation, while ammonia MBE generates dots by the droplet epitaxy process, which we discuss below for the example of dot growth in lattice-matched systems. (We take up the topic of MBE growth of these materials in our next chapter.)

All these methods of dot formation had a common feature, the involvement of strain, and, though we may still be some way from a full understanding of the precise mechanism, it appeared that strain was an essential requirement. This was unfortunate because it seemed to rule out the possibility of growing dots in lattice-matched systems such as AlGaAs/GaAs, which were rather better understood than the alternatives involving In. It came as something of a surprise, therefore, when a group of workers at the National Research Institute for Metals in Tsukuba, Japan, discovered a novel technique for doing just that. It involved a modified MBE technique called 'droplet epitaxy' and it all began with the growth of InSb dots on CdTe substrates. It had become evident from the early days of MBE that, if Group III elements were deposited in the absence of a group V flux, the result would be the formation of Group III droplets (Foxon et al. 1973). It was this, of course, that led to the standard method of growing GaAs under As-rich conditions—drops, at that time, being regarded as weeds in the MBE garden of Eden. However, on the basis that a weed is only the right plant in the wrong place, the Japanese group recognised that, in 1991, a droplet in the right place could result in a flower of great beauty. Koguchi et al. (1991) used an InSb substrate on which they grew an initial layer of InSb, followed by 200 nm of CdTe (at 200 °C). This formed the substrate for deposition of In droplets, using an In flux of 4×10^{18} m^{-2}, again at 200 °C, for 30 s. After deposition of these In droplets, an Sb flux of 1×10^{18} m^{-2} was supplied for 100 s to turn the In droplets into InSb dots with dimension of $200 \times 150 \times 70$ nm^3 as measured by electron microscopy. As CdTe has a lattice constant close to that of InSb and a much larger band gap, the hope was that this might form the basis for a lattice-matched quantum dot system but it soon became apparent that the technique of droplet epitaxy might be of much wider application.

Two years later the same group (Koguchi and Ishige 1993) applied it to growing GaAs dots with dimensions of $43 \times 25 \times 15$ nm^3 by the artefact of first treating a GaAs surface with sulphur, in order to satisfy surface bonds. Thus, when a Ga beam impinged on the surface at 200 °C, Ga droplets formed and, when this was followed by an As flux, these were converted to GaAs dots, the sulphur layer preventing any two-dimensional growth. A control sample, not treated with sulphur, showed conventional growth of a 2D GaAs film by the process of 'migration-enhanced epitaxy' (Horikoshi et al. 1986) which we discussed in Chapter 5 (Section 5.2). The sulphur treatment employed consisted of dipping the sample in an ammonium sulphide solution, which meant taking it from the MBE machine but Koguchi et al. (1993) modified the procedure by exposing the cleaned GaAs surface to sulphur vapour in the sample introduction chamber and repeated the rest of the process, though the dots they produced were rather larger ($70 \times 70 \, nm^2$ in area). They pointed out the possibility of enclosing the dots in AlGaAs barrier material but this final stage was only completed in the year 2000 (Watanabe et al. 2000). The delay appeared to be caused by the poor quality of the earlier dot material, as characterised by high resistivity and short minority-carrier lifetime, possibly as a result of the incorporation of sulphur atoms or excess As atoms in the dots. The solution involved doing away with the sulphur treatment and replacing it with carefully chosen growth conditions for the Ga droplets, the As 'conversion' process and for the AlGaAs cladding. Let us look at the details.

Watanabe et al. first grew a GaAs buffer layer on a (001) substrate, followed by a $1.5 \, \mu m$ thick $Al_{0.3}Ga_{0.7}As$ layer at 580 °C. They then turned off the As beam and grew Ga droplets at 200 °C. These were converted to GaAs dots by irradiation with an As_4 flux and it was this process which was critical to success, its being necessary to use a very high As flux ($\sim 4 \times 10^{-5}$ Torr BEP) and a low temperature (150 °C) in order to avoid 2D growth. They argued that the latter proceeded by (thermally activated) Ga atom migration from the Ga droplets to the AlGaAs surface (hence the low temperature), and a high As flux was necessary to 'mop up' the Ga atoms in the droplets before they could migrate. Having successfully produced the required dots, it was then necessary to deposit a containing layer of AlGaAs and this was accomplished in two stages, an initial 10 nm layer grown at 150 °C, followed by a 90 nm layer at 580 °C, topped off with a 10 nm GaAs cap. The low-temperature growth of the initial AlGaAs was important to prevent As dissociation from the dots, leading to further 2D growth. Altogether a complicated process but electron microscope images showed the existence of pyramid-shaped dots with sizes of the order of $10 \times 15 \times 5 \, nm^3$, and photoluminescence spectra contained broad (~ 100 meV) emission lines at photon energies consistent with these dimensions. We should probably caution that the essential measurements of both growth temperature and As beam flux are very much machine dependent but there can be no doubt that this Japanese work opened the way to the growth of self-assembled quantum dots in strain-free material systems.

Finally, we come to what may be termed the Rolls Royce of quantum dot manufacture, based on the concept of 'cleaved edge overgrowth' (CEO), which offered the ultimate control over dot size and spatial distribution. Two obvious disadvantages with self-assembled dots were the facts that: (1) dot size was dependent on some ill-understood relation with a complex of growth parameters and (2) there was a statistical distribution of sizes which resulted in inhomogeneously broadened emission lines with $\Delta h\upsilon \sim 100$ meV, compared to the natural width of perhaps a few tens of microelectron volts. In a sense, this made a mockery of the almost atomic-scale capability of MBE to control film thickness and, a return to this fundamental mode of thinking implied that it must surely be possible to do better. It was just(!) a matter of turning the growth process on its side so that the control afforded by MBE could be applied in first two and finally in three dimensions. The basic idea came (inevitably?) from Bell Labs, being first demonstrated by Loren Pfeiffer et al. (1990) (we referred to it briefly at the beginning of this section). They showed that it was possible, in principle, to grow a quantum well structure on the (001) surface of GaAs, then to cleave it in a (110) plane and grow a second such structure at right angles, so that the coincidence of the two wells would constitute a well-controlled quantum wire. And the beauty of it was that, not only would the wire be totally enclosed in a 'skin' of AlGaAs with high-quality interfaces, but it would be of constant width throughout its length. And, with hindsight, we now see that it was only a matter of repeating the process with a second cleave in a (–110) plane to arrive at a perfect dot! All its dimensions would now be under MBE control, and the statistical problems would be things of the past. The secret lay in finding a method of growing high-quality GaAs and AlGaAs on (110) planes, an achievement previously beyond the capabilities of MBE growers, worldwide—indeed, very few had even tried it. In fact, the 'secret' was no more than a matter of careful selection of growth conditions—an As_4 BEP of 1.6×10^{-5} Torr, a growth temperature of 480 °C–500 °C, a growth rate of 0.5 MLs^{-1}— and the use of two-dimensional electron gas mobility (see Section 6.6) as a monitor of film quality. Oh, and there was the little matter of perfecting a technique for accurately cleaving GaAs samples in the MBE vacuum system! It sounds easy enough when summarised thus glibly but it represented a crucial step forwards, as we shall shortly appreciate.

The Bell group went on to apply their new-found technique to the growth of high-quality quantum wires (Goni et al. 1992) but the final link in the chain, leading to equally high-quality quantum dots, grew out of a postdoctoral visit to Murray Hill by W. Wegscheider, who, on his return to the Technical University of Munchen, proceeded to apply it to dots (Wegscheider et al. 1997, 1998). Not only could dots be grown with accurately controlled dimensions but with similarly well-controlled separation. By basing the growth procedure first on a multi-quantum well, it was possible to form a line of dots and, by varying the well separations, to adjust the dot separation. It was thus possible to explore the coupling between dots—to explore, as the authors graphically put it, the 'way an artificial molecule, characterised by the existence of bonding and antibonding states, can be assembled from two . . . artificial atoms'. All this was achieved by detailed study of micro-PL and micro-PLE spectra at low temperatures from quantum wells, wires and dots. Excitation light was focussed onto the sample by way of a microscope objective lens, and the same lens was used to collect the resulting emission before feeding it into the spectrometer. It was thus possible to relate particular emission lines (some as sharp as 40 μeV wide) to specific regions of the sample. The detail was complex and we make no attempt to reproduce it here (the original papers are well worth perusing)—from the MBE viewpoint the important fact is that here, at last, was a technique for forming a perfectly regular array of quantum dots with dimensions which were fully under the control of the grower. It surely represents a triumph of the crystal growers' art.

Overall, then, it is clear that MBE-grown quantum dots have established, for themselves, a vital niche in the edifice that has been erected on the foundation of LDS and we shall refer to them again when we discuss electro-optic devices in Chapter 10. However, the time has now come to transfer our attention to yet another remarkable field of endeavour which also depended on MBE to provide minutely controlled structures—that of the two-dimensional electron gas.

6.6   The Two-Dimensional Electron Gas (2DEG)

In 1978, Ray Dingle and collaborators at Bell Labs, having been responsible for the initial studies of optical properties of quantum wells, continued their innovative streak by inventing what rapidly became known as the 2DEG, a structure employing modulation doping to obtain much improved electrical conductivity in GaAs films. If we consider, for a moment, bulk samples of n-type GaAs and ask what limits their conductivity, we immediately come upon the frustrating fact that, as the free-carrier density is increased by increasing the donor density, the electron mobility is reduced by ionised impurity scattering. Each free electron implies an ionised donor, and the resulting Coulomb interaction inevitably limits the electron mobility, particularly at low temperatures. There seemed to be no way of avoiding it—until Dingle et al. (1978) demonstrated their genius for lateral thinking. Their innovation depended on the fact that, if the donor atoms could be confined to the barrier regions of an AlGaAs/GaAs superlattice, the resulting free electrons would drop into the GaAs conduction band, where they would be spatially separated from their parent donors, thus reducing the unwanted Coulomb interaction. What was more, they further improved their scheme by depositing Si donors only in the central region of the AlGaAs, leaving undoped spacer layers 6 nm wide and which increased the donor–electron separation. Once again, MBE came into its own, allowing accurate choice of the precise distribution of donor atoms—not only could the Si atoms be deposited exactly where required, the relatively low growth temperature militated against subsequent diffusion. Room temperature mobilities of modulation-doped samples with n ~ 10^{23} m^{-3} were measured to be nearly a factor of 2 greater than corresponding bulk MBE films of GaAs but, as anticipated, the effect was even greater at lower temperatures. Peak mobility (at T ~ 100 K) was 1.5 m^2 V^{-1} s^{-1} for the MD samples, compared with about 0.6 m^2 V^{-1} s^{-1} for a bulk sample, and varying the thickness of the GaAs over the range 10 – 45 nm made little difference. Not that this performance was significantly better in absolute terms than bulk samples grown by LPE and VPE but it was surely significant that the MD samples showed a marked improvement over corresponding bulk MBE samples—this was, after all, the first demonstration of modulation doping—and things could only get better!

As we shall see in a moment, mobilities did indeed get better—very much better—but first the Bell team demonstrated that a similar two-dimensional electron gas could be associated with a single AlGaAs/GaAs heterojunction, rather than a multi-quantum well system (Störmer et al. 1979a, b). Again, Si dopant atoms were incorporated in the AlGaAs while the resulting free electrons transferred into the GaAs where they were confined within an approximately triangular quantum well, as shown in Figure 6.23. In this case, the confining well, which resulted from interface-band bending, was slightly narrower than those used in the earlier multilayer structures, being typically of the order of 10 nm (though dependent on the details of doping, aluminium content, etc. of the

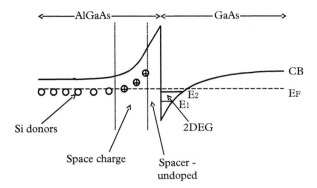

Figure 6.23 *Conduction-band energy of an AlGaAs/GaAs heterojunction, showing the formation of a 2DEG. Electrons from the silicon donors in the AlGaAs transfer to the GaAs and are trapped close to the interface, where they fill all confined states up to the Fermi level E_F. Two confined energy states, E_1 and E_2 are indicated.*

structure involved—see e.g. the discussion in Weisbuch and Vinter 1991, Section II.8). This was the structure which was to set the standard for a great many 2DEG (and two-dimensional hole gas—2DHG) studies to be reported over the next three decades—it consisted of about 3 μm of undoped GaAs grown on a semi-insulating substrate, followed by 3 μm of $Al_xGa_{1-x}As$ (x = 0.25 – 0.3, doped with 10^{24} m^{-3} Si) and a thin cap layer of undoped GaAs (0.02 μm) which was fully depleted of carriers. Störmer et al. were particularly concerned with identifying the two-dimensional nature of the 'electron gas' by means of the Shubnikov–de Haas effect and cyclotron resonance measurements. The former is a magneto-quantum-effect which results in an oscillatory variation in sample resistance as a function of applied magnetic field. It is closely related to the quantum Hall effect (which we discuss below) and, as outlined in Box 6.5, it reflects periodic transitions of the electron Fermi level between Landau levels as the magnetic field is increased. Störmer et al. observed two Shubnikov–de Haas frequencies which demonstrated the electron occupation of two confined states within the quantum well and from which they estimated the corresponding areal densities as 1.4×10^{16} m^{-2} and 2.0×10^{15} m^{-2}, respectively, in reasonable agreement with the total density of 1.2×10^{16} m^{-2} obtained from Hall measurements. To prove that they were, indeed, measuring a two-dimensional gas, they explored the effect of changing the angle between the field direction and the normal to the interface plane. In the usual three-dimensional case, there is no change in the oscillations as a function of angle, whereas, for the 2D case, because the gas is already quantised by the electrostatic well, only the normal component of magnetic field is relevant and this gives rise to a cos θ dependence on angle. Cyclotron resonance absorption was consistent with an electron effective mass some 12% larger than the value of $0.067m_0$ appropriate to GaAs, an enhancement they attributed to non-parabolicity of the GaAs conduction band (the electron Fermi level lying 140 meV above the GaAs band edge at the interface; see e.g. Nelson 2001).

Box 6.5 Shubnikov–de Haas oscillations

In this box we offer a simplified account of the Shubnikov–de Haas effect (for a more rigorous account, see e.g. Davies 1998, Section 6.4). This reveals itself in the shape of an oscillatory variation of the electrical resistance of a metal or semiconductor as a function of an applied magnetic field B. First, we should recognise that the resistance depends on B because of (in classical terms) the spiral trajectory of electrons when subject to crossed electric and magnetic fields. The larger B is, the greater is the perturbation of electron motion from its field-free condition but this does not, of course, explain the observed oscillations. To understand these, we are obliged to look into the quantum description of the electron's motion.

In the two-dimensional case, the magnetic field B (perpendicular to the plane of the gas) splits electron band states into Landau levels, which are sharply defined because the electrons are fully confined—in the z-direction by the quantum well and in the x–y plane by the magnetic field—each level corresponding to a specific circular orbit of radius a. Thus, the energies are given by

$$E_n = (n + \tfrac{1}{2})(h/2\pi)\omega_c \quad (n = 0, 1, 2, 3, - - -) \tag{B6.31}$$

the radii by

$$a_n = [(2n + 1)\, h/2\pi eB]^{1/2} \tag{B6.32}$$

and the cyclotron frequency by

$$\omega_c = eB/m_e \tag{B6.33}$$

Figure 6.24 represents a plot of Landau level energies as a function of B, for the case of electrons in GaAs, having an effective mass of $m_e = 0.067m_0$.

Another important property of the Landau levels is the fact that their degeneracy (the number of allowed electron states per square metre) increases with magnetic field according to

$$N_s = eB/h \tag{B6.34}$$

In terms of the magnetic field B in Tesla, we can then write $N_s = 2.42 \times 10^{14}\, B$ m^{-2} and we see that the n_s electrons (per square metre) will be distributed over a number of Landau levels and, as B changes, this distribution will also change. (Note that the measurements are made at liquid helium temperatures, where $kT \ll (h/2\pi)\omega_c$.)

It is this redistribution which is responsible for the Shubnikov–de Haas effect. Whenever an electron makes a transition from one Landau level to another, its orbit changes and this, in turn, changes its contribution to the magnetoresistance. The question of interest is how to relate the measured oscillation frequency to the other parameters such as effective mass, free-carrier density, etc. Let us try to make this clear. We need to calculate the differences between adjacent values of B at which these 'jumps' between Landau levels occur. Imagine for a moment that the applied field B = 10T, for which case each Landau level has a degeneracy of $N_s = 2.42 \times 10^{15}$ m^{-2} and that the density of free electrons $n_s = 10^{16}$ m^{-2}, as was the case in the sample measured by Störmer et al. Clearly, the four lowest Landau levels will be filled but with a small overspill into the fifth level. Now suppose that B is gradually reduced. At a value of B = 8.26 T, we have $5 \times 2.42 \times 10^{14}\, B = 10^{16}$, implying that all five levels will be full and any further reduction in B will start to populate a sixth level. Further reduction will fill

continued

Box 6.5 *continued*

this sixth level at B = 6.89 T and so on, as B is decreased further, the condition for the 'jumps' being simply that

$$B_i = (1/i)(h/e)n_s \qquad (B6.35)$$

where i is an integer (see Figure 6.24). Thus, the difference between adjacent values of $1/B_i$ (which defines the period of the oscillation in magnetoresistance) is $\Delta(1/B) = (e/h)n_s^{-1}$ and n_s can then be calculated as follows:

$$n_s = (e/h)/\Delta(1/B) \qquad (B6.36)$$

from which n_s can easily be determined. Note that there is no involvement of effective mass or any other material parameter.

   This account suggests that the plot of R vs (1/B) should consist of a series of spikes, rather than a roughly sinusoidal form but this is the result of idealising the Landau levels as being infinitely sharp. In real samples, we must always expect a degree of disorder, which broadens the levels inhomogeneously and rounds the spikes into a smoothly varying waveform. None-theless, the period is still given by equation (B6.36), and a value for n_s thereby obtained. However, we should end with a note of caution—the above account makes the assumption that the Landau levels are spin split, as is the case for large magnetic fields. At lower fields, where each level contains both up and down spins, the density of states is doubled and the formula for n_s becomes

$$N_s = 2(e/h)/\Delta(1/B) \qquad \text{(low magnetic fields)} \qquad (B6.37)$$

Needless to say, this complication can lead to a deal of confusion—you have been warned!

Störmer et al. (1979b) reported yet another interesting observation, namely, that the measured electron-sheet density could be very significantly modified by shining visible light onto the sample at low temperature. This persistent photoconductivity was thought to result from the deep level (known as a DX centre) associated with Si doping of AlGaAs. Electrons trapped on these centres could be transferred into the 2DEG when they were released by light absorption. For example, a free-carrier density of $n_s = 1.1 \times 10^{16}$ m^{-2} at 4 K in the dark was increased to 1.6×10^{16} m^{-2} under illumination. The effect could be reversed by raising the sample temperature above 100 K and the cycle repeated reproducibly. It was an effect which would be much utilised in exploring the properties of many 2DEG samples by subsequent authors, when it was noted (see e.g. Hiyamizu et al. 1983) that the electron mobility also increased, in sympathy with the electron density—ionised impurity scattering decreases with increasing free-carrier velocity which, in turn, depends on the electron Fermi energy.

   Perhaps the obvious next step in exploring two-dimensional effects in AlGaAs/GaAs heterostructures would be concerned with a 2D hole gas, and the Bell group duly obliged (Störmer and Tsang 1980). By doping the AlGaAs p-type with beryllium and leaving the GaAs undoped, it was hoped to generate a hole gas at the interface in a manner directly

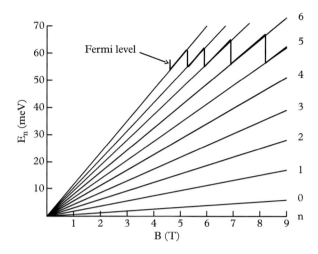

Figure 6.24 *Plot of Landau level energies E_n in an AlGaAs/GaAs 2DEG as a function of magnetic field B. The electron occupation of the levels is indicated for the case where $n_s \sim 10^{16}$ m^{-2} in terms of the Fermi level position.*

parallel to the 2DEG described above. They recognised, however, that it might not be quite so straightforward, on account of the smaller step in the valence band (still believed to be only 15% of the band-gap difference) and the much greater acceptor binding energy, compared to the donor binding energy appropriate to the 2DEG. Once again, MBE was chosen as the preferred method of achieving a clean interface and accurate placement of the Be acceptors, a choice which proved satisfyingly successful—interpretation of Hall measurements was complicated by parallel conduction in the AlGaAs but low-temperature Shubnikov–de Haas oscillations proved the existence of a 2DHG. The hole mobility at $T = 5\,K$ was measured as 0.17 m^2 V^{-1} s^{-1}, compared with typical values of 3 m^2 V^{-1} s^{-1} for 2DEG structures (in 1980).

The way was now open for MBE growers around the world to compete in a frenzy of activity aimed at optimising the quality of 2DEG and 2DHG structures, quantified in terms of the measured low-temperature electron or hole mobilities. First, let us concentrate on the more popular 2DEG (its mobility would always be greater than that of the corresponding 2DHG on account of the electron's smaller effective mass). As shown quite dramatically in Figure 6.6 of Orton (2004), 4.2 K mobilities climbed from roughly 1 m^2 V^{-1} s^{-1} in 1978 to nearly 1000 m^2 V^{-1} s^{-1} in 1989, when progress abruptly slowed. From then on to the present day (2012), life became very much harder—current performance is characterised by a 1 K mobility of $3.5 \times 10^3 m^2$ V^{-1} s^{-1} and the quest for the holy grail of 10 000 still looks like being a long haul. This flurry of activity during the 1980s was quite remarkable, involving six different countries and upwards of ten laboratories. It was also of interest to note the dominance of industrial laboratories over academic establishments in what was essentially an academic exercise, the driving force being largely that of the exciting new physics inherent in studies of the quantum

Hall effect. It was certainly here that the need for ever greater electron mobility could be seen to advantage, while the oft-repeated maxim that electronic devices only have a future when they work at room temperature may have been temporarily overlooked (as we shall see in a later chapter, this is a slight exaggeration—the high electron transistor worked very well at room temperature, though its mobility was extremely modest, by comparison with those being spoken of here!). The reason for this odd mismatch, as we discussed earlier (Section 6.1), was probably the fact that MBE machines were expensive and, initially, only industrial concerns were prepared to invest in them (on the basis of Al Cho's pioneering demonstration of MBE-grown device structures). However, before looking more closely into the subtleties of MBE growth and the amazing number of modifications which made possible the achievement of $\mu_e = 1000 \, \text{m}^2 \, \text{V}^{-1} \, \text{s}^{-1}$, it would be well to outline the nature of the stimulus.

The quantum Hall effect was discovered, as it happened, at the very beginning of the III-V MBE boom in 1980 but perversely enough (!) in silicon (von Klitzing et al. 1980). It concerned a two-dimensional electron gas but one confined at the interface between a sample of p-type silicon and the oxide layer of what was, by then, the long-established MOS transistor. By applying a positive voltage to a gate electrode evaporated on top of the oxide film, it was possible to induce a sheet of free electrons at the interface and thereby control conductivity between source and drain electrodes alloyed into the silicon. The key to the transistor's success was the very low density of interface states associated with the thermally grown oxide. In order to study the detailed behaviour of the 2DEG, von Klitzing et al. chose to define a conventional Hall bar so as to examine the Hall voltage as a function of magnetic field applied normal to the interface. Measurements were made at a temperature of 1.5 K and at magnetic fields up to 18 T. For small values of field they observed the normal behaviour—the Hall resistance $R_H = V_H/I_x$ varied linearly with magnetic field—but at certain points, characterised by

$$V_H/I_x = h/ie^2 \qquad (6.18)$$

where i is an integer, they observed plateaux in the R_H vs B_z characteristic. The remarkable feature of these plateaux was their being independent of sample shape, oxide thickness or gate voltage and (as later work has confirmed) showing a value of $R_H = 25812.807/i \, \Omega$ to an accuracy of about one part in a billion. This latter observation stimulated an immediate interest in its application as a resistance standard (see Chapter 9) but, for the moment, we shall concentrate on the exciting new physics involved.

It was hardly surprising that, once this remarkable result was published, those concerned to develop 2DEG structures in AlGaAs/GaAs should attempt to reproduce the effect in their own samples and, what was more, they were soon able to report how much easier it was to demonstrate well-resolved plateaux than had been the case for the silicon MOS devices (Tsui and Gossard 1981). MOS studies required temperatures as low as 2 K and magnetic fields greater than 15 T while AlGaAs/GaAs structures showed the effect at 4.2 K and 4.2 T. The reason, as Tsui and Gossard emphasised, was the much smaller electron effective mass in GaAs ($0.067m_0$) compared with Si ($0.19m_0$) but, to understand this, demands some knowledge of the effect, itself. We can discover a sufficient understanding by following the argument presented in Box 6.5 to explain the Shubnikov–de Haas effect (see Figure 6.24). As explained in Orton (2004) Section 6.2, the Hall plateaux occur at the precise values of magnetic field at which the electron Fermi level jumps between Landau levels, that is, the field at which specific levels are just filled, and these are given by

$$B_i = hn_s/ie \qquad (6.19)$$

Thus, they depend on the 2D electron density n_s, which can be controlled by the doping level in the AlGaAs, by the width of the spacer layer, or by illumination (or by the gate voltage in the MOS structure). In principle, then, it appears that the effect can be made to occur at low magnetic fields by choosing a low value of n_s. However, two comments are in order: first, we note that small values of B result in small Landau level splittings ΔE and this demands low temperatures to satisfy the condition that $kT \ll \Delta E$; second, the concept of a sharply defined Landau level requires that

$$\mu B \gg 1 \qquad (6.20)$$

so, the smaller B is, the larger must be the electron mobility. Quantitatively, if B = 1T, the mobility would need to be $10 \text{ m}^2 \text{ V}^{-1} \text{ s}^{-1}$, a value only just within practical reach at the time (the best sample used by Tsui and Gossard had a mobility of $7.9 \text{ m}^2 \text{ V}^{-1} \text{ s}^{-1}$) and certainly not available in silicon. So, for any particular value of B, the smaller effective mass in GaAs results in both an increase in ΔE and an increase in μ, whence the comment above.

At this juncture, it would be well to emphasise that the integer quantum Hall effect is not only a remarkable occurrence but also demands a high degree of subtlety in its explanation (a thoughtful account is given by Davies 1998, Chapter 6). We make no pretence of providing that here but one particular feature should be made clear—it is that the effect can only be seen in samples which show random imperfections, resulting in inhomogeneous broadening of the Landau levels and (even more important) in a degree of localisation of the electron states. While it may be necessary for the electron mobility to be high, well-developed Hall plateaux can only be anticipated from samples which depart significantly from perfection. There was really no need for any high-mobility chase at all (!)—not, that is, until the Bell researchers made an even more remarkable discovery in the shape of the fractional quantum Hall effect. At the back of their minds was a hope that they might find evidence for the existence of a so-called Wigner crystal in the 2DEG associated with an AlGaAs/GaAs interface. The idea that an electron fluid held at low temperature might 'freeze' to produce a crystal lattice had been predicted by Wigner in 1934 and an example had been discovered experimentally in 1979 for the case of an electron gas on the surface of liquid helium. The condition to be satisfied for crystallisation to occur is that the electron's kinetic energy should be less than its potential energy and, as the effect of a high magnetic field is to quench the kinetic energy, it made good sense to look for Wigner crystallisation in a 2DEG sample in the quantum limit—that is when all the electrons are in the lowest Landau level. Also, as the potential energy increases with electron density (see MacKinnon 2001), it would be helpful to choose a sample with relatively low free-carrier density. Thus, Tsui et al. (1982) chose an AlGaAs/GaAs modulation-doped sample with $n_s \sim 1 \times 10^{15} \, \mathrm{m}^{-2}$ and examined the region of the Hall resistance vs magnetic field characteristic in the region above the i = 1 Hall plateau. To their surprise, they discovered a further series of plateaux corresponding to *fractional* filling factors. Features with a value of i = 1/3 were very clear but there was evidence for i = 2/3 also. It was clear that these could not be explained on a model based on the integer QHE, and an intriguing field of speculation was thereby thrown open to the world of semiconductor physics. Here, if it were needed, was a valid stimulus to the achievement of 2DEGs with ever higher electron mobilities.

As we have seen already, success came rapidly—over the decade between 1978 and 1988 4.2 K mobilities rose from $1 \, \mathrm{m^2 \, V^{-1} \, s^{-1}}$ to the dizzy heights of almost $1000 \, \mathrm{m^2 \, V^{-1} \, s^{-1}}$. Compare this with the peak mobility of $30 \, \mathrm{m^2 \, V^{-1} \, s^{-1}}$ for the very best bulk n-type GaAs. How was this achieved? The key was to understand the nature of the various scattering mechanisms that limit mobility in different structures. As background, we note that in lightly doped bulk GaAs two processes are of particular importance—at temperatures above 100 K, polar optic-phonon scattering dominates while, at temperatures below 30 K, ionised impurity scattering takes over, limiting the 4.2 K value to approximately $10 \, \mathrm{m^2 \, V^{-1} \, s^{-1}}$. In the region of peak mobility, acoustic scattering plays a role, as shown in Figure 6.25 from Stillman and Wolfe (1976). Clearly, the major success in fabricating 2DEG structures was the minimisation of ionised impurity scattering by separating free electrons from their parent donors but, as always, the devil is in the detail. We should be clear precisely what limited μ_e at each stage of the chase, not forgetting

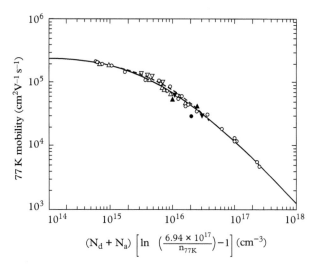

Figure 6.25 *Plot of electron mobility in a high purity sample of GaAs as a function of temperature, compared with calculation of the contributions from the different scattering mechanisms. (From Stillman and Wolfe 1976, courtesy Elsevier)*

that well-established calculations of the contributions from different mechanisms in three dimensions are frequently modified when applied to a two-dimensional electron gas. In particular, acoustic phonon scattering shows an approximately T^{-1} dependence on temperature in a 2D system (cf. $T^{-3/2}$ in 3D) and ionised impurity scattering becomes independent of temperature in 2D (cf. $T^{3/2}$). Thus, high-mobility samples show three clearly resolved regimes, representing polar optic-phonon scattering above about 50 K, acoustic phonon scattering between 2 K and 50 K and ionised impurity scattering below 2 K (see Figure 6.26). This offers a clear aid to understanding and helps to confirm the contribution of various sample changes to the measured mobility. Nor should we overlook the value of free-carrier density n_s—the challenge thrown up by the study of the fractional quantum Hall effect is one of growing samples with maximum mobility at *low* values of n_s, in order to persuade all the carriers into the lowest Landau level at relatively modest values of magnetic field.

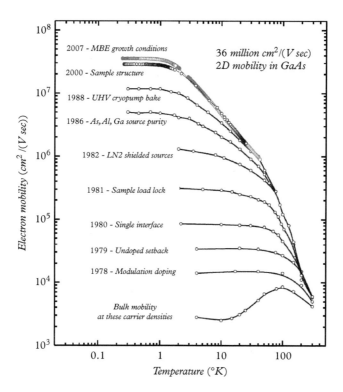

Figure 6.26 *Plot of measured electron mobilities in 2DEGs grown by MBE over the period 1978 to 2007. Samples with the highest mobilities show clearly three temperature regimes, corresponding to polar optic-phonon scattering, acoustic phonon scattering and remote ionised impurity scattering. (From Schlom and Pfeiffer 2010, courtesy Macmillan)*

The obvious starting point for any attempt to optimise 2DEG mobilities was the minimisation of ionised impurity scattering by the donor atoms in the AlGaAs and this involved the introduction of the so-called spacer layer—a region of undoped material separating the donors from the interface. Interestingly, such a layer (6 nm wide) was introduced in the very first 2DEG structures (the MD superlattices used by Dingle et al. 1978) but was apparently missing from the single heterostructures used by Störmer et al. (1979a, b) and by Tsui and Gossard (1981), then re-introduced by Tsui et al. (1982) (50 nm wide) but its effect was never discussed in detail by any of these authors. The first serious attempt to investigate the dependence of mobility on such material parameters appears to have been that reported by Drummond et al. (1981) of the University of Illinois (Al Cho, never long parted from MBE innovative venture, was a visiting professor at the time). As they pointed out, a number of parameters could be expected to influence 2DEG properties: the AlAs mole fraction x, the doping level in the AlGaAs, the width of the spacer layer and the thickness of the doped layer. Of greatest significance, from their work, was the spacer-layer thickness, which they found to be optimum at a value of about 5 nm (though they only measured mobility at 300 and 77 K). In a later paper (Drummond et al. 1982) they obtained values of $\mu \sim 20\, m^2\, V^{-1}\, s^{-1}$ at 10 K for an optimum spacer layer of 6 nm.

The Illinois group had certainly started an interesting exploration of sample properties and their influence on 2DEG sheet density and mobility but their conclusion that a 6 nm spacer layer optimised mobility was soon challenged. Hiyamizu et al. (1983), of Fujitsu, found the optimum thickness to lie near 20 nm and observed electron mobilities as high as 125 $m^2\, V^{-1}\, s^{-1}$ at 5 K in the dark and 212 $m^2\, V^{-1}\, s^{-1}$ after illumination (well over a million in the 'old' units!). This obviously represented a major step forwards and supported the authors' claim that they had achieved much improved interface quality, together with reduced background impurity levels in their GaAs. The importance of these aspects was confirmed by work at the German Post Office in Darmstadt (Weimann and Schlapp 1985), a study which showed that an MBE growth temperature between 620 °C and 640 °C, which was known to result in optimum GaAs quality, also resulted in optimum 2DEG mobilities. Weimann and Schlapp also showed that mobility improved in samples with thicker GaAs layers, the purity of the GaAs apparently improving as the layer thickness was increased. They achieved a 2DEG mobility of 200 $m^2\, V^{-1}\, s^{-1}$ (after illumination) with a spacer thickness of 33 nm. The importance of ensuring optimum GaAs quality was confirmed by our own work (Harris et al. 1986), which demonstrated a steady improvement in mobility with time after reloading source cells. The background impurity level of bulk GaAs films was used as a monitor, and the 2DEG mobility at 4 K improved in sympathy, reaching a value of 300 $m^2\, V^{-1}\, s^{-1}$ after some 60 growth runs. There was a well-defined peak in mobility for a spacer-layer thickness of about 50 nm and evidence of improved mobility resulting from increased thickness of the doped AlGaAs. Unusually, the experimental results showed a marked tendency to *exceed* theoretical predictions!

This Japanese/European gauntlet was then taken up by the old firm at Bell, English et al. (1987), who responded with a report of a 2DEG mobility of 500 $m^2\, V^{-1}\, s^{-1}$ at a somewhat lower sheet-carrier density of $1.6 \times 10^{15}\, m^{-2}$, though, in fairness, we should note that this mobility was measured only at temperatures below 1 K—the 4 K value was close to 300 m^2 V^{-1} s^{-1}. They attributed their success to two factors: a novel method of incorporating Si donors in the AlGaAs, and the introduction of a GaAs/Al$_{0.3}$Ga$_{0.7}$As multilayer prior to growing the undoped GaAs. The former followed the idea originally proposed by Wood et al. (1980)—Si being deposited in a sequence of sheets, each providing ~10^{15} electrons per square metre in the GaAs. These sheets were deposited during growth interrupts, the lack of a Ga beam thereby minimising the proportion of Si atoms incorporated on As sites and acting as acceptors. This, again, was intended to reduce remote ionised impurity scattering. The multilayer sequence was incorporated in order to trap background impurities at each of the interfaces. The paper certainly included one significant advance, in providing details of the temperature dependence of mobility over the whole temperature range from 300 K down to 0.3 K, demonstrating distinct regimes corresponding to polar optic-phonon, acoustic phonon and ionised impurity scattering, the latter region, characterised by a temperature-independent mobility, being reached only at temperatures below 1 K. Clearly, any meaningful comparison between 'competing' research groups should ideally be made at these extremely low temperatures.

The next move in this intriguing game of molecular beam 'chess' came from France. Etienne and Paris (1987) of CNRS, Bagneux employed yet another Si doping technique involving just two planar-doped layers, one near the surface of the 2DEG structure, which accounted for surface depletion charge, and a second, some 80–100 nm from the AlGaAs/GaAs interface and which supplied the 2DEG. They reported μ = 370 m^2 V^{-1} s^{-1}, for n$_s$ = 1.8 × 10^{15} m^{-2} at T = 1.5 K, even though their samples were grown very soon after recharging their source cells. They emphasised the importance of including the value of n$_s$ by proposing a figure-of-merit as μ/n$_s^{3/2}$ on the basis that ionised impurity scattering can be expected to vary approximately as n$_s^{3/2}$. However, some care was obviously necessary in accepting this approach—the assumption that low-temperature mobility was in all cases limited by ionised impurity scattering could be open to question, as made clear by the Philips group (Foxon et al. 1989). All the improvements achieved up to this point had referred to samples showing values of n$_s$ in the range 1.5 – 3.0 × 10^{15} m^{-2} though, as we noted in our introduction to this mobility chase, it was desirable to achieve high mobility combined with low carrier density. The Philips group then made their second major contribution by demonstrating 1.5 K mobilities as high as 1000 m^2 V^{-1} s^{-1} in samples with sheet-carrier density as low as 2 × 10^{14} m^{-2} and showed convincing evidence that, even at 1.5 K, the dominant scattering was by acoustic phonons. Once again, they attached great importance to improving background purity levels, employing an AlAs/GaAs superlattice as a pre-layer to trap impurity atoms (similarly to the procedure adopted by English et al.). Their other principal innovation was to reduce the doping level in the AlGaAs from 1–2 × 10^{24} m^{-3}, as reported in the majority of earlier papers, to 1 × 10^{23} m^{-3}, while increasing its *thickness* to 200 nm. It was presumably this reduced doping level that reduced ionised impurity scattering to such an extent that acoustic phonon scattering dominated down to 1.5 K.

Yet again the ball was back in the Bell Labs court and yet again they responded vigorously (Pfeiffer et al. 1989), reporting a mobility as high as 1170 m^2 V^{-1} s^{-1} at temperatures below 1 K, the value remaining constant as a function of temperature in this very low-temperature regime—limited, therefore, by ionised impurity scattering? On the other hand, the corresponding carrier density was 2.4×10^{15} m^{-2}, five times larger than the best Philips sample. Pfeiffer et al. took considerable pains to improve the level of background impurities in their MBE vacuum system and claimed a system pressure of 1.5×10^{-12} Torr prior to growth. They used an AlGaAs/GaAs multilayer to clean up the GaAs layer and also followed the Etienne and Paris doping method, involving two distinct regions of Si monolayers in the AlGaAs doped layer, the spacer layer being 70 nm thick. This achievement of mobilities in excess of 1000 m^2 V^{-1} s^{-1} represented the end of the 'easy' advances—from this point on the going was to become very much harder, as witnessed by the fact that the best results reported to date (some 20 years later) lie in the region of 3500 m^2 V^{-1} s^{-1} (Umansky et al. 2009). It would appear that the field is now restricted to three research groups: Loren Pfeiffer at Princeton University, Vladimir Urmansky at the Weizmann Institute in Israel and Werner Wegscheider at ETH Zurich. Perhaps, by the time this book is published, one of them will have reached the next major landmark of 10 000 m^2 V^{-1} s^{-1} —we wait with bated breath! However, no matter where the race eventually comes to an end, the achievements of MBE growers in improving AlGaAs/GaAs 2DEGs to the degree described here represents a well-earned feather in their collective hats—of that there can be little doubt. By way of a summary, we reproduce in Figure 6.26 a collection of experimental results published by Schlom and Pfeiffer (2010) and which provides a graphic account of the manner in which 2DEG mobilities improved between 1978 and 2007. It also makes clear the different temperature regimes corresponding to the various scattering mechanisms.

Nor is this quite the end of the present story. We need to round it off by briefly considering two further aspects: those of interface scattering and of the corresponding developments in the two-dimensional hole gas. The fact that interface quality might affect low-temperature mobility came to the fore quite early in the story when the Illinois group (Morkoc et al. 1981) made 2DEG structures employing the so-called inverse interface—the GaAs being grown on top of the doped AlGaAs. Measured electron mobilities were notably lower than those recorded for 2DEGs at 'normal' interfaces—by roughly a factor of 50. There were two possible causes: silicon and impurity segregation from the AlGaAs to the interface, and interface roughness. Various attempts to improve the situation yielded only modest success until 1988, when Shtrikman et al. (1988) of IBM took up the challenge. They employed low growth temperatures (600 °C) to minimise impurity segregation, and a sequence of growth interrupts to smooth the interface and thus achieved an electron mobility as high as 46 m^2 V^{-1} s^{-1} at a carrier density of 2×10^{15} m^{-2}. The fact that samples grown without interrupts showed mobilities of 15 m^2 V^{-1} s^{-1} suggested just how important was the quality of the interface. Further improvements were obtained by Sajoto et al. (1989), who replaced the AlGaAs layer with an undoped AlAs/GaAs superlattice (grown with interrupts at each interface) and *induced* the 2DEG by applying a gate voltage. This involved growing the whole structure on a conducting substrate, rather than the SI substrates used by others. Their best 4 K mobility was measured as 200 m^2 V^{-1} s^{-1}, still about a factor of 4 lower than that obtained on 'normal' interfaces but a very considerable improvement on earlier 'inverse' interfaces. It is worth remembering that 2DEGs formed in multi-quantum well structures inevitably involve 'inverse' interfaces and we should note yet another result from Bell Labs in which Pfeiffer et al. (1992) recorded low-temperature mobilities of 400 m^2 V^{-1} s^{-1} in such structures. Two questions arise: first, what is the role of inter-face scattering in limiting mobilities in the best 'normal' structures?—the observation of temperature-independent mobilities at very low temperatures does not rule out a role for interface quality but the extent of its contribution is still unclear; second, what is the future for induced 2DEGs?—the possibility of doing away with specific dopant spe-cies offers obvious advantages. An interesting example has been described by Harrell et al. (1999) of Cambridge University, as they show that induction may have particu-lar advantages for samples with low carrier densities. They achieved a low-temperature mobility of 400 m^2 V^{-1} s^{-1} at n$_s$ = 4 × 10^{15} m^{-2}, identical to that measured on a doped sample, but found μ = 100 m^2 V^{-1} s^{-1} at n$_s$ = 1 × 10^{14} m^{-2}, many times greater than for a corresponding doped sample.

Finally, we should say a word about 2D hole gases. As noted above, Störmer and Tsang (1980) demonstrated the possibility of a 2DHG in a 'normal' AlGaAs/GaAs structure but with a rather modest mobility of 0.17 m^2 V^{-1} s^{-1}. There was clearly scope for improvement! Indeed, not only did Wang et al. (1986) achieve considerably greater mobilities, they did it in a particularly interesting manner, by growing on the (311)A surface of GaAs. On this orientation, Si impurities tend to occupy As sites and act as ac-ceptors, rather than following the donor behaviour exclusively found on (100) surfaces. Their best low temperature (i.e. T < 1 K) hole mobility was 38 m^2 V^{-1} s^{-1} at a hole dens-ity of 1 × 10^{15} m^{-2}, about a factor of 10 lower than the best electron mobilities available in 1986 but this was probably largely down to the larger hole effective mass. Fascin-ating insight has been thrown on this question by experiments performed on undoped structures in which the same interface could be populated by either electrons or holes, according to the sign of gate voltage used (Chen et al. 2012). In all cases measured, hole mobility was less than that of electrons by a factor between 7 and 10 (for the same carrier density), an observation in fair agreement with values calculated for a combin-ation of ionised impurity and interface roughness scattering. Peak mobilities lay at about 400 and 80 m^2 V^{-1} s^{-1} for electrons and holes, respectively. Perhaps the most interesting aspect of these gate-controlled structures is the possibility of creating both electron and hole gases within a distance of the order of 10 nm, so close that their separation is sig-nificantly less than the distance between electrons (or holes) within the respective layers. This opens the way to all manner of exciting physics (see the review by Das Gupta et al. 2011) but it would take us well out of our comfort zone to go into detail. We leave it at that!

However, we should certainly be at fault if we failed to outline *some* of the excit-ing new physics that has emerged as a result of the MBE growers' ability to provide high-quality 2DEG samples. We have already referred to the fact that, when Tsui et al. (1982) stumbled upon the fractional quantum Hall effect, they were actually looking for evidence of Wigner crystallisation of the 2D electrons. It soon became apparent that in this ultimate quantum limit (filling factor less than unity) the electrons behave as a correlated liquid—in other words, electron–electron interactions become dominant over electron interactions with external fields. Theoretical prediction then suggested that, as the applied magnetic field was increased, at a filling factor somewhere between 1/3 and 1/9, this liquid could be expected to crystallise into a two-dimensional solid. Two practical problems made it difficult for the experimentalist to demonstrate its existence: first, it required both very high magnetic fields and very low temperatures; second, it was not immediately obvious how to show unequivocally that such a solid state had been achieved. The first such evidence came from France in a joint project between the MBE growers at CNRS Bagneux and the magnetic resonance experts at the Atomic Energy Commissariat, Saclay (Andrei et al. 1988). The samples made use of the double planar-doping technique first employed by Etienne and Paris (1987) with free-carrier densities between 4.6×10^{14} and 1.2×10^{15} m^{-2}, and electron mobilities of about 100 m^2 V^{-1} s^{-1}. Measurements were made at temperatures between 50 mK and 250 mK and at magnetic fields of up to 28 T. The method used to demonstrate crystallisation involved measuring the RF absorption spectrum at frequencies in the range 50–1400 MHz, which demonstrated rigidity to sheer—in other words, the generation of longitudinal phonons. Raising the temperature or decreasing the magnetic field led to melting, with the loss of the characteristic resonance lines, enabling the authors to plot out the solid–liquid phase diagram, a splendid vindication of many years of theoretical stimulus.

A second area of interest for the semiconductor physicist was concerned with the phenomenon of ballistic transport. The extremely high mobilities made available in 2DEGs implied unusually long mean-free-paths λ and allowed measurements on electrons free from inelastic scattering. The mean-free-path is defined as

$$\lambda = \tau v_T \qquad (6.21)$$

where v_T is the electron thermal velocity and τ is the scattering time. Bearing in mind that the mobility μ is related to τ according to

$$\mu = e\tau/m_e \qquad (6.22)$$

and that, in a degenerate electron gas, the thermal energy $E_T = m_e v_T^2/2$ must be identified with the Fermi energy E_F, we have

$$\lambda = (\mu/e)(2m_e E_F)^{1/2} \qquad (6.23)$$

from which it follows that, for a mobility of 1000 m^2 V^{-1} s^{-1} and a typical value of $E_T = 30$ meV, $\lambda = 150\,\mu$m, which is much longer than typical dimensions that can be defined photolithographically. It was thus possible to study electrons in solids under similar conditions to those appropriate to high-vacuum conditions.

What did this mean in practice? It meant, for example, that electrons could be confined within a waveguide defined by a pair of parallel gates and allowed a demonstration of the Aharonov–Bohm effect (see Orton 2004 Section 6.3). It also offered the opportunity for study of quantum effects in point contacts such as that shown in Figure 6.26. Two groups published at almost the same time (van Wees et al. 1988; Wharam et al. 1988), both reporting that, in ballistic transport through a narrow channel between a

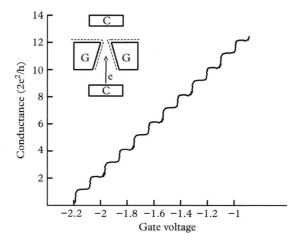

Figure 6.27 *Typical experimental data for the conductance of a quantum point contact, plotted in units of (2e²/h). The geometry of the experiment is shown as an inset, the gates G defining the aperture through which electrons flow between the ohmic contacts C.*

pair of gates, the conductance was quantised in units of $2e^2/h$. A typical set of results is shown in Figure 6.27, where the channel conductance is plotted in units of $(2e^2/h)$ as a function of gate voltage. As the gate voltage is made more negative, thus expanding the depletion regions under the gate electrodes, we should obviously expect to see a gradual reduction in conductance but it is less obvious why this should occur in regular steps. The explanation centres on the fact that electrons passing through the constriction must satisfy quantum conditions similar to those imposed on electrons in deep quantum wells, where their wave functions are constrained to fall to zero at the sides of the wells (see Box 6.4). By a parallel argument to the one used there, we find that the y-component of the k-vector must satisfy the condition

$$k_y = i\pi/W \tag{6.24}$$

where W is the width of the constriction, and i is an integer. Because the magnitude of the total k-vector k_F is fixed by the Fermi energy E_F,

$$k_F = (2m_e E_F)^{1/2}/(h/2\pi) \tag{6.25}$$

This also implies a quantisation of the x-component k_x in the direction of current flow and it follows from this that the conductance will be quantised in units of $2e^2/h$—the argument is clearly presented by Johnson (2001). As further examples of the use of quantum point contacts in scientific experimentation, we might cite their use as emitters or collectors in experiments with ballistic electrons. Thus, an electron beam could be injected into an electric-field-free region by one point contact, bent into a circular path by a magnetic field and collected at a second contact. It was also found possible to focus electron beams with specially shaped gate electrodes in very much the manner of focussing light with dielectric lenses. For a comprehensive account of these so-called 'mesoscopic' systems, see the review by Beenakker and van Houten (1991).

Another fascinating set of experiments made possible by the availability of high-mobility 2DEG samples was concerned with single electron charging of quantum dots. In this instance, the dots were formed by a series of gate electrodes defined by electron lithography, an example being provided in Figure 6.28. Applying appropriate voltages to the various electrodes set up barriers which isolated the central dot region from the rest of the 2DEG. Typically, such dots were of order 0.8 μm in diameter, very much larger than the self-organised dots we discussed earlier but small enough to present a tiny capacitance $C \sim 10^{-17} - 10^{-16}$ F; small enough, in fact, that the amount of energy required to charge it with a single electron $e^2/2C \gg kT$ at the working temperature (most experiments were performed at milli-Kelvin temperatures). Provided that the barriers between the dot and electrodes 1 and 2 were thin enough, they acted as tunnelling barriers which allowed electrons from the left-hand side 2DEG region to tunnel onto the

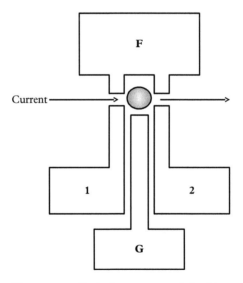

Figure 6.28 *Typical gate pattern, defined by electron-beam lithography, used to form a single quantum dot with dimension of order 0.8 μm. Electron current from left to right may flow through the tunnel barriers between gates 1 and F and between 2 and F. Gate G controls the Coulomb blockade regime in which current is limited by the charging energy required to place a single electron on the dot.*

dot and then to tunnel off again into the right-hand side. However, such processes were limited by the fact that the required charging energy could not be supplied as thermal energy—in fact it could not be supplied at all—and the charging process was effectively blocked, a situation referred to as a 'Coulomb blockade'. However, if a positive voltage is applied to the gate electrode G so as to induce a charge on the dot just equal to the electron charge e, then *one* electron could tunnel onto the dot. It was then free to tunnel to the 'drain' reservoir and then another electron could tunnel from the 'source' region and so on. But only one electron could tunnel at a time, in marked contrast to the situation in a typical FET, where literally thousands of electrons flow together from source to drain.

An interesting modification to this basic experiment was described by Kouwenhoven et al. (1991) in which they controlled the tunnelling processes by applying two RF signals (of frequency f) to the tunnel barriers, with π phase difference between them. Thus, the first phase of the cycle consisted of an enhanced probability for tunnelling through barrier 1, while tunnelling out through barrier 2 was barred—only one electron could flow in because the Coulomb blockade prevented any ensuing flow. In the second phase of the cycle, the first barrier was raised while the second one was lowered, allowing the electron to flow from the dot into the 'drain' but barring any further electron flow onto the dot. Thus, at the end of the cycle, one electron had flowed through the system and it was then primed to repeat the process at a rate determined by the RF frequency applied. The operation was likened to a *turnstile* in which one person at a time could pass through. Overall, a current I given by

$$I = ef \tag{6.26}$$

flowed from source to drain, and the source–drain current–voltage characteristic showed a sequence of steps in current, each step corresponding to this single electron per RF cycle flow.

Needless to say, the above example was merely one of many applications of this clever scientific toy and many others were to follow. We shall make mention of two of them: the work of the English Cambridge group (Tewordt et al. 1992), who studied tunnelling onto a quantum dot as a function of magnetic field, and that of the American Cambridge group (aided and abetted by the University of California; Waugh et al. 1995), who demonstrated coupling between a sequence of dots. Waugh et al. used an array of 14 gates on an AlGaAs/GaAs 2DEG to form 3 quantum dots of dimension $0.5 \times 0.8\,\mu m$ and studied the current–voltage characteristic of the series combination as the coupling was varied by application of gate voltages. Coupling between dots gave rise to splitting of the Coulomb blockade conductance pulses—into two when only two dots were measured and into three when all three dots were used. Tewordt et al. formed their quantum dots by growing an AlGaAs/GaAs RTD (resonant tunnel diode) and defining a 200 nm diameter device by plasma etching. The magnetic field was applied along the direction of current flow (i.e. normal to the growth plane) and used to study tunnelling through zero-dimensional states in the quantum dot. We make no pretence of providing details, and even less of understanding—our purpose should be clear; we simply wish to outline some of the exciting new physics which accrued from the development of very high-mobility 2DEG samples by the world's many MBE growers. From the moment when Tsui, Störmer and Gossard reported the discovery of the fractional quantum Hall effect, the world of the condensed matter physicist has blossomed in so many different directions as to overwhelm the imagination. But none of it would have been possible without the materials scientist. We can surely feel proud of our contribution.

....

REFERENCES

Andrei, E Y, Deville, G, Glattli, D C, Williams, F I B, Paris, E and Etienne, B (1988) Observation of a magnetically induced Wigner solid. Phys Rev Lett 60, 2765.
Arakawa, Y and Sakaki, H (1982) Multidimensional quantum well laser and temperature dependence of its threshold current. Appl Phys Lett 40, 939.
Asada, M, Miyamoto, Y and Suematsu, Y (1986) Gain and the threshold of three-dimensional quantum-box lasers. IEEE J Quantum Electron QE22, 1915.
Asada, M, Suzuki, S and Kishimoto, N (2008) Resonant tunneling diodes for sub-terahertz and terahertz oscillators. Jap J Appl Phys 47, 4375.
Beenakker, C W J and van Houten H (1991) 'Quantum transport in semiconductor nanostructures' in 'Solid State Physics', Vol 44 (eds H Ehrenreich and D Turnbull), Academic Press, New York, p 1.
Berger, P R, Chang, K, Bhattacharya, P, Singh, J and Bajaj, K K (1988) Role of strain and growth conditions on the growth front profile of $In_xGa_{1-x}As$ on GaAs during the pseudomorphic growth regime. Appl Phys Lett 53, 684.
Bimberg, D, Grundmann, M and Ledentsov, N N (1999) 'Quantum Dot Heterostructures', Wiley, Chichester.
Björk, M T, Ohlsson, B J, Sass, T, Persson, A I, Thelander, C, Magnusson, M H, Deppert, K, Wallenberg, L R and Samuelson, L (2002) One-dimensional heterostructures in semiconductor nanowhiskers. Appl Phys Lett 80, 1058.
Chang, K H, Bhattacharya, P K and Gibala, R (1989) Characteristics of dislocations at strained heteroepitaxial InGaAs/GaAs interfaces. J Appl Phys 66, 2993.
Chang, L L, Esaki, L, Howard, W E and Ludeke, R (1973a) The growth of a GaAs–GaAlAs superlattice. J Vac Sci Technol 10, 11.
Chang, L L, Esaki, L, Howard, W E, Ludeke, R and Schul, G (1973b) Structures grown by molecular beam epitaxy. J Vac Sci Technol 10, 655.
Chang, L L, Esaki, L and Tsu R (1973c) Resonant tunneling in semiconductor double barriers. Appl Phys Lett 24, 593.
Chen, J C H, Wang, Q, Klochan, O, Micolich, A P, Hamilton, A R, Das Gupta, K, Sfigakis, F, Richie, D A, Reuter, D, Wieck, A D and Hamilton, A R (2012) Fabrication and characterization of ambipolar devices on an undoped AlGaAs/GaAs heterostructure. Appl Phys Lett 100, 052101.
Cirlin, G E, Dubrovskii, V G, Soshnikov, I P, Sibirev, N V, Sansonenko, Y B, Bouravleuv, A D, Harmand, J C and Glas, F (2009) Critical diameters and temperature domains for MBE growth of III–V nanowires on lattice mismatched substrates. Phys Stat Sol RRL 3, 112.
Das Gupta, K, Croxall, A E, Waldie, J, Nicholl, C A, Beere, H E, Farrer, I, Ritchie, D A and Pepper, M (2011) Experimental progress towards probing the ground state of an electron-hole bilayer by low-temperature transport. Adv Cond Matter Phys 2011, ID 727958.

Davies, J H (1998) 'The Physics of Low-Dimensional Semiconductors', Cambridge University Press, Cambridge.

Dawson, P, Duggan, G, Ralph, H I, Woodbridge, K and 't Hooft, G W (1985) Positions of the sub-band minima in GaAs–(AlGa)As quantum well heterostructures. Superlattice Microst 1, 231.

Dawson, P, Moore, K J, Duggan, G, Ralph, H I and Foxon, C T B (1986b) Unambiguous observation of the 2s state of the light-and heavy-hole excitons in GaAs-(AlGa) As multiple-quantum-well structures. Phys Rev B 14, 6007.

Dawson, P, Moore, K J and Foxon, C T (1987) 'Photoluminescence Studies of Type II GaAs/AlAs Quantum Wells Grown by MBE' in 'Quantum Well and Superlattice Physics, Proc SPIE 792' (ed G H Doehler, J N Schulman), p 208.

Dawson, P, Moore, K J and Foxon, C T (1988) Effects of electronic coupling on the band alignment of thin GaAs/AlAs quantum-well structures. Phys Rev B 38, 3368.

Dawson, P, Wilson, B A, Tu, C W and Miller, R C (1986a) Staggered band alignments in AlGaAs heterojunctions and the determination of valence-band offsets. Appl Phys Lett 48, 541.

Dingle, R (1975) 'Confined Carrier Quantum States in Ultrathin Semiconductor Heterostructures' in 'Festkörperprobleme XV' (ed H J Queisser), Pergamon-Vieweg, Braunschweig, p 21.

Dingle, R, Gossard, A C and Wiegmann, W (1975) Direct observation of superlattice formation in a semiconductor heterostructure. Phys Rev Lett 34, 1327.

Dingle, R, Störmer, H L, Gossard, A C and Wiegmann, W (1978) Electron mobilities in modulation-doped semiconductor heterojunction superlattices. Appl Phys Lett 33, 665.

Dingle, R, Wiegman, W and Henry, C H (1974) Quantum states of confined carriers in very thin $Al_xGa_{1-x}As$-GaAs-$Al_xGa_{1-x}As$ heterostructures. Phys Rev Lett 33, 827.

Dodson, B W and Tsao, J Y (1987) Relaxation of strained-layer semiconductor structures via plastic flow. Appl Phys Lett 51, 1325.

Döhler, G H (1972) Electrical and optical properties of crystals with "nipi-Superstructure" Phys Stat Sol (b) 52, 79, 533.

Döhler, G H, Künzel, H, Olego, D, Ploog, K, Ruden, P, Stolz, H J and Abstreiter, G (1981) Observation of tunable band gap and two-dimensional subbands in a novel GaAs superlattice. Phys Rev Lett 47, 864.

Döhler, G H and Ploog, K (1979) Periodic doping structure in GaAs. Prog Crystal Growth Charact 2, 145.

Döhler, G H, Tsu, R and Esaki L (1975) A new mechanism for negative differential conductivity in superlattices. Sol St Commun 17, 317.

Drummond, T J, Kopp, W, Keever, M, Morkoc, H and Cho, A Y (1982) Electron mobility in single and multiple period modulation-doped (Al, Ga) As/GaAs heterostructures. J Appl Phys 53, 1023.

Drummond, T J, Morkoc, H and Cho, A Y (1981) Dependence of electron mobility on spatial separation of electrons and donors in $Al_xGa_{1-x}As$/GaAs heterostructures. J Appl Phys 52, 1380.

Dubrovskii, V G, Cirlin, G E, Soshnikov, I P, Tonkikh, A A, Sibirev, N V, Samsonenko, Yu B and Ustinov, V M (2005) Diffusion-induced growth of GaAs nanowhiskers during molecular beam epitaxy: theory and experiment. Phys Rev B 71, 205325.

Duggan, G (1985) A critical review of heterojunction band offsets. J Vac Sci Technol B 3, 1224.

Duggan G and Ralph, H I (1987) 'Exciton Binding Energy In Type-II GaAs-AlAs Quantum Well Heterostructures' in 'Quantum Well and Superlattice Physics, Proc SPIE 792' (ed G H Doehler, J N Schulman), p 147.

Eaglesham, D J and Cerullo, M (1990) Dislocation-free Stranski-Krastanow growth of Ge on Si(100). Phys Rev Lett 64, 1943.

Elman, B, Koteles, E S, Melman, P, Jagannath, C, Lee, J and Dugger, D (1989) *In situ* measurements of critical layer thickness and optical studies of InGaAs quantum wells grown on GaAs substrates. Appl Phys Lett 55, 1659.

English, J H, Gossard A C, Störmer, H L and Baldwin, K W (1987) GaAs structures with electron mobility of 5×106 cm2/V s. Appl Phys Lett 50, 1826.

Esaki, L and Chang, L L (1976) Semiconductor superfine structures by computer-controlled molecular beam epitaxy. Thin Solid Films 36, 285.

Esaki, L and Tsu, R (1970) Superlattice and negative differential conductivity in semiconductors. IBM J Res Dev 14, 61.

Etienne, B and Paris, E (1987) Two-dimensional electron gas of very high mobility in planar doped heterostructures. J Phys France 48, 2049.

Fewster, P F, Andrew, N L and Curling, C J (1991) Interface roughness and period variations in the AlGaAs system grown by molecular beam epitaxy. Semicond Sci Technol 6, 5.

Foxon, C T, Harris, J J, Hilton, D, Hewett, J and Roberts, C (1989) Optimisation of (Al,Ga)As/GaAs two-dimensional electron gas structures for low carrier densities and ultrahigh mobilities at low temperatures. Semicond Sci Technol 4, 582.

Foxon, C T, Harvey, J A and Joyce, B A (1973) The evaporation of GaAs under equilibrium and non-equilibrium conditions using a modulated beam technique. J Phys Chem Solids 34, 1693.

Franchi, S, Trevisi, G, Seravalli, L and Frigeri, P (2003) Quantum dot nanostructures and molecular beam epitaxy. Prog Cryst Growth Charact Mater 47, 166.

Gammon, D, Snow, E S, Shanabrook, B V, Katzer, D S and Park, D (1996) Fine structure splitting in the optical spectra of single GaAs quantum dots. Phys Rev Lett 76, 3005.

Goldstein, L, Glas, F, Marzin, J Y, Charasse, M N and Le Roux, G (1985) Growth by molecular beam epitaxy and characterization of InAs/GaAs strained-layer superlattices. Appl Phys Lett 47, 1099.

Goni, A R, Pfeiffer, L N, West, K W, Pinczuk, H U and Störmer, H L (1992) Observation of quantum wire formation at intersecting quantum wells Appl Phys Lett 61, 1956.

Gossard, A C (1982) 'Molecular Beam Epitaxy of Superlattices in Thin Films' in 'Treatise on Materials Science and Technology, Volume 24: Preparation and Properties of Thin Films' (ed K N Tu and R Rosenberg, Academic Press, New York, p 13.

Gossard, A C, Petroff, P M, Weigmann, W, Dingle, R and Savage, A (1976) Epitaxial structures with alternate-atomic-layer composition modulation. Appl Phys Lett 29, 323.

Greene, R L, Bajaj, K K and Phelps, D E (1984) Energy levels of Wannier excitons in GaAs-$Ga_{1-x}Al_xAs$ quantum-well structures. Phys Rev B 29, 1807.

Grundmann, M, Christen, J, Ledentsov, N N, Bohrer, J, Bimberg, D, Ruvimov, S S, Werner, P, Richter, U, Gosele, U, Heydenreich, J, Ustinov, V M, Egorov, A Y, Zhukov, A E, Kop'ev, P S and Alferov. Zh I (1995) Ultranarrow luminescence lines from single quantum dots. Phys Rev Lett 74, 4043.

Guha, S, Madhukar, A and Rajkumar, K C (1990) Onset of incoherency and defect introduction in the initial stages of molecular beam epitaxical growth of highly strained $In_xGa_{1-x}As$ on GaAs(100). Appl Phys Lett 57, 2110.

Harrell, R H, Thompson, J H, Ritchie, D A, Simmons, M Y, Jones G A C and Pepper, M (1999) Very high quality 2DEGS formed without dopant in GaAs/AlGaAs heterostructures. J Cryst Growth 201, 159.

Harris, J J, Foxon, C T, Lacklison, D E and Barnham, K W J (1986) Scattering mechanisms in (Al, Ga)As/GaAs 2DEG structures. Superlattice Microstruct 2, 563.

Hiyamizu, S, Saito, J, Nanbu, K and Ishikawa, T (1983) Improved electron mobility higher than 106 cm2/Vs in selectively doped GaAs/N-AlGaAs heterostructures grown by MBE. Jap J Appl Phys, 22, L609.

Horikoshi, Y, Kawashima, M and Yamaguchi, H (1986) Low-temperature growth of GaAs and AlAs-GaAs quantum-well layers by modified molecular beam epitaxy. Jap J Appl Phys 25, L868.

Ilegems, M (1977) Beryllium doping and diffusion in molecular-beam epitaxy of GaAs and $Al_xGa_{1-x}As$. J Appl Phys 48, 1278.

Jensen, L E, Björk, M T, Jeppesen, S, Persson, A I, Ohlsson, B J and Samuelson, L (2004) Role of surface diffusion in chemical beam epitaxy of InAs nanowires. Nano Lett 4, 1961.

Johnson, E A (2001) 'Electrons in Quantum Semiconductor Structures: An Introduction' in 'Low-Dimensional Semiconductor Structures' (ed K Barnham and D Vvedenski), Cambridge University Press, Cambridge, p 56.

Joyce, P B, Krzyzewski, T J, Bell, G R, Joyce, B A and Jones, T S (1998) Composition of InAs quantum dots on GaAs(001): direct evidence for (In,Ga)As alloying. Phys Rev B 58, R15981.

Joyce, B A, Sudijono, J L, Belk, G J, Yamaguchi, H, Zhang, X M, Dobbs, H T, Zangwill, A, Vvedensky, D D and Jones, T S (1997) A scanning tunneling microscopy-reflection high energy electron diffraction-rate equation study of the molecular beam epitaxial growth of InAs on GaAs(001),(110) and (111)A–quantum dots and two-dimensional modes.. Jap J Appl Phys 36, 4111.

Joyce, B A and Vvedensky D D (2004) Self-organized growth on GaAs surfaces. Mat Sci Eng Reports 46, 127.

Kapon, E, Hwang, D M and Bhat, R (1989) Stimulated emission in semiconductor quantum wire heterostructures. Phys Rev Lett 63, 430.

Kapon, E, Tamargo, M C and Hwang, D M (1987) Molecular beam epitaxy of GaAs/AlGaAs superlattice heterostructures on nonplanar substrates. Appl Phys Lett 50, 347.

Kawabe, M, Kondo, M, Matsuura, N and Yamamoto, K (1983) Photoluminescence of $Al_xGa_{1-x}As/Al_yGa_{1-y}As$ multiguantum wells grown by pulsed molecular beam epitaxy. Jap J Appl Phys 22, L64.

Koguchi, N and Ishige, K (1993) Growth of GaAs epitaxial microcrystals on an S-terminated GaAs substrate by successive irradiation of Ga and As molecular beams. Jap J Appl Phys 32, 2052.

Koguchi, N, Ishige, K and Takahashi, S (1993) New selective molecular-beam epitaxial growth method for direct formation of GaAs quantum dots. J Vac Sci Technol B 11, 787.

Koguchi, N, Takahashi, S and Chikyow, T (1991) New MBE growth method for InSb quantum well boxes. J Cryst Growth 111, 688.

Koshiba, S, Noge, H, Akiyama, H, Inoshita, T, Nakamura, Y, Shimitzu, A, Nagamune, Y, Tsuchiya, M, Kano, H and Sakaki, H (1994) Formation of GaAs ridge quantum wire structures by molecular beam epitaxy on patterned substrates. Appl Phys Lett 64, 363.

Kouwenhoven, L P, Johnson, A T, van der Vaart, N C, Harmans, C J P M and Foxon, C T (1991) Quantized current in a quantum-dot turnstile using oscillating tunnel barriers. Phys Rev Lett 67, 1626.

Kurtenbach, A, Eberl, K and Shitara, T (1995) Nanoscale InP islands embedded in InGaP. Appl Phys Lett 66, 361.

Lee, H G, Jeon, H C, Kang, T W and Kim, T W (2001) Gallium arsenide crystalline nanorods grown by molecular-beam epitaxy. Appl Phys Lett 78, 3319.

Leon, R, Petroff, P M, Leonard, D and Fafard, S (1995) Spatially resolved visible luminescence of self-assembled semiconductor quantum dots. Science 267, 1966.

Lewis, B F, Lee, T C, Grunthaner, F J, Madhukar, A, Fernandez, R and Maserjian, J (1984) RHEED oscillation studies of MBE growth kinetics and lattice mismatch strain-induced effects during InGaAs growth on GaAs (100). J Vac Sci Technol B 2, 419.

Liu, H C and Sollner, T C L G (1994) 'High Frequency Resonant-Tunneling Devices' in 'Semiconductors and Semimetals, Vol 41: High Speed Heterostructure Devices' (ed R Willardson, R Kiehl, T G Sollner, A Beer and E Weber), Academic Press, New York, p 359.

MacKinnon, A (2001) 'Localization and Quantum Transport' in 'Low-Dimensional Semiconductor Structures' (ed K Barnham, and D Vvedensky, Cambridge University Press, Cambridge, p 149.

Mariette, H (2005) Formation of self-assembled quantum dots induced by the Stranski–Krastanow transition: a comparison of various semiconductor systems. Comptes Rendu Physique 6, 23.

Marzin, J-Y, Gerard, J-M, Izrael, A, Barrier, D and Bastard, G (1994) Photoluminescence of single InAs quantum dots obtained by self-organized growth on GaAs. Phys Rev Lett 73, 716.

Mathews, J W and Blakeslee, A E (1974) Defects in epitaxial multilayers: I. Misfit dislocations. J Cryst Growth 27, 118.

Meynadier, M H, Delalande, C, Bastard, G, Voos, M, Alexandre, F and Lievin, J L (1985) Size quantization and band-offset determination in GaAs-GaAlAs separate confinement heterostructures. Phys Rev B 31, 5539.

Miller, R C, Gossard, A C, Kleinman, D A and Munteanu, O (1984a) Parabolic quantum wells with the GaAs-Al$_x$Ga$_{1-x}$As system. Phys Rev B 29, 3740.

Miller, R C, Kleinman, D A and Gossard, A C (1984b) Energy-gap discontinuities and effective masses for GaAs-Al$_x$Ga$_{1-x}$As quantum wells. Phys Rev B 29, 7085.

Moore, K J, Dawson, P and Foxon, C T (1986) Observation of luminescence from the 2 s heavy-hole exciton in GaAs-(AlGa) As quantum-well structures at low temperature. Phys Rev B 14, 6022.

Moore, K J, Dawson, P and Foxon, C T (1987) Optical studies of type I and type II recombination in GaAs-AlAs quantum wells. J Phys Colloques 48, C5-525.

Moore, K J, Duggan, G, Dawson, P and Foxon, C T (1988) Short-period GaAs-AlAs superlattices: optical properties and electronic structure. Phys Rev B 38, 5535.

Morkoc, H, Drummond, T J, Thorne, R E and Kopp, W (1981) Mobility enhancement in inverted Al$_x$Ga$_{1-x}$As/GaAs modulation doped structures and its dependence on donor-electron separation. Jap J Appl Phys 20, L913.

Nelson, J (2001) 'Electronic States and Optical Properties of Quantum Wells' in 'Low-Dimensional Semiconductor Structures' (ed K Barnham and D Vvedensky), Cambridge University Press, Cambridge, p 180.

Nötzel, R and Ploog K H (1992) Surface and interface ordering on non-(100)-oriented GaAs substrates. J Vac Sci Technol B 10, 2034.

Nötzel, R and Ploog, K H (1993) Direct synthesis of semiconductor quantum-wire and quantum-dot structures. Adv Mater 5, 22.

Novikov, B V, Serov, S Yu, Filosofov, N G, Shtrom, I V, Talalaev, V G, Vyvenko, O F, Ubyivovk, E V, Samsonenko, Yu B, Bouravleuv, A D, Soshnikov, I P, Sibirev, N V, Cirlin, G E and Dubrovskii, V G (2010) Photoluminescence properties of GaAs nanowire ensembles with zincblende and wurtzite crystal structure. Phys Stat Sol RRL 7, 175.

Ohlsson, B J, Björk, M T, Magnusson, M H, Deppert, K, Samuelson, L and Wallenberg, L R (2001) Size-, shape-, and position-controlled GaAs nano-whiskers. Appl Phys Lett 79, 3335.

Ohlsson, B J, Björk, M T, Persson, A I, Thelander, C, Wallenberg, L R, Magnusson, M H, Deppert, K and Samuelson, L (2002) Growth and characterization of GaAs and InAs nano-whiskers and InAs/GaAs heterostructures. Physica E 13, 1126.

Orton, J (2004) 'The Story of Semiconductors' Oxford University Press, Oxford.

Orton, J W, Fewster, P F, Gowers, J P, Dawson, P, Moore, K J, Dobson P J, Curling, C J, Foxon, C T, Woodbridge, K, Duggan G and Ralph, H I (1987) Measurement of 'material' parameters in multi-quantum-well structures. Semicond Sci Technol 2, 597.

People, R and Bean, J C (1985) Erratum: calculation of critical layer thickness versus lattice mismatch for Ge_xSi_{1-x}/Si strained-layer heterostructures [Appl. Phys. Lett. 47, 322 (1985)]. Appl Phys Lett 47, 322.

Persson, A I, Fröberg, L E, Jeppesen, S, Björk, M T, and Samuelson, L (2007) Surface diffusion effects on growth of nanowires by chemical beam epitaxy. J Appl Phys 101, 034313.

Petroff, P M and DenBaars, S P (1994) MBE and MOCVD growth and properties of self-assembling quantum dot arrays in III-V semiconductor structures. Superlattice Microst 15, 15.

Pfeiffer, L N, West, K W, Eisenstein, J P, Baldwin, K W and Gammel, P (1992) Multiquantum well structure with an average electron mobility of 4.0×10^6 cm^2/V s. Appl Phys Lett 61, 1211.

Pfeiffer, L, West, K W, Störmer, H L and Baldwin, K W (1989) Electron mobilities exceeding 10^7 cm^2/V s in modulation-doped GaAs. Appl Phys Lett 55, 1888.

Pfeiffer, L N, West, K W, Störmer, H L, Eisenstein, J P, Baldwin, K W, Gershoni, D and Spector, J (1990) Formation of a high quality two-dimensional electron gas on cleaved GaAs. Appl Phys Lett 56, 1697.

Ploog, K (1987) Delta-(°-) doping in MBE-grown GaAs: concept and device application. J Cryst Growth 81, 304.

Ploog, K, Fischer, A and Künzel, H (1979) Improved p/n junctions in Ge-doped GaAs grown by molecular beam epitaxy. Appl Phys 18, 353.

Ploog, K, Fischer, A and Künzel, H (1981a) The use of Si and Be impurities for novel periodic doping structures in GaAs grown by molecular beam epitaxy. J Electrochem Soc 128, 400.

Ploog, K, Künzel, H, Knecht, J, Fischer, A and Döhler G H (1981b) Simultaneous modulation of electron and hole conductivity in a new periodic GaAs doping multilayer structure. Appl Phys Lett 38, 870.

Rajkumar, K C, Kaviani, K, Chen, P, Madhukar, A, Rammohan, K and Rich, D H (1993) One-step in-situ quantum dots via molecular beam epitaxy. J Cryst Growth 127, 863.

Raymond, S, Fafard, S, Charbonneau, S, Leon, R, Leonard, D, Petroff, P M and Merz, J L (1995) Photocarrier recombination in $Al_yIn_{1-y}AsAl_xGa_{1-x}As$ self-assembled quantum dots. Phys Rev B 52, 17238.

Sajoto, T, Santos, M, Heremans, J J, Shayegan, M, Heiblum, M, Weckwerth, M V and Meirav, U (1989) Use of superlattices to realize inverted GaAs/AlGaAs heterojunctions with low-temperature mobility of 2×10^6 cm^2/V s. Appl Phys Lett 54, 840.

Sakaki, H (1980) Scattering suppression and high-mobility effect of size-quantized electrons in ultrafine semiconductor wire structures. Jap J Appl Phys 19, L735.

Schiff, L J (1955) 'Quantum Mechanics' (second edition), McGraw-Hill, New York.

Schlom, D G and Pfeiffer, L N (2010) Oxide electronics: upward mobility rocks! Nat Mater 9, 881.

Schubert, E F, Horikoshi, Y and Ploog, K (1985) Radiative electron-hole recombination in a new sawtooth semiconductor superlattice grown by molecular-beam epitaxy. Phys Rev B 32, 1085.

Shamirzaev, T S, Nenashev, A V, Gutakovskii, A K, Kalagin, A K, Zhuravlev, K S, Larsson, M and Holtz, P O (2008) Atomic and energy structure of InAs/AlAs quantum dots. Phys Rev B 78, 085323.

Shtrikman, H, Heiblum, M, Sco, K, Galbi, D E and Osterling, L (1988) High-mobility inverted selectively doped heterojunctions. J Vac Sci Technol B 6, 670.

Snyder, C W, Mansfield, J F and Orr, B G (1992) Kinetically controlled critical thickness for coherent islanding and thick highly strained pseudomorphic films of $In_xGa_{1-x}As$ on GaAs (100). Phys Rev B 46, 9551.

Snyder, C W, Orr, B G, Kessler, D and Sander, L M (1991) Effect of strain on surface morphology in highly strained InGaAs films. Phys Rev Lett 66, 3032.

Solner, T C L G, Goodhue, W D, Tannenwald, P E, Parker, C D and Peck, D D (1983) Resonant tunneling through quantum wells at frequencies up to 2.5 THz. Appl Phys Lett 43, 588.

Stillman, G E and Wolfe, C M (1976) Electrical characterization of epitaxial layers. Thin Solid Films 31, 69.

Störmer, H L, Dingle, R, Gossard, A C, Wiegmann, W, and Sturge, M D (1979a) Two-dimensional electron gas at differentially doped $GaAs–Al_xGa_{1-x}As$ heterojunction interface. J Vac Sci Technol 16, 1517.

Störmer, H L, Dingle, R, Gossard, A C, Wiegmann, W and Sturge, M D (1979b) Two-dimensional electron gas at a semiconductor-semiconductor interface. Sol St Commun 29, 705.

Störmer, H L and Tsang, W T (1980) Two-dimensional hole gas at a semiconductor heterojunction interface. Appl Phys Lett 36, 685.

Tanaka, M and Sakaki, H (1987) Atomistic models of interface structures of $GaAs–Al_xGa_{1-x}As$ (x = 0.2–1) quantum wells grown by interrupted and uninterrupted MBE. J Cryst Growth 81, 153.

Tanaka, M, Sakaki, H and Yoshino, J (1986) Atomic-scale structures of top and bottom heterointerfaces in $GaAs–Al_xGa_{1-x}As$ (x=0.2-1) quantum wells prepared by molecular beam epitaxy with growth interruption. Jap J Appl Phys 25, L155.

Tewordt, M, Martin-Moreno, L, Law, V J, Kelly, M J, Newbury, R, Pepper, M, Ritchie, D A, Frost, J E F and Jones A C (1992) Resonant tunneling in an $Al_xGa_{1-x}As$/GaAs quantum dot as a function of magnetic field. Phys Rev B 46, 3948.

Tonkikh, A A, Cirlin, G E, Dubrovskii, V G, Sibirev, N V, Soshnikov, I P, Samsonenko, Yu B, Polyakov, N K and Ustinov, V M (2006) Influence of MBE growth conditions on the surface morphology of Al(Ga)As nanowhiskers. Phys Stat Sol (a) 203, 1365.

Tsang, W T and Cho A Y (1977) Growth of $GaAs-Ga_{1-x}Al_xAs$ over preferentially etched channels by molecular beam epitaxy: a technique for two-dimensional thin-film definition. Appl Phys Lett 30, 293.

Tsu, R and Esaki, L (1973) Tunneling in a finite superlattice. Appl Phys Let 22, 562.

Tsui, D C and Gossard A C (1981) Resistance standard using quantization of the Hall resistance of $GaAs-Al_xGa_{1-x}As$ heterostructures. Appl Phys Let 38, 550.

Tsui, D C, Störmer, H L and Gossard, A C (1982) Two-dimensional magnetotransport in the extreme quantum limit. Phys Rev Lett 48, 1559.

Umansky, V, Heiblum, M, Levinson, Y, Smet, J, Nübler, J and Dolev, M (2009) MBE growth of ultra-low disorder 2DEG with mobility exceeding 35×10^6 cm^2/V s. J Cryst Growth 311, 1658.

Van Der Merwe, J H (1962) Crystal interfaces. Part ii. Finite overgrowths. J Appl Phys 34, 123.

van Wees, B J, Kouwenhoven, L P, van Houten, H, Beenakker, C W J, Mooij, J E, Foxon C T and Harris J J (1988) Quantized conductance of magnetoelectric subbands in ballistic point contacts. Phys Rev B 38, 3625.

Von Klitzing K, Dorda, G and Pepper, M (1980) Realization of a resistance standard based on fundamental constants. Phys Rev Lett 45, 494.

Wagner, R S and Ellis, W C (1964) Vapor-liquid-solid mechanism of single crystal growth. Appl Phys Lett 4, 89.

Wang, T and Forchel, A (1998) Growth and optical investigation of strain-induced AlGaAs/GaAs quantum dots using self-organized GaSb islands as a stressor. Appl Phys Lett 73, 1847.

Wang, W I, Mendez, E E, Iye, Y, Lee, B, Kim, M H and Stillman, G E (1986) High mobility two-dimensional hole gas in an $Al_{0.26}Ga_{0.74}As$/GaAs heterojunction. J Appl Phys 60, 1834.

Wasilewski, Z R, Dion, M M, Lockwood, D J, Poole, P Streater, R W and Springthorpe, A J (1997) Composition of AlGaAs. J Appl Phys 81, 1683.

Watanabe, K, Koguchi, N and Gotoh, Y (2000) Fabrication of GaAs quantum dots by modified droplet epitaxy. Jap J Appl Phys 39, L79.

Waugh, F R, Berry, M J, Mar, D J, Westervelt, R M, Campman, K L and Gossard, A C (1995) Single-electron charging in double and triple quantum dots with tunable coupling. Phys Rev Lett 75, 705.

Wegscheider, W, Schedelbeck, G, Abstreiter, G, Rother, M and Bichler, M (1997) Atomically precise GaAs/AlGaAs quantum dots fabricated by twofold cleaved edge overgrowth. Phys Rev Lett 79, 1917.

Wegscheider, W, Schedelbeck, G, Bichler, M and Abstreiter, G (1998) Atomically precise quantum dots fabricated by two-fold cleaved edge overgrowth: from artificial atoms to molecules. Physica E 3, 103.

Weimann, G and Schlapp, W (1985) Molecular beam epitaxial growth and transport properties of modulation-doped AlGaAs-GaAs heterostructures. Appl Phys Lett 46, 411.

Weisbuch, C and Vinter, B (1991) 'Quantum Semiconductor Structures: Fundamentals and Applications', Academic Press, San Diego.

Wharam, D A, Thornton, T J, Newbury, R, Pepper, M, Ahmed, H, Frost, J E F, Hasko, D G, Ritchie, D A and Hones, G A C (1988) One-dimensional transport and the quantisation of the ballistic resistance. J Phys C 21, L209.

Wood, C F C, Metze, G, Berry, J and Eastman, J (1980) Complex free-carrier profile synthesis by "atomic-plane" doping of MBE GaAs. J Appl Phys 51, 383.

Wu, Z H, Mei, X Y, Kim, D, Blumin, M and Ruda, H E (2002) Growth of Au-catalyzed ordered GaAs nanowire arrays by molecular-beam epitaxy. Appl Phys Lett 81, 5177.

Wu, Z H, Mei, X, Kim, D, Blumin, M, Ruda, H E, Liu, J Q and Kavanagh, K L (2003) Growth, branching, and kinking of molecular-beam epitaxial ⟨110⟩ GaAs nanowires. Appl Phys Lett 83, 3368.

Wu, Z H, Sun, M, Mei, X Y and Ruda, H E (2004) Growth and photoluminescence characteristics of AlGaAs nanowires. Appl Phys Lett 85, 657.

Xie, Q, Madhukar, A, Chen, P and Kobayashi, N P (1995) Vertically self-organized InAs quantum box islands on GaAs(100). Phys Rev Lett 75, 2542.

Yoshita, M, Okada, T, Akiyama, H, Okano, M, Ihara, T, Pfeiffer, L N and West, K W (2012) Quantitative absorption spectra of quantum wires measured by analysis of attenuated internal emissions. Appl Phys Lett 100, 112101.

Zhang, J, Dawson, P, Neave, J H, Hugill, K J, Galbraith, I, Fawcett, P N and Joyce, B A (1990) Influence of growth interruption on inverted interface quality in single AlAs-GaAs quantum wells grown by molecular beam epitaxy. J Appl Phys 68, 5595.

Zhang, Y, Guan, M, Liu, X and Zeng, Y (2011) Dependence of the electrical and optical properties on growth interruption in $AlAs$/$In_{0.53}Ga_{0.47}As$/InAs resonant tunneling diodes. Nanoscale Res Lett 6, 603.

7

Nitrides, Phosphides, Antimonides and Bismides

7.1 Introduction

The excitement associated with the development of low-dimensional structures based on the remarkably well-matched combination of AlAs and GaAs has enticed us away from our duty, as historians of MBE, to provide an account of the broad spectrum of applications of this growth technique to semiconductor materials in general. Granted that the previous chapters have certainly emphasised the AlGaAs system to the almost total exclusion of all others, we are now duty-bound to restore the balance and take up the story of MBE's no mean contribution to growing films of many alternative materials. In doing so, we shall use the earlier account of GaAs and AlAs growth as a basis, and emphasise similarities and (more importantly) differences between these archetypical materials and their counterparts from other parts of the periodic table. But, before venturing into the realms of II-VI, IV-VI compounds and the long-established world of Si and Ge epitaxy (see Chapter 8), we shall begin by examining the other family members within the III-V group of compounds, beginning at the bottom of the Group V column with nitrides and climbing, via the intermediate phosphides, to reach the heights occupied by the antimonides. We shall find much to comment on in the manner of different growth behaviour and of different material properties. However, the literature, particularly in the case of nitride materials, is enormous and we can do no more than select a few representative examples of appropriate behaviour.

7.2 Nitrides

The nitrogen atom differs significantly from the other Group V elements in regard to its small covalent radius (0.7 Å compared with 1.10 Å for P, 1.18 Å for As, and 1.36 Å for Sb) and the chemical stability of its compounds (bond energies being 2.88 eV for AlN, 2.24 eV for GaN, and 1.93 eV for InN—compared e.g. with 1.63 eV for GaAs—see Table 7.1). One consequence is that the nitrides tend to crystallise in the hexagonal

Table 7.1 *Bond energies in eV for III-V compounds (from Harrison 1980).*

	N	P	As	Sb
Al	2.88	2.13	1.89	1.74 (est.)
Ga	2.24	1.78	1.63	1.48
In	1.93	1.74	1.55	1.40

Note that some authors use a unit of 'bond energy per atom' which, for tetrahedral bonding, implies numbers four times greater than those quoted above.

wurtzite form, rather than the more familiar zinc blende structure common to other III-V compounds. They are of particular interest in so far as they offer a range of direct band gaps from 0.7 eV (1.8 μm) in the infrared, for InN, through 3.4 eV (360 nm) for GaN, and out to 6.2 eV (200 nm) in the ultraviolet for AlN and the fact that this covers the whole of the visible spectrum is clearly of considerable practical importance. The history of their evolution is an interesting one (see e.g. our earlier review, Orton and Foxon 1998). It began in the early 1970s with the HVPE growth of good-quality epitaxial films of GaN on sapphire substrates, thus illustrating one of the problems inherent in nitride epitaxy, the lack of lattice-matched substrates. (The other difficulty concerns the supply of active nitrogen to the growth surface, which we shall consider in a moment.) The first bulk crystals of GaN only became available during the 1990s and then only on a relatively small scale. The extremely high melting temperature and high nitrogen vapour pressure involved rendered conventional Bridgman or Czochralski growth impractical, and the only viable technique appeared to be solution growth from liquid gallium at temperatures of 1500 °C and nitrogen overpressures of 10 kbar. Success in this difficult enterprise was confined to the group at the High Pressure Research Centre in Warsaw led by Professor Porowski, and a limited supply of GaN substrates became available from them during the early years of the new millennium. Nevertheless, most workers were obliged to compromise with 'foreign' substrates such as sapphire, silicon carbide, silicon, gallium arsenide, gallium phosphide or zinc oxide, and this fact implied that nitride epitaxy must find ways of coping with seriously lattice-mismatched substrates. The material also suffered from what, for many years, appeared an insurmountable difficulty in regard to p-type doping. However, this was finally overcome and the 1990s witnessed dramatic success in the development of high-brightness blue and green LEDs and blue injection lasers, as reported by Japanese groups led by Isamu Akasaki at Nagoya University and by Shuji Nakamura at the Nichia Chemical Company (see e.g. Akasaki and Amano 1994; Nakamura and Fasol 1997). (The saga of Nakamura's remarkable achievement and subsequent sad departure from Nichia has been catalogued in detail in Bob Johnstone's (2007) book and the recent award of Nobel prizes to Akasaki, Amano and Nakamura will probably be familiar to most of our readers.) Nor must we overlook the important development of high-power microwave HEMTs and HBTs based on AlGaN/GaN heterostructures (McCarthy et al. 2001; Pengelly et al. 2012). Nevertheless, in writing about MBE, we should acknowledge that the majority of commercial devices based on nitrides are, today, grown by MOVPE. It behoves us, therefore, in this section to concentrate on those aspects of nitride growth where MBE has made a particular contribution, one of which is undoubtedly its contribution to the understanding of the basic physical and electronic properties of these materials, while a second concerns the growth of cubic phase material.

The MBE story also begins in Japan. As early as 1975, Yoshida et al. (1975) at the Electrotechnical Laboratory reported the use of reactive evaporation to grow AlN films on sapphire substrates for use in surface acoustic wave devices. A beam of aluminium was generated by evaporation from a beryllia crucible, while they solved the problem of providing active nitrogen by using thermally decomposed ammonia gas, which involved substrate temperatures in the range 1000 °C–1400 °C. Their vacuum system was pumped by an oil-diffusion pump to a background pressure of 10^{-7} Torr, while during growth the pressure of ammonia gas rose to about 10^{-4} Torr so one may reasonably question whether this could be referred to as MBE growth. Nevertheless, it resulted in transparent, smooth, crack-free, single crystal films with electrical resistivities greater than 10^{11} Ωm. Reflection electron diffraction showed the films to be oriented crystallographically with respect to the substrate—for (0001) surfaces the c-axis of the AlN lay perpendicular to the substrate, while on (01–12) surfaces it was oriented approximately 28° from the normal. Four years later the same group (Yoshida et al. 1979) marginally refined their technique, using electron beam evaporation of aluminium, together with similar ammonia gas flow rates and substrate temperatures, though introducing a significant modification by referring to it as 'reactive molecular beam epitaxy'. Film quality was unchanged. Probably of greater significance was the observation that growth occurred under ammonia-rich conditions, the arrival rate of NH_3 molecules at the substrate being five times that of Al. This suggested a degree of similarity to the arsenic-rich conditions by now well established for MBE growth of GaAs but whether this corresponded to *nitrogen*-rich conditions depended, of course, on the (unknown) ammonia cracking efficiency. They presented data showing that growth of stoichiometric AlN required larger NH_3 fluxes at lower substrate temperatures, implying that the efficiency increased with increasing temperature, thus explaining, perhaps, why they had chosen such a high growth temperature in the first place. After a further four years they also reported the growth of GaN films at substrate temperatures of 700 °C using very much the same apparatus (Yoshida et al. 1983). Hall measurements showed these samples to be heavily n-type $\left(n \sim 10^{25} - 10^{26} \text{ m}^{-3}\right)$ with rather poor electron mobilities of about $3 \times 10^{-3} \text{m}^2 \text{V}^{-1} \text{s}^{-1}$, while PL measurements at 77 K showed a sharp emission line at 360 nm (3.44 eV). This is some 60 meV below the band gap of GaN, consistent with the high n-type doping level. Clearly, this could not be seen as high-quality material; however, the work did establish one important fact: films grown on pre-deposited AlN layers showed much improved mobilities and very much more efficient luminescence than similar films grown directly on sapphire surfaces, a fact which turned out to be key to much future work on nitride films. The group also demonstrated the growth of films of AlGaN over the whole composition range, which showed a slight degree of departure from Vegard's Law, nor was the band gap quite a linear function of composition (Yoshida et al. 1982).

Without doubt, this approach to gas source MBE can only be described as 'primitive' but it, at least, signposted the way to more sophisticated developments. So far as we can ascertain, the first attempt to grow nitride films in a conventional UHV MBE system was undertaken at the French CNET laboratory in Paris (Alexandre et al. 1982). The motivation was very different—namely the growth of AlN on GaAs as an insulator in a metal/insulator/semiconductor (MIS) structure. The AlN films were therefore grown on top of n-type GaAs films, both films being grown at the same temperature of 550 °C, on conventionally prepared GaAs substrates. Following the lead given by the Japanese group, during AlN growth the ratio of NH_3 molecules to Al atoms arriving on the GaAs surface was approximately ten to one. The AlN films showed excellent insulating properties but the AlN/GaAs interface behaved in less than ideal fashion, showing strong Fermi-level pinning due to a trapping level 0.8 eV above the GaAs valence band, seen in several other GaAs MIS structures. No follow-up work appears to have been reported so we must conclude that this was seen as an insurmountable difficulty.

In 1991, the Yoshida group, by now using a UHV system, found that GaN films would not grow on GaAs using ammonia and introduced the notable innovation of providing nitrogen in the form of dimethylhydrazine (DMHy), which decomposes more readily (their growth temperature being 660 °C; Okamura et al. 1991). They also discovered a significant difference between growth on the (001) and (111) surfaces—the latter generating wurtzite GaN, while the former resulted in cubic (i.e. zinc blende) material. The existence of cubic GaN as a stable structure had earlier been established by Paisley et al. (1989) using plasma-assisted MBE and we shall have more to say about both aspects in a moment but first we should complete our brief survey of work with gas sources.

A major activity developed at the CNRS laboratory in Valbonne towards the end of the 1990s which began (Mesrine et al. 1998) by addressing the question of 'ammonia efficiency' in the growth of GaN epitaxial films—as they commented, it was remarkable that such a study had not been undertaken earlier. Their results are summarised in Figure 7.1, where it is clear that for substrate temperatures below 500 °C ammonia is not decomposed at all. Cracking efficiency increases rapidly up to 600 °C but then appears to level off at about 4% but our feeling is that this is an artefact of the fact that GaN starts to decompose at temperatures above 600 °C. One would surely expect cracking efficiency to be thermally activated and to continue increasing at temperature above this. In practice, growth temperatures for the nitrides reflect the compromise between the need for Group III atom surface mobility (which encourages the use of high substrate temperature) and nitrogen evaporation rate (which follows the opposite trend)—thus, InN is grown typically at 500 °C–600 °C, GaN at 700 °C–800 °C and AlN at 850 °C–900 °C, temperatures increasing according to chemical stability. It is necessary to supply large ammonia fluxes to counter this, typically 50 sccm for GaN and even higher for InN (200 sccm) and such high fluxes suggest the likelihood of intermolecular scattering, with possibility of gas phase reactions (though, to our knowledge, this has never been reported). One problem which certainly does occur concerns the large amount of ammonia trapped on liquid-nitrogen-cooled cryopanels and consequent high vapour pressures on warming the system after growth, which requires special cells for the Group III elements—hot-lipped cells for Ga and In, and cold-lipped cells for Al.

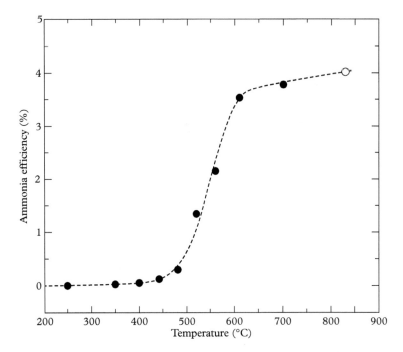

Figure 7.1 *Experimental data on the efficiency with which ammonia is cracked on a GaN surface with pre-deposited Ga. The open symbol refers to an experiment in which both NH_3 and Ga were supplied to the surface. (From Mesrine et al. 1998.) Reproduced with permission from Mesrine, M, Grandjean, N and Massies, J (1998) Appl Phys Lett 72, 350. Copyright 1998, AIP Publishing LLC.*

Even with special care being taken to warm the system slowly, high pumping capacity is an essential requirement.

Returning now to the Valbonne programme, which was concerned with luminescence behaviour of MBE-grown GaN/AlGaN QWs, GaN/AlN QDs and InGaN/GaN QWs grown on sapphire substrates, we may note that it demonstrated several highly significant features of nitride structures (Grandjean et al. 2001). The Group III atoms were supplied from solid sources, and the nitrogen by thermal cracking of ammonia. Low-temperature PL measurements were made using either a HeCd laser (3.8 eV) or a frequency-doubled argon ion laser (5.1 eV) as excitation source. Luminescence transition energies for GaN/Al$_{0.17}$Ga$_{0.83}$N QWs showed a strong variation with well width and, in particular, values significantly below that for GaN films grown on sapphire. This apparent anomaly results from the so-called quantum-confined Stark effect (QCSE) due to the build-up of a large electric field within the wells which distorts the confining potential (see Figure 7.2a). Electrons are confined towards one side of the well, holes to the other, giving rise to a downward shift in transition energy which increases with well

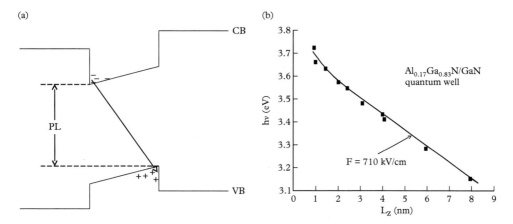

Figure 7.2 *(a) Schematic diagram to show the quantum-confined Stark effect when an electric field is applied across a quantum well. (b) Experimental data on PL energies from a GaN/Al$_{0.17}$Ga$_{0.83}$N quantum well as a function of well thickness. The line represents values calculated for the case of an internal electric field of 710 kV/cm. (Following Grandjean et al. 2001).*

width. At the same time, the overlap of electron and hole wave functions is reduced, with consequent increase in radiative lifetime. Two contributions to the electric field occur, resulting from spontaneous polarisation of the material, together with piezoelectric polarisation (due to the difference in lattice parameters), and calculation of the magnitude of the total field in a quantum well is complicated by the difficulty in knowing the distribution of strain between well and barrier. (We should note here that GaN and AlN do not follow the example of the arsenides, the in-plane lattice parameters differing by 2.5% (cf. 0.1% for the arsenides).) Figure 7.2b provides a comparison between measured and calculated PL energies assuming an electric field within the wells of 710 kV cm^{-1}. In fact, this represented a relatively small field (values as large as 10 MVcm^{-1} having been estimated theoretically (Bernardini et al. 1997)) because, in this case, most of the strain was in the barrier material, rather than in the well.

From a practical viewpoint, InGaN quantum wells were of considerably greater interest because these could be used to generate visible light, though their growth implied significant changes in growth conditions. The greater evaporation rate of In demanded a growth temperature as low as 500 °C but it was also found necessary to increase the ammonia flow rate to maintain two-dimensional growth. At rates much less than 200 sccm, a 2D–3D transition occurred after the deposition of some 4–6 monolayers, in very much the same manner as found for InAs QD growth on GaAs (see Section 6.5). Using the higher flow rate, excellent In$_{0.2}$Ga$_{0.8}$N wells were grown, showing good spatial uniformity and no evidence of the InN/GaN phase separation which had been encountered in some instances by other workers. Varying the well width from 3 nm to 5.5 nm resulted in PL emission energies covering the whole visible spectrum from blue to red and, once again, these energies implied the existence of a huge electric field (2.5 MVcm^{-1}) in the well regions. Note that the field not only shifts the photon energy but also makes a major contribution to line width by way of small random variations in well width—the observed line widths of less than 100 meV at 10 K confirmed the high degree of uniformity achieved in these structures. Nevertheless, temperature-dependent behaviour strongly suggested that the excitons involved were localised at temperatures below 250 K, possibly as a result of small variations in composition of the alloy material or, and we feel more probably, as a result of monolayer variations in well width (Phil Dawson, private communication). In the event, such localisation probably turned out to be advantageous in improving luminescence efficiency—the much reduced oscillator strength resulting from the QCSE (Figure 7.2) favours non-radiative recombination at dislocations or other defects, while localisation prevents excitons from exploring their quantum well environment to discover such recombination sites. Interestingly, these InGaN/GaN QWs could be utilised in making white-light LEDs by the simple artifice of combining different well widths in the same device.

Returning to the question of three-dimensional growth, we note that both the GaN-on-AlN and InGaN-on-GaN systems can also be used to grow quantum dots. The Valbonne group demonstrated AFM images of GaN islands and obtained room-temperature luminescence over the range 2.0 to 3.0 eV (roughly 600 nm to 400 nm). The lattice mismatch of 11% between GaN and InN is considerably larger than the 6% difference for the corresponding arsenides so the nitrides can be expected to show similar behaviour. They found that indium fractions greater than about 12% favoured the SK growth mode and that $In_{0.15}Ga_{0.85}N$ dots generated room-temperature luminescence between 2.6 and 3.1 eV, the QCSE reduction in oscillator strength militating against the achievement of longer wavelengths. It would be necessary to increase the In fraction to obtain luminescence in the green and red spectral regions.

As we have hinted, thermal cracking of ammonia does have its disadvantages as a method of supplying active nitrogen to the growth surface, and later workers have preferred the alternative of electrically induced activation of the chemically inactive N_2 molecule. Again, a Japanese group appears to have been early into the field. Sato and Sato (1989) pointed the way by introducing an RF coil into their evaporation system and activating a N_2 beam in situ with power at 13.56 MHz. Using indium evaporated from a heated crucible, they grew films of InN on GaAs substrates at a rate of 0.25 μm/h. The first true MBE film growth using plasma-activated nitrogen was reported from Bob Davis' group at Raleigh, North Carolina (Paisley et al. 1989, 1990), a technique which became known as 'plasma-assisted MBE' (PAMBE; occasionally as 'plasma-enhanced MBE'). Active nitrogen was produced by passing N_2 through a glass tube in which a microwave discharge was generated at a frequency of 2.45 GHz, a system which, in its turn, came to be known as an 'electron cyclotron resonance' (ECR) source. Using a conventional gallium effusion cell and a high-temperature boron cell, they grew films of GaN and BN on β-SiC (a cubic form), though their growth rates were very low—of the order of 100–200 Å/h. BN is of current interest because it crystallises in a layered structure which provides a reasonable match to that of graphite and could be used as a substrate for the growth of graphene but Paisley et al. were concerned to grow cubic BN as a possible substrate for diamond epitaxy. These attempts were rewarded with, at best, polycrystalline material but their GaN films turned out to be of interest in so far as they were zinc blende, rather than wurtzite in form, possibly as a consequence of the use of a cubic substrate. We shall take up the subject of cubic GaN in a moment but first we want to pursue the question of plasma sources.

The next important step was the use of a 13.56 MHz inductively coupled RF plasma source by Hoke et al. (1991) which allowed them to grow wurtzite films of InN on GaP and GaN on GaAs at growth rates of 0.25 µm/h. Interestingly, an experiment to grow GaN with simultaneous As_2 and plasma nitrogen beams resulted in the growth of pure GaN, illustrating the greater chemical reactivity of active nitrogen, compared with arsenic. Alternative nitrogen sources have included ion sources (Powell et al. 1993; Rubin et al. 1994) and supersonic jets (Sellidj et al. 1996) but the emphasis has been largely on the use of RF plasma sources, with which growth rates have increased to the satisfactory value of over 1 µm/h (Li et al. 2003). There has, however, been some controversy over the precise nature of the species which is active in nitride film growth. It is clear from studies of optical emission spectra and ion currents from different RF plasma sources (see e.g. Blant et al. 2000; Grant et al. 2007) that they contain excited molecular nitrogen, neutral atomic nitrogen and ionised species. It is obviously desirable that the energy of ions arriving at the growth surface be moderated so as to minimise damage to the film and Blant et al. showed that, for the two RF sources they investigated, conditions could be found which reduced the ion content to negligible proportions, with concomitant improvement in film quality. Li et al. were able to show that their growth rate was proportional to the atomic nitrogen concentration and they also showed that probably all the N_2 in the source was decomposed but that most of it recombined before emerging from the source. There is, however, equally convincing evidence that, in some cases, growth results from incidence of excited N_2 species (Oye et al. 2007) and likewise that ECR sources may achieve film quality every bit as good as those grown with the use of RF sources (Korakakis et al. 1998). We can do no more than reserve judgement.

One point, however, we *can* make with confidence—the growth of nitrides by PAMBE differs significantly from that of other III-V materials in so far as it should occur close to stoichiometry, rather than under Group V-rich conditions. Growth under moderately nitrogen-rich conditions results in discontinuous films with poor electron mobilities (Heying et al. 2000), while use of extremely nitrogen-rich conditions produces so-called nanocolumns (Ristic et al. 2008). Under gallium-rich conditions, when Ga droplets form, surfaces are atomically flat but electron mobilities are significantly lower than those found in samples grown close to stoichiometry. Heying et al. define a narrow 'intermediate regime' which depends on substrate temperature, as shown in Figure 7.3. A sample grown at 720 °C on a GaN substrate and represented by the open circle in the figure was recorded to have a room-temperature carrier concentration of 3×10^{22} m^{-3}, and electron mobility as high as 0.119 m^2 V^{-1} s^{-1}, the highest value reported at that time and confirming the importance of carefully optimised growth conditions. Its significance is clear from comparison with the data collected in Figure 7.4, taken from Orton and Foxon (1998).

This brings us to the interesting topic of cubic nitrides which have very largely been grown by MBE. While the nitrides certainly prefer to crystallise in the wurtzite

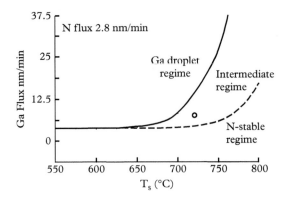

Figure 7.3 *Plasma-assisted GaN growth conditions appropriate to the three distinct regimes: 'Ga droplet', 'intermediate' and 'N-stable', as reported by Heying et al. (2000).*

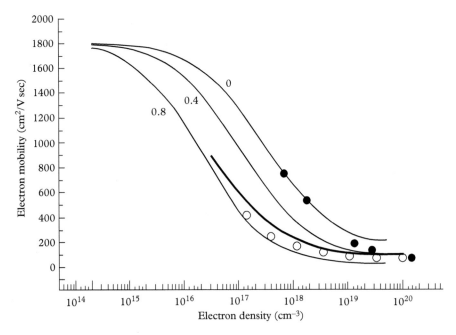

Figure 7.4 *Experimental and theoretical values of room-temperature electron mobility in GaN as a function of electron density. The heavy curve represents the best data for WZ GaN available in 1998, the open circles, the best data from 1973, and the light full curves, the calculations by Rode for three values of the compensation ratio* $C = N_A/N_D$. *The full circles represent unusually good data for ZB GaN from Kim et al. (1994). (From Orton and Foxon 1998, courtesy Institute of Physics).*

phase, this differs from the more usual zinc blende structure only in the arrangement of next-nearest neighbour atoms and we might reasonably expect not only that metastable cubic phase material could exist but that its properties might differ only slightly from those of the wurtzite phase. Indeed, such seems to be the case—all three compounds, InN, GaN and AlN, have been grown epitaxially as cubic crystals, though, for a long time, only GaN was studied in any great detail. One important point to note, however, is the fact that the high electric fields present in wurtzite heterostructures are absent in the case of cubic materials grown on (100) surfaces and this gives them a potential advantage in many practical situations. Higher electron mobilities and saturation drift velocities also offer worthwhile advantages in high-power field effect transistors.

The first report of MBE growth of cubic GaN appeared at the end of the 1980s (Paisley et al. 1989), thin films (50–170 nm) being deposited by PAMBE on (100) oriented β-SiC substrates at temperatures in the range 500 °C–700 °C. The lattice parameter was measured as $a_0 = 4.54$ Å, which is in excellent agreement with that calculated for zinc blende material from the known values of covalent radii (4.53 Å). During the first half of the 1990s a number of other reports appeared (see Orton and Foxon 1998) in which MBE was used to deposit cubic GaN films on a variety of substrates, such as (100) GaAs, MgO and β-SiC, suggesting, perhaps, that MBE had some magic ingredient which favoured cubic over hexagonal crystal growth. However, it gradually became clear that the important parameter was one of low substrate temperature—MBE just happened to employ the appropriate range of temperatures. It also appeared necessary to use substrates with cubic symmetry, even though there was little agreement in lattice parameter with the grown film; but it then transpired that cubic GaN could also be grown on (111) GaAs (which 'should' have favoured hexagonal growth). Finally, to add further confusion, it was discovered that the presence of a secondary As beam could also influence the growth mode on (100) GaAs. Several workers found it possible to obtain films containing mixed phases, XRD measurements showing wide ranging ratios of hexagonal to cubic material. Mullhauser (1998) at the Paul Drude Institute in Berlin made a careful study of the PAMBE growth conditions required to produce cubic GaN on GaAs (001) substrates and showed the importance of the initial nitridation stage. It was important that this occurred slowly and he found it necessary to use an As flux to prevent premature nitridation. A low growth rate (~0.05 ML/s), high V:III ratio (~4) and a low growth temperature (590 °C) were essential for effective nitridation. At this stage RHEED patterns consisted of spots appropriate to three-dimensional GaN growth, together with GaAs spots, the latter gradually disappearing when 1.4 ML of GaN had been deposited. This nucleation stage was terminated after 5–7 ML, the substrate temperature being raised to 640 °C–650 °C and the V/III ratio reduced to unity for continuing stable growth at about 130 nm/h, while GaN RHEED streaks replaced the spotty patterns seen during the nitridation stage. Such was the care needed to obtain phase-pure zinc blende epitaxial films, films which maintain a crystallographic relationship with the GaAs substrate, even though there exists a 25% difference in lattice parameter. However, even this exemplary study may not represent the last word—Kim et al. (2003), for example, using PAMBE to grow GaN on GaP (100), claimed that, in their case, it was necessary to use Ga-rich conditions during the low-temperature nucleation stage if phase-pure cubic films were to materialise. A particularly interesting comment comes from a Hong Kong group (Shi et al. 2006) who actually grew cubic GaN on hexagonal GaN. They, too, are adamant that the growth of the cubic phase requires not only a low temperature (420 °C) but also Ga-rich conditions. In our own work we have relied on the use of an additional As beam to encourage the nucleation of the cubic phase. Perhaps we might simply (though unsatisfactorily) conclude that there is overwhelming evidence to the effect that pure cubic material can indeed be grown but that the precise conditions for its growth are a matter for careful exploration in the particular MBE machine involved and depend, of course, on the chosen substrate.

Though, initially, cubic GaN dominated the experimental scene, it is now clear that both InN and AlN can also be grown in zinc blende form by MBE, opening the possibility (and, indeed, the practice) of quantum well and quantum dot studies based on cubic materials. Again, a wide range of substrates has been employed. For example, InN has been grown on GaAs (Tabata et al. 1999), r-plane sapphire (Hsiao et al. 2008), MgO (Inoue et al. 2008), zirconia (Nakamura et al. 2007) and β-SiC (Schörmann et al. 2006), AlN on MgO (Kakuda et al. 2012), β-SiC (Schupp et al. 2010) and Si (Thompson et al. 2001). More recently, zinc blende AlN/GaN superlattices (Mietze et al. 2011) and quantum dots (Bürger et al. 2012) have also been grown by PAMBE—the degree of sophistication is clearly increasing purposefully and we can look forward to serious applications in the not-too-distant future.

Having looked briefly at the growth of these zinc blende materials, what can be said of their properties? Perhaps, the first generalisation one might essay is that, with the important exception of polarisation fields, they behave in rather similar fashion to their hexagonal counterparts. However, there are differences. Their band gaps are, in all cases slightly smaller and β-AlN, (like its cohorts AlAs and AlP) is probably indirect. For convenience, we have collected data on Group III nitride band gaps in Table 7.2 (Casallas-Moreno et al. 2013; Mourad 2013). Perhaps the most interesting aspect of the zinc blende nitrides lies in their electrical properties. Two aspects stand out: those of electron mobility and of p-type doping. The electron mobility of ZB GaN has been shown to be much superior to that of the corresponding WZ phase, as can be seen from Figure 7.4, samples grown by Kim et al. (1994) probably representing the best quality material available today. Another interesting development was the p-type co-doping of ZB GaN with Be and O by Brandt et al. (1996) which resulted in much improved p-type conductivity—a matter of some practical importance for the performance of

Table 7.2 *Band gaps (in eV) of Group III nitrides (300 K).*

	Hexagonal	Cubic
InN	~0.6	~0.55
GaN	3.44	3.23
AlN	6.20	5.50 (indirect)

high-brightness LEDs and laser diodes. Because of the large difference in band gaps between the two constituents, AlN/GaN quantum wells offer the possibility for inter-subband transitions in the near-infrared, as demonstrated by Mietze et al. (2011), while AlN/GaN quantum dots show strong cathodoluminescence emission at photon energies in the range 3.6–3.7 eV (Bürger et al. 2012).

Finally, we should consider another interesting feature of nitride growth, that of nanocolumns (otherwise referred to as nanowhiskers or nanowires). We have already discussed these in some detail in connection with the growth of GaAs and InAs in Chapter 6 (Section 6.5). As noted earlier (Ristic et al. 2008), growth under highly nitrogen-rich conditions produces not continuous films but a dense accumulation of tiny columns with diameters measured in nanometres, an example from our own work (Foxon et al. 2009) being shown in Figure 7.5. As can be seen, there is a fairly wide distribution of sizes, 'diameters' varying from about 20 nm to 120 nm—though the clear evidence for hexagonal cross-sections renders the word 'diameter' merely impressionistic. The first reference to such columnar growth in GaN films appeared in Japan towards the end of the 1990s (Yoshizawa et al. 1997) and interest has burgeoned with the realisation that such columns are free of defects and may be used in numerous ways to generate a range of device functions. A glance at the earlier literature also reveals that, as we saw earlier, such structures were by no means new—similar columns had been seen on silicon in the 1960s and a model to explain their growth formulated by Wagner and Ellis (1964). This they christened as the 'vapour–liquid–solid (VLS)' mechanism which depends on the presence on the surface of a liquid drop of metal (such as gold, platinum or nickel), the size of the column being determined by the size of the drop. Si atoms arriving at the surface were preferentially incorporated into the drop and a silicon column would grow with a metal drop atop. During the 1990s a number of multiwire structures were grown using III-V materials (see references in Dubrovskii et al. 2005), and the concept of self-organised wires came into being (i.e. there was no requirement for a pre-deposited metal drop). Such were the GaN nanocolumns reported by Yoshizawa et al., grown on (0001) sapphire substrates by RF-PAMBE, using a single crystal AlN pre-layer, grown at 500 °C. GaN was then deposited at 800 °C with various N:Ga flow ratios (at constant Ga flux), the c-axis of the columns lying consistently normal to the plane of the substrate. In all cases the growth rates were determined by the nitrogen supply—in other words, growth occurred under Ga-rich conditions. Column diameters lay in the range 50 nm to 150 nm depending on growth conditions, as did the areal density. The fact that the column diameter reduced as the RF power was reduced suggested that the kinetic energy of the active nitrogen species played a role in determining its migration length on the growth surface.

It was obviously very tempting to suppose that GaN nanocolumns grew in the VLS mode, where the metal droplets were simply gallium (rather than Au, Ni, etc.) but this, like so many temptations, proved illusory. Ristic et al. (2008) reported detailed experiments which refuted such a hypothesis—they showed that Ga droplets played no part whatsoever in column growth and proposed a completely different model based on Ga diffusion. According to this, nucleation occurs as a direct result of lattice mismatch, via a Volmer–Weber growth mechanism, the excess nitrogen preventing coalescence of the

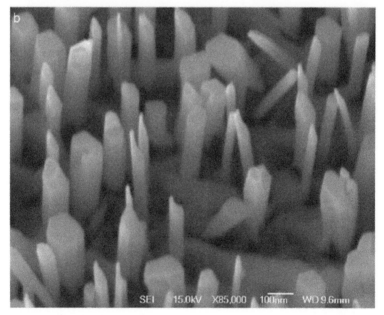

Figure 7.5 *Two examples of GaN nanocolumns grown by RF plasma-assisted MBE. In (a) the substrate was rotated at 20 rpm, while in (b) it was stationary. Samples were grown on sapphire substrates with AlN buffer layers. The images were produced by SEM with a magnification of approximately 80 000. (From Foxon et al. 2009, courtesy Elsevier.)*

nascent nanocolumns. Growth continues by way of direct Ga incorporation on top of the columns and by diffusion of Ga atoms from the base of the columns to the growing surfaces. At the substrate temperature of 700 °C used in their experiments, Ga surface diffusion may be adequate for such a mechanism, even though growth takes place under nitrogen-rich conditions. However, as pointed out by Foxon et al. (2009), yet another feature of MBE growth may well play an even more important role. In many MBE chambers Ga and N fluxes impinge on the growth surface at complementary angles, as suggested by Figure 7.6, implying that, while Ga and N atoms arrive simultaneously on the top surface of the column, such is not the case for the sides. (Interestingly, while vertical growth occurs by conventional MBE, sideways growth involves MEE.) This alone suggests a strong tendency for columns to grow in height, while maintaining an approximately constant diameter. What is more, there is also a simple geometric factor which results in significantly greater flux densities arriving on the top surface than on the sides. In the case discussed by Foxon et al. this ratio was estimated to be about 5 or 6. An interesting feature of this geometrical model concerns the effect of substrate

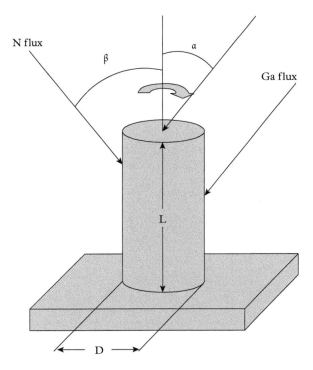

Figure 7.6 *Schematic diagram to illustrate the geometry of nanocolumn growth in a typical MBE machine. The Ga and N beams are incident at angles of α and β to the normal, respectively; thus, growth on the side walls is relatively low compared with that on top of the column. (From Foxon et al. 2009, courtesy Elsevier.)*

rotation, there being a significant difference in shape between columns grown with and without rotation (see Figure 7.5(b)). Such considerations do not, of course, deny the possibility of Ga diffusion to the top surfaces and we might add that the relatively small N flux reaching the sides of the columns would serve to encourage such diffusion, so a complete growth model should probably include both aspects. What is undoubtedly clear is that interesting possibilities for device applications are offered by this unusual phenomenon of nanocolumnar growth. Enough about nitrides—let us now look at the corresponding phosphides.

7.3   Phosphides

Both GaP and GaAsP played a major role in the early years of visible LED development during the 1960s. $In_{0.48}Ga_{0.52}P$, lattice-matched to GaAs, showed early promise as a red emitter and, with the addition of aluminium, allowed both orange and yellow light to be generated with modest efficiency. However, growth of all these materials relied on VPE and LPE—it was not until 1970 that MBE came to be seriously considered. Arthur and LePore (1969) reported MBE growth of GaAsP and GaP, Cho (1970) grew GaP on CaF_2, and Cho et al. (1970) grew GaP on GaP, while Gonda and Matsushima (1976) studied the composition of GaAsP as functions of substrate temperature and flux ratios, all of which demonstrated MBE's capability to make a serious contribution. Similarly, increasing interest in InP as a microwave device material during the 1970s led to the early study of its MBE growth by Farrow (1974). It was clear that MBE could produce material of reasonable quality but in no case was there any serious follow-up—VPE and LPE offered success with a degree of simplicity and financial circumspection which apparently outweighed the sophistication inherent in the MBE method. There were also genuine disadvantages to the use of MBE which sprang from difficulties in providing a beam of pure P_2.

   Much of the early work on phosphide growth made use of a P_4 beam from a conventional cell containing red phosphorus but, as with arsenide growth based on As_2, it was felt preferable to use P_2. Three alternative methods of generating the P_2 beam are available: use of a solid source such as polycrystalline GaP (though some, e.g. Yamane et al. (2009), preferred InP), use of red phosphorus in a valved cracker cell (Wicks et al. 1991) and use of phosphine gas, also via a cracker cell. As is well known, PH_3 is an extremely toxic gas and many growers decline to run the risk inherent in its use, though this is not the only danger to life and limb associated with phosphide growth. Fire due to spontaneous combustion of white phosphorus deposited within the MBE chamber is an ever-present concern. It is advisable to minimise the level of such deposition by removing it at weekly intervals, carefully warming the system cold traps to transfer the phosphorus into the pumping system and, if there is any need to open the chamber to atmosphere, by overnight baking to ensure no condensed phosphorus remains when air enters. Such precautions apply, of course, no matter what P_2 source is employed. While the GaP source is certainly the safest and most convenient to use, there will always be issues of purity depending on the supply of phosphine, red phosphorus or polycrystalline GaP and these may change with time as manufacturers improve their procedures. It should also be borne in mind that typical crackers are less than 100% efficient in generating P_2—a few percent of P_4 molecules will usually be present in the beam.

Having said this, we should note that not all MBE growers persuaded themselves to eschew phosphide growth. In fact, MBE played an important role, on the one hand, in clarifying the mysterious behaviour of the AlGaInP quaternary system during the 1980s and 1990s, while on the other, still being exploited in the demanding task of growing device-quality GaP films on silicon substrates. The former was of critical importance in the development of visible LEDs and lasers, the latter may well be crucial to the integration of silicon electronic with GaP optoelectronic functions. We should examine both aspects. (Figure 7.7 provides data concerning the relevant band gaps and lattice constants.)

The story of AlGaInP is a fascinating one—it went something like this. The first red LEDs (and, indeed, the first red laser) were made from $GaAs_{1-x}P_x$, an alloy characterised by a crossover from direct to indirect gap at x = 0.45 corresponding to a band gap of 1.95 eV. Thus the shortest wavelength emission with even modest efficiency lay at about 650 nm. It was comfortably into the red region of the visible spectrum but that was all—there was no hope of generating orange, yellow or green light. There were also efficiency problems which arose from the lattice mismatch between the alloy and its GaAs substrate. Then along came $Al_xGa_{1-x}As$, closely lattice-matched (and therefore considerably more efficient) but also exhibiting a crossover which limited useful emission to the red part of the spectrum. Into this frustrating maelstrom (*c.*1970) came GaInP, yet another alloy showing a band-gap crossover but this time at a photon energy of

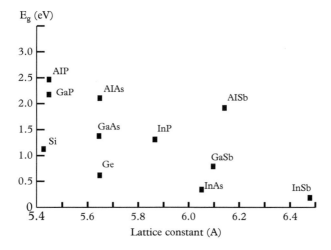

Figure 7.7 *Band gaps and lattice constants for a range of III-V compounds, together with Si and Ge for comparison. The antimonides, with InAs, form a group of materials with high lattice constants.*

2.25 eV, corresponding to a wavelength of 550 nm, in the green, and close to the point of maximum eye sensitivity. Here was surely hope for better things—though it had to be recognised that the alloy composition which was lattice-matched to GaAs, that is, $Ga_{0.52}In_{0.48}P$, had a somewhat smaller band gap, close to 2.0 eV, thus limiting *efficient* emission to about 620 nm, still well short of (or, more correctly, 'long of') the highly desirable green region. Could it be that the gods of enlightenment were determined to limit mankind's visual aspirations strictly to the red?—well, no, there was to be yet another throw of the dice. By replacing some of the gallium atoms with aluminium, it was possible to increase the direct band gap, while retaining lattice match. However, it was known that $Al_{0.52}In_{0.48}P$ was indirect, raising the question as to just how much Al could be introduced before reaching yet another annoying crossover? The answer was bedevilled by serious difficulties in growing adequate quality films and it was here that MBE was to make its most valuable contribution. Let us examine these crystal growth problems in a little more detail.

MBE was first applied to the growth of GaInP, lattice-matched to GaAs, in 1979 at the Philips Lab in Redhill (Scott and Roberts 1979; Roberts et al. 1981), followed shortly afterwards by the group at NTT, Tokyo (Kawamura et al. 1981). In both cases the phosphorus was supplied as P_4 from a red phosphorus cell, substrate temperatures being 400 °C–520 °C in the former case, and 500 °C–580 °C in the latter. Substrates were (100) GaAs. Considerable care was required to obtain lattice-matched material (x = 0.52), particularly at the higher substrate temperatures, where the more volatile Group III species (In) tended to desorb preferentially. According to Kawamura et al., mirror smooth surfaces only resulted provided the normal-to-the-plane lattice constants matched within a few parts in a thousand. They used a pair of Group III cells and a premixed Ga/In cell together with a separate In cell to compensate for the gradually decreasing In beam intensity. Roberts et al. used separate Ga and In cells, which gave rise to a composition gradient across the substrate surface, as revealed by a corresponding variation in PL energy of some tens of milli-electron volts. The observed line widths were much greater than anticipated for purely thermal broadening, possibly an indication of In/Ga segregation, which had been observed earlier in samples grown by VPE and LPE. One positive observation was the lack of misfit dislocations in 1 μm thick films, thought to be a consequence of the low growth temperatures used in MBE.

Just one year later, the NTT group (Asahi et al. 1982) reported MBE growth of the lattice-matched quaternary, AlGaInP which, as already noted, offered the possibility of wider direct gap material than GaInP, highly desirable for both visible LED and laser development. This was an important step in so far as, apparently, it could not be grown by either LPE or VPE and had only been synthesised somewhat unreliably by a modified Bridgman method. Optical absorption measurements showed a linear variation of the Γ gap with Al fraction and a Γ–X crossover at an Al fraction of 0.35 (band gap 2.3 eV). An interesting observation was the fact that, even at Al fractions of 50%, the films were stable in water (in marked contrast to the behaviour of AlP), a feature favourable to the technology of DH laser making.

Later work at the Hewlett-Packard labs in which lattice-matched AlGaInP was grown by MOVPE suggested that the Γ–X crossover occurred at an Al fraction close to 0.7, roughly double that found by Kawamura and Nagai(!), thereby emphasising the need for extreme care in assessing this complex material. A definitive study was undertaken at the University of Sheffield in 1994 (Mowbray et al. 1994) including a complete range of samples from $Ga_{0.52}In_{0.48}P$ to $Al_{0.52}In_{0.48}P$, grown by solid source MBE. Measurements of PL and PLE showed a linear variation of the direct gap (at T = 4 K) given by

$$E_g(x) = 1.979 + 0.704x \, (eV) \tag{7.1}$$

(where x is the fraction of Al substituted for Ga), together with a Γ–X crossover at x = 0.50 ± 0.02 (roughly midway between the earlier estimates!). Coincidentally, a similar study was under way at the Sharp Laboratories of Europe (Najda et al. 1995) based on samples grown by gas source MBE. Their results agreed fairly closely but showed a small departure from linearity, the direct band gap being given by

$$E_g(x) = 2.014 + 0.499x + 0.16x^2 \, (eV) \tag{7.2}$$

while the direct–indirect crossover occurred 'near x ~ 0.55'. The Sharp researchers referred to earlier work which showed that the optical properties of GaInP could be significantly affected by segregation of the Group III atoms—the resulting 'ordered' structure having a slightly smaller band gap than the random alloy. Following this earlier data, they therefore employed GaAs substrates deliberately off-cut by 15° towards the nearest (111)A plane, though they also found that such effects were essentially confined to GaInP (note the smaller band gap measured by the Sheffield group, who used normally aligned substrates). Perhaps remarkably, the presence of Al appeared to inhibit ordering. These conclusions, concerning direct and indirect band gaps, were later confirmed by a detailed study of MBE-grown samples with x close to the crossover value (Dörr et al. 1998). Thus, the maximum photon energy for efficient luminescence (at room temperature) could be found at about 2.2 eV (λ = 560 nm) in the green spectral region, permitting a range of red, orange, yellow and green LEDs to be developed. And, as we shall see in Chapter 10, this material system was to prove highly appropriate for the development of red DH lasers. MBE growers could therefore feel a degree of satisfaction in helping to resolve the uncertainties associated with these 'difficult' materials.

We now turn attention to the continuing efforts to grow high-quality films of GaP on silicon substrates, with a view to combining electronic and optical functions on a single chip. GaP was, of course, one of the first light-emitting semiconductors to be developed, flourishing during the 1970s as a result of two important developments. First, Czochralski-grown bulk crystals became available to serve as substrates for, second, LPE film growth. The result was the development of material of such excellent quality that efficient red and green electroluminescence was obtained in spite of GaP having an indirect gap. MBE played no part in all this but came seriously into the picture when it became clear that there might be considerable commercial interest in growing 'optical' films on silicon substrates. An early proposal, for example, was concerned with the idea of optical coupling between high-speed silicon circuits. Attempts were therefore made to grow GaAs on Si, with a view to making infrared emitters but this was a task beset by problems of gross lattice mismatch and polar incompatibility—silicon being monatomic and GaAs diatomic meant there was a mismatch in step heights, resulting in the formation of antiphase domains. GaP, on the other hand, is favoured by a lattice constant differing from that of silicon by only 0.37% (compare, for example, the 'well-matched' AlAs/GaAs discrepancy of 0.16% and the 'seriously mismatched' GaAs/GaP difference of nearly 4%). Needless to say, the antiphase domain problem still exists but that can be minimised by growing on off-axis substrates—usually 4–6° off the (100) plane towards the (011) plane—the idea being to persuade the silicon surface to adopt double rather than single atomic steps.

In passing, it is of interest to note the wide range of motives for growing GaP on silicon. In addition to the optoelectronic functions already mentioned, these range from high-temperature electronics (GaP's large band gap being critical to this possibility), the use of much larger and cheaper substrates for conventional LEDs (though the strongly absorbing nature of silicon in the visible spectral region is an obvious *disadvantage*), the use of GaP as a birefringent material for non-linear optics (its high thermal conductivity is seen as vital, where the need for high power levels is paramount) and the use of a GaP emitter in a GaP–Si heterojunction bipolar transistor and as a 'first stage' in the growth of GaAs on Si for the development of AlGaAs/GaAs DH lasers. Much work has therefore been undertaken to improve the crystal perfection of the resulting GaP films and much of this has involved MBE growth, on account of its well-established capabilities for monitoring the structure of the first few monolayers, in particular the use of RHEED to distinguish 2D from 3D growth mechanisms. The use of sophisticated deposition methods, such as atomic layer epitaxy (ALE) and migration enhanced epitaxy (MEE) is also of value in controlling the properties of the nucleation layers. Unsurprisingly, the quality of subsequent films is frequently found to depend strongly on the nature of the early growth.

Probably the first attempt to grow GaP on Si by solid source MBE was reported in the early days of MBE by Gonda et al. (1978). They grew on substrates misaligned by 5° from the (111) plane in the direction of the [100] axis (cleaned by ion sputtering followed by thermal annealing at 800 °C), and obtained streaky RHEED patterns, showing good-quality GaP films. They also noted only a small auto-doping of the films by Si from the substrate—much lower than films grown by MOVPE—on account of the lower substrate temperatures used (580 °C, compared with 1050 °C). Interesting though this was, realistic device applications dictate the use of (100) substrates in line with standard silicon technology. Rather few further papers appeared during the 1980s and 1990s, though we should certainly refer to the work of Wright et al. (1984) who were interested in developing HBTs. Because of the critical nature of the GaP–Si interface in this application, they gave considerable attention to the preparation of the substrate, passivating the surface with a thin layer of oxide which was then removed by exposure to a Ga beam at a temperature of 800 °C—though care was necessary to avoid Ga droplet formation which led to pits in the silicon surface. They studied growth on several different surfaces, concluding that (211)-oriented Si gave films free from antiphase domains and of much higher quality than obtainable on other orientations. A nucleation layer of GaP was obtained by first depositing a monolayer of Ga, then supplying P_2 (from a GaP cell) while monitoring the film with RHEED. Subsequent GaP quality was found to be optimised by growing at temperatures near 580 °C with low P_2:Ga ratios, though not so low as to produce Ga droplets.

The subject seemed to take on an enhanced urgency following the millennium. Sadeghi and Wang (2001) grew on tilted (100) substrates by solid source MBE, using red phosphorus to provide a P_2 beam by way of a valved cracker. They employed ALE to deposit a nucleation layer at an optimum temperature of 500 °C, followed by MBE growth of 1 μm-thick films which showed minimum surface roughness when grown also at 500 °C. However, overall crystal quality, as revealed by high resolution XRD, suggested an optimum temperature of 600 °C. Yu et al. (2004), using a GaP cell to provide P_2, grew on 4° misaligned (100) substrates to form single-phase layers. In contrast to the work of Wright et al., they formed a nucleation layer by first depositing a monolayer of phosphorus, before turning on the Ga beam, though they agreed on the merit of a low P_2:Ga ratio—values between 2 and 3 being optimum. Their interest was in using GaP as a non-linear optic material, for which they required an orientation-patterned structure, vertical domains differing in crystal orientation. It was therefore important to study the phase of the grown films as a function of growth parameters and they concluded that substrate temperature was of greatest significance. Growing at 500 °C resulted in high-quality, single-phase material, while at low temperatures a different phase, rotated by 90° about the (001) axis predominated. Grassman et al. (2009) concentrated on the elimination of structural defects from their GaP films. To ensure that their GaP was grown on a clean surface, they first grew a thin layer of silicon, followed by an anneal at 800 °C to produce the required double-height steps, then used MEE to grow a nucleation layer, initiated with Ga, to avoid the surface roughening associated with the interaction of phosphorus and silicon. This latter step was performed at 350 °C and then the temperature was raised to 545 °C for film growth, first by MEE and then by conventional MBE. Resulting films were shown to be free of antiphase domains, stacking faults and microtwins—the only remaining defects were the misfit dislocations expected from the small discrepancy in lattice constant. Putuato et al. (2009) introduced a source of atomic hydrogen as an aid to thorough cleaning of their substrates and employed a version of MEE in which the Ga was supplied in pulses, together with a continuous flow of P_2 (from an InP source). Evidence from XRD suggested that the first 0.1 μm of GaP, when grown with atomic hydrogen, was nearly pseudomorphic. It seems that the initial stages of layer growth are critical to success and Yamane et al. (2009) further illustrated this thesis in describing their own attempts to achieve smooth, defect-free epitaxy. They also grew GaP by solid source MBE, using a polycrystalline InP cell to provide P_2. They were ultimately concerned to grow the lattice-matched ternary GaPN (to which we shall make reference later) and regarded GaP as an intermediary between Si and GaPN, so wished to grow GaP layers less than the critical thickness (50 nm) with smooth, defect-free surfaces. MEE growth with an initial monolayer of phosphorus resulted in pits in both the GaP and the Si substrate which resulted from melt-back etching by Ga droplets. The mechanism is intriguing—the initial monolayer of phosphorus atoms bonds to the substrate in such a way as to form a chemically inert surface so that, when Ga atoms are supplied, they simply coalesce to form droplets which then dissolve not only the phosphorus but also the silicon surface atoms. Increasing the substrate temperature from 440 °C to 540 °C increased the size of the pits on account of the increased surface diffusion of Ga atoms giving rise to larger droplets. The solution was found to lie in replacing the initial MEE growth with two monolayers of GaP grown by conventional MBE. These layers grew three-dimensionally (as witnessed by RHEED spots) but subsequent MEE growth became two-dimensional (RHEED streaks) and resulted in smooth GaP surfaces (as shown by AFM). An example of the use of gas source MBE for growing GaP on Si is provided by a group from the Chinese Academy of Sciences in Beijing (Yu et al. 2010), while the study of antiphase domains was still seen as of importance by a widespread collaborative French group (Letoublon et al. 2011).

One might, perhaps, summarise the present situation by saying that the level of desire to obtain device quality layers of GaP on silicon substrates is matched only by the obvious difficulty in achieving success. Considerable progress has obviously been made but commercial exploitation seems still some way away. We can only await further developments with interest.

7.4   Antimonides

Turning, now, to antimonides, we should note that, as with the phosphides, MBE was brought to bear on their growth relatively early in its history. In 1977 Leo Esaki's IBM group reported the MBE growth of heterojunctions between $In_{1-x}Ga_xAs$ and $GaSb_{1-y}As_y$ (Sakaki et al. 1977) and a year later a group at Waseda University in Tokyo also described MBE growth of the alloy GaSbAs (Yano et al. 1978). The IBM team employed the same growth chamber they had earlier used to grow GaAs/AlAs superlattices. It simply required the incorporation of four effusion cells containing In, Ga, Sb and As. They grew on (100) GaAs, InAs and GaSb surfaces at temperatures in the range 450 °C to 600 °C. This choice of materials is interesting in so far as InAs and GaSb are nearly lattice-matched (see Figure 7.7) and, by adding a small amount of arsenic to GaSb, a perfect match can (in principle) be obtained, and this can be maintained by adding controlled amounts of gallium to the InAs, together with more arsenic to the GaSbAs. According to Sakaki et al., the necessary relationship is

$$y = 0.918x + 0.082 \tag{7.3}$$

It is of further interest for the growth of heterostructures on account of the extreme type II band alignment—the conduction band of InAs actually lying below the valence band of GaSb (see Figure 7.8). One unusual feature of this alignment was the demonstration of p-GaSbAs/n-InGaAs heterojunctions which showed *ohmic* I-V characteristics—as pointed out by Sakaki et al., it leads to carrier accumulation at the interface, rather than the more familiar depletion layers which characterise rectifying characteristics. Of even greater interest, perhaps, was the behaviour of InAs–GaSb superlattices, reported by Chang et al. (1979). These consisted of some 2 μm of material made up from various thicknesses of the two constituents, that of the InAs varying between 9 nm and

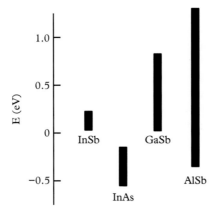

Figure 7.8 *Band alignment for InAs and selected antimonides.*

100 nm, while the GaSb thickness was comparable (though not so well controlled). Earlier calculations (Sai-Halasz et al. 1978) of the band structure predicted a change from semiconductor to semimetal behaviour as the layer thicknesses were increased, the critical thickness of InAs being roughly 10 nm, and their experimental results bore out this expectation dramatically, the measured free-carrier density increasing by a factor of about 20 as the InAs thickness increased from 9 nm to 15 nm. (Note that samples with InAs thickness somewhat less than this critical value behave as narrow gap semi-conductors and have recently been studied as possible replacements for CdHgTe IR detectors—see e.g. Kroemer 2004.) Yano et al. grew micron-thick films of GaSbAs on (100) GaAs substrates at temperatures of 500 °C–550 °C and with V:III ratios close to 3. They measured their electrical properties and, as later investigators were to con-firm, they found that undoped layers were p-type, while Te proved to be a satisfactory n-type dopant (both Si and Sn, the preferred MBE donors in AlGaAs films, behave as acceptors in GaSb and AlSb).

Perhaps a natural development from the use of GaSb, the inclusion of AlSb (with a similar lattice constant—see Figure 7.7) was proposed in an invited paper given at the MBE-CST-2 conference in Tokyo (Chang 1982). The idea put forward here concerned what was referred to as a 'polytype superlattice' which included the three materials AlSb, GaSb and InAs. However, a somewhat less exotic application concerns the use of AlSb as a barrier in AlSb–InAs quantum wells. AlSb is characterised by a much larger, though indirect, band gap (1.6 eV) and this, together with its band alignment (Figure 7.8) offers the possibility of very deep conduction-band quantum wells, when combined with thin InAs layers, an opportunity explored in a series of papers from Herb Kroemer's group at the University of California (Tuttle et al. 1989, Tuttle et al. 1990, Bolognesi et al. 1992). The principal motivation for this work lay in the high electron mobility available in InAs, which, combined with a well depth of 1.35 eV, offered exciting possibilities for high-frequency HEMTs operating at low power levels (Bennett et al. 2005).

Samples were grown by MBE, using As_4 (later As_2) and Sb_4 beams on SI GaAs substrates, considerable care being taken to mitigate the deleterious effect of the 8% lattice mismatch between GaAs and AlSb. Typically, a 1 μm thick GaSb buffer layer, followed by a 2 μm AlSb buffer (grown at 530 °C), followed, in turn, by a ten period (2.5 nm + 2.5 nm) GaSb/AlSb superlattice (grown at 500 °C) was used to obtain a smooth base for the growth of the quantum well. This latter (also grown at 500 °C) consisted of a 12 nm InAs well sandwiched between two 20 nm AlSb barriers, capped by a 5 nm GaSb layer to protect the AlSb from attack by airborne water vapour. (It is worth commenting that, in spite of these efforts, misfit dislocation densities of 10^{11} m^{-2} were indicated by TEM measurements.) Sheet electron densities of order 10^{16} m^{-2} were measured, even without deliberate doping, these being explained as due to the presence of deep donor levels within the barriers. Room-temperature mobilities of 2.5 m^2 V^{-1} s^{-1} were typical, rising to a temperature-independent value of the order of 30 m^2 V^{-1} s^{-1} at 50 K, this being interpreted as being limited by interface roughness scattering. Measurements on samples with various well widths, in the range L_z = 5 nm to 15 nm, showed mobilities varying as L_z^6, in agreement with theoretical predictions by Gold (1987) (and, incidentally, with earlier measurements on AlGaAs/GaAs QWs).

One of the more intriguing features of this AlSb/InAs QW system is the fact that there are neither common anions nor common cations at the interfaces, which makes it possible to engineer two quite different types of interface. These may be either InSb-like or AlAs-like, according to the manner in which the molecular beams are switched (see Figure 7.9). For example, in growing the lower interface, the Al beam may be shuttered off before the Sb beam, the latter being kept open for a 'soak' time of (say) 15 s before the In beam is switched on. After a further time sufficient to deposit a single monolayer of InSb, the As beam is finally turned on to grow the InAs layer. This procedure clearly produces an InSb interface, the AlAs interface being grown in complementary fashion. Tuttle et al. (1990) grew all possible combinations of these interfaces and found a very marked effect on carrier density and mobility. In particular, an AlAs bottom interface resulted in severe degradation of mobility, while an AlAs top interface also degraded mobility but to a much lesser extent. Highest values resulted from using InSb interfaces at both top and bottom of the well. However, sheet carrier densities were considerably enhanced by using AlAs bottom interfaces, an observation explained in terms of an As$_{Al}$ anti-site defect at the interface, acting as a donor. (The sequence of beam switching implied that at the bottom interface the AlSb surface was 'soaked' in As, while this was not the case at the top interface.) In fact, such a defect was deliberately engineered within the upper barrier layer, at a distance of 10 nm from the interface and resulted in modulation doping of the well in very much the same manner as done conventionally by monolayer impurity doping. It would appear that such a defect at the interface would also explain the severe degradation in mobility—the implicit charge density being a major source of electron scattering. In passing, we might note that low-temperature mobilities close to 100 m^2 V^{-1} s^{-1} have been measured by several workers (see Kroemer 2004) on carefully optimised samples, a remarkable achievement in a relatively less well developed material system.

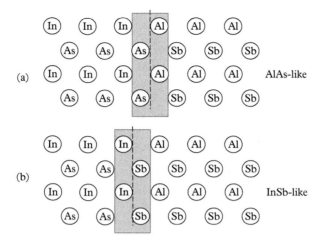

Figure 7.9 *The two possible interface structures for the AlSb/InAs heterojunction, showing in (a) an AlAs-like and in (b) an InSb-like interface (as discussed by Tuttle et al. 1990).*

Subsequent work has often concentrated on AlGaSb–InAs–AlGaSb QWs (with some 20% Al), the motivation being to reduce the degree of mismatch between well and barrier and to avoid the complication of deliquescence associated with pure AlSb. For example, Behet et al. (1998) grew such structures on GaAs substrates, using a set of pre-layers very similar to that devised by Tuttle et al. (1989) and measured 2DEGs with sheet electron densities of 8×10^{15} m^{-2} and mobilities of 2.8 m^2 V^{-1} s^{-1} at room temperature and 40 m^2 V^{-1} s^{-1} at 20 K. There was no need to use a GaSb capping layer but, as discussed by Kroemer (2004), the use of a different capping layer influences the free-carrier density in the well because GaSb cap layers exhibit high densities of surface states $\left(\gg 10^{16} \text{ m}^{-2} \right)$ from which electrons drain into the well. We shall have more to say about the antimonides when we discuss devices in our later chapters—for the moment it is probably sufficient to acknowledge that recent progress in their growth and application has been relatively rapid and there must be hope of their supplanting the better established materials in several specific cases.

7.5 Highly Mismatched Alloys

Our final topic in this brief survey of 'other III-V compounds' concerns the fascinating question of the so-called highly mismatched alloys, those alloys containing an element like N, which is much smaller than Ga, Al, As, P, etc., or an element like Bi, which is much larger. (Because the first such alloys to be investigated were those containing small amounts of N atoms, they were initially known as 'dilute nitrides' but the term 'highly mismatched alloys' is clearly of more general applicability.) In both cases the size mismatch made it difficult to introduce the 'outcast' species into the alloy while also leading to some major surprises in electronic properties. Previous experience with alloys such as GaAsP and GaAsSb led one to anticipate an approximately linear variation of band gap between the two extreme members (for example GaAs and GaP), with very little bowing and this inveigled one into anticipating similar behaviour for alloys such as GaAsN or GaPN. The former material, for example, looked promising as a potential direct gap alloy covering the whole of the visible spectrum—but it was not to be! The problem with nitrogen is not only that it is physically much smaller than the other Group V anions but that it also differs considerably in its electronegativity and this led to some dramatic and quite unexpected results. In particular, the introduction of N into GaAs produced a marked *reduction* in band gap, an observation that came as a complete surprise to semiconductor theorists in the early 1990s. At the same time, it proved extremely difficult to persuade more than a few percent of N atoms to enter the GaAs lattice and somewhat similar comments, with regard to both aspects, turned out to be applicable to GaPN. More recently, studies of InPN, GaSbN, GaBiN suggest that this extreme bowing of the band gap is a general phenomenon whenever N is combined with other Group V anions. It has proved to be both a curse and a benefaction, as we shall see in the following account but, first, we should remind ourselves of the relative simplicity of controlling alloy composition in the case of the 'straightforward' ternaries.

MBE growth of alloys like GaAsP or InGaAsP proceeds on the basis of two concepts: the relative sticking coefficients and the relative evaporation rates of the two anions. For example, if we consider GaAsP or InGaAsP at normal growth temperatures, the sticking coefficient of As is much greater than that of P so the composition can be adjusted by limiting the flow of As relative to the Group III flux while supplying excess P. At higher temperatures one must also take account of the fact that the evaporation rate of P is greater than that of As. A similar situation holds for GaAsSb, the sticking coefficient of Sb being much greater than that of As. Generally speaking, however, composition control is relatively straightforward over the whole range. The same cannot be said for the dilute nitrides.

The first MBE studies were undertaken at the University of Illinois and were concerned with incorporating modest amounts of N into GaP films grown on (100) GaP substrates (Baillargeon et al. 1992a, b). The material GaP/N had been well known since the 1960s, forming the basis of both green and yellow LEDs, and the nature of the isoelectronic N centre in GaP was well understood. It formed a highly localised level below the X-conduction-band minimum, isolated N centres producing green emission, and N–N pairs producing yellow emission with surprisingly high efficiencies (given that GaP was an indirect material), the high degree of localisation implying a wide spread in momentum, thereby giving the radiative transition an element of directness. Samples for LEDs had been grown by VPE and by LPE but the amounts of N which could be incorporated were extremely small, being less than about 2×10^{26} m^{-3} (0.8%). There was an obvious interest in the incorporation of larger amounts, with the possible formation of a genuine GaP$_{(1-x)}$N$_x$ alloy and in this they were certainly successful, obtaining samples with up to 3.5% N. Growth took place at 680 °C, and nitrogen was supplied by injecting NH$_3$ and PH$_3$ through a heated, tantalum-based, high-pressure gas-injection cell which generated the dimer PN (as well as P_2 and N_2) and it was this species which proved vital. Their photoluminescence results showed the effective band gap to take a dramatic downwards plunge as the concentration of N increased and Liu et al. (1993) confirmed this by measuring optical absorption on these same samples. They explained their results on the basis of the electronegativity theory of Van Vechten, deriving an expression for the indirect X band gap $E_g(x)$,

$$E_g(x) = 2.268 + 1.732x - 14.103x\,(1-x)\,\text{eVs} \tag{7.4}$$

which offered a modestly acceptable fit to their data. It also predicted the occurrence of negative gaps in the region of x = 0.3 to x = 0.7 (see Figure 7.10), should it prove possible to grow samples with such large amounts of nitrogen.

At much the same time, the NTT group in Tokyo, using a plasma-assisted MOVPE process (active nitrogen being provided by a microwave plasma source) reported very similar effects for $GaAs_{1-x}N_x$ alloys with 0 < x < 0.015 (Weyers et al. 1992, Weyers and Sato 1993). They estimated that (over this range of x) the alloy band gap varied as

$$E_g(x) = E_g(\text{GaAs}) - \alpha x \text{ with } \alpha = 12 \text{ eV} \tag{7.5}$$

While disappointing anyone hoping for a convenient range of visible-light-emitting materials, their observation certainly excited the interest of those looking for a GaAs-based technology for 'optic-fibre' wavelength ($\lambda \sim 1.5 \ \mu$m) lasers. On the basis of equation (7.5), it would only be necessary to incorporate approximately 5% N in order to

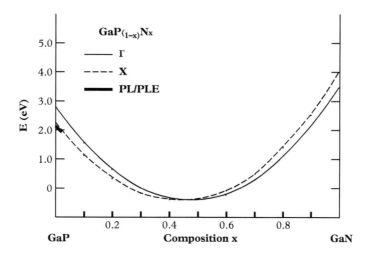

Figure 7.10 *Band gap vs composition for the alloy $GaP_{(1-x)}N_x$. The thinner lines represent theoretical results of the electronegativity theory, while the PL and PLE experimental data from Baillargeon et al. (1992a) and Liu et al. (1993) for P-rich samples is shown as a heavy line.*

reach this wavelength—though it was clearly not going to be an easy matter to persuade GaAs to accept even this modest infiltration of N impurity atoms. The key, based on Weyers and Sato (1993), appeared to be to grow layers at reduced temperature and this suggested the use of MBE.

The first reports (that we are aware of) of MBE growth came from Hitachi (Kondo et al. 1994), using an RF plasma nitrogen source in a gas source MBE machine, arsenic being supplied in the form of arsine. The Hitachi interest was in possible integration of optoelectronic devices with silicon so they were aiming to incorporate 20% N to achieve lattice-matched material. However, they chose, in the first instance, to experiment with GaAsN growth on (100) GaAs substrates as an easier option. Samples grown at a substrate temperature of 500 °C contained up to 1.5% N and showed even greater band-gap reduction than found by their NTT rivals, quoting a value of $\alpha = 18$ eV for the appropriate parameter in equation (7.5). They accounted for the difference in terms of erroneous estimation of the N content by NTT–XRD measurements must be corrected for strain if they are to be relied on for this purpose. (This alloy GaAsN is of great interest in relation to near-infrared laser development and we shall look at it again in Chapter 10.)

Perhaps the next development of interest came in 1996 when Bi and Tu (1996) reported the growth of InPN films (with up to 1% N) on (100) InP substrates by gas source MBE, supplying phosphorus by thermal cracking of PH_3 and active nitrogen from an Oxford Applied Research RF source. Again, the alloy band gap showed a sharp reduction with increasing N content, the appropriate constant ($\alpha = 21$ eV) in equation (7.5) being even larger than for either the GaPN or the GaAsN case. It certainly began to look as though this band-gap reduction by the incorporation of N was a general phenomenon. Also in 1996 came the first attempts to study the N-rich side of the GaPN alloy by Gonda's group at Osaka University (Iwata et al. 1996, 1997; Kuroiwa et al. 1997). They grew samples on sapphire substrates and on (111)A GaAs by gas source MBE, using a cyclotron resonance radical cell to supply N and included an ion removal system consisting of a pair of magnets located at the cell exit, which resulted in much improved quality GaN layers. For small P concentrations they found strong band-gap bowing but at 1.5% P there was evidence of phase separation into P-rich and N-rich regions.

The year 1999 saw an important breakthrough in theoretical understanding. As we have already hinted, the prediction of negative band gaps by the electronegativity theory had given cause for concern and there was probably something of a collective sigh of relief when Shan et al. (1999) proposed an alternative—the band anti-crossing (BAC) model, which gradually became accepted as explaining band bowing in all the so-called strongly mismatched alloys (we discuss it in detail in Box 7.1). They first applied it to GaAsN where the N atoms give rise to localised states some 0.3 eV above the bottom of the GaAs conduction band and where the bowing effect results from repulsion between these states and the delocalised GaAs CB states. The greater is the concentration of N atoms, the stronger is the repulsion, accounting for the sharp reduction in band gap. A year later they applied it further to the P-rich side of the GaPN alloy (Shan et al. 2000). As we noted above, the N atom creates a localised state just below the X CB minimum in GaP and this shifts to lower energy as the N concentration is increased.

This downward trend is, again, explained by the mutual repulsion of CB and N states, while the strongly localised nature of the N states results in a change from indirect to direct band gap. They grew a range of samples on GaP substrates with up to 3% N content by gas source MBE and measured the variation of the N-induced absorption edge as a function of x. Their measured value of the parameter α in equation (7.5) was 15 eV, very similar to the corresponding value in the GaAsN alloy.

Box 7.1 Band-gap bowing

It is frequently found that binary alloy band gaps do not vary exactly linearly with composition, a plot of $E_g(x)$ against x tending to bow slightly downwards. If we take the example of the $In_{1-x}Ga_xP$ alloy, the Γ gap varies with x as shown in Figure 7.11 and can be represented by the equation

$$E_g(x) = E_g(InP) + ax - bx(1-x)$$
$$= 1.35 + 1.45x - 0.8x(1-x)eV$$

(B7.1)

The parameter b is known as the bowing parameter, the term bx(1x) being maximum at $x = \frac{1}{2}$. In this case, the maximum bowing is 0.2 eV. Note that this expression represents the form given by the Van Vechten electronegativity theory of band bowing and we can expect

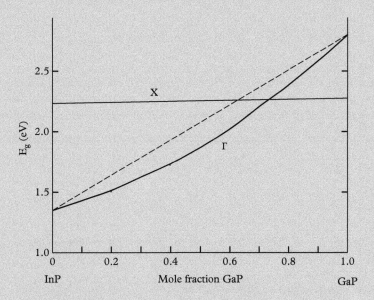

Figure 7.11 *Plot of energy gap (E_g) vs mole fraction of GaP(x) in the alloy $In_{(1-x)}Ga_xP$. The direct gap Γ shows a degree of bowing which is maximum at x = 0.5 and equal to 0.2 eV.*

continued

Box 7.1 *continued*

the bowing to be very much larger for ternary alloys containing the strongly electronegative N atom (see e.g. Baillargeon et al. 1992a).

In the case of the compounds $GaAs_{1-x}N_x$ and $GaP_{1-x}N_x$, the bowing parameter is so large as to dominate their behaviour, resulting in a *decrease* in band gap for small values of x. Note that (for GaAsN), when x is small, we can write equation (B7.1) approximately as

$$E_g(x) = E_g(GaAs) + [a - b]x \qquad (B7.2)$$

So, when b > a, $E_g(x)$ decreases with increasing x, with a slope given by (a − b). Knowing the band gaps of the end members allows us to evaluate the variable a, and a measurement of the downward slope (a − b) gives a value for the bowing parameter b. Thus, $E_g(GaAs)$ = 1.43 eV, $E_g(GaN)$ = 3.25 eV (for the zinc blende form) so a = 3.25 − 1.43 = 1.82 eV. Taking the measured slope for small values of x as −18 eV (Kondo et al. 1994), we find the bowing parameter b = 19.8 eV. The maximum bowing, at x = 1/2, is therefore about 5 eV, consistent with a negative band gap between x = 0.09 and x = 0.82.

While the electronegativity theory appears to explain the sharp decrease in band gap for small values of x, its prediction of negative gaps is not supported by experimental evidence and we must look for an alternative model to explain the behaviour of these strongly mismatched alloys containing nitrogen. The band anti-crossing (BAC) model first proposed by Shan et al. (1999) appears to be much more successful in explaining the available data and we shall attempt to provide a simple account in the rest of this box.

In the case of the alloy GaAsN, the N atoms form localised levels which lie within the GaAs conduction band roughly 0.3 eV above the bottom of the band. The theory assumes that the interaction of GaAs band states and N localised states can be treated as a perturbation V_{MN} and that its effect is to mix the two states and push them apart. In particular, this results in the conduction band being pushed downwards by the presence of the resonant N levels, thus accounting for the observed band-gap reduction. Standard perturbation theory gives the effective energies of the two resulting states in terms of the following determinantal equation:

$$\begin{vmatrix} E - E_M & V_{MN} \\ V_{MN} & E - E_N \end{vmatrix} = 0 \qquad (B7.3)$$

where E_M and E_N are the unperturbed energies of the GaAs CB edge and the N level relative to the top of the valence band. The two solutions of the problem can be written as

$$E_\pm = \left[E_N + E_M \pm \left\{ (E_N - E_M)^2 + 4V^2_{MN} \right\}^{1/2} \right] / 2 \qquad (B7.4)$$

And, knowing the values of E_N and E_M, one can immediately calculate the extent of the gap reduction in terms of the interaction potential V_{MN}. From equation (B7.4), it is easy to show that, for small values of V_{MN}, the GaAs band gap will shrink according to

$$E_g = E_M - V^2_{MN}/(E_N - E_M) \qquad (B7.5)$$

which implies, from the observed linear reduction of $E_g(x)$, that V_{MN} must vary as $x^{1/2}$. Indeed, this is precisely what Lindsay and O'Reilly (1999) predicted on the basis of a tight-binding calculation. They concluded that

$$V_{MN} = \beta x^{1/2} \text{ with } \beta = 1.50 \text{ eV} \tag{B7.6}$$

which gives the band-gap reduction as

$$E_g = E_M - \beta^2 x/(E_N - E_M) \tag{B7.7}$$

Thus, the slope of the E_g vs x line (which we call α) is $b^2/(E_N - E_M)$ and, taking the slope to be 18 eV and $(E_N - E_M) = 0.3$ eV we find $\beta = 2.3$ eV, somewhat at variance with Lindsay and O'Reilly's estimate.

Shan et al. (1999) employed a rather different approach, measuring the pressure-dependence of the $E_{\pm}$ levels for three GaInAsN samples (the small amount of In was added to obtain an improved lattice match with the GaAs substrate), demonstrating their anti-crossing behaviour and deriving values for V_{MN} from their data. They found, as expected, that V_{MN} did, indeed, increase with increasing N content and they quoted a value of $V_{MN} = 0.12$ eV for x = 0.009 (their sample with smallest value of x). This gives $\beta = 1.26$ eV, in reasonable agreement with Lindsay and O'Reilly but in rather poor agreement with the experimental data on E_g vs x!

In the case of the dilute GaPN alloy, they (Shan et al. 2000) fitted their absorption-edge data to the same theoretical model (with a 'new' CB state emerging from the interaction between the N state and the original GaP CB) and obtained a value of $\beta = 3.05$ eV, corresponding to a slope of E_g vs x as $\alpha = 15$ eV, which agrees well with other published data.

So far so good—but this model covered no more than a tiny fraction of (E_g vs x) space. The next move was concerned with the other extreme—that is the nitrogen-rich region of the GaAsN alloy. In this case, the As atom is responsible for a donor-like state approximately 0.6 eV above the GaN valence-band edge and one might therefore expect that it could be treated in very much the same manner as for phosphorus-rich GaPN. This, indeed, proved to be the case—Yu et al. (2009) showed how to treat this according to the BAC model, though, because of the greater complexity of the valence band, it involved an 8 *times* 8 determinant, rather than the simple 2×2 version appropriate to the conduction-band examples. This meant that it was no longer possible to derive a simple algebraic expression for the energy levels but diagonalisation of the 8×8 determinant revealed the generation of a new valence band just above the GaN valence-band edge, and the alloy band gap corresponded to transitions between this and the original GaN-like CB. Fitting this model calculation to their experimental absorption data gave a value for the interaction potential of $V = \beta x^{1/2} = 0.75 x^{1/2}$ eV (roughly one-quarter of the value appropriate to the earlier CB calculations).

Finally, there arose the question of how to treat the remaining 80% of the range, for which there was no experimental data! Lacking any detailed theory, they adopted a linear interpolation method, according to the following equation:

$$E_g(x) = x E_g(\text{N-rich}) + (1 - x) E_g(\text{As-rich}) \tag{B7.8}$$

where x is the fraction of N in the GaAsN alloy. It is this expression which was used to obtain the BAC curve in Figure 7.12. It is, of course, fitted to experimental data at both ends, while

continued

Box 7.1 *continued*

relying on equation (B7.8) to interpolate between. In marked contrast to the electronegativity theory, it predicts a minimum band gap of 0.7 eV at x ~ 0.2. The spectre of negative band gaps has disappeared.

A great deal of interest has been shown in attempts to develop and control both GaAsN and GaPN layers—we can do no more than look very briefly at some of the outcomes. Most of the data collected up to this point had been concerned with samples having no more than about 3% of the minority atom and there was clearly a need to extend this if such material was to have useful application. A first step was taken by the Gonda group (Asahi et al. 2000), who studied the effect of substrate temperature on the gas source MBE growth of GaN with added P. They showed that, when the temperature was reduced from 680 °C to 570 °C, the P content increased monotonically to a level of 8%. Phase separation, which, at the higher temperatures, occurred at P concentrations as low as 1.5%, was suppressed. Another important step was taken by Furakawa et al. (2002)in demonstrating the plasma-assisted MBE growth of GaPN films on (100) Si substrates. Incorporating a thin GaP buffer layer, they obtained material lattice-matched to Si within 0.05% (less than the mismatch between GaAs and AlAs), showing no evidence of misfit or threading dislocations. Several groups noted the improvement in crystalline quality, when using an RF plasma source, if N ions were removed from the beam with some form of ion collector. Doping of such films was clearly of importance for device applications and both Mg (Mitsuyoshi et al. 2010) and C (Liu et al. 2011) p-type doping was demonstrated in MBE growth, with, in both cases, hole densities greater than 10^{25} m^{-3} being achieved. A final point of interest in this somewhat random walk through what is by now a considerably dense thicket of literary growth is provided by the report of MBE growth of AlPN (with 3.3% N), lattice-matched to Si (Kawai et al. 2011). Such material appears ideal as cladding for GaPN quantum wells on silicon substrates.

The perceptive reader will no doubt be wondering whether experimental evidence ever did accrue for any of the strongly mismatched alloys, across the whole composition range. It had to wait until 2009. For a long time it seemed that it would be impossible to grow such samples due to the existence of large miscibility gaps but, all of a sudden, for the case of the GaAs–GaN system, the gap closed—MBE, being kinetically controlled, could ignore the restraints of miscibility! We have already noted the work of Asahi et al. on GaPN, where the use of reduced substrate temperature allowed up to 8% of N to be incorporated—the same approach worked even more effectively for GaAsN. By growing at temperatures as low as 100 °C, Sergei Novikov in Nottingham found it possible to grow samples over a large fraction of the composition range and these allowed band-gap measurements which could be compared with the BAC theory discussed in Box 7.1 (Yu et al. 2009). Films were grown by plasma-assisted MBE on sapphire substrates at temperatures between 100 °C and 550 °C and, starting from the GaN end of the range, the

As fraction increasing monotonically from 2% to 70% as the temperature was lowered. Samples with up to 15% As were crystalline, as were a pair of samples with about 80% As, while those in the range 20% to 70% were amorphous—nevertheless, the measured absorption edges all appeared equally steep and there was a monotonic change in band gap as the composition was varied. The results are plotted in Figure 7.12, where they are compared with the prediction of the BAC model. The model agrees remarkably well with measurements on crystalline samples, while those which are amorphous show gaps somewhat larger than predicted, though we may take solace in the well-known fact that amorphous silicon shows a gap considerably larger than that measured on crystalline material. It would nevertheless be dangerous to push this comparison too far—amorphous silicon shows an absorption edge far less steep than its crystalline counterpart, while, as we noted already, this is certainly not the case for the GaAsN samples measured here. An important comment concerning Figure 7.12 is to note that at the GaN end of the range the effective band gap is some 0.7 eV lower than the accepted gap for GaN. This is consistent with the earlier observation that As-doped GaN samples grown by MBE in Nottingham showed a strong blue emission at photon energy of 2.7 eV (Winser et al. 2000), suggesting that As impurities give rise to a level 0.7 eV above the GaN VB and it is this level which forms the effective VB in the alloy, the repulsive interaction between it and GaN VB states accounting for the severe band bowing. In this respect, it parallels the case of the 'new' CB introduced by N atoms in GaP. The band-gap data plotted for the GaAs end of the composition range in Figure 7.12 are taken from the paper by Uesugi et al. (1999) as it probably represents the best data available at the time. They grew samples on (100) GaAs substrates by MOMBE in the temperature range 520 °C –570 °C and achieved N concentrations up to 4.5%.

So much for the case of alloys containing an unusually *small* element—what of the opposite situation where the 'intruder' is much *larger* than its neighbours? This is the case represented by the alloy GaAsBi. The covalent radius of Bi is 1.46 Å, compared with those for As (1.18 Å) and P (1.10 Å). While the compound GaBi appears not to have been made, if it could be made, this would imply a lattice constant of $a_0 = 6.3$ Å, close to that of InSb at the right hand margin of Figure 7.7. In practice, only modest quantities of Bi have been incorporated in GaAs and, once again, it has a dramatic effect on the band gap, reducing it from 1.43 eV to just under 0.8 eV when 10% of Bi is introduced (Sweeney et al. 2011). This is of particular significance in so far as it represents a wavelength of 1.55 μm, corresponding to the minimum loss in the ubiquitous fibre-optic cable—it offers a tantalising vision of a GaAs-based technology for lasers and photodetectors at these communication wavelengths. What is more, the inclusion of N in the mismatched mix promises to achieve such a result while minimising the amount of strain—N and Bi both reduce the gap while they pull in opposite directions with regard to their effect on lattice parameter.

However, the incorporation of Bi in III-V compounds did not begin with GaAsBi but with the alloy InSbBi. As early as 1982 a group at Westinghouse in Pittsburgh used solid source MBE to grow films of InSb containing a few percent of Bi, their motivation being that of developing far-infrared detectors for the 8–12 μm band (Noreika et al. 1982, 1983). Bi was evaporated from a conventional effusion cell, and growth took place on

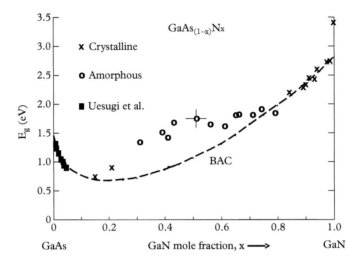

Figure 7.12 *Experimental and theoretical plots of band gap (E_g) vs GaN mole fraction (x) for MBE-grown GaAs$_{(1-x)}$N$_x$, taken from Yu et al. (2009). Crystalline samples are represented by crosses, and amorphous samples by open circles. Data at the GaAs-rich end of the scale is taken from Uesugi et al. (1999). The dashed curve represents the prediction of the band anti-crossing model (BAC) (see Wu et al. 2004).*

InSb or GaAs substrates at temperatures in the range 295 °C–380 °C. Substitutional (?) incorporation of Bi under In-rich conditions increased to a maximum of about 3% as the substrate temperature was reduced—under Sb-rich conditions only interstitial Bi could be detected. The wavelength corresponding to the band gap of the 3% alloy (0.124 eV) was close to 10 μm, thus showing promise for the development of detectors for the 8–12 μm band. Assuming a linear decrease of band gap with Bi fraction, the parameter α defined in equation (7.5) was 3.2 eV. In common with other large atoms such as Pb and Sn, Bi tended to segregate on the surface of the film but this apparently had no deleterious effect on the alloy film quality. However, He-ion channelling showed that most of the Bi in the film was present on non-substitutional sites (see discussion in Farrow 1983) which accounts for the relatively small value of α referred to above (in contrast to values of the order of 15–20 eV for alloys such as GaAsN). It probably also accounts for the somewhat diffuse nature of the absorption edge in the InSbBi alloy films.

Of greater promise was the combination of Bi with GaAs, an alloy first grown by MBE in 2003 (Francoeur et al. 2003; Tixier et al. 2003). Films of thickness 100–300 nm were grown on (100) GaAs substrates at temperature of 380 °C and with a V:III ratio close to stoichiometry—only thus was it possible to incorporate significant levels of Bi. Samples with up to 3.6% Bi showed a rapid decrease in band gap with increasing Bi content, both PL and modulated reflectance measurements yielding a value of α = 8.3 eV.

Films grew pseudomorphically to the GaAs substrates, the resulting tetragonal distortion splitting the light- and heavy-hole states by up to about 20 meV. Both XRD and PL measurements suggested that the films suffered from a degree of structural disorder, consistent with an extrapolated lattice mismatch between GaAs and GaBi of some 12%. In addition, the relatively low substrate temperature no doubt, resulted in less than ideal thermal diffusion coefficients for the constituent atomic species. The inverse dependence of Bi content on growth temperature T_S was further pursued by Bertulis et al. (2006), increasing the Bi content to 5% when $T_S = 280\,°C$, the corresponding band gap being slightly less than 0.9 eV (wavelength ~1.4 μm). The BAC theory has been applied to explain this band-gap reduction by Alberi et al. (2007).

More recently, the group under Steve Sweeney at the University of Surrey have pointed out another important property of the $GaAs_{1-x}Bi_x$ alloy (Sweeney et al. 2011). Not only does the incorporation of Bi effect a rapid decrease of band gap but it also results in a corresponding increase in the spin-orbit splitting of the valence-band states. Based on Figure 7.13, which shows experimental data on the variation of E_g and Δ_{SO} with Bi content, it is apparent that Δ_{SO} exceeds E_g for values of x greater than 9%, and this is predicted to be of major importance for the performance of 1.55 μm lasers. The temperature dependence of threshold current in InP-based devices can be attributed to the occurrence of Auger recombination processes, involving transitions between the split-off and the heavy-hole band—but such processes are not allowed when $\Delta_{SO} > E_g$. We leave further discussion to Chapter 10, where we consider optical devices.

A final brief comment is in order concerning the very highly mismatched alloy $GaN_{1-x}Bi_x$, this being the extreme example of mismatched anions. By the use of

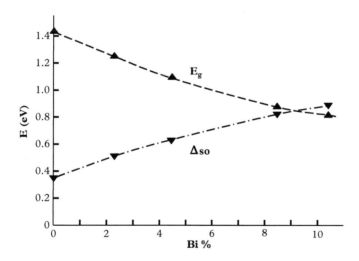

Figure 7.13 *Experimental data showing the variation of band-gap and spin-orbit splitting for the alloy GaAsBi, as a function of the percentage of Bi. When more than 9% Bi is incorporated, $\Delta_{SO} > E_g$.*

low-temperature MBE, films have been grown on sapphire substrates with values of x as large as 0.11 (Levander et al. 2010). The material turned out to be amorphous and the corresponding absorption edges were relatively diffuse but it was clear that the effective band gap decreased strongly with increasing Bi content, the value at x = 0.11 being close to 1.2 eV (an average value of $\alpha = 20$ eV, as expected, much larger than that measured on $GaAs_{1-x}Bi_x$).

Before leaving the topic, it might be appropriate here to comment on this question of low-temperature growth. In fact, there are many examples in the literature concerning the difficulty of incorporating significant amounts of physically large atoms—one thinks, for instance, of Sn and Pb as dopants in GaAs. In high-temperature growth, these atoms are mobile enough to segregate to the growth surface, where they feel most comfortable. In contrast, at low temperature, being much less mobile, they are simply buried within the growing film, which grows faster than they can diffuse. Needless to say, this is far from being in the best interest of crystal quality and the MBE expert must, of course, work hard to achieve a satisfactory compromise between adequate dopant incorporation and structural perfection. The case of the *small* atom N is somewhat subtler in so far as there is no obvious physical reason why it should be difficult to incorporate it in a significantly *larger* lattice site. However, the size mismatch does imply weaker binding, making it prone to thermal emission from the site and subsequent diffusion to the surface. The mechanism may be different but the result is the same—large amounts of N can be incorporated in GaAs (for example) only at low substrate temperatures. With that, we shall put the III-V compounds to rest—but rest assured that we shall wrest even greater delight from their performance in numerous device applications in Chapters 9 and 10—now for new territory.

REFERENCES

Akasaki, I and Amano H (1994) Widegap column-III nitride semiconductors for UV/blue light emitting devices. J Electrochem Soc 141, 2266.

Alberi, K, Dubon, O D, Walukiewicz, K M, Yu, K, Bertulis, K and Krotkus, A (2007) Valence band anticrossing in $GaBi_xAs_{1-x}$. Appl Phys Lett 91, 051909.

Alexandre, F, Masson, J M, Post, G and Scavennec, A (1982) A1N/GaAs structures grown by molecular beam epitaxy for metal/insulator/semiconductor devices. Thin Solid Films 98, 75.

Arthur, J R and LePore, J J (1969) GaAs, GaP, and $GaAs_xP_{1-x}$ epitaxial films grown by molecular beam deposition. J Vac Sci Technol 6, 545.

Asahi, H, Kawamura, Y and Nagai, H (1982) Molecular beam epitaxial growth of InGaAlP on (100) GaAs. J Appl Phys 53, 4928.

Asahi, H, Tampo, H, Hiroki, H Asami K and Gonda, S (2000) Gas source MBE growth of GaN-related novel semiconductors. Matl Sci and Eng B 75, 199.

Baillargeon, J N, Cheng, Hofler, G E, Pearah, P J and Hsieh, K C (1992a) Luminescence quenching and the formation of the $GaP_{1-x}Nx$ alloy in GaP with increasing nitrogen content. Appl Phys Lett 60, 2540.

Baillargeon, J N, Pearah, P J, Cheng, K Y, Hofler, G E and Hsieh, K C (1992b) Growth and luminescence properties of GaP:N and $GaP_{1-x}N_x$. J Vac Sci Technol B 10, 829.

Behet, M, Nemeth, S, De Boeck, J, Borghs, G, Tummler, J, Woitok, J and Guerts, J (1998) Molecular beam epitaxy and characterization of heterostructures for magnetic sensing applications. Semicond Sci Technol 13, 428.

Bennett, B R, Magno, R, Boos, J B, Kruppa, W and Ancona, M G (2005) Antimonide-based compound semiconductors for electronic devices: a review. Solid St Electron 49, 1875.

Bernardini, F, Fiorentini, V and Vanderbilt, D (1997) Phys Rev B 57, R9435.

Bertulis, K, Krotkus, A, Aleksejenko, G, Pacebutas, V, Adomavicius, R, Molis, G and Marcinkevicius, S (2006) GaBiAs: a material for optoelectronic terahertz devices. Appl Phys Lett 88, 201112.

Bi, W G and Tu, C W (1996) N incorporation in InP and band gap bowing of InN_xP_{1-x}. Appl Phys Lett 80, 1934.

Blant, A V, Hughes, O H, Cheng, T S, Novikov, S V and Foxon, C T (2000) Nitrogen species from radio frequency plasma sources used for molecular beam epitaxy growth of GaN. Plasma Sources Sci Technol 9, 12.

Bolognesi, C R, Kroemer, H and English, J H (1992) Well width dependence of electron transport in molecular-beam epitaxially grown InAs/AlSb quantum wells. J Vac Sci Technol B 10, 877.

Brandt, O, Yang, H, Kostial, H and Ploog, K H (1996) High p-type conductivity in cubic GaN/GaAs(113)A by using Be as the acceptor and O as the codopant. Appl Phys Lett 69, 2707.

Bürger, M, Schupp, T, Lischka, K and As, D J (2012) Cathodoluminescence spectroscopy of zinc-blende GaN quantum dots. Phys Stat Sol (c) 9, 1273.

Casallas-Moreno, Y L, Pérez-Caro, M, Gallardo-Hernández, S, Ramírez-López, M, Martínez-Velis, I, Escobosa-Echavarría, A and López-López, M (2013) Study of structural properties of cubic InN films on GaAs(001) substrates by molecular beam epitaxy and migration enhanced epitaxy. J Appl Phys 113, 214308.

Chang, L L (1982) 'Polytype Heterostructures' in 'Collected Papers of MBE-CST-2 Conference, Tokyo, 1982', p 57.

Chang, L L, Kawai, N, Sai-Halasz, G A, Ludeke, R and Esaki, L (1979) Observation of semiconductor-semimetal transition in InAs-GaSb superlattices. Appl Phys Lett 35, 939.

Cho, A Y (1970) Epitaxial growth of gallium phosphide on cleaved and polished (111) calcium fluoride. J Appl Phys 41, 782.

Cho, A Y, Panish, M B and Hatashi, I (1970) 'Molecular beam epitaxy of GaAs, $Al_xGa_{1-x}As$ and GaP' in 'Proceedings of the Third International Symposium on Gallium Arsenide and Related Compounds' (ed K Paulus), IOP, Aachen, p 18.

Dörr, U, Schwartz, W, Wörner, A, Westphäling, R, Dinger, A, Kalt, H, Mowbray, D J, Hopkinson, M and Langbein, W (1998) Optical properties of $(Al_xGa_{1-x})_{0.52}In_{0.48}P$ at the crossover from a direct-gap to an indirect-gap semiconductor. J Appl Phys 83, 2241.

Dubrovskii, V G, Cirlin, G E, Soshnikov, I P, Tonkikh, A A, Sibirev, N V, Samsonenko, Yu B and Ustinov, V M (2005) Diffusion-induced growth of GaAs nanowhiskers during molecular beam epitaxy: theory and experiment. Phys Rev B 71, 205325.

Farrow, R F C (1974) Growth of indium phosphide films from In and P_2 beams in ultra-high vacuum. J Phys D: Appl Phys 7, L121.

Farrow, R F C (1983) The stabilization of metastable phases by epitaxy. J Vac Sci Technol B 1 222.

Foxon, C T, Novikov, S V, Hall, J L, Campion, R P, Cherns, D, Griffiths, I and Khongphetsak, S (2009) A complementary geometric model for the growth of GaN nanocolumns prepared by plasma-assisted molecular beam epitaxy. J Cryst Growth 311, 3423.

Francoeur, S, Seong, M J, Mascarenhas, A, Tixier, S, Adamcyk, M and Tiedje, T (2003) Band gap of $GaAs_{1-x}Bi_x$, $0 < x < 3.6\%$. Appl Phys Lett 82, 3874.

Furukawa, Y, Yonezu, H, Ojima, K, Samonji, K, Fujimoto, Y, Momose, K and Aiki, K (2002) Control of N content of GaPN grown by molecular beam epitaxy and growth of GaPN lattice matched to Si(100) substrate. Jap J Appl Phys 41, 528.

Gold, A (1987) Electronic transport properties of a two-dimensional electron gas in a silicon quantum-well structure at low temperature. Phys Rev B 35, 723.

Gonda, S and Matsushima, Y (1976) Effect of substrate temperature on composition ratio x in molecular-beam-epitaxial $GaAs_{1-x}P_x$. J Appl Phys 47, 4198.

Gonda, S, Matsushima, Y, Mukai, S, Makita, Y and Igarashi, O (1978) Heteroepitaxial growth of GaP on silicon by molecular beam epitaxy. Jap J Appl Phys 17, 1043.

Grandjean, N, Damilano, B and Massies, J (2001) Group-III nitride quantum heterostructures grown by molecular beam epitaxy. J Phys: Condens Matter 13, 6945.

Grant, V A, Campion, R P, Foxon, C T, Lu, W, Chao, S and Larkins, E C (2007) Optimization of RF plasma sources for the MBE growth of nitride and dilute nitride semiconductor material. Semicond Sci Technol 22, 15.

Grassman, T J, Brenner, M R, Rajagopalan, S, Unocic, R, Dehoff, R, Mills, M, Fraser, H and Ringel, S A (2009) Control and elimination of nucleation-related defects in GaP/Si(001) heteroepitaxy. Appl Phys Lett 94, 232106.

Harrison, W A (1980) 'Electronic Structure and the Properties of Solids'. W H Freeman and Co, San Francisco, p 176.

Heying, B, Smorchkova, I, Poblenz, C, Elsass, C, Fini, P, Den Baars, S, Mishra, U and Speck, J S (2000) Optimization of the surface morphologies and electron mobilities in GaN grown by plasma-assisted molecular beam epitaxy. Appl Phys Lett 77, 2885.

Hoke, W E, Lemonias, P J and Weir, D G (1991) Evaluation of a new plasma source for molecular beam epitaxial growth of InN and GaN films. J Cryst Growth 111, 1024.

Hsiao, C -L, Liu, T -W, Wu, C -T, Hsu, H -C, Hsu, G -M, Chen, L -C, Shiao, W -Y, Yang, C C, Gallstrom, A, Holtz, P -O, Chen C -C, and Chen, K -H (2008) High-phase-purity zinc-blende InN on r-plane sapphire substrate with controlled nitridation pretreatment. Appl Phys Lett 92, 111914.

Inoue, T, Iwahashi, Y, Oishi, S, Orihara, M, Hijikata, Y, Yaguchi, H and Yoshida, S (2008) Photoluminescence of cubic InN films on MgO (001) substrates. Phys Stat Sol (c) 5, 1579.

Iwata, K, Asahi, H, Asami, K and Gonda, S (1996) Gas source molecular beam epitaxial growth of $GaN_{1-x}P_x$ (x/Leq 0.015) using ion-removed electron cyclotron resonance radical cell. Jap J Appl Phys 35, L1634.

Iwata, K, Asahi, H, Asami, K and Gonda, S (1997) Gas source MBE growth of GaN rich side of $GaN_{1-x}P_x$ using ion-removed ECR radical cell. J Cryst Growth 175/176, 150.

Johnstone, B (2007) 'Brilliant—Shuji Nakamura and the Revolution in Lighting Technology', Prometheus Books, New York.

Kakuda, M, Makino, K, Ishida, T, Kuboya, S and Onabe, K (2012) MBE growth of cubic AlN films on MgO substrate via cubic GaN buffer layer. Phys Stat Sol (c) 9, 558.

Kawai, T, Yamane, K, Furukawa, Y, Okada, H and Wakahara, A (2011) Growth of AlPN by solid source molecular beam epitaxy. Phys Stat Sol (c) 8, 288.

Kawamura, Y, Asahi, H and Nagai, H (1981) Molecular beam epitaxial growth of undoped low-resistivity $In_xGa_{1-x}P$ on GaAs at high substrate temperatures (500–580 C). Jap J Appl Phys 20, L807.

Kim, J G, Frenkel, A C, Liu, H and Park, R M (1994) Growth by molecular beam epitaxy and electrical characterization of Si-doped zinc blende GaN films deposited on β-SiC coated (001) Si substrates. Appl Phys Lett 65, 91.

Kim, M -H, Juang, F S, Hong, Y G, Tu, C W and Park, S -J (2003) Single-crystal zincblende GaN grown on GaP (1 0 0) substrate by molecular beam epitaxy. J Cryst Growth, 251, 465.

Kondo, M, Uomi, K, Hosomi, K and Mozume, T (1994) Gas-source molecular beam epitaxy of GaN_xAs_{1-x} using a N radical as the N source. Jap J Appl Phys 33, L1056.

Korakakis, D, Ludwig, K F and Moustakas, T D (1998) X-ray characterization of GaN/AlGaN multiple quantum wells for ultraviolet laser diodes. Appl Phys Lett 72, 1004.

Kroemer, H (2004) The 6.1 Å family (InAs, GaSb, AlSb) and its heterostructures: a selective review. Physica E 20, 196.

Kuroiwa, R, Asahi, H, Iwata, K, Kim, S, Noh, J, Asami, K and Gonda, S (1997) Gas source molecular beam epitaxy growth of GaN-rich side of GaNP alloys and their observation by scanning tunneling microscopy. Jap J Appl Phys 36, 3810.

Letoublon, A, Guo, W, Cornet, C, Boulle, A, Veron, M, Bondi, A, Durand, O, Rotret, F, Detraese, C, Chevalier, N, Bertru, N and Le Corre, A (2011), J Cryst Growth 323, 409.

Levander, A X, Yu, K M, Novikov, S V, Tseng, A, Foxon, C T F, Dubon, O D, Wu, J and Walukiewicz, W (2010) $GaN_{1-x}Bi_x$: extremely mismatched semiconductor alloys. Appl Phys Lett 97, 141919.

Li, T, Campion, R P, Foxon, C T, Rushworth, S A and Smith, L M (2003) MOMBE growth studies of GaN using metalorganic sources and nitrogen. J Cryst Growth 251, 499.

Lindsay, A and O'Reilly E P (1999) Theory of enhanced bandgap non-parabolicity in GaN_xAs_{1-x} and related alloys. Sol St Commun 112, 443.

Liu, X, Bishop, S G, Baillargeon, J N and Cheng, K Y (1993) Band gap bowing in $GaP_{1-x}N_x$ alloys. Appl Phys Lett 63, 208.

Liu, Z, Kawanami, H and Sakata, I (2011) Carbon doping for p-type GaP and GaPN in molecular beam epitaxy using carbon tetrabromide source. Phys Stat Sol (c) 8, 285.

McCarthy, L S, Smorchkova, I P, Xing, H, Kozodoy, P, Fini, P, Limb, J, Pulfrey, D L, Speck, J S, Rodwell, J W, DenBaars, S P and Mishra, U K (2001) GaN HBT: toward an RF device. IEEE Trans Electon Devices 48, 543.

Mesrine, M, Grandjean, N and Massies, J (1998) Efficiency of NH_3 as nitrogen source for GaN molecular beam epitaxy. Appl Phys Lett 72, 350.

Mietze, C, DeCuir, E A, Manasreh, M O, Lischka, K and As, D J (2011) Intrasubband transitions in cubic AlN/GaN superlattices for detectors from near to far infrared. Phys Stat Sol (c) 8, 1204.

Mitsuyoshi, S, Umeno, K, Furukawa, Y, Urakami, N, Wakahara, A and Yonezu, H (2010) Electrical and luminescence properties of Mg-doped p-type GaPN grown by molecular beam epitaxy. Phys Stat Sol (c) 7, 2498.

Mourad, D (2013) Tight-binding branch-point energies and band offsets for cubic InN, GaN, AlN, and AlGaN alloys. J Appl Phys 113, 123705.

Mowbray, D J, Kowalski, O P, Hopkinson, M, Skolnik, M S and David, J P R (1994) Electronic band structure of AlGaInP grown by solid-source molecular-beam epitaxy. Appl Phys Lett 65, 213.

Müllhäuser, J R (1998) 'Properties of Zincblende GaN and (In, Ga, Al) N Heterostructures grown by Molecular Beam Epitaxy'. Doctoral dissertation, Humboldt-Universität zu Berlin, Mathematisch-Naturwissenschaftliche Fakultät I.

Najda, S P, Kean, A H, Dawson, M and Duggan, G (1995) Optical measurements of electronic bandstructure in AlGaInP alloys grown by gas source molecular beam epitaxy. J Appl Phys 77, 3412.

Nakamura, S and Fasol, G (1997) 'The Blue Laser Diode', Springer, Berlin.

Nakamura, T, Tokumoto, Y, Katayama, R, Yamamoto, T and Onabe, K (2007) RF–MBE growth and structural characterization of cubic InN films on yttria-stabilized zirconia (001) substrates. J Cryst Growth 301, 508.

Noreika, A J, Greggi, J, Takei, W J and Francombe, M H (1983) Properties of MBE grown InSb and $InSb_{1-x}Bi_x$. J Vac Sci Technol A 1, 558.

Noreika, A J, Takei, W J, Francombe, M H and Wood, C E C (1982) Indium antimonide-bismuth compositions grown by molecular beam epitaxy. J Appl Phys 53, 4932.

Okamura, H, Misawa, S and Yoshida, S (1991) Epitaxial growth of cubic and hexagonal GaN on GaAs by gas-source molecular-beam epitaxy. Appl Phys Lett 59, 1058.

Orton, J W and Foxon C T (1998) Group III nitride semiconductors for short wavelength light-emitting devices. Rep Prog Phys 61, 1.

Oye, M M, Mattord, T J, Hallock, G A, Bank, S R, Wistey, M A, Reifsnider, J M, Ptak, A J, Yuen, H B, Harris, J S and Holmes, A L (2007) Effects of different plasma species (atomic N, metastable N_2^*, and ions) on the optical properties of dilute nitride materials grown by plasma-assisted molecular-beam epitaxy. Appl Phys Lett 91, 191903.

Paisley, M J, Sitar, Z, Posthill, J B and Davis, R F (1989) Growth of cubic phase gallium nitride by modified molecular-beam epitaxy. J Vac Sci Technol A 7, 701.

Paisley, M J, Sitar, Z, Yan, B and Davis, R F (1990) J Vac Sci Technol B 8, 323.

Pengelly, R S, Wood, S M, Milligan, J W, Sheppard, S T and Pribble, W L (2012) A review of GaN on SiC high electron-mobility power transistors and MMICs. IEEE Trans Microw Theory Tech 60, 1764.

Powell, R C, Lee, N E, Kim, Y W and Greene, J E (1993) Heteroepitaxial wurtzite and zinc-blende structure GaN grown by reactive-ion molecular-beam epitaxy: growth kinetics, microstructure, and properties. J Appl Phys 73, 189.

Putuato, M A, Bolkhovityanov, Y B, Vasilenko, A P and Gutakovskii, A K (2009) Crystal perfection of GaP films grown on Si substrates by solid-source MBE with atomic hydrogen. Semiconductors 43, 1235.

Ristic, J, Calleja, E, Fernandez-Garrido, S, Cerutti, L, Trampert, A, Jahn, U and Ploog, K H (2008) On the mechanisms of spontaneous growth of III-nitride nanocolumns by plasma-assisted molecular beam epitaxy. J Cryst Growth 310, 4035.

Roberts, J S, Scott, G B and Gowers, J P (1981) Structural and photoluminescent properties of $Ga_xIn_{1-x}P$ ($x\approx0.5$) grown on GaAs by molecular beam epitaxy. J Appl Phys 52, 4018.

Rubin, M, Newman, N, Chan, J S, Fu, T C and Ross, J T (1994) p-type gallium nitride by reactive ion-beam molecular beam epitaxy with ion implantation, diffusion, or coevaporation of Mg. Appl Phys Lett 64, 64.

Sadeghi, M and Wang, S (2001) Growth of GaP on Si substrates by solid-source molecular beam epitaxy. J Cryst Growth 227–228, 279.

Sai-Halasz, G A, Esaki, L and Harrison, W A (1978) InAs-GaSb superlattice energy structure and its semiconductor-semimetal transition. Phys Rev 18, 2812.

Sakaki, H, Chang, L L, Ludeke, R, Chang, C-A, Sai-Halasz, G A and Esaki, L (1977) $In_{1-x}Ga_xAs$-$GaSb_{1-y}As_y$ heterojunctions by molecular beam epitaxy. Appl Phys Lett 31, 211.

Sato, Y and Sato, S (1989) Growth of InN on GaAs substrates by the reactive evaporation method. Jap J Appl Phys 28, L1641.

Schörmann, J, As, D J, Lischka, K, Schley, P, Goldhahn, R, Li, S F, Löffler, W, Hellerich, M and Kalt, H (2006) Molecular beam epitaxy of phase pure cubic InN. Appl Phys Lett 89, 261903.

Schupp, T, Rossbach, G, Schley, P, Goldhahn, R, Roppischer, M, Esser, N, Cobet, C, Lischka, K and As, D J (2010) MBE growth of cubic AlN on 3C-SiC substrate. Phys Stat Sol (a) 207, 1365.

Scott, G B and Roberts, J S (1979) 'Gallium Arsenide and Related Compounds' IOP Conference Series 45, 181.

Sellidj, S, Ferguson, B A, Mattord, T J, Streetman, B G and Mullins, C B (1996) Growth of GaN on sapphire (0001) using a supersonic jet of plasma-generated atomic nitrogen. Appl Phys Lett 68, 3314.

Shan, W, Walukiewicz, W, Ager, J W III, Haller, E E, Geisz, J F, Friedman, D J, Olson, J M and Kurtz, S R (1999) Band anticrossing in GaInNAs alloys. Phys Rev Lett 82, 1221.

Shan, W, Walukiewicz, W, Yu, K M, Wu, J, Ager, J W, Haller, E E, Xin, H P and Tu, C W (2000) Nature of the fundamental band gap in GaN_xP_{1-x} alloys. Appl Phys Lett 76, 3251.

Shi, B M, Xie, M H, Wu, H S, Wang, N and Tong, S Y (2006) Transition between wurtzite and zinc-blende GaN: an effect of deposition condition of molecular-beam epitaxy. Appl Phys Lett 89, 151921.

Sweeney, S J, Batool, Z, Hild, K, Jin, S R and Hosea, T J C (2011) 'The Potential Role of Bismide Alloys In Future Photonic Devices' in '13th International Conference on Transparent Optical Networks' (ed M Jaworski and M Marciniak), National Institute of Telecommunications, Warsaw, p 1.

Tabata, A, Lima, A P, Teles, L K, Scolfaro, L M R, Leite, J R, Lemos, V, Schottker, B, Frey, T, Schikora, D and Lischka, K (1999) Condensed matter: structural, mechanical, thermal, and atomic transport properties–structural properties and Raman modes of zinc blende InN epitaxial layers. Appl Phys Lett 74, 362.

Thompson, M P, Auner, G W, Zheleva, T S, Jones, K A, Simko, S J and Hilfiker J N (2001) Deposition factors and band gap of zinc-blende AlN. J Appl Phys 89, 3331.

Tixier, S, Adamcyk, M, Tiedje, T, Francoeur, S, Mascarenhas, A, Wei, P and Schiettekatte, F (2003) Molecular beam epitaxy growth of $GaAs_{1-x}Bi_x$. Appl Phys Lett 82, 2245.

Tuttle, G, Kroemer, H and English, J H (1989) Electron concentrations and mobilities in AlSb/InAs/AlSb quantum wells. J Appl Phys 65, 5239.

Tuttle, G, Kroemer, H and English J C (1990) Effects of interface layer sequencing on the transport properties of InAs/AlSb quantum wells: evidence for antisite donors at the InAs/AlSb interface. J Appl Phys 67, 3032.

Uesugi, K, Morooka, N and Suemune, I (1999) Reexamination of N composition dependence of coherently grown GaNAs band gap energy with high-resolution x-ray diffraction mapping measurements. Appl Phys Lett 74, 1254.

Wagner, R S and Ellis, W C (1964) Vapor-liquid-solid mechanism of single crystal growth. Appl Phys Lett 4, 89.

Weyers, M and Sato, M (1993) Growth of GaAsN alloys by low-pressure metalorganic chemical vapor deposition using plasma-cracked NH_3. Appl Phys Lett 62, 1396.

Weyers, M, Sato, M and Ando, H (1992) Red shift of photoluminescence and absorption in dilute GaAsN alloy layers. Jap J Appl Phys 31, L853.

Wicks, G W, Koch, M W, Varriano, J A, Johnson, F G, Wie, C R, Kim, H M and Colombo, P (1991) Use of a valved, solid phosphorus source for the growth of $Ga_{0.5}In_{0.5}P$ and $Al_{0.5}In_{0.5}P$ by molecular beam epitaxy. Appl Phys Lett 59, 342.

Winser, A J, Novikov, S V, Davis, C S, Cheng, T S, Foxon, C T and Harrison I (2000) Strong blue emission from As doped GaN grown by molecular beam epitaxy. Appl Phys Lett 77, 2506.

Wright, S L, Kroemer, H and Inada, M (1984) Molecular beam epitaxial growth of GaP on Si. J Appl Phys 55, 2916.

Wu, J, Walukiewicz, W, Yu, K M, Denlinger, J D, Shan, W, Ager, J W, Kimura, A, Tang, H F and Kuech, T F (2004) Valence band hybridization in N-rich $GaN_{1-x}As_x$ alloys. Phys Rev B 70, 115214.

Yamane, K, Kobayashi, T, Furukawa, Y, Okada, H, Yonezu, H and Wakahara, A (2009) Growth of pit-free GaP on Si by suppression of a surface reaction at an initial growth stage. J Cryst Growth 311, 794.

Yano, M, Suzuki, Y, Ishii, T, Matsushima, Y and Kimata, M (1978) Molecular beam epitaxy of GaSb and $GaSb_xAs_{1-x}$. Jap J Appl Phys 17, 2091.

Yoshida, S, Misawa, S, Fujii, Y, Takada, S, Hayakawa, H, Gonda, S and Itoh, A (1979) Reactive molecular beam epitaxy of aluminium nitride. J Vac Sci Technol 16, 990.

Yoshida, S, Misawa, S and Gonda, S (1982) Properties of $Al_xGa_{1-x}N$ films prepared by reactive molecular beam epitaxy. J Appl Phys 53, 6844.

Yoshida, S, Misawa, S and Gonda, S (1983) Epitaxial growth of GaN/AlN heterostructures. J Vac Sci Technol B 1, 250.

Yoshida, S, Misawa, S and Itoh, A (1975) Epitaxial growth of aluminum nitride films on sapphire by reactive evaporation. Appl Phys Lett 26, 461.

Yoshizawa, M, Kikuchi, A, Mori, M, Fujita, N and Kishino, K (1997) Growth of self-organized GaN nanostructures on Al_2O_3(0001) by RF-radical source molecular beam epitaxy. Jap J Appl Phys 36, L459.

Yu, J, Chen, B, Yu, Z and Wang, Q (2010) GaP/Si heterostructures grown by GS-MBE. Proc SPIE 3551, 9.

Yu, K M, Novikov, S V, Broesler, R, Demchenko, I N, Denlinger, J D, Liliental-Weber, Z, Luckert, F, Martin, R W, Walukiewicz, W and Foxon, C T (2009) Highly mismatched crystalline and amorphous $GaN_{1-x}As_x$ alloys in the whole composition range. J Appl Phys 106, 103709.

Yu, X, Kuo, P S, Ma K, Levi, O, Fejer, M and Harris, J S Jr (2004) Single-phase growth studies of GaP on Si by solid-source molecular beam epitaxy. J Vac Sci Technol B 22, 1450.

8

II-VI and IV-VI Compounds and Si/Ge

8.1   II-VI Compounds

The so-called II-VI compounds represent a limited set from the totality of those possible. In particular, the sulphides, selenides and tellurides of zinc, cadmium and mercury have been studied for a long time, as has zinc oxide. With the exception of CdTe and the mercury chalcogenides, all these materials have direct band gaps covering the visible and UV spectral regions and they are well known in the form of phosphors, with application to fluorescent lighting and to the now-old-fashioned TV tubes. It goes without saying that their ability for efficient generation of visible luminescence made them prime targets for possible LEDs and LDs and they were the subject of many studies designed to elucidate the fine detail of optical emission spectra. As early as the 1950s there was much speculation as to the nature of so-called activator centres responsible for specific emission lines (Kröger and Vink 1954) and a particularly significant development saw the analysis of donor–acceptor pair luminescence in ZnS (Williams 1960). Indeed, it is clear from a brief perusal of the book by Aven and Prener (1967) that the II-VI compounds were under intense study during the 1960s by luminescence, electron spin resonance and electronic transport techniques and that this was made possible by the development of various crystal growth methods. Bulk crystals of millimetre-to-centimetre size were grown by vapour transport or from the melt, both techniques involving temperatures of the order of 1000 °C. Detailed studies of exciton spectra were undertaken and considerable advances made in the understanding of recombination mechanisms. However, the further application to visible LEDs and LDs came up against the brick wall represented by an apparently total impossibility of achieving both n- and p-type doping. While ZnTe could be doped only p-type, the remaining compounds took the contrary position and steadfastly refused to allow hole conduction. The consensus of opinion favoured an explanation in terms of compensating defect centres which were thermodynamically inevitable but that did not totally inhibit crystal growers' attempts to outwit nature's inherent bias. After all, it was also possible that the true explanation depended on the presence of unplanned and unrecognised impurity atoms and, if this were so, it should be possible to eliminate them. Much would depend on the nature of the crystal growth processes. However, before discussing specific growth methods, we should first note the important fact that some of these compounds crystallise in the cubic (zinc blende) phase, others in the hexagonal (wurtzite) phase—indeed one or two can be persuaded to adopt either structure, depending on the details of the deposition method. Thus, CdS, CdSe and ZnO insist on being hexagonal but CdTe and ZnTe prefer to be cubic, while ZnS and ZnSe are happy to exist in either form (Kasap and Capper 2007). Finally, we should note that more recently both Mg and Be have been incorporated in II-VI alloys.

As noted above, the early work had been done on bulk crystals, grown at high temperatures, which were far from perfect and it was clear that some form of epitaxy might offer improved control of crystal quality and, of particular significance for the incorporation of lattice defects, growth at much lower temperatures. However, this immediately raised the question of what to use as substrates. One obvious possibility, made clear by Figure 8.1, was the use of GaAs substrates for the growth of ZnSe epitaxial films, the lattice mismatch being only 0.25% (cf. e.g. the 0.15% difference between AlAs and GaAs). Indeed, the majority of workers preferred this to the use of the poorer quality bulk ZnSe crystals then available. Films were grown by both VPE and LPE during the early 1970s and by MBE and MOVPE in the second half of the decade, resulting in a gradual improvement in crystal quality as evidenced by the sharpness of X-ray diffraction rocking curves and of low-temperature photoluminescence lines. Indeed, PL spectra provided not only a measure of overall quality but also direct evidence for the presence of various unintentional donor and acceptor species (Werkhoven et al. 1981; Dean et al. 1983). However, the struggle to obtain p-type ZnSe with useful conductivity had to wait until the early 1990s, when the use of active nitrogen doping during MBE growth led to the achievement of hole densities on the order of 10^{24} m^{-3}. Similarly,

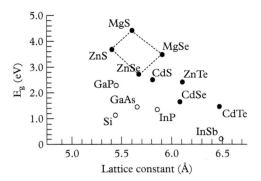

Figure 8.1 *Lattice parameters and corresponding band-gap energies for a number of zinc blende II-VI compounds, showing their relation to some readily available substrate materials. Note that the group linked by dotted lines has been of particular importance in the development of green laser diodes.*

highly conducting n-type films of ZnTe were achieved by both MBE and MOVPE in the mid 1990s. Dogged persistence paid off and the doom-mongers were once again proved wrong. Hopes were therefore high when, in 1991, initial success was achieved in the demonstration of a green injection laser based on ZnSe, a story we shall be happy to elaborate in the following pages. Even the very-wide-gap materials ZnS and ZnO were eventually doped p-type, though the net acceptor levels achieved were somewhat lower and the precise conduction mechanisms are still somewhat controversial.

Having said this, we must also recognise the importance of the ternary alloy HgCdTe (MCT) in the context of far-infrared detection. As HgTe is a semimetal (i.e. it has a negative band gap), its combination with CdTe (Eg = 1.475 eV at room temperature) made possible the growth of material with direct band gaps covering both the 3–5 μm (0.25 – 0.4 eV) and 8–14 μm (0.1 eV–0.15 eV) infrared bands vital for effective night-vision systems. Far-infrared detection, once the realm of the lead chalcogenides and indium antimonide, was rapidly colonised by MCT during the 1970s as a result of the latter's much greater flexibility and adaptability. Initially, detectors were made from bulk crystals grown by the Bridgman method but increasing demands on accurate composition and uniformity soon led to the introduction of epitaxy. CdTe substrates were available in modest sizes and with acceptable perfection, the fact that HgTe and CdTe happened to be well matched implying a satisfactory lattice match to MCT over the whole composition range. VPE took the stage in 1971, and LPE in 1975, both showing promise; but both were seriously challenged during the 1980s, first by MOVPE and then by MBE. Both these latter techniques possessed the advantage of low growth temperatures which allowed accurate formation of quantum well structures, unspoiled by Cd–Hg interdiffusion. They also demonstrated the ability for the controlled n- and p-type doping required for photovoltaic detectors, while the development of large-area staring arrays of diodes, allied to silicon CCD signal processing, suggested the need for heteroepitaxy, a demand met with some degree of success by the MBE process during the new millennium. The dramatic contrast between the early night-vision systems dependent on mechanical scanning technology and today's staring arrays, combined with electronic scanning, represents one of the major successes of optoelectronic innovation, and MBE can claim much of the credit.

It obviously makes sense to treat the two topics of wide- and narrow-gap materials separately, so we shall do that. With regard to the former, our approach will be to regard ZnSe and ZnTe as prototypical compounds, therefore treating them in some detail, while providing rather sketchier accounts of the other materials. Here we go!

So far as we are aware, the first application of MBE to the growth of wide-gap II-VI compounds occurred as early as 1975, when Smith and Pickhardt (1975) of the Perkin Elmer Corporation reported the growth of CdSe, CdTe, ZnSe and ZnTe on a variety of substrates, including GaAs (100), GaAs (110), CdS, CdSe, CaF$_2$, and BaF$_2$. MBE was in its infancy and Perkin Elmer were considering the possibility of launching their commercial machine so this can probably be seen as part of their development programme. They also saw these II-VI compounds as having a future in integrated optics, being transparent from visible wavelengths to beyond 10 μm. Molecular beams were generated from separate Knudsen cells containing the pure elements Cd, Zn, Se and Te, thus allowing easy control of the VI:II ratio at the substrate. Beam intensities and growth rates were measured by an in situ quartz crystal monitor. The authors made the important point that the vapour pressures of these various elements are considerably higher than those of the Group V elements which determine re-evaporation rates from the growing films and this implies the use of much lower growth temperatures. Optimum substrate temperatures were found to be 450 °C for ZnSe, 350 °C for ZnTe, and 300 °C for CdTe.

Having set the scene, the Perkin Elmer group appears to have faded into the background, the running being taken up in Japan at the Electrotechnical Laboratory in Tokyo (Yao et al. 1976, 1977) and at the Tokyo Institute of Technology (Kitagawa et al. 1980). Concentrating on the growth of ZnSe and ZnTe from elemental sources on (100) GaAs, they explored the question of what controls growth rate. Keeping the substrate temperature constant at 350 °C, they varied the Se:Zn and Te:Zn ratios between limits of 10:1 and 1:10, showing that growth was, in all cases, dominated by the flux of the minority component. Thus, for example, at low Se fluxes ($F_{Zn} \gg F_{Se}$) ZnSe growth was limited by F_{Se}, while at high Se fluxes it saturated, behaviour which could be explained on the basis of a simple model in which Se atoms bond only to surface Zn atoms and Zn only to surface Se, while excess atoms immediately evaporate from the surface. Building on this early progress, they then followed the example set by concurrent investigations of III-V material growth, by studying RHEED spectra (Yao et al. 1986, 1990). In the first paper they reported the first observations of RHEED oscillations from II-VI materials— during the growth of ZnSe on (100) GaAs at 350 °C and only when the Se:Zn flux ratio was close to 3:1. These conditions related to the question of stoichiometry and in the second paper they flushed out the details of the surface phase diagram. A zinc-stabilised surface is characterised by a C(2 × 2) RHEED pattern, a Se-stabilised surface by a (2 × 1) pattern and Figure 8.2 shows the demarcation between the two regimes as a function of substrate temperature. Because Se is more volatile than Zn, it is necessary to

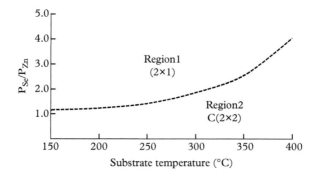

Figure 8.2 *Surface phase diagram for ZnSe as reported by Yao et al. (1989). Region 1 represents the (2 × 1) surface reconstruction characteristic of a Se-stabilised surface, region 2 the C(2 × 2) reconstruction of a Zn-stabilised surface.*

increase the Se:Zn ratio to maintain stoichiometry at higher substrate temperatures. It is probably worth noting that, in the growth of ZnTe, Yao et al. (1989) observed RHEED oscillations with up to 150 periods, indicative of excellent two-dimensional growth.

Two topics are of particular interest in the MBE struggle to tame ZnSe, those concerning, on the one hand, the GaAs/ZnSe interface and, on the other, effective p-type doping. Both are discussed at length in the review by Han and Gunshor (1997) in 'Semiconductors and Semimetals' and we acknowledge their excellent account in aiding our own presentation (there is also much else of relevance in this volume). As already mentioned, many ZnSe growers preferred to use (100) GaAs substrates, in preference to bulk ZnSe, on account of its better quality, well-established cleaning procedure and larger size. There was also a feeling that any subsequent commercial application would benefit from the fact that GaAs was both more readily available and considerably cheaper. While recognising the relatively small lattice mismatch, we should not overlook two other features of the GaAs/ZnSe interface—the change of valence militates against a flat, abrupt interface, implying a degree of intermixing of the constituents, while both Zn and Se are dopants in GaAs and Ga and As are similarly dopants in ZnSe. Though such complications may have little consequence for the subsequent growth of micron-thick films of ZnSe, they do present problems in regard to the electrical behaviour of the contact between the pair. Not only can there be consequences for the conduction and valence-band offsets but also for the interface state density. Both are important if the interface forms part of an electrical contact to a device such as an LED or LD, and the second is vital to the use of ZnSe as an insulator in a GaAs MISFET (metal-insulator field effect transistor), the lack of a high-quality native oxide making it imperative that some alternative be found.

Success in forming a well-controlled interface depended on the initial growth of a GaAs epilayer in a dedicated III-V chamber, followed by vacuum transfer to the II-VI chamber. The GaAs layer was first pre-heated to produce an As-deficient surface, as indicated by appropriate RHEED patterns. Four different patterns were observed, in ascending order of As-deficiency, $C(4 \times 2)$, (4×2), (4×6) and (4×3), this latter differing from previously known GaAs reconstructions and thought to represent a Se-modified surface, due to the presence of residual Se in the II-VI chamber. In fact, it was most clearly apparent when a Se molecular beam was directed at the GaAs surface during heat treatment. It has been speculated that this corresponded to the formation of a ZB Ga_2Se_3 monolayer but, no matter what the precise detail, it became clear that this treatment led to the minimisation of the surface state density at values similar to those obtainable in the GaAs/AlGaAs system. As anticipated, measurement of band offsets also revealed considerable variation, depending on the nature of the GaAs surface reconstruction.

From the viewpoint of film quality, the nucleation stage of ZnSe growth is of vital importance. The development of the green laser diode demonstrated a clear link between device operational lifetime and the presence of dark-line defects, originating from extended defects (stacking faults and dislocations) in ZnMgSSe alloy films, and such defects depended on the nature of the initial ZnSe growth. RHEED studies showed that the variation of the GaAs epilayer stoichiometry could be used to control the ZnSe growth mode, a pretreatment with Se leading to 3D island growth, while Zn treatment resulted in 2D layer-by-layer growth, this latter leading to much reduced defect densities (see e.g. Gaines et al. 1993). Factors such as substrate cleaning, In-free substrate mounting, reduction of background contamination, etc., allied with Zn pretreatment and the use of low-temperature MEE (migration-enhanced epitaxy—see Section 5.2) reduced defect levels to less than 10^8 m^{-2}. Sadly, this considerable achievement was still not adequate to allow useful device lifetimes, the ease with which defects propagate in II-VI materials being an intrinsic property of the relatively weak bonding in ionic compounds.

Turning now to the topic of p-type doping of ZnSe, the obvious first question had to be that of choosing a suitable acceptor. In principle, any Group I element (Li, Na, K, Rb), substituted for Zn or any Group V element (N, P, As, Sb) substituted for Se might be expected to act as acceptors. However, there are complications which make such a simplistic approach less than totally reliable and suggest we look a *little* more carefully at doping in general, both n- and p-type (though this is no place for a thorough discussion of a rather complex subject!—see e.g. Van de Walle 1997). Knowing that ZnSe can readily be doped n-type, it may be well to start with this. Typical Zn-site donors are Al and Ga, and typical Se-site donors are Cl and I. For example, Cl doping, which appears to be most successful, results in a shallow donor with activation energy $E_D = 26$ meV (compared with an estimated hydrogenic value of 26 meV, based on an electron effective mass $m_e = 0.17m_0$ and a relative dielectric constant of $\varepsilon = 9.2$). The largest free-electron density achieved (Zhu et al. 1992b) appears to be $n \sim 3 \times 10^{26}$ m^{-3}, which represents a singularly high value. On the face of it, this is very satisfactory but, from the MBE viewpoint, it is perhaps rather less so—Cl is hardly a gas to be welcomed into a UHV system. Are there more acceptable alternatives? A variety of measurements on other donors (e.g. Merz et al. 1972) showed that Al, Ga, In, F and I all provide shallow donor levels with activation energies in the range 26–30 meV, though the level of free carriers generated is significantly lower. Ga-doping achieves no more than $n = 5 \times 10^{23}$ m^{-3} while Al does somewhat better at $n = 5 \times 10^{24}$ m^{-3} (Oh et al. 2003) and I achieves 1×10^{25} m^{-3} (Yoshikawa et al. 1989). Two factors limit the practical free-carrier level, that of dopant solubility and, more seriously, that of complex formation. In several cases, free-carrier density increases linearly with dopant at low doping levels, then saturates (or even decreases again) at high levels due to the formation of an acceptor-like complex of a donor atom with a lattice vacancy. For example, Ga in ZnSe is thought to form a complex such as $(Ga_{Zn} + V_{Zn})$. In other words, n-type doping in ZnSe follows simple expectations rather well at low doping levels but not always when the doping level exceeds 10^{24} m^{-3}.

In the case of acceptors, there is even greater difficulty in obtaining useful hole densities, partly because acceptor ionisation energies are much larger (the hydrogenic value being close to 100 meV), meaning that they are far from fully ionised at room temperature. But, more importantly, there are difficulties in incorporating some acceptor impurities due to low solubility and in other cases (e.g. P and As) dopants give rise to deep levels. Thus, P doping gives rise to a shallow acceptor approximately 90 meV above the valence-band maximum but at high doping levels, this is dominated by a second level some 550 meV above it (Yao and Okada 1986). The most studied acceptors, prior to the successful use of N have been Li and Na. Li forms a shallow acceptor with

$E_A = 115\,meV$. However, it is a small atom, which means not only that it is unstable but that it is readily incorporated as an interstitial donor ($E_D = 26\,meV$) as well as a substitutional acceptor (Neumark 1988). Thus, the maximum hole density obtainable with Li lies in the range $1-3 \times 10^{23}\,m^{-3}$ (Zhu et al. 1992a), too small for many practical applications. Na, with an acceptor level 124 meV above the valence-band maximum, is larger but, unfortunately, it is too large to fit easily into the ZnSe lattice. In fact, its solubility turns out to be some 10^3 times smaller than that of Li. (K and Rb are larger still and can be expected to show even smaller solubilities.)

The remaining hope of achieving acceptable p-type conductivity lay with the N acceptor, though it was not immediately clear just how such acceptors were to be produced. That N atoms generated a shallow acceptor level had been known for some time. As early as 1983 Dean et al. (1983) had identified a luminescence line associated with a donor–acceptor transition in N-doped ZnSe and, by analysing the spectrum, had arrived at an activation energy $E_A = 111 \pm 1\,meV$, slightly less than that appropriate to Li. Their samples (grown by LPE) did not show any indication of p-type conduction but this confirmation of acceptor behaviour provided encouragement for others to pursue the quest via alternative growth methods. As we have already suggested, the low substrate temperatures associated with MBE growth offered hope of minimising the concentration of compensating defects and this no doubt stimulated efforts to achieve practical doping levels.

The breakthrough in p-type doping of ZnSe probably came towards the end of the year 1990 when Park et al. (1990) described the use of a free radical source (Oxford Applied Research) to provide a nitrogen beam for doping ZnSe-on-GaAs films—though it didn't happen overnight. There had been several earlier attempts to achieve p-type doping by MBE and one of the more exotic (which came from the Sony Corporation in Yokohama) depended on the (somewhat surprising) use of oxygen as an acceptor (Akimoto et al. 1989). These authors described (an admittedly primitive) p–n junction diode grown on a Si-doped GaAs substrate. The n-type dopant was Ga, which limited the electron density to $4 \times 10^{23}\,m^{-3}$, while the maximum hole density achieved was p = $1.2 \times 10^{22}\,m^{-3}$. Nevertheless, p–n junction luminescence was observed at a wavelength close to 450 nm in the blue spectral region. The next report came from the 3M Company in St Paul, Minnesota (Haase et al. 1990), using MBE to grow Li-doped material but the maximum hole density was only $8 \times 10^{22}\,m^{-3}$. It was soon followed by the rather different approach of using MOMBE to grow N-doped films, the N acceptors being provided by cracking NH_3 (Taike et al. 1990). Hall effect measurements showed hole densities as high as $5.6 \times 10^{23}\,m^{-3}$, though estimates of the N concentration suggested $[N] \sim 10^{25}\,m^{-3}$, yet another example of the compensation phenomena which seemed to dog all attempts at achieving high hole conductivity.

The paper by Park et al. represented a successful collaborative programme between the University of Florida and the 3M Company. Layers were grown in a custom-designed MBE chamber at substrate temperature of 275 °C and with a Se:Zn flux ratio of 2:1. RHEED spectra were characteristic of a slightly Se-stabilised surface (see Figure 8.2). The free radical source was known to produce a beam containing some fraction of atomic nitrogen, together with a much larger flux of N_2 molecules and studies of PL spectra from samples grown with ($N + N_2$) and those grown with only N_2 showed clear evidence that the p-type doping resulted from atomic nitrogen. Capacitance–voltage profiling allowed measurement of net acceptor density, the maximum value achieved being 3.4×10^{23} m^{-3} and, though this could be regarded as no more than modest, it was nevertheless adequate for them to make p–n junction LEDs which emitted in the blue region of the visible spectrum at room temperature. The p-type ZnSe was grown on a p$^+$ GaAs substrate, followed by a Cl-doped n-type layer (Cl being obtained from a ZnCl$_2$ effusion cell), and contact made between the GaAs and a lithographically defined metal film on top of the structure. Though acting as a useful demonstrator, this device suffered from the presence at the GaAs/ZnSe interface of a large valence-band step which resulted in a large voltage drop—typical operating voltages being 13 V, compared to the expected value of about 3 V. It nevertheless provided the stimulus to the later development of both LED and LD devices, as we shall see in the following. Finally, in 1991 Qiu et al. (1991) reported a more detailed study of N doping in which they obtained hole densities as high as 1×10^{24} m^{-3}. However, it was clear that this required an N concentration [N] > 10^{25} m^{-3}—once again, high doping was beset by compensation difficulties.

The quest for high-brightness LEDs and for laser diodes operating in the green–blue spectral region had a long history, effectively dating back to 1907, and the first report of semiconductor electroluminescence (from SiC) by Captain Henry Joseph Round of the Marconi Company. By the end of the 1980s efficient red LEDs had been developed based on LPE growth of AlGaAs but the shorter wavelengths were still not well catered for. Some considerable effort to control the growth of the quaternary AlGaInP pushed the colour range into the orange–yellow regions but it was clear that some major breakthrough was needed if the whole of the visible spectrum was to be covered, and—those who wait patiently for London buses will recognise the feeling—not one but two such breakthroughs materialised together during the early years of the 1990s. By 1995, Nakamura at the Nichia Chemical Company had demonstrated high-brightness blue and green LEDs based on InGaN/AlGaN quantum wells, while the II-VI compounds were represented by DH structures involving ZnCdSe (blue) and ZnSeTe (green). What was more, these were rapidly followed by LDs based, on the one hand, on the same nitride materials and, on the other, on the ZnMgSSe alloy system which we shall discuss in a moment. It may be disappointing for those responsible for the II-VI device development that nitride devices have demonstrated much superior reliability and have therefore established a dominant commercial position but we should certainly not overlook the scientific and technical merits of the II-VI structures produced. As we shall see, MBE played an important role in their emergence.

The primitive LEDs referred to above were inefficient because they lacked carrier confinement and, therefore, following the precedent set by III-V light emitters, it was natural for anyone designing visible LEDs to look to a similar technology. This explains both the use of DH structures to confine the recombining free carriers and the much improved performance which resulted. It was exemplified by the work of Eason et al. (1995), a collaboration between the North Carolina State University and the Eagle Pitcher Company. The latter grew the ZnSe substrates, while the former made the LEDs. The advantage of using ZnSe, rather than GaAs, substrates was that it avoided the problem of mismatch in thermal expansion coefficients between the two materials and it also minimised reabsorption of light, as GaAs absorbs strongly in the visible region. The authors described the MBE growth of green and blue LEDs with external efficiencies of 5.3% and 1.3%, respectively. The green device, using a simple DH structure, emitted a relatively broad line peaking at 512 nm, the active layer being $ZnSe_{0.9}Te_{0.1}$, while its blue counterpart, making use of a $Zn_{0.9}Cd_{0.1}Se$ MQW, emitted a much narrower line at 489 nm. The performance of these devices compared well with that of current nitride LEDs but suffered from premature degradation, operating lifetimes being on the order of a few hundred hours.

Returning to purely scientific matters, we should note one interesting aspect in regard to the band gap of the $ZnSe_{0.9}Te_{0.1}$ alloy. A linear interpolation between the gaps of the extreme members suggests a value of E_g = 2.66 eV (λ = 466 nm) whereas the reality is that this system provides yet another example of large bowing (see Section 7.5), and the actual value is close to 2.4 eV ($\lambda = 517$ nm) which explains the apparent anomaly. Figure 8.3, taken from the work of Larach et al. (1957), shows not only that such bowing occurs but also that it has been known about for rather a long time! (Later work by Seong et al. (1999) has confirmed the reduction in band gap when up to 17% Se is incorporated in ZnSeTe alloy, though they confined their attention entirely to the ZnTe-rich region.) Note, by the way, that the amount of bowing is not symmetrical about the 50% composition—in other words, equation (B7.1) no longer applies for these examples of very large bowing. Walukiewicz et al. (2000) have seized the opportunity to examine the bowing effect in this alloy system in terms of their band anti-crossing model (which we described in Section 7.5 with respect to GaAsN). They, too, considered only the ZnTe-rich end of the composition range and found good agreement with their model, based on the pressure dependence of the conduction-band edge (compare similar measurements

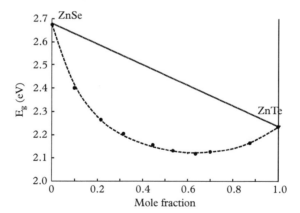

Figure 8.3 *Experimental data on the variation of energy gap for the alloy ZnSeTe, showing a very large degree of bowing. Measurements of diffuse reflection and diffuse transmission were made on polycrystalline samples. (Data taken from Larach et al. 1957).*

on GaAsN). According to their thesis, the Se atom, being considerably more electro-negative than Te (just as N is more electronegative than As), forms a localised energy level which interacts strongly with the extended states of the ZnTe conduction band—note, too, the case of oxygen-doped ZnSe, mentioned above, where it is clear that O forms a shallow donor state. It seems likely that such bowing effects are extremely wide-spread, though we should be careful not to suppose them to be omnipresent—the band gap of the ZnSSe alloy, for example, shows very good linearity across the full compos-ition range. No doubt deeper understanding will gradually accrue—we shall comment no further here.

Having reached this point, it was natural for II-VI workers to take the next step and attempt to develop a technology for making laser diodes. However, this was far from trivial, bearing in mind the need for both optical and carrier confinement, which required not only films with different band gaps but also different refractive indices. First attempts made use of ZnCdSe active regions, ZnSe barriers and ZnSSe optical confining layers (Haase et al. 1991; Jeon et al. 1991). In band-gap terms, this was a sensible enough selection but it suffered from the lack of lattice-matching (only ZnSSe with 6% S could be matched accurately to GaAs) and the net result was a plague of misfit dislocations which (as was well known from the early days of AlGaAs/GaAs laser development) led to rapid degradation under operating conditions. The solution involved the introduction of a new alloy system $Zn_xMg_{1-x}S_ySe_{1-y}$ which, as is clear from Figure 8.1, can be lattice-matched to GaAs (or ZnSe) by appropriate choice of the ratios x and y. Its use was first described by Okuyama et al. (1991) of the Sony Corporation, Yokohama—Sony made a determined effort to develop a viable II-VI laser (see Ishibashi 1996). They grew it by MBE at substrate temperatures between 210 °C and 250 °C, and it was then taken up by many others. As we explain in more detail in Box 8.1, the introduction of Mg provides a choice of band gaps all the way from the 2.8 eV of $ZnS_{0.06}Se_{0.94}$ to the 4.1 eV of $Zn_{0.18}Mg_{0.82}S$ using alloys lattice-matched to GaAs (see Figure 8.4). As also demonstrated by Okuyama et al., the refractive index decreases as the band gap increases, making the wide-gap alloy suitable as an optical confining layer in a DH laser. This was, without doubt, a valuable step forward but a number of difficulties remained when it came to developing controlled growth of these various alloys by MBE.

Box 8.1   ZnMgSSe

The zinc blende alloy $Zn_xMg_{1-x}S_ySe_{1-y}$ is of considerable importance in the development of green and blue laser diodes in II-VI materials because it can be lattice-matched to GaAs, while readily providing band gaps in the range 2.9 eV to 3.5 eV. In this box we look briefly at some of the details, aiming at plausibility rather than at any degree of rigour! In fact, our approach may well remind the reader of Thurber's famous put-down of wine snobbery: 'It's only a naive, do-mestic Burgundy, with little or no breeding, but I think you may be amused by its pretension.' Amusing as it may seem, we hope that what follows may nevertheless prove helpful.

We should first point out that the end members MgS and MgSe both crystallise in the rock salt structure (like MgO) so there has been no reliable measurement of either their ZB lattice parameters or their band gaps. Such values as appear in the literature are the result of extrapolations (see e.g. Jobst et al. 1996). With regard to lattice parameters, we shall adopt an approach based on published values of covalent radii as tabulated in Chapter 3 of Kittel's (1967) iconic 'Introduction to Solid State Physics'. Thus, we use the following: Mg, 1.42 Å; Zn, 1.31 Å; S, 1.04 Å; and Se, 1.14 Å, where we have taken the liberty of increasing the Mg radius from 1.40 to 1.42 Å on the basis that this is necessary in order to explain the *measured* lattice parameter of MgTe (which *does* crystallise in the ZB form). Our argument is based on the idea (assuming Vegard's Law to hold) that we can estimate the average bond length in terms of an appropriate average of the above radii. Thus,

$$\begin{aligned}<B> &= xR_{Zn} + (1-x)R_{Mg} + yR_{S} + (1-y)R_{Se} \\ &= x(R_{Zn} - R_{Mg}) + y(R_{S} - R_{Se}) + R_{Mg} + R_{Se}\end{aligned}$$

(B8.1)

The condition that the alloy be lattice-matched to GaAs can be expressed by equating $<B>$ to the bond length in GaAs, which is given (for the ZB structure) by $a_0 \times 0.433 = 2.45$ Å. Hence,

$$<B> = 2.45 = -0.11x - 0.10y + 2.56$$

or

$$0.11x + 0.10y = 0.11$$

which, to a fair approximation, has the form

$$x + y \sim 1$$

(B8.2)

Needless to say, this represents no more than a general guide to the precise lattice-matching condition (we don't know the values of covalent radii with sufficient accuracy to do any better)

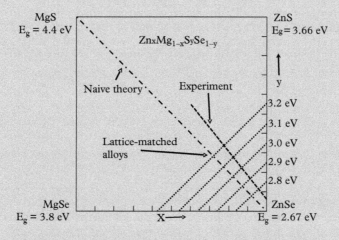

Figure 8.4 *Simple linearised theory to predict the band gap of $Zn_xMg_{1-x}S_ySe_{1-y}$ alloys, lattice-matched to GaAs substrates.*

continued

Box 8.1 *continued*

---

but gives a fair idea of what to expect. Equation (B8.2) is plotted as the dot–dash line in Figure 8.4. The question remaining is one of estimating corresponding band gaps and we hope to achieve that aim in a similarly cavalier fashion!

Again, referring to Figure 8.4, we see that the band gaps of MgSe and ZnS are closely similar and, for simplicity, we shall take them to be equal at $E_g \sim 3.7$ eV. If we also take the band gap of ZnSe to be ~ 2.65 eV, so that $[E_g(\text{ZnS})—E_g(\text{ZnSe})] \sim [E_g(\text{MgSe})—E_g(\text{ZnSe})] \sim 1.0$ eV, we can express the values of E_g for $\text{ZnS}_y\text{Se}_{1-y}$ with $y = y_1$ and for $\text{Mg}_{1-x}\text{Zn}_x\text{Se}$ with $x = x_1 = (1-y_1)$ by the simple equation

$$E_g(y_1) = E_g(1-x_1) = 2.65 + y_1 = 2.65 + (1-x_1) \qquad \text{(B8.3)}$$

If we then assume that all band gaps along the line $x_1 - y_1$ are equal, we see that the ZnMgSSe alloy with $y = y_1/2$ and $(1-x) = (1-x_1)/2$ has this same value of E_g. This enables us to plot lines of constant band gap which intersect at right angles with the line representing the lattice-matching condition. Thus, for values of y up to 0.5, lattice-matched alloys are available with gaps up to 3.15 eV (corresponding to a wavelength of 390 nm in the UV). In this rather crude approximation, we can write

$$E_g(y) = 2.67 + 1.87y \, \text{eV} \qquad \text{(B8.4)}$$

Finally, we should return to reality by noting the position of experimental data as reported by Okuyama et al. (1991, 1993)—see the dashed line in Figure 8.4. Given the lack of 'breeding' in our simple theory, the discrepancies are surprisingly little. In fact, it would be easy enough to modify the choice of covalent radii to shift the dot–dash line to coincide with the dashed line but that would be stretching 'pretention' a little too far. Our main aim here was simply to make plausible the correlation between elementary ideas and experimental data.

Typical practice was to use source cells of Zn, Mg, Se and ZnS, the latter being preferred to a simple S source because of the very low temperature required for the latter. The lower the cell temperature, the greater the demands on temperature control, which are, in turn, complicated by thermal emission from other hot bodies within the chamber, such as the substrate. In fact, substrate temperature T_S is a particularly important parameter in determining alloy composition not only because of its effect on cell performance but also because sticking coefficients of the various elements vary with T_S. Here again, the fact that II-VI compounds are grown at low temperatures makes for greater difficulty in achieving adequate stabilisation. As estimated by Han and Gunshor (1997) in their review, a change of substrate temperature by 5 °C could be expected to produce a 10% change in sulphur fraction when growing S-containing alloys. Another problem stemmed from the fact that both MgS and MgSe crystallise in the rock salt form and this results in their alloys having a tendency to 'flip' between ZB and WZ, leading to the generation of stacking faults, an established source of laser degradation. Yet another difficulty concerned p-type doping. Active N doping of ZnMgSSe proved only moderately successful in yielding p-type conductivity, as the thermal activation energy of the acceptors appeared to be significantly larger than was the case for N-doped ZnSe, and this was of practical significance in considerably increasing contact resistance in LEDs and LDs. Further investigation showed that the 'culprit' took the form of a deep level due to the formation of a DX-like centre, similar to that observed in n-type AlGaAs when the Al concentration exceeded 20% (Han et al. 1994). As the authors comment, on account of its association with an acceptor impurity, this should presumably be referred to as an AX centre. In the case of ZnSe the centre lies within the valence band but, as the band gap is increased by adding Mg and S, it emerges into the gap and acts as gradually deepening trapping level.

Such difficulty in achieving low-resistivity p-type cladding layers has hampered the development of LDs and it is of interest to mention (however briefly) the use of ZnTe as an aid to the formation of low-resistance contacts. As we said earlier, ZnTe can readily be doped p-type—though it was only with the arrival of active nitrogen that really high doping levels ($N_a - N_d > 10^{25}$ m^{-3}) could be attained in MBE growth. The question arose, therefore, whether an effective contact could be made between p-type ZnTe and p-type ZnSe (or related alloys). A simple abrupt contact proved impracticable on account of a large step between the valence bands, so it was found necessary to employ a carefully graded ZnSeTe layer and, because it proved difficult to maintain accurate control over the Se/Te ratio, the final solution consisted of a graded superlattice in which the width of the ZnSe and ZnTe layers was varied smoothly from end to end. Thus was the operating voltage of LDs brought down from something like 20 V in the early structures to a highly presentable 4 V. We should also point out another ingenious application of ZnTe as a means of doping ZnSe p-type. Jung et al. (1997) incorporated delta-doped layers of ZnTe:N in MBE-grown ZnSe, thus producing net hole densities as high as 7×10^{24} m^{-3}, while modifying the overall properties of the layers only slightly. They claimed that the ZnTe:N was confined to single monolayers, each one being separated by approximately ten monolayers of ZnSe.

Finally, we might note that the difficulty in doping ZnTe n-type was finally overcome in MBE growth by using Cl atoms (from a ZnCl$_2$ cell) (Tao et al. 1994) but even then the free-electron density reached only 3×10^{22} m^{-3}. Seven years later (Chang et al. 2001), by the use of Al donors raised this to 4×10^{24} m^{-3}, a figure which is on the borderline of being practically useful. The fact that the first MBE growth of ZnTe had been reported as early as 1976 (Yao et al. 1976) illustrates the magnitude of the difficulty in obtaining adequate control over growth conditions. Indeed, the natural reluctance of wide-gap II-VI compounds to remain monotypic was not to be brushed lightly to one side. As can be seen from Figure 8.1, there was no obvious lattice-matched substrate available for ZnTe epitaxy and various compromise solutions have been adopted. Ge (111), Ge (100), GaAs (100) and bulk ZnTe substrates have all been tried and it may be significant that the best n-type doping result was achieved on ZnTe substrates. However, the search for more practical substrates goes on, as illustrated by the recent application of GaSb to the role (Zhang et al. 2012). We shall leave the matter there, and move on to consider the MBE growth of the two wide-gap materials ZnS and ZnO.

Serious attempts to grow ZnS by MBE first appeared during the 1980s (see e.g. Yokoyama et al. 1986; Yokoyama and Ohta 1986; Benz et al. 1988). Because of the reasonably close lattice match to GaP (0.75%), this has been widely used as substrate, though GaAs and bulk ZnS have also been used, the former because of its higher quality, and the latter because of its perfect match! Even Si has been called into play, though preferably with a buffer layer of CaF_2 to avoid anti-phase domain formation (Tong et al. 1996). Samples have been grown using either ZnS (Kanie et al. 1994) or S effusion cells, though in the latter case, because of the unusually high vapour pressure of sulphur, it was necessary to design a special cell with a post-heater at the orifice (see Figure 8.5; Cook et al. 1992; Yoneta et al. 1993). ZnS, in its ZB configuration, has a band gap of 3.66 eV (at room temperature), which makes it a candidate for electroluminescent devices operating in the blue and UV spectral regions and this led to the familiar struggle to produce both n-type and p-type conductivities. Though MBE has demonstrated the ability to grow films showing excellent low-temperature PL spectra, dominated by free and bound exciton lines (Tran et al. 1997), it has singularly failed in its mission to achieve significant p-type doping. Having led the charge to obtain low-resistivity p-type conductivity in ZnSe, one may be forgiven for having anticipated similar success with ZnS but, in the event, this was achieved not by MBE but by good old-fashioned VPE. Iida et al. (1989) produced p-type films on (100) GaAs with hole densities as high as 6×10^{24} m^{-3} by supplying both Zn and NH_3 to the growth surface, the NH_3 to provide N acceptors, the Zn to saturate Zn sites, to persuade the N to fill S sites and also to dissuade Ga (from the GaAs substrate) from filling Zn sites (where it would act as a donor). The growth temperature was 670 °C—so much for the thesis that the low growth temperatures inherent in MBE would solve these doping problems! Similarly, Svob et al. (2000) obtained hole densities of 10^{24} m^{-3} with MOVPE growth, again using N doping,

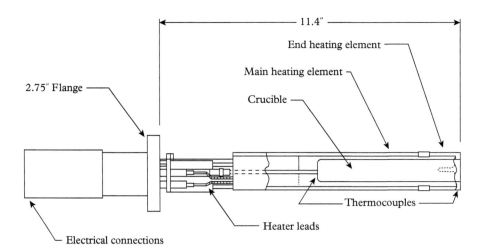

Figure 8.5 *Schematic diagram of the sulphur effusion cell used by Cook et al. (1992) for the MBE growth of ZnS on GaP substrates. The cell has two independently controlled temperature zones, one the main heater and one at the outer end to prevent clogging of the orifice. (From Cook et al. 1992.) Reproduced with permission from Cook, J W, Eason, D B, Vaudo, R P and Schetzina, J F (1992) J Vac Sci Technol B 10, 901. Copyright 1992, AIP Publishing LLC.*

though only after a post-growth anneal to remove compensating hydrogen atoms. They also achieved a hole density of 8.8×10^{23} m^{-3} by diffusing Li into an undoped sample. The best p-type doping achieved with MBE appears to be that reported by Jung et al. (1997) using their ZnTe:N delta-doping procedure which worked so well in ZnSe. In the case of ZnS, though, they managed hole densities no higher than 5×10^{23} m^{-3}. Efforts are still being made, as exemplified by Ichino et al. (2004) who grew Li-doped ZnS on GaP using Zn, S and Li effusion cells. Their low-temperature PL spectra showed excellent exciton spectra with narrow line widths but the best hole conductivity corresponded to $(N_A - N_D) = 3 \times 10^{21}$ m^{-3}. More recently, they had little more success with P doping and with (N + Ag) co-doping (Ichino et al. 2008, 2010). One can only conclude that, even after all these years of effort, the doping of wide-gap II-VI compounds is still far from understood.

While the interest in ZnS was generally quite modest, there has been a veritable explosion in that concerning ZnO. Zinc oxide is a material with a long history. It is a well-known additive in the production of concrete, figures large in the pharmaceutical industry and is also added to the rubber used to make car tyres, while in the electronics industry it has, for some time, been recognised as a useful transparent conducting thin film, in the style of indium tin oxide. It is also frequently used for acoustic surface wave devices. However, its use as a semiconductor per se is of considerably more recent vintage. The development of bulk single crystals during the 1960s and 1970s allowed accurate measurement of its crystal structure (WZ), band gap (E_g = 3.37 eV at room temperature) and exciton binding energy (E_X = 60 meV), the latter being significant in implying the existence of free excitons at room temperature. This and the 'UV' band gap made ZnO yet another material of interest for blue light emitters—in many ways it compares very closely with GaN (though GaN has a much smaller exciton binding energy E_X = 26 meV)—but it, too, proved very difficult to dope p-type. N-type doping could readily be achieved by substituting Al, Ga or In on the Zn site, yielding free-electron densities as high as 10^{26} m^{-3} but p-type conductivity demanded considerably more effort. It was only following the sad demise of ZnSe LEDs and LDs in the 1990s that serious attention was given to the development of ZnO, in the hope that it might prove more resilient, and, as in many other instances, MBE was to play an important role. Indeed, the surge of interest in ZnO was quite remarkable—according to Claus Klingshirn (in Klingshirn et al. 2010), the then current annual publication rate for ZnO papers exceeded 2000 and the total literary volume amounted to some 26 000 papers in all. What is more, no less than six full length books on ZnO were published between 2006 and 2012, a clear illustration of the seductive influence attaching to this 'new' semiconductor material. Needless to say, we make no claim to rival them—we shall simply try to highlight a few important developments.

In considering the possibility of ZnO epitaxy, two questions were paramount: what to use as substrate and how to supply the oxygen. That bulk ZnO crystals were available we have already noted but, in the early days, these tended to be rather small, of questionable quality and (typically of any innovative technology in its early days) relatively expensive. Consequently, many growers favoured alternatives such as sapphire or silicon carbide (widely used in GaN epitaxy), GaAs or silicon, in spite of the large lattice mismatch and, in the case of sapphire, significant Al diffusion into the grown layer. An interesting alternative which did not suffer from the former disadvantage ($\Delta a/a_0 = 0.09\%$) was ScAlMgO (or SCAM) which has an appropriate hexagonal structure and, though brittle and far from easy to use, has indeed been used by several groups (Ohtomo et al. 1999; Tsukazaki et al. 2004). In all such examples, it must be recognised that some form of buffer layer is essential to ultimate success, MgO being a case in point, yielding $MgAl_2O_4$ (spinel) on substrates containing Al. More recently, the quality and size of bulk ZnO crystals has improved sufficiently to encourage several workers to return to using them—the jury is still out, therefore, on this fundamental choice, though it is clear that, while SCAM has been of value in developing ZnO epitaxy, it cannot be regarded as having a commercial future. With regard to the oxygen source, given the rather low beam equivalent pressures used in MBE growth, simply supplying a beam of O_2 proved too inefficient for useful growth rates—it would be necessary to supply oxygen in a more reactive form. Atomic oxygen from a plasma activated source, such as developed for nitrogen doping of Zn chalcogenides was an obvious possibility, though, as demonstrated in the case of doping, it would be desirable to remove ionised species. Alternatives were ozone or hydrogen peroxide.

We believe the first demonstration of ZnO MBE growth to be that described by Johnson et al. (1996) of North Carolina State University. Ironically, perhaps, they were concerned primarily with GaN growth and made use of ZnO merely as a compliant buffer layer between it and both sapphire and SiC substrates. It was convenient, at least in the case of SiC, because it showed good conductivity ($n = 9 \times 10^{24}\,m^{-3}$, $\mu = 0.026\,m^2\,V^{-1}\,s^{-1}$) and a small conduction-band offset, thus allowing low-resistance electrical contact to a GaN device. Johnson et al. were already well versed in the use of an RF plasma source for supplying atomic nitrogen for the growth of GaN—it was a simple matter to add a second source to provide atomic oxygen. Noteworthy was the use of RHEED to demonstrate two-dimensional growth of both ZnO and GaN. As with so much II-VI growth, the centre of interest then moved to Japan. Chen et al. (1997, 1998) and Segawa et al. (1997) reported MBE growth of ZnO on c-plane (0001) sapphire substrates using, in the first case, a plasma oxygen source, in the second, laser ablation of a ZnO source, allied with the introduction of pure oxygen gas. Typical growth temperatures were 550 °C, significantly higher than the 300 °C–400 °C used for the zinc chalcogenides, a reflection of the greater stability of oxides. (It was this, of course, which stimulated hope of greater longevity in possible ZnO laser diodes.) The importance of the nucleation stage was emphasised by both groups, an initial clean-up phase using just the oxygen beam at 600 °C being key. RHEED studies showed that growth began in a two-dimensional form but gradually changed to 3D island growth, resulting in a hexagonal island structure in the final layer. XRD and PL measurements suggested material quality to be good while Segawa et al. found persuasive evidence for the predicted room temperature recombination via free excitons at a wavelength of 375 nm (3.30 eV), as shown in Figure 8.6. They also observed laser action in some of their films when excited at high intensity. So, the prognosis was good—but there was still the spectre of p-type doping to be laid.

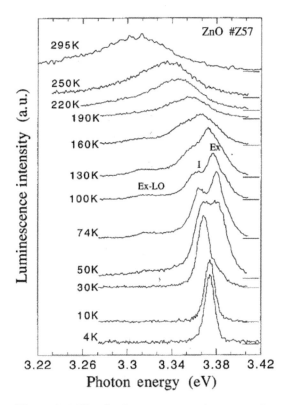

Figure 8.6 *Photoluminescence spectra from a sample of ZnO grown by laser MBE on a sapphire substrate, the emission being excited by the 325 nm line of a He–Cd laser. At low temperatures emission is dominated by a bound exciton transition, while above 100 K free exciton emission (E$_X$) appears and shifts to lower energy in sympathy with the temperature dependence of the ZnO band gap. As can be seen, this emission line persists up to room temperature, its line width resulting from thermal broadening. (From Segawa et al. 1997, courtesy Wiley.)*

Various attempts were made to achieve p-type doping round about the turn of the century, success being somewhat variable (see Look 2005) but the work of Look et al. (2002) provided a degree of confidence in the possibility for controlled and reproducible behaviour—it also suggested a desirable degree of understanding. They grew films on bulk ZnO substrates by plasma-enhanced MBE at a substrate temperature of 525 °C. Doping was achieved by supplying nitrogen to the plasma source, along with the essential oxygen. Hall measurements were consistent with a hole density of 9×10^{22} m^{-3}, in satisfactory agreement with calculations based on SIMS measurements of the nitrogen concentration of order 10^{25} m^{-3} and an acceptor energy of approximately 200 meV. They concluded their letter with the words 'these results suggest that a p–n homojunction can be produced in ZnO'. Could such prognostication be proved correct? In the case of GaN, one remembers, no sooner was p-type conductivity a reality (in 1989) than high-brightness LEDs and LDs took the startled world of electronics by storm. Could ZnO follow suit in the new millennium? Would the II-VI teams pursue their holy grail with similar confidence and drive? The answer, alas, must be a qualified 'no'. If one looks at some of the papers published in 2012, one is struck by introductory passages such as 'ZnO has an exciton binding energy of about 60 meV which *may* (our italics) lead to the realisation of more efficient light emitting devices' (Zhao et al. 2012). Ten years after the initial breakthrough, one was left with the same rather empty feeling—what went wrong?

Perhaps we should first explain why we suggested a 'qualified' negative answer to our question. In fact, during the period 2006–2008, a number of p–n junction diodes *were* reported by a group from the University of California and both UV LEDs and LDs, together with UV photodetectors, demonstrated (Mandalapu et al. 2006; Chu et al. 2008a, b; Mandalapu et al. 2008). Figure 8.7 provides an example of the electroluminescence spectrum from one of their homojunction diodes, showing near-band-edge

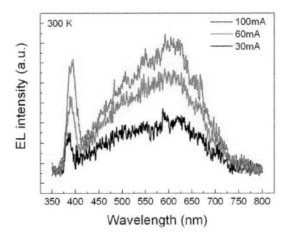

Figure 8.7 *Electroluminescence spectra at different injection levels from a ZnO homojunction diode grown by MBE on a Si substrate. The n- and p-type dopants were Ga and Sb, respectively. The spectra show near-band-edge emission at 390 nm, together with an intense, broad emission from defect levels in the ZnO film. (From Chu et al. 2008a.) Reproduced with permission from Chu, S, Lim, J H, Mandalapu, L J, Yang, Z, Li, L and Liu, J L (2008a) Appl Phys Lett 92, 15210. Copyright 1992, AIP Publishing LLC.*

emission at 390 nm, though somewhat swamped by deep-level green emission, resulting from defects in the ZnO film. We should take the opportunity here of mentioning that ZnO can be combined with either MgO or CdO to form quantum well structures—this, indeed, was a feature of the laser diode work reported by Chu et al. All these structures were grown on Si (111) substrates by plasma-assisted MBE but p-type material relied on Sb doping, rather than the more commonly used N—the same group had earlier demonstrated that Sb-doped films were capable of low-resistivity hole conduction, though some doubt has since been expressed concerning the reliability of their Hall measurements (Janotti and Van de Walle 2009). Indeed, such doubts seemed to proliferate within the II-VI community. Hall measurements were shown to suffer from the fact that Hall voltages tend to be very small (because hole mobilities are very low) and from the likelihood of carrier trapping in ZnO surface states (Look 2005). Similarly, there appears to be considerable difficulty in understanding the behaviour of acceptor atoms. For example, after many years of study, it is still a matter of controversy whether N behaves as a simple shallow acceptor (though with a relatively large ionisation energy ~ 200 meV) or whether it produces acceptor behaviour only in conjunction with some native defect. The precise nature of compensating species is similarly still disputed. Needless to say, the understanding of alternative acceptors such as As, P and Sb (on O sites) is fraught with even greater uncertainty. For example, it is clear that the luminescence emitted by the ZnO LEDs referred to above is generated within the Sb-doped material (the electron density on the n-side of the junction being much greater than the p-side hole density) and the emission spectra are of relatively poor quality. In particular, there is a large component of green emission from undesirable deep levels.

Similar comments are in order concerning the efforts of a Japanese group who struggled over many years to achieve very modest emission efficiency from ZnO LEDs (Tsukazaki et al. 2004; Nakahara et al. 2010). They preferred the use of N as acceptor species and developed a novel growth procedure for its incorporation into their samples, grown by laser MBE on SCAM substrates. This involved high substrate temperatures (of the order of 950 °C) to achieve smooth, defect-free layers, interspersed with much lower temperature phases (typically 450 °C) for N doping (N was not incorporated at 950 °C because of its high volatility). In this manner they achieved high-quality films with nitrogen concentrations ranging from 10^{24} to 10^{26} m^{-3}, though the LED reported by Nakahara et al. had to make do with a p-type doping level of only 10^{22} m^{-3}. However, in their more recent work these authors came down strongly in favour of growing by plasma MBE on ZnO substrates, which they regard as much more commercially viable. They grew at high temperature but on Zn-polar, rather than the more common O-polar, surfaces and have also been successful in employing NH$_3$ as a source of nitrogen. Not only were films of improved structural quality, but they were also characterised by long recombination lifetimes (Takamizu et al. 2008).

Looking at the literature in general, there is much evidence pointing to the fact that many ZnO films contain high densities of defects, even to the extent that they are polycrystalline. Possibly, high-temperature growth will prove successful in eradicating such defects but we must leave the matter there. The reader looking for further detail on the mysteries of ZnO should consult the works by Look (2005), Janotti and Van de Walle (2009) or Klingshirn et al. (2010).

Thus, it is clear that a proper understanding of semiconducting ZnO is still a long way away and this must, at least in part, explain the apparent lack of progress in device development. While a huge effort is in hand to *understand* the many vagaries of ZnO behaviour, there appears to be rather little devoted to device work—it is noteworthy that very few ZnO papers emanate from industrial laboratories. The emphasis is heavily biased towards universities, and the aim of university research is, of course, primarily to improve understanding so this is precisely what seems to be happening. Progress may be slow but it is crystal clear that ZnO is enormously more complicated than are silicon or the III-V materials—that is, on the one hand, a blight but, on the other, it provides just the sort of challenge that attracts the academic researcher. In the meantime, any serious challenge to InGaAlN as visible and UV light emitter may have to wait a long time yet.

Our final topic in this section will be concerned with a quite different spectral region, that of the mid infrared, and with a semiconductor alloy which has achieved widespread practical success. Thermal imaging as a means of 'seeing in the dark' (see Orton 2004, Chapter 9) came into prominence during the First World War, based, at that time on a near infrared detector material thallium sulphide. The need for longer wavelength detection was militated by the fact that thermal radiation from moderately 'warm' targets peaked in the wavelength region 3–10 μm, while atmospheric absorption showed two windows at approximately 3–5 μm and 8–14 μm. Various detector materials were investigated with a view to utilising these windows, including the lead chalcogenides, InSb and extrinsic materials such as Au-doped Ge but none of these offered the flexibility required to provide an accurate wavelength match, while extrinsic materials lacked the desired sensitivity. What was required was a direct-gap semiconductor whose band gap could be varied in a controlled manner and this implied the use of an appropriate ternary compound. Two alternatives presented themselves, the IV-IV/VI compounds, which we shall discuss in the next section, and the II-II/VI compound $Hg_{1-x}Cd_xTe$ (widely known as MCT) which was the brainchild of W D Lawson and co-workers at RRE Malvern (Lawson et al. 1959). This latter system offers direct band gaps all the way from zero (HgTe being a semimetal) to 1.6 eV (T = 77 K), with the added advantage of an almost constant zinc blende lattice constant—$\Delta a/a_0 = 3 \times 10^{-3}$. MCT can therefore be seen as the II-VI equivalent of AlGaAs! It shows an almost exactly linear variation of band gap with composition (see Figure 8.8) and, in particular, the compositions x = 0.2 and x = 0.3 possessed band gaps appropriate for the 8–14 μm and 3–5 μm windows, respectively (see e.g. Kruse 1981).

The growth of high-quality single crystals of MCT was always going to be a matter of considerable difficulty, given the extreme toxicity of all three constituents and the high vapour pressure of Hg at typical growth temperatures. A general overview is available in Orton (2004), Chapter 9, while a detailed discussion can be found in Micklethwaite (1981). As might be expected, bulk crystals came first, though suffering from problems of composition control and uniformity. The availability of CdTe substrates, which were transparent to the radiation of interest, served as a stimulus to the development of epitaxial techniques, close-spaced epitaxy being pioneered at the French CNRS laboratory

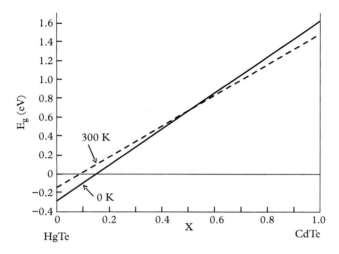

Figure 8.8 *The direct energy gap of the alloy Hg$_{1-x}$Cd$_x$Te as a function of composition for two temperatures, 300 K and 0 K. (Following Kruse 1981.)*

towards the end of the 1960s, while VPE made an appearance round about 1970, followed by LPE from about 1975. (What was more, CdZnTe substrates were also available which offered the possibility of a perfect lattice match to MCT.) Growth temperatures were, in all cases, of the order of 500 °C, and films suffered from non-uniformity due to Hg diffusion from film to substrate and Cd diffusion from substrate to film. Achieving precise control of composition became something of an art and there was clearly a need for epitaxial growth at significantly lower substrate temperatures. This was provided almost simultaneously in 1981 by MOVPE (Irvine and Mullin at RSRE, Malvern) and by MBE (Faurie and Million at LETI, Grenoble).

Prior to the first attempts to grow MCT by MBE, Farrow et al. (1979) made a study of its decomposition under Knudsen effusion conditions, using the method of modulated beam mass spectrometry, originally applied by Foxon et al. (1974) to the evaporation of GaAs (see Section 4.2). They found that, at temperatures up to 100 °C, the vapour pressure of Hg followed the equilibrium vapour pressure but above 100 °C it increased more slowly due to the limited diffusion rate of Hg to the MCT surface, leaving a region depleted of Hg. This result emphasised that for MBE growth at temperatures above 100 °C it would be necessary to supply excess Hg to compensate for the evaporation loss. Faurie and Million (1981a) first demonstrated the MBE growth of CdTe on CdTe substrates, using a single CdTe effusion cell and found it possible to obtain smooth surfaces at temperatures right down to room temperature. They then went on (Faurie and Million 1981b) to grow MCT from three cells, Cd, Te and HgTe, showing that the film composition was determined by the Cd:Te flux ratio (the Hg flux being much higher than that of Cd, in accord with the Farrow MBMS studies). Compositions in the range x = 0 – 0.35 were achieved at substrate temperatures as low as 100 °C, films being monocrystalline and mirror-smooth up to 6 μm in thickness. Here, then, appeared to be the basis for a much more reliable method of growing uniform, well-controlled films for thermal imaging. There were, however, hints of disquiet. Farrow et al. (1981), of RSRE, drew attention to the somewhat variable quality of CdTe substrate crystals, due to a pronounced tendency for twinning and the formation of low-angle grain boundaries. They, therefore, preferred to grow on the much better quality InSb which was currently available (with lattice parameter differing by only 0.05% from that of CdTe). What was more, InSb was commercially available with a diameter of 40 mm, compared to the 15–30 mm of CdTe. They reported the MBE growth of CdTe films on (001) InSb substrates from a single CdTe effusion cell and found that films grown in the temperature range 150 °C–220 °C were free of low-angle grain boundaries and accurately lattice-matched to the substrate. Here was an alternative approach which offered an advantage possibly crucial to the development of high-quality imaging systems. It would be interesting to follow future developments.

Just one year later the French authors reported a second important advance (Faurie et al. 1982). Sticking with CdTe substrates, they grew CdTe–HgTe multilayers at temperatures in the range 120 °C–200 °C, using a CdTe cell for the growth of CdTe and separate Te and Hg cells for HgTe. Crystal quality, as monitored by in situ RHEED measurements, was considerably better for HgTe than for CdTe and improved markedly with increasing growth temperature. SIMS and AES analysis showed that interdiffusion of Cd and Hg was less than the resolution of these techniques (~5 nm) and a collaboration with colleagues at CNRS Paris showed that these 200-period structures behaved as quasi zero-energy-gap semiconductors and provided the first determination of band offsets between the constituent compounds (Guldner et al. 1983). Another interesting contribution from the LETI group concerned the electron mobility in n-type MCT films (Faurie and Million 1982). Films grown at 110 °C showed poor mobilities, characteristic of carrier scattering at grain boundaries, whereas films grown at 180 °C showed much higher values, close to those measured on samples grown by alternative methods. Further work on MCT films demonstrated that both n- and p-type films could be grown to order by controlling the Hg flux, thus facilitating the fabrication of a p–n junction photodetector (Faurie et al. 1983). Key to success was the use of a special Hg effusion cell which could be introduced into the growth chamber as and when required (it was not possible to leave it there permanently because the Hg pressure at room temperature was incompatible with accepted MBE practice). The RSRE group (Dean et al. 1984) countered by describing an unusual luminescence process (associated with an extended defect) observed in MBE-CdTe grown on InSb substrates, again demonstrating the exciting new physics associated with this material system. More importantly, however, theoretical studies suggested that CdTe–InSb superlattices might offer a flexible approach to long-wavelength devices and the RSRE group set out to investigate the practical possibilities of growing them by MBE (Williams et al. 1985). The difficulty was one of finding a compromise between conditions appropriate to the growth of InSb (which was normally grown at temperatures in the range 250 °C–450 °C) and CdTe (150 °C–220 °C). They showed that it was possible, using a two-step growth technique, to grow CdTe at temperatures as high as 310 °C and that there was no evidence for interdiffusion between adjacent layers. The way to practical superlattices was clearly open. In the meantime, however, the initiator of the RSRE work, Robin Farrow had moved on and was now (Farrow 1985) investigating a similar approach at the Westinghouse laboratory in Pittsburgh, while his LETI counterpart, Jean-Pierre Faurie, had followed him across the Atlantic to the University of Illinois.

The Illinois group proceeded to study the growth of CdTe and MCT on differently oriented CdTe substrates (Sivananthan et al. 1986), showing that, for growth of CdTe at low temperatures (150 °C), growth rates were essentially the same on (111)A, (111)B and (100) faces, while at higher temperatures this was far from true. Not only did the rate decrease significantly but the different surfaces behaved very differently. Excess Cd flux was necessary to maintain growth on (111)A, and excess Te on (111)B but on (100) there was no evidence that the rate could be influenced by excess of either flux. Clearly, the growth chemistry was more complex for the II-VIs than for the more familiar III-V compounds, the differences in vapour pressure between anion and cation being much greater for the III-Vs. For MCT the growth rate was limited at high temperature by the Hg flux and, to maintain single crystal growth, the ratios of Hg flux on the various surfaces should be in the ratio 1.0:4.4:9.0 for the three surfaces (111)B, (100) and (111)A, respectively, these ratios being the result of the different bonding arrangement on the different surfaces. We might note that the need for a lower Hg flux for growth on (111)B proved influential in its choice by future workers.

Possibly of much greater long-term interest, the Illinois group also reported the MBE growth of CdTe and MCT on both GaAs and Si (100) substrates (Faurie et al. 1986; Sporken et al. 1989). It was already clear that the future of far-infrared imaging would probably depend on integrating focal plane MCT detectors with Si integrated circuit signal processing technology so this growth problem was of some considerable importance. That it might well be a 'problem' could readily be inferred from a knowledge of the 19% lattice mismatch between the two materials, so no one was anticipating an easy ride, and no one was surprised when the anticipated difficulty turned into reality. After all, crystal growers had been battling for some years to persuade GaAs to grow prettily on Si substrates. In the event, Sporken et al. found it possible to grow (111)B and (100) CdTe on (100) Si, the former directly, the latter by interposing a buffer layer of ZnTe. They then demonstrated that MCT could be grown successfully on top of the (111)B CdTe. Here was evidence that the mission was not quite 'impossible' but difficulties there certainly were. In fact, the (111)B films tended to grow with two domains, rotated by 90° with respect to one another and it was necessary to use substrates misoriented by 8° towards [011] to avoid this, though it was far from clear whether this angle was optimum. It was also necessary to establish just how good was the quality of the resulting MCT films.

The early 1990s saw a considerable burgeoning of interest in the MBE growth of MCT and it became clear that two major issues stood out, those of controlling crystal quality and doping. We have already referred to the tendency for CdTe to form twins and, as MCT was usually grown either on bulk CdTe (occasionally CdZnTe) or on CdTe buffer layers on heterogeneous substrates, it too suffered from the same problem. The situation was well summarised by Faurie et al. (1991). The control problem was complicated by the fact that not all twins were revealed by in situ RHEED measurements, though they showed up well enough in ex situ SEM or electron channelling patterns. It was vital to control substrate preparation, buffer layer growth and MCT growth conditions, key issues being those of substrate temperature ($\pm 1\,°C$) and of Hg flux ($\pm 1\%$), both extremely difficult targets to aim for. For example, the use of a conventional back-surface thermocouple proved inadequate—it was, of course, the *front* surface which had to be monitored—and a front-surface couple was incompatible with substrate rotation. Faurie et al. compromised with the combination of a back-surface couple together with a front-surface IR pyrometer reading, while control of Hg flux was achieved with a specially designed 'constant level' cell. We should note that the presence of twins not only had a major influence on the etch-pit count of MCT layers but also their electrical properties. Twinned layers showed higher p-type background levels, though, surprisingly, their hole mobilities were also higher—difficult to understand!

Yanka et al. (1991) of GE Syracuse proposed an alternative approach to twin reduction, combining photo-assisted growth (using an argon-ion laser) with compositionally modulated structures. Photo-assisted MBE (PAMBE)—not to be confused with plasma-assisted MBE(!) had earlier been used to control the doping of CdTe layers by Bicknell et al. (1986), and the Syracuse workers then demonstrated its application to MCT growth. (While it is not altogether clear exactly how it works, it certainly appears to improve crystal quality!) In addition, their introduction of thin layers of CdTe at regular intervals during MCT growth showed evidence that it could block dislocations. However, they too emphasised the importance of accurate control of substrate temperature and Hg flux—it was never likely that minor modification of the basic technique would find a way round these fundamental requirements. Taking a very different approach, Benz et al. (1991) addressed flux control by the use of chemical beam epitaxy instead of or together with conventional effusion cells. In particular, their precursors were pre-cracked to produce atomic Cd or Te, the latter in order to enhance the Hg sticking coefficient, always an important issue in attempts to produce high-quality MCT. A constant Hg flux was provided by a specially designed Hg-pressure-controlled vapour source. Electron mobilities in n-type MCT were found to be comparable with the best values reported at the time, while those measured on HgTe layers grown by CBE were considerably greater than their MBE counterparts. Nevertheless, it would appear that CBE has not been taken up in the long-term development of commercial MCT devices, as we shall see in Chapter 10. However, the importance of controlling Hg flux was widely appreciated, another example of a pressure-controlled Hg cell being described by Haseyama et al. (1991) of the Tokyo Institute of Technology (see Figure 8.9). Finally, and further emphasising the Japanese contribution to a rapidly expanding technology, we should also note the work of Sasaki et al. (1995) of NEC, who concentrated on the problem of controlling the substrate temperature accurately. Using a long-wavelength pyrometer, they observed significant decrease in temperature during growth of CdTe buffer layers on GaAs substrates—values depended on sample size and surface roughness but were as large as $39\,°C$ in some cases. However, using this data, they were then able to compensate for temperature changes during MCT growth and thereby achieve improved crystallinity and etch-pit density.

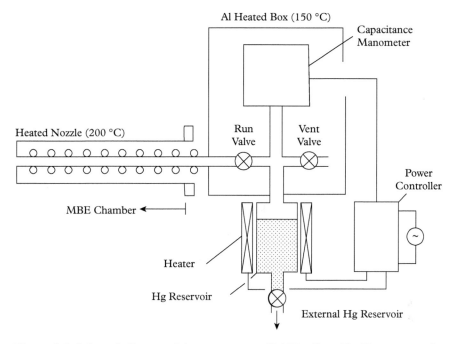

Figure 8.9 *Schematic diagram of the pressure-controlled Hg cell used by Haseyama et al. (1991) for the growth of MCT. The Hg pressure on the upstream side of the run valve is measured by the capacitance manometer and controlled via the feedback loop. Hg is supplied to the substrate via a heated nozzle which extends 45 cm into the chamber. Measured Hg flux was found to be an accurately linear function of Hg pressure. Copyright 1991, The Japan Society of Applied Physics.*

Finally, as we noted earlier, the difficulties in achieving good crystallinity were matched only by those of controlling n- and p-type doping. Originally, this was left to the vagaries of Hg flux control, its being well established that Hg vacancies were responsible for p-type conductivity. Low-temperature (T < 300 °C) annealing in a mercury atmosphere could then be used to convert crystals to n-type. As emphasised by Capper (1991), who reviewed the whole question of MCT doping, and by Boukerche et al. (1988), who summarised the situation with respect to MBE-grown material, the situation is extremely complex. Factors which influence conductivity type include substrate temperature, substrate orientation, Hg flux and composition (i.e. the Cd:Hg ratio) and, in many cases, the outcome is very sensitive to subtle changes. Clearly, a technology as important as that of thermal imaging could not be left hanging on such a knife-edge, a situation made all the more critical when it became clear that the future lay with large arrays of uniform p–n junction diodes. It was essential that both n- and p-type doping should be under the control of specific donor and acceptor impurity species which could be introduced in measured quantity and which would remain stable under device processing. Would it be possible to find well-behaved donors and acceptors which could be relied on to provide efficient and reproducible conductivity control?

The first report (by the Illinois group) of extrinsic doping concerned the use of In as an n-type dopant in MCT layers grown on (111) CdTe/(100) GaAs substrates. They achieved doping efficiencies between 10% and 60%, though there was evidence of an undesirable memory effect in the MBE system. The first successful p-type doping was demonstrated by Wroge et al. (1988) of the McDonnell Douglas research laboratories, using Sb in a CdTe/HgTe superlattice, but the process was not reproducible. Similarly, As could act as an acceptor when occupying a Te site but this appeared to happen only at relatively high temperatures (T ~ 400 °C)—when introduced at normal growth temperatures of ~190 °C it was incorporated on a metal site and acted as a donor. Li was shown to be an efficient p-type dopant but, as in other II-VI compounds, it proved to be a fast diffuser, with a tendency to segregate on the surface of the film, and therefore it was of little practical value. Ag, on the other hand, not only worked as a p-type dopant but showed very little tendency to move from its appointed lattice site. Later studies of In doping seemed to minimise the importance of the memory effect and it was seen to be a reliable donor species. Si, too, could be used as a donor. It *was* possible, then, to achieve reasonably well-controlled doping, both n-type and p-type but only under carefully controlled growth conditions. Clearly, MCT places demands on the crystal grower such that those of us fortunate enough to be 'brought up' on III-V compounds can only shudder to contemplate.

In summary, one can sympathise with those who carry the burden of responsibility for producing large-area, high-quality heteroepitaxial films of MCT (their difficulties are legion), while at the same time admiring them for the tremendous progress already made. We shall take up their story again in Chapter 10.

8.2 IV-VI Compounds

Having just discussed MCT, the material of preference for infrared night viewing sensors, it now seems appropriate to take up the story of its chief competitor in the shape of the IV-VI compounds. The first point to note is that, whereas MCT was a single alloy system, the IV-VI compounds cover a wide range of materials. The second point, that the IV-VIs were developed primarily with a rather different objective in view, that of pollution monitoring. The ability to measure the IR absorption spectrum of a gas allows it to be identified with a high degree of certainty and that requires not only sensors but sources. Because many applications depend on detecting very small concentrations of gaseous impurities, it is important to employ tuneable, single-wavelength, high-power sources, and the IV-VIs are notable for their use as infrared lasers as well as photodetectors. The first IV-VI injection laser dates from 1964, when Butler et al. (1964) of Lincoln Labs reported laser action in PbTe ($\lambda = 6.5$ nm) and PbSe ($\lambda = 8.5$ nm) at a temperature of 12 K. They used bulk Bridgman-grown crystals and formed the necessary p–n junction by annealing them in a Pb-rich atmosphere to effect a type change from p to n. The three lead chalcogenides, PbS, PbSe and PbTe, however, had been in use as IR detectors

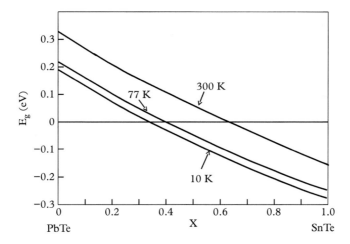

Figure 8.10 *Band gap of the alloy Pb$_{1-x}$Sn$_x$Te as a function of composition. (Following Lovett (1977), Figure 7.22.)*

since the 1930s and were already well established as sensors for the spectral region 3–8 μm. Two motives stimulated the further development of material technology: the need for detectors sensitive in the 8–14 μm window, and the need for precisely controlled laser wavelengths over the whole of the mid infrared spectrum. The necessary initiative came from the group of Ivars Melngailis at the MIT Lincoln Laboratory in the form of two ternary compounds, PbSnTe and PbSnSe. In somewhat the same manner as was true of the HgCdTe system, the SnTe and SnSe ends of the range were characterised by negative band gaps so variation of the Pb:Sn ratio allowed a choice of direct band gaps (at 77 K) between 0.0 and approximately 0.2 eV (see Figure 8.10), thus covering the wavelength range 5 μm to (notionally) infinity. (Note that, though E$_g$ increases with increasing temperature, extending the cut-off wavelength to about 4 μm at room temperature, these early devices tended to operate only at very low temperatures.)

The way was obviously clear for the development of both sources and sensors in this important mid infrared wavelength range but there was equally clearly a need for the greater sophistication associated with epitaxial growth. Sensors in the form of photoconductors demanded material with well-controlled, low doping levels which could not be obtained reproducibly using stoichiometry control. On the other hand, injection lasers and photovoltaic detectors required both n- and p-type doping at similarly well-controlled levels, which implied the incorporation of donor and acceptor atoms. Early lasers were also characterised by high-threshold currents, typical of homojunction devices, and, following the experience with GaAs lasers, it was clear that it would be necessary to develop heterostructure technology. This, again, implied some form of epitaxy. LPE had worked for GaAs and AlGaAs and it also worked in the first instance for IV-VI devices (see e.g. Horikoshi 1985) but many of those concerned with more sophisticated materials preferred the greater control inherent in MBE growth, as we shall shortly see. However, before taking up the MBE story, we should first comment on some of the material peculiarities of these IV-VI compounds.

It is interesting to recognise that the lead salts appeared on the semiconductor scene well before the III-V compounds and were discussed in some detail by R. A. Smith in his seminal book 'Semiconductors' published in 1959 (Smith 1959). They are characterised by being ionically bonded, crystallising in the octahedral rock salt structure and, though having direct band gaps, these occur at the L points of the Brillouin zone, rather than at the zone centre. Unusually, their band gaps *increase* with increasing temperature and there is a somewhat surprising sequence of band gap in relation to chemical property— the values (at room temperature) being $E_g = 0.42$ eV (PbS), 0.27 eV (PbSe) and 0.31 eV (PbTe) (Lovett 1977).

As we saw in Chapter 2 (Section 2.4), an early form of MBE was used to grow IV-VI compounds during the 1970s, the formative years of MBE (see e.g. Holloway et al. 1970 and, for an overall summary, Holloway and Walpole 1980). The base pressure in the vacuum chamber may have been somewhat higher than that taken for granted in a modern MBE system, and the effusion cells may have been cruder in design than those used today but there can be no doubting the value of the work described. Not only were the three lead chalcogenides grown but also the ternaries PbSnTe, PbSnSe and PbSeTe. Both n-type and p-type doping was established (with Bi and Tl, respectively), free-carrier mobilities were found to be equal to the values measured on the best bulk crystals then available, and films were shown to be relatively free of small-angle grain boundaries. Both photoconductive and photovoltaic IR detectors were demonstrated with detectivities of the order of $D^* = 10^{11}$ cm Hz$^{1/2}$ W^{-1}, and heterostructure injection laser diodes demonstrated with threshold currents as low as 100 Acm^{-2} at temperatures below 77 K. These various successes were, at least in part, due to the choice of substrate. Earlier work had employed alkali halides but much improved lattice-matching was achieved by growing on (111) cleaved BaF$_2$ and SrF$_2$ (for photodetectors) or on PbTe (for laser diodes). In their summary, Holloway and Walpole pointed out that there was room for much more work to correlate crystal perfection (native defects, dislocations, impurities, interface states) with device performance but, considering all this had been achieved by about 1978, one can only admire their pertinacity. We might, for instance, recall that the separate confinement double heterostructure (SCDH) GaAs laser diode only came into prominence in the mid 1970s and the fibre-optic communications laser not until the 1980s. What is more, the vital MBE pioneering work of Arthur and Cho at Bell Labs also dates from this very same period, 1970–1975. It is only predated by Bruce Joyce's study of silicon deposition.

So, what of modern MBE? As was true of the II-VI compounds, the Perkin Elmer research team of Smith and Pickhardt (1976, 1978) was probably first in applying the subtle skills of modern MBE to growing lead salts. They reported the growth of PbTe, PbSe, PbSnTe and PbSnSe on chemically polished (100) BaF$_2$ substrates at substrate temperatures in the region of 350 °C–450 °C. Though it was possible to effect n- or p-type doping (at a level of $\sim 10^{23}$ m^{-3}) by control of stoichiometry, they preferred the much better control offered by impurity doping, using, the same two dopants Bi and Tl preferred by Holloway and Walpole. Sources were either elemental or binary compounds, such as PbTe, which were known to evaporate congruently. Fluxes were measured by means of a quartz crystal monitor. They were able to obtain free-carrier densities greater than 10^{25} m^{-3} and measured the diffusion behaviour of Bi and Tl, demonstrating that abrupt junctions could be grown (with widths less than 0.2 μm). (We might note here that donor and acceptor behaviour in these compounds differs from that in the majority of semiconductors on account of their very large dielectric constants.) On the hydrogen model, the ionisation energies of shallow donors or acceptors are given by

$$E_{D,A} = 13.6 \times (m_{e,h}/m_0)(\varepsilon/\varepsilon_0)^{-2} \, eV \qquad (8.1)$$

where the effective masses are small (of the order of 0.04–0.08) and the relative dielectric constants are unusually large (of the order of 200–500 at room temperature, even larger at low temperatures), resulting in values for $E_{D,A}$ of the order of μeV. (It has even been suggested that donor and acceptor levels actually lie within the appropriate bands.) One consequence is that free-carrier densities show no freeze-out at low temperatures. The work of Holloway et al. at the Ford Motor Company and Walpole at MIT, together with that of Smith and Pickardt at Perkin Elmer represented a sound basis for future progress—it remains for us to explore the motivation for and the nature of that progress.

There were probably three driving forces spurring future development, those of the need for greater purity, better crystal quality and the integration of infrared function with silicon data processing. More specifically, there was a clear requirement for accurately lattice-matched structures to minimise dislocation densities and, in particular, to facilitate the growth of optimised laser diodes. The introduction of BaF$_2$ substrates had already proved valuable in improving the performance of photoconductive detectors but, on the other hand, an injection laser diode should ideally be 'built' on a conducting substrate. The obvious choice was the readily available PbTe. It would then be necessary to grow lattice-matched films of different band gap and refractive index in the interest of producing separate confinement, double heterostructures and we should remember that there was no exemplary AlGaAs-like material with which to achieve this—the IV-VI compounds were not blessed with a pair of naturally lattice-matched materials. A degree of initiative would be essential to solving this particular problem. Then, finally, it became apparent during the early 1980s that the future of IR imaging would demand the integration of sensors with Si (or just possibly GaAs?) signal processing, so growth of high-quality films on Si (or GaAs) substrates would be an important field of endeavour.

Let us first take up the question of lattice-matched materials suitable for the growth of SCDH lasers. Assuming the use of a conducting PbTe substrate, the requirement was for materials with larger band gaps which crystallised in the rock salt structure and which could be matched to PbTe. These would form the carrier confining and light confining layers and would also offer the possibility of growing quantum well active regions. The work of Dale Partin and colleagues at the General Motors Research Laboratory in Warren, Michigan well illustrates the nature of the challenge (Partin 1984; Goltsos et al. 1985; Partin 1991). They used a modern ion-pumped MBE system to grow a number of new ternary and quaternary compounds which could be lattice-matched to PbTe by suitable choice of composition. While the 'classic' alloys PbSnTe and PbSnSe, formed by replacing some fraction of Pb atoms by Sn, result in reduced band gaps, the need for increased gaps can be met by substitution with elements such as Ge, Cd, Ca, Sr, Yb, Eu, etc. An early example of this was reported by the Ford group (Holloway and Jesion 1982) who studied PbCaS and PbSrS. The latter alloy was shown to exist as a single phase over the full composition range, with a band gap increasing from 0.4 eV to 4.6 eV (though far from linearly). At least for compositions of up to 20% Sr, PbSrS could be doped both n- and p-type with Bi and either Ag or Tl, respectively. For lattice-matching to PbTe, however, their General Motors rivals chose PbEuSeTe (see Figure 8.11) which could be grown from the following selection of effusion cells: PbTe plus Te to effect accurate stoichiometry, together with PbSe to control lattice match and a separate Eu cell to provide the required band-gap enhancement. Growth temperatures were in the range 300 °C–400 °C, low enough to minimise interdiffusion of the various layers, a feature well illustrated by their demonstration of PbTe–PbEuSeTe quantum wells. Well widths down to 5 nm were shown to generate PL emission in good agreement with that expected from square wells, line widths being consistent with well width fluctuations of only 1.5 monolayers (note that quantum confinement effects in these structures are much enhanced by the small effective masses of both electrons and holes, m_e, $m_h \sim 0.04\ m_0$—see Box 6.4). This alloy system enabled them to grow DH lasers with much reduced threshold currents, operating up to higher temperatures than had been possible hitherto but we shall leave the details until discussing optical devices in Chapter 10. However, we might make one relevant comment: it is interesting to note that these alloys show a marked effect on free-carrier mobility due to alloy scattering. Typically, the addition of just 10% Eu, Ca or Sr is sufficient to reduce electron mobility by a factor of twenty (Partin 1991) and this limits the change of band gap available in practice to a few tenths of an electron volt.

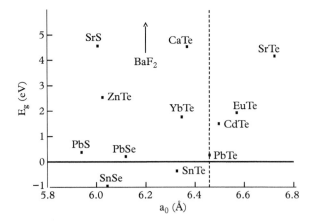

Figure 8.11 *Band gaps and lattice parameters for a range of compounds used in IV-VI lasers and IR detectors.*

One important material property which we have so far taken rather for granted concerns band offsets. It is easy to suppose that a heterojunction such as that between PbTe and PbEuSeTe provides confinement of both electrons and holes but there is no good reason to assume this without careful investigation, an investigation made all the more difficult by the rather small band gaps involved. A number of appropriate studies based on MBE-grown heterostructures showed that most of the relevant combinations do, indeed, result in type I alignment (typical of the AlAs/GaAs system), a good example being provided by the work of Simma et al. (2012) of the Johannes Kepler University in Linz, Austria. They grew a wide range of PbSe/PbEuSe quantum well structures with well widths varying from 3.4 nm to 14.2 nm and Eu content varying between 7% and 13% and measured the appropriate band offsets using temperature-modulated differential transmission spectroscopy (DTS). If our own experience can be taken as typical, we surmise that there may be readers unfamiliar with this technique so we have taken the liberty of providing Box 8.2 to assist with its understanding but, in a sentence, it is a refined method of measuring the interband optical transmission energies corresponding to the confined quantum well states. What Simma et al. did was to measure these energies and compare them with values calculated on the envelope function approximation, using the band offset as an unknown variable. Excellent agreement was found when the ratio $\Delta E_c : \Delta E_g = 0.45 \pm 0.15$. In other words, the conduction- and valence-band wells were nearly equal in all their samples and were independent of temperature between 20 K and 300 K, such information being important for the understanding of infrared laser performance, particularly with regard to operating temperatures.

Box 8.2 Temperature-modulated differential transmission spectroscopy

---

In this box we enlarge somewhat on the account given in the text concerning measurement of optical transmission spectra on PbSe/PbEuSe quantum wells. Two issues require comment: first, we offer an explanation of the measurement technique, and, second, we discuss the fact that each spectral feature appears as a doublet.

Idealised transmission spectra in the region of a quantum well band edge show abrupt steps due to the nature of the 2D density of states (as shown in Figure 6.19). Thus, the transmitted intensity (t) for the n =1 and n =2 transitions is shown schematically in Figure 8.12(a) as a function of photon energy (E)—it drops abruptly at the energy corresponding to the absorption edge and then remains constant before showing a second downward step at the second edge. In practice, detection of such steps is often difficult because the thickness of absorbing material in quantum wells is small. It is therefore helpful to employ a differential technique to improve signal-to-noise ratio. This is illustrated in Figure 8.12(b) for a single step. It involves measuring the transmission intensities t_1 and t_2 at two different temperatures T_1 and T_2 a few degrees apart and subtracting one value of t from the other. Thus,

$$\Delta t = (t_2 - t_1) \tag{B8.5}$$

continued

Box 8.2 *continued*

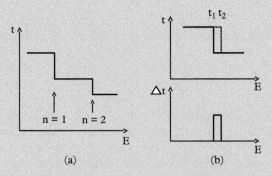

(a)   (b)

Figure 8.12 *Schematic diagram to illustrate the use of temperature-modulated transmission spectroscopy in measuring the confined energy levels in a PbSe/PbEuSe quantum well. In (a) is shown an idealised step in optical transmission (t) as a function of photon energy (E) due to the nature of the 2D density of states. Part (b) shows plots of t_1 and t_2 corresponding to two slightly different sample temperatures T_1 and T_2. Taking the difference $(t_2 - t_1)$, the resulting spectrum has the form of a rectangular peak, as seen in the lower part of (b).*

As can be seen from Figure 8.12(b), a plot of Δt vs E is represented by a rectangular peak. Needless to say, in practice, the measured spectrum suffers a degree of rounding which results in a spectral line similar in appearance to a typical emission line in PL spectroscopy, this being much easier to detect than a not very well-defined step.

Bearing in mind that the band gaps of the IV-VI compounds increase with increasing temperature, t_2 corresponds to the higher of the two temperatures and therefore $T_2 > T_1$. The two temperatures are chosen so that the energy difference between the two absorption edges is approximately 1 meV and this represents a limit to the available spectral resolution. The technique seems to have been first used by Dai et al. (1998) to observe excitonic transitions in InSb quantum wells. Note that most authors use the symbol 'T' to represent transmission—not, of course, to be confused with 'T' for temperature! They also plot a normalised differential transmission ($\Delta T/T$) but this follows the same shape as ΔT itself.

Experimental data presented by Simma et al. (2012) (and, incidentally, by several other workers, measuring similar QW structures) show closely spaced pairs of lines, the higher energy line having an intensity roughly three times that of the lower. This is a characteristic of the IV-VI compounds and arises because their direct energy gaps lie at the L points of the Brillouin zone rather than at the zone centre as is the case for GaAs and many other semiconductors. In measuring transmission, light propagates normal to the sample surface, which, in this case was a (111) surface. Thus, light travels parallel to one of the [111]-type directions while making an angle of 70.5° to the other three, so the spectra are split into two regimes, one for the single longitudinal valley and one for the three oblique valleys. The splitting is a consequence of the differing effective masses associated with the longitudinal and oblique valleys and the intensity ratio follows from the fact that there are three oblique valleys to only one longitudinal.

This, however, was not the whole story. In earlier experiments (Simma et al. 2009) the Linz group had shown that band offsets could be sensitive to strain in the quantum well material. They grew MQW structures in which PbSe wells were confined between PbEuSeTe barriers and, because the band gap of the PbEuSeTe is largely determined by the Eu content while its lattice parameter is largely determined by the Te content, it was possible to vary the strain while keeping the band-gap step essentially constant. Films were grown on (111) BaF_2 substrates, the first 3–4 μm being PbEuSeTe, acting as a pseudo-substrate, followed by the MQW structure with 7.7 nm wells and 29 nm barriers. A range of samples was selected with Te content varying from 0% to 22%, as a means of varying the tensile strain on the well material and the optical transmission edges measured as in the above paragraph. An increase in strain resulted in an increase in the PbSe band gap, so the transmission edge shifted to higher photon energy, until, at about 1% strain, the corresponding DTS peak gradually disappeared. This was due to the holes no longer being confined within the well material, the band alignment having changed from type I to staggered type II (see Figure 8.13) and the electron/hole wave function overlap being sharply reduced. Thus the oscillator strength of the band-to-band

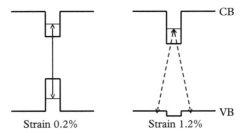

Strain 0.2%   Strain 1.2%

Figure 8.13 *Schematic diagram of band energies for a PbSe/PbEuSeTe quantum well under different strain conditions. At low values of strain, the band alignment is type I whereas, for values of strain greater than about 1%, the alignment becomes staggered type II, and holes are no longer confined in the well. The interband oscillator strength is therefore much reduced as compared with the type I alignment.*

transition was similarly sharply reduced and the transition could no longer be detected, even with the sophistication provided by the temperature-modulation technique. As a final comment, we might simply note that such strain-induced changes of alignment take on a much enhanced significance in these alloys on account of their small energy gaps.

Following this interest in quantum wells, it is scarcely surprising to find that MBE growers have also explored both 1D and 0D structures in IV-VI materials. MBE-grown PbTe nanowires were reported in 2010 by a group in Warsaw (Dziawa et al. 2010). These were grown by the Au-catalysed VLS method on (111)B GaAs substrates (see our earlier discussion of nanowires in Section 6.5). These wires grew normal to the GaAs surface, crystallised in the rock salt structure and tapered in diameter from (typically) 90 nm at the base to 60 nm at the top. Unlike many other such wires, they were free from stacking faults. The same group later demonstrated Bi-catalysed PbTe wires on KCl (001) substrates which were uniform in diameter (d ~ 50–200 nm) with no detectable twinning (Volobuev et al. 2011). (These two modest papers, amounting to a total of 9 pages, may be seen as remarkable for their including no less than 20 different authors—materials research becoming the province of groups, rather than of individuals.)

Probably of greater significance has been the foray into self-assembled quantum dot growth by the Stranski–Krastanow mode, such as employed in the fabrication of InGaAs dots on GaAs substrates (see Section 6.5). The first report of such 0D structures in the IV-VI compounds appeared in 1998 from the same Linz group (Pinczolits et al. 1998) in the form of PbSe dots on (111) PbTe substrates. Following the growth of a uniform wetting layer of approximately 2 ML thickness, PbSe islands developed in the shape of well-characterised tetrahedra (a result of the threefold symmetry associated with the (111) surface) and grew to heights of between 6 and 20 nm, with surface densities of $(1-20) \times 10^{14}\,\text{m}^{-2}$. Size distributions lay in the range $\pm 15\%$ to $\pm 25\%$. Two comments are appropriate: first, as emphasised by Springholz and Bauer (2013) in their review of IV-VI compounds, these PbSe dots are under tensile strain, unlike the majority of earlier dot structures which suffer compressive strain—though this is of no consequence for the growth dynamics because the energies involved do not depend on the sign of the strain; second, the use of (001) PbTe substrates produced no dots at all, strain relief being achieved simply by the generation of misfit dislocations.

Quantum dots must, in general, be encapsulated in a suitable capping layer to stabilise them from unwanted attention from the environment, so, in the case discussed above, it was logical to grow a further layer of PbTe to complete the structure. This, however, led to serious compromising of dot properties. STM studies showed that there was considerable intermixing of PbSe and PbTe, resulting in severe truncation of the tetrahedra. Indeed, after deposition of only 4 nm of PbTe the overall surface was found to be perfectly flat, while synchrotron scattering and X-ray diffraction measurements showed that the dots consisted of PbSeTe, rather than the PbSe originally grown. Interestingly, a very different outcome resulted if the capping layer was composed of the ternary PbEuTe (with approximately 8% Eu). In this case, the tetrahedral shape was retained for capping layers up to 10 nm in thickness and the intermixing of Se and Te was considerably inhibited (by the presence of a EuTe film). Nevertheless, careful studies of the dot composition showed it to vary from 55% Se at the base to 100% at the tip, an observation which we might refer back to our discussion of InAs dots in GaAs confining layers in Section 6.5. There too, careful analysis of AFM and STM data indicated that nominal InAs dots actually contained large amounts of GaAs (though varying with growth temperature).

Finally, we should provide a brief account of the work performed to investigate two and three-dimensional ordering of quantum dots. Springholz and Bauer (2013) describe in detail their studies on PbSe/PbEuTe dots. As shown by earlier measurements on InAs/GaAs dots, for example, repeated layers of dots may show strong correlation due to the strain fields existing throughout such structures but the Linz group were able to demonstrate an interesting range of behaviours dependant on growth conditions, dot size, layer spacing, number of layers, etc. which provide a fascinating insight into what one may call 'dot lattices', or, indeed, 'dot superlattices'. As might well be anticipated, varying the spacing between dot layers changes the coupling between layers and these particular studies show a clear demarcation between three regimes: (I) closely spaced layers (10–40 nm) result in strong coupling, characterised by vertical alignment between the dots in adjacent layers; (II) spacings between 40 and 55 nm result in a face-centred cubic (fcc) lattice, characterised by a hexagonal in-plane arrangement of dots—this hexagonal lattice being increasingly sharply defined as the number of layers increases; (III) in the case of spacing greater than 60 nm there is no correlation between layers, each layer consisting of a random arrangement of dots. These observations are also sensitive to other growth parameters such as dot size and growth temperature. Thus, when dots are smaller than 3 ML in height, the associated strain fields are too small to produce any ordering but when they are between 4 and 6 ML a good fcc lattice is formed, while for dots greater than 6 ML thick there is a tendency for the formation of clusters ('superdots') with a strong vertical alignment. Low growth temperatures (320 °C–340 °C) produce a high density of small dots and no correlation, intermediate temperatures (360 °C–380 °C) result in dots of about 9 nm in height, with excellent hexagonal ordering, while values between 380 °C and 400 °C produce large dots (height ~16 nm), correspondingly large strain and vertical alignment. There is much detail in the original papers which we, unfortunately, have no room to include. Hopefully, this brief summary will not only stimulate the interested reader to refer to Springholz and Bauer's excellent review but will further illustrate the virtue of MBE growth in providing a vast range of well-controlled samples.

For our part, we must turn attention to a quite different area of study. Though, in the early stages of IV-VI epitaxy, it had been common practice to grow on non-lattice-matched substrates, as we saw above, efforts to achieve lattice-matched DH laser structures proved the value of such procedures. On the other hand, the need to combine narrow-gap photodetectors with silicon signal processing and the relative lack of success in combining the two functions in hybrid fashion led to a serious attempt during the 1990s and the new millennium to grow high-quality material on silicon. What was more, the detectors took the form of photodiodes, rather than photoconductors on account of the latter's slow response times (the very high values of dielectric constant associated with the lead chalcogenides playing a major role in increasing device capacitance). This, in turn, required well-controlled n- and p-type doping at typical levels of 3×10^{23} m^{-3}.

The prospect for growing high-quality IV-VI layers on silicon substrates appears bleak when one considers the considerable mismatch not only in lattice parameter but also in thermal expansion coefficients, as shown in Table 8.1. Thus, the lattice constants of PbTe and Si differ by 19% but their thermal expansion coefficients β differ by almost a factor of 2. Assuming a growth temperature of 400 °C suggests that cooling a PbTe layer from growth temperature to room temperature will result in a strain of 0.65%, not huge but sufficient to generate dislocations in a relatively 'soft' material.

First successful attempts to grow lead salts on (111) silicon were reported by Zogg and Hüppi (1985). They employed a CaF_2/BaF_2 buffer to help bridge the lattice constant gap, first in the form of a graded $(CaBa)F_2$ layer but later as separate CaF_2 (~10 nm) plus BaF_2 (~200 nm) layers deposited at a substrate temperature of ~700 °C. Epitaxy of CaF_2 on Si was already well established, as was the growth of lead salts on (111) BaF_2. The resulting PbSe films were smooth, apart from a pattern of misfit dislocation glide steps along <110> directions (see discussion in Zogg 2012). Free-carrier mobilities were comparable to those measured on bulk PbSe samples. While it would have been preferable to grow on (001) Si to comply with standard IC processing, it turned out that such layers tended to crack rather badly—a consequence of inhibited dislocation glide. In a later paper (Zogg et al. 1991) the Swiss group reported the growth of PbTe, PbSSe, PbEuSe and PbSnSe on (111) Si which had already been processed to provide appropriate signal processing for sensor arrays. This involved growing the fluoride buffer layers at 450 °C and the lead salts at 300 °C–450 °C. X-ray rocking curves showed line widths of 150 arc sec or better. Arrays of photovoltaic detectors were fabricated successfully but we shall leave discussion of the device aspects until Chapter 10.

An interesting alternative to the use of fluoride buffers was described by Taylor et al. (2009), who grew (211) PbSnSe films on (211) Si, using a $5-7\,\mu m$ layer of ZnTe as buffer. Though ZnTe has the zinc blende structure and PbSnSe rock salt, they share a common (211) symmetry and show only a small lattice mismatch (see Figure 8.11). The (211) orientation is also advantageous for dislocation glide, so thermal strain can readily be relieved. The ZnTe layers were grown in one MBE machine, cleaned with Br_2/methanol and transferred to a second machine for the growth of the lead salt films.

Table 8.1 *Lattice parameters relevant to growth of IV-VI compounds on silicon.*

	$a_0(\text{Å})$	$\beta(K^{-1} \times 10^{-6})$
PbTe	6.462	19.8
PbSe	6.124	19.4
CaF_2	5.463	19.1
BaF_2	6.200	18.8
Si	5.431	2.6

Because of the near lattice-matching, dislocation densities were unusually low (down to 1.2×10^{10} m^{-2}), film surfaces were smooth and X-ray rocking curves showed line widths of 200 arc sec.

We might make one final comment concerning the doping required to form p–n junctions for these detectors. The use of Bi and Tl as n- and p-type dopants, respectively has been mentioned earlier but further detail can be derived from a paper by Partin et al. (1989) which reports doping studies on PbTe, using Bi_2Te_3 and Tl_2Te as dopant sources. To be effective, both dopants must substitute for Pb atoms, so it is useful to employ their tellurides, rather than simple elemental sources—a slight excess of tellurium, tending to generate cation vacancies, could be seen as encouraging the dopants to occupy the appropriate sites. Partin et al. showed that hole densities as high as 10^{26} m^{-3} could be obtained, though Tl tended both to segregate at the growing surface and to diffuse appreciably at such high concentrations. At concentrations of the order of 10^{25} m^{-3} both Bi and Tl behaved well at growth temperatures up to 370 °C. The authors concluded that diffusion would certainly not be a problem for devices based on micron-sized dimension, such as photodetectors, though it might possibly be a problem in the case of quantum well devices. Certainly, the problem of achieving both n- and p-type doping, which so plagued the II-VI compounds, plays no role in the case of IV-VI materials.

8.3   Silicon/Germanium

Silicon epitaxy dates from the beginning of the 1960s and we might (provocatively?) introduce the topic by asking the simple question 'why?'. Silicon came to displace germanium as the preferred material for transistors round about the middle of the 1950s but struggled at first to compete at the high frequency end of the radio-frequency spectrum. However, the introduction of planar technology at the end of the decade, together with the invention of the integrated circuit in 1958 appeared to have sealed germanium's fate (until the introduction of the Si/Ge bipolar transistor in 1987, to which we shall return) and it could all be done by diffusing donors and acceptors into *bulk* silicon slices. Unlike the situation in the field of compound semiconductors, the quality of bulk silicon crystals seemed to be such as to obviate the need for epitaxy. The total ionised impurity concentration in high-quality Bridgman-grown silicon lay in the range of 10^{17}–10^{18} m^{-3}, in contrast to that in bulk GaAs, for example, of a few times 10^{22} m^{-3}, and this left more than adequate scope for the diffusion of base and emitter layers to form the archetypal bipolar transistor. There were, nevertheless, two difficulties. First, it was required that a low-resistance contact be made to the collector and this could most easily be done by growing a low-doped epitaxial collector film on a highly conducting substrate. Second, it became clear that the collector-base junction breakdown voltage was too low to allow the rapid carrier sweep-out required for high frequency operation and this could be improved by introducing a low-doped film into the base, adjacent to the collector. These demands were initially met by the vapour phase method of hydrogen reduction of silicon tetrachloride, introduced by Henry Theuerer of Bell Labs in 1961 (Theuerer 1961):

$$\mathrm{SiCl_4 + 2H_2 = Si + 4HCl} \qquad\qquad (8.2)$$

While various alternatives have been tried, this process has been widely used in both bipolar and CMOS technology ever since. It takes place at a typical substrate temperature of 1270 °C. However, this relatively high temperature inevitably implies a degree of interdiffusion of dopant atoms across structural boundaries and, where particularly sharp interfaces are required, it is clearly advantageous to look for a low-temperature process such as offered by vacuum evaporation.

Silicon films had been grown by vacuum deposition from the early 1960s but most of this work made use of glass chambers evacuated by oil-diffusion pumps (see references in the summary paper by Bean 1981) and resulted in film quality which scarcely competed with that available from CVD processes. It was only during the 1970s, when UHV conditions were widely introduced, that material quality came to challenge that of conventional epitaxial films. Impurity-induced defects were eliminated, films were free of stacking faults, with etch-pit densities as low as 10^6 m^{-2} and it was only then that the potential for growing well-controlled lightly doped material could be realised. At the same time, one is once again faced with the somewhat vexed question as to whether the process could strictly be regarded as MBE?

In Chapter 2 we saw that the earliest use of molecular beams to study semiconductor deposition was indeed concerned with the homo-epitaxy of silicon. This work at the Plessey laboratory in Caswell evolved from a developing interest in the growth of Si films for integrated circuits, being concerned to gain a better understanding of silicon deposition from the pyrolysis of silane. Though it was clearly a study of fundamentals, rather than an attempt to grow useful films, it could perhaps be seen as the true beginning of silicon MBE. The date was 1966. However, in that form, it could never be seen as a practical growth technique. Central to the philosophy of the work was the need to avoid gas phase reactions and this demanded the use of a Knudsen cell to generate a truly molecular beam of silane. Deposition rates were therefore strictly limited and it was never possible that this apparatus could have been used to grow practical silicon films. Indeed, this leads to the heart of the matter—how can one generate a beam suitable for real film growth? The problem here is, of course, silicon's high melting point, it being necessary to heat it to temperatures of order 1400 °C in order to generate a sufficient atom flux for practical growth. Various methods were used in the early days of 'vacuum sublimation growth' but gradually, it became clear that electron-beam evaporation was most convenient and this demanded a careful source design in order to avoid the evaporation of impurity atoms from the crucible and other non-silicon components, a typical structure being shown in Figure 8.14. An important feature is the fact that only the centre of the silicon sample is heated by the electron beam, thus avoiding the inadvertent evaporation of impurity atoms from other parts of the structure. Also, the hot filament is arranged to lie beneath the silicon, which shields it from the substrate so as to avoid tungsten contamination of the growing film. Without delving into greater detail, it is clear that such a source is very far from being a Knudsen cell, and the resulting flux is probably not a true molecular beam, in the sense that the molecules (strictly atoms in this case) do not interact with each other. However, such criticism can also be applied

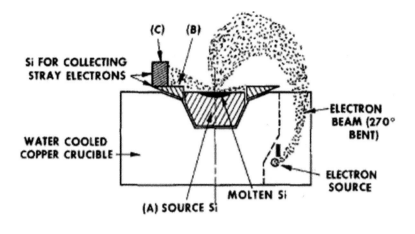

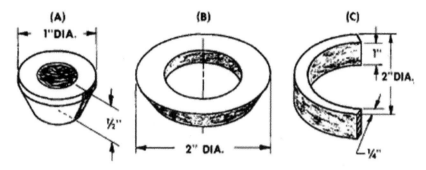

Figure 8.14 *An example of a silicon electron-beam evaporation source for MBE growth of silicon films. Note the silicon doughnut section acting as a screen to prevent the sputtering of copper atoms from the crucible by stray electrons. Only the central portion of the source material is melted, while the hot tungsten filament is screened from line-of-sight to the substrate. (From Ota 1977.) Copyright 1977, The Japan Society of Applied Physics.*

to many modern MBE effusion cells and we can only conclude that, in practical terms, it really doesn't matter. So long as a pure flux of the appropriate species reaches the substrate surface, epitaxial growth will always be determined by surface chemistry—the precise details of the impinging beam are secondary and the designation 'MBE' should be seen as convenient, rather than as descriptive.

While the Plessey work was concerned to study the interaction of SiH_4 with a clean silicon surface, other contemporary efforts were aimed more directly at understanding the growth of silicon films from Si beams generated by EB evaporation (Jona 1966; Abbink et al. 1968) or from a heated Si filament (Widmer 1964). These workers, too, appreciated the importance of studying atomically clean surfaces and therefore made use of stainless-steel UHV systems pumped by ion, Ti sublimation and sorption, rather than oil-diffusion pumps. Background pressures were in the region of 10^{-10} Torr. Both Si (111) and Si (100) bulk crystals were used as substrates. Widmer used TEM to study the effects of varying substrate preparation and growth temperature, found it possible to grow high-quality films at temperatures as low as 550 °C and emphasised the importance of working in a UHV environment. Jona used LEED to show that epitaxy occurred for temperatures above about 400 °C and studied the various surface structures which occur as amorphous films are recrystallised. Abbink et al. were concerned with growth mechanisms on (111) surfaces, employing a remarkably sophisticated electron microscope replication technique, capable of resolving single monolayer steps. They were able to distinguish island growth and step growth and determined a value of surface diffusion coefficient at a substrate temperature of 800 °C. Our concern here is that such studies were under way, rather than with the fine details of their conclusions. They were clearly forerunners of what came to be a highly practical method of depositing silicon films at modest growth temperatures during the 1970s.

Returning, then, to the historical development of silicon MBE, we might note that one of its strongest protagonists was the Bell Laboratories at Murray Hill; hardly surprising, perhaps, in view of the major contribution made by Cho and Arthur to III-V MBE during the early 1970s. That Bell were deeply involved in silicon technology, following their invention of the transistor, goes without saying and the example of III-V MBE 'just down the corridor' naturally led to similar pioneering efforts in the field of silicon epitaxy, as evidenced by a group of papers towards the end of the 1970s (Becker and Bean 1977; Ota 1977; Bean 1978). These papers were primarily concerned with the aspiration towards growing lightly doped films on highly conducting substrates and the accurate control of the dopant species to be employed. Bakeable, cryopumped, stainless-steel growth chambers were typical of those use for III-V MBE, apart from the introduction of an electron-beam source (which allowed growth rates of about 1 μm per hour). On the other hand, dopants were provided courtesy of conventional effusion sources. Growth temperatures were chosen to minimise diffusion of dopant atoms, which meant that T_S had to be kept below about 950 °C and could be as low as 500 °C. However, some dopants tended to segregate on the growth surface at low temperatures so practical substrate temperatures were typically in the range 600 °C to 900 °C but well below the 1270 °C appropriate to CVD.

As with all attempts to grow high-quality crystalline films, careful preparation of the substrate was an imperative, and, as with GaAs (see Cho and Arthur 1975), the 'trick' lay in creating a thin layer of oxide on the etch-polished surface. This had the effect of passivating the surface while the substrate was manipulated into the UHV environment. It could then be removed, either by heating to 1200 °C for 1 min or by argon beam etching at room temperature, followed by thermal annealing at 800 °C–900 °C for 10 min to remove inherent damage. The surface structure was checked by RHEED and cleanliness by AES. The (100) surface consistently showed a (2×1) reconstruction, while a faint trace of carbon could frequently be detected by AES. For completeness, we should add that a similarly cleaned Si (111) surface showed (Jona 1966) the 'classic' Si (7×7) reconstruction that we discussed in Chapter 4 (Section 4.3). The importance of substrate quality was well illustrated by measurement of dislocation densities on both substrate and film. Densities of the order of 10^7 m^{-2} were typical but, significantly, it was found that values measured on the grown films were identical with those measured on the corresponding substrates. Clearly, the growth process was not responsible for any increase. TEM studies on the films showed no evidence for stacking faults.

Moving on now to the doping behaviour, the first question to be answered concerned the choice of dopants. The preferred species were, of course, P or As for n-type doping and B for p-type but there were problems with both of these in MBE growth. Boron has an extremely high melting temperature (2300 °C), making it very difficult to evaporate it from a conventional effusion cell, while P and As, at the opposite extreme, having high vapour pressures, tend to contaminate the MBE chamber and cause unacceptably high levels of background doping. One way to circumvent such difficulties lay in the use of silicon sources pre-doped with the appropriate species but this obviously lacked flexibility. Some compromise candidates were clearly to be preferred, namely, Sb for n-type and either Ga or Al for p-type.

Sb doping was studied by Ota (1977) and by Bean (1978). The Sb beam was generated from a standard effusion cell, which allowed the flux to be determined from measured cell temperature. It was then possible to investigate the Sb sticking coefficient as a function of substrate temperature, by comparison with measured doping levels. Unfortunately, between the two papers there were rather severe differences which could be associated with difficulties in accurate measurement of the Sb source temperature, but it was certainly clear that the sticking coefficient decreased strongly with increasing growth temperature over the range 600 °C to 900 °C, being of the order of 0.1 at 600 °C and falling to the order of 0.02 at 900 °C. In practical terms, this could be utilised as a means of controlling doping level, it apparently being much easier to control substrate temperature than that of the effusion cell. Doping levels from 2×10^{20} m^{-3} to 6×10^{24} m^{-3} were measured on both (100) and (111) substrates but of greater significance was the observation of sharp doping profiles, which were important for the growth of microwave diodes and transistors. However, at growth temperatures below 850 °C there was evidence of Sb surface segregation which tended to smear such profiles so it was necessary to grow at $T_S = 900\,°C$. Under these conditions there was no evidence of Sb diffusion nor of out-diffusion of As from the substrate, so here was clear evidence of the advantage of MBE's low growth temperature, compared to that of the CVD process.

Becker and Bean (1977) undertook a similar investigation of p-type doping with Ga and Al. Again they found problems with surface segregation, in the case of Al over the whole range of growth temperatures but with Ga only for growth temperatures below 550 °C. Though, as for the n-type dopant Sb, the Ga sticking coefficient varied strongly with substrate temperature (being thermally activated with activation energy 1.5 eV), it was possible to vary the doping level in a controlled manner between 10^{20} m^{-3} and 10^{24} m^{-3} and to achieve very sharp doping profiles. Typical growth temperatures lay in the region of 700 °C. Hole mobilities were close to those measured on bulk silicon samples. An n$^+$ – p diode formed by growing a p-layer on an n$^+$ substrate showed excellent diode characteristics, with forward current I given by

$$I = I_0 \exp (eV/nkT) \tag{8.3}$$

with n = 1.08, characteristic of diffusion-limited current. Al doping resulted in severely distorted doping profiles as a result of surface segregation, ruling it out as a plausible replacement for the commonly used B acceptor.

Ota (1979) is also credited with the development of ion-beam doping as a means of countering the high vapour pressures of As and P. By first ionising them, then forming them into a low energy (~600 eV) ion beam, these elements could be implanted into the silicon substrate to a small depth, just sufficient to avoid surface segregation and to considerably reduce any tendency for them to re-evaporate from the growing surface. As a result it was possible to achieve doping levels higher than 10^{24} m^{-3}, not possible with Sb doping (though see the results quoted by AEG Telefunken in the paragraph following). Because of the low energy employed, damage to the silicon film was minimised and, because of the ease with which the beam intensity could be measured, it was also possible to use computer control to generate pre-programmed doping levels. This latter facility was incorporated into a magnificent new MBE machine, which proved how seriously Bell regarded silicon MBE as a method of growing new device structures (Bean and Sadowski 1982; see Figure 8.15). It featured a sample load lock capable of dealing

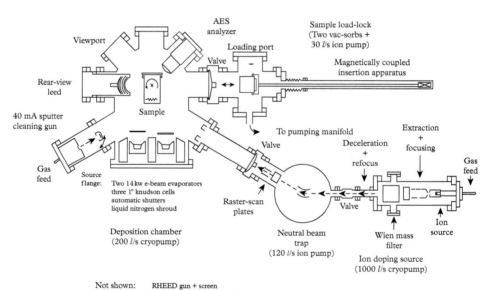

Silicon MBE system

(Vertical cross section, scale: |———|10 cm)

Figure 8.15 *Schematic diagram of Si MBE machine described by Bean and Sadowski (1982). Reproduced with permission from Bean, J C and Sadowski, E A (1982) J Vac Sci Technol 20, 137. Copyright 1982, AIP Publishing LLC.*

with 4-inch diameter substrates, a sputter cleaning gun, EB sources for Si and for metals (such as Co or Ni), and conventional Knudsen cells for dopants, together with an ion-beam source. Fluxes could be measured and regulated to an accuracy of ~1%, while substrate rotation allowed uniformity of deposition to better than 1% across a three inch wafer. Also included were facilities for LEED, RHEED, AES and residual gas analysis. Sample loading, deposition and unloading took no more than 45 min—as many as 8 runs could be completed within a normal working day. Compared with III-V MBE growth, the obvious difference was in the growth rate, which (according to Bean and Sadowski) could exceed $1/3$ μm/min (20 μm/h), some 20 times faster than typical GaAs rates. Such a dramatic difference tempts one to ask 'why?' The answer lies, of course, with the use of the EB source but probably also includes subtleties such as the fact that only one type of atom is concerned, making the surface chemistry correspondingly simpler. Albeit, such growth rates certainly improved the ability of Si MBE to challenge conventional CVD methods in the development of device technology. Other aspects were represented by the use of metal beams to deposit silicide layers and a capability to grow silicon films on insulating substrates such as sapphire and spinel (Bean 1981). In much the same manner as shown by Cho for III-V epitaxy, the Bell researchers had already demonstrated their ability to grow a range of Si device structures, such as microwave mixer diodes, varactor diodes, PIN diodes and both bipolar and MOS transistors.

It would, of course, be misguided to imply that only Bell Labs were keen to exploit the virtues of MBE to grow Si device structures. Silicon was already of worldwide interest and both AEG Telefunken and Hitachi declared their interest in MBE growth at about this time. Konig et al. (1979) reported on their own version of a would-be commercial MBE machine, employing Sb doping at levels between 3×10^{20} m^{-3} and 2×10^{25} m^{-3}. They confirmed the very sharp doping profiles observed by the Bell group and introduced a significant modification of their own with regard to substrate cleaning. They showed that heat-cleaning at 900 °C for 5 min was adequate to allow satisfactory film growth, rather than the previously used temperatures in the region of 1200 °C. This was designed to facilitate growth on pre-processed device layers, without damaging features already incorporated. Interestingly, they also referred to earlier work in which this machine had been used to grow monocrystalline SiGe superlattices as early as 1975, some ten years before the world became generally aware of the device possibilities associated with this material combination. Six years later (Kasper and Wörner 1985) they described the use of this machine in fabricating a high-speed integrated circuit, thus confirming the seriousness of their intent. Katayama et al. (1979) of Hitachi, in Tokyo described the use of MBE to make an n-MOS transistor on a p-Si substrate, using As as the n-type dopant. They claimed field-effect mobilities comparable with those measured on conventional bulk silicon structures. The same group (Ishizaka et al. 1982) later reported a technique for thermal cleaning of silicon substrates at the surprisingly low temperature of 750 °C, again with a view to combining MBE growth with silicon device processing. In an invited paper at MBE-CST-2 held in Tokyo in 1982 Yasuhiro Shiraki (1982) summarised recent developments in Japan, emphasising the development of 'Si-dedicated MBE machines which fit LSI processes', together with considerable emphasis on the application to device development. He also referred to important advances in ion doping achieved by Sugiura of ECL (Sugiura 1980) who obtained accurate control of Sb doping over the range $n = 10^{21} - 10^{26}$ m^{-3}. The ULVAC group (Japanese MBE machine manufacturers) had also looked at the use of ionised Si beams for Si MBE, finding a significant lowering of the growth temperature (Saitoh et al. 1981). Without spelling out all the details, it was clear that the potential of Si MBE was widely appreciated within the Japanese research community. On the other hand, looking at the subject subjectively for a moment, we note that there was no activity within the British Isles until 1985, coincident with the UK LDS Initiative (see Box 6.1 in Chapter 6). This was consistent with the remarkable lack of any coherent programme of silicon research within the UK, following the catastrophe of unhappy decision-making within the British Post Office during the 1950s and 1960s (see Orton 2004, p. 97).

Moving on, then, to the middle of the 1980s, we are fortunate that John Bean, who had been so largely responsible for the initial surge into Si MBE, was moved to write a brief review of progress up to the year 1986 (Bean 1987). Apparatus, substrate preparation, doping, epitaxial metals, heteroepitaxy and SiGe all come in for comment but we shall be less expansive in our treatment. One aspect which stands out, when one compares the situation with III-V MBE, is the emphasis on developing machines and techniques appropriate to commercial exploitation. Silicon technology was, of course, already well developed and it was clear that, if MBE were to establish a foothold on its rapidly shifting terrain, this would demand significant developments in convenience, throughput and reliability. What was more, such features were needed urgently. Substrate preparation was therefore subject to widespread study, particularly with a view to reducing cleaning temperatures below 800 °C to allow film growth on processed wafers. The Ishizaki procedure, referred to above, was further examined, as was the technique of ion-beam etching. Japanese workers proposed the use of a silicon beam to reduce the surface oxide to the monoxide SiO, which is highly volatile and therefore capable of removal at lower temperatures. A problem faced by everyone (in III-V MBE also) was the removal of residual carbon from the substrate surface, a problem dealt with by Tatsumi of NEC by the use of an ozone treatment. Very low defect densities could be achieved on (100) surfaces, though the (111) surface proved slightly less tractable.

Bearing in mind the undesirable variation of sticking coefficient of Al, Ga and In with growth temperature, efforts were made to provide a viable source of boron for p-type doping. Kubiak et al. from the City of London Polytechnic described a high-temperature Knudsen cell and used it to dope MBE films at levels between about 10^{21} m^{-3} and 10^{24} m^{-3}. Happily, the B sticking coefficient showed no variation with temperature. An alternative approach came from Ostrom and Allen at UCLA, using either B_2O_3 or B-doped Si, which could be evaporated at much lower temperatures than elemental B. They obtained doping levels in the range 2×10^{23} m^{-3} to 2×10^{26} m^{-3} with good control and no evidence of oxygen incorporation in the grown films. The technique of ion doping, involving, as it did, mass selection, focussing, raster scanning and differential pumping was seen as unduly complicated and work was therefore aimed at simplifying it. The groups at AEG Telefunken and at the City of London Polytechnic both succeeded in doping Si films simply by applying a bias to the substrate. It was speculated that a fraction of the Si atoms evaporated from the EB source were ionised and that these knocked dopant atoms on the film surface deeper into the film, thus preventing their re-evaporation. There was apparently no need to ionise the donors themselves, thus saving considerably in machine complexity. In another, somewhat bizarre initiative, Kubiak et al. uncovered an unexpected contribution to B doping from the use of boro-silicate glass viewports in their MBE chamber—replacing these with quartz reduced the effect by a factor of 50!

At this point we should make a brief pause to take note of the various applications of Si MBE to the better understanding of Si surfaces and Si crystal growth. In Chapter 4 (Section 4.4) we described the application (during the 1980s) of scanning tunnelling microscopy (STM) to GaAs surface studies so it should come as no surprise that similar work was concerned with silicon. Combined with measurement of RHEED images and RHEED intensity oscillations, STM contributed much towards the understanding of the fundamentals of Si crystal growth and we should be very remiss in ignoring the fact, as a result of our concentration on more 'practical' aspects of Si MBE. The reader with an interest in such fundamental studies will find much of interest in the NATO conference proceedings edited by Max Lagally (1990), a specially relevant paper being that by Sakamoto et al. on p. 263, which is also concerned with the growth of Ge on the Si (001) surface.

Leaving aside any further detail, it is without doubt that one of the most significant developments during the 1980s was the extension of the MBE technique to growing strained layers of SiGe on Si substrates, and, once again, Bell Labs was at the centre of the activity. Not only was this material system of great scientific interest but it also offered scope for the development of 2DEG structures and (somewhat later) hetero-junction bipolar transistors (HBTs). As we saw above, there had been earlier work at AEG in the mid 1970s which led to the growth of Si/SiGe superlattices and there had also been even earlier studies of evaporated films of Ge on Si substrates (e.g. Cullis and Booker 1971) but the understanding of such heterostructures took a sudden leap forward in the years 1984 et seq. It began in January 1984 when Bean et al. (1984b) published the first of a rich seam of 'Applied Physics Letters', reporting the growth of SiGe films on (100) Si substrates, cleaned by argon-ion sputtering and an 800 °C–850 °C anneal. They first deposited a 100 nm Si film to ensure a reproducible starting condition and then followed it with an appropriate SiGe film. Growth temperatures varied from 400 °C to 750 °C, and thicknesses from 10 nm to 250 nm. The mismatch between Ge and Si lattice constants is 4.2% so the initial SiGe film is always under in-plane compressional stress and, according to various well-known theories (such as that of Mathews and Blakeslee), one must expect pseudomorphic growth up to some critical thickness h_c, before the formation of misfit dislocations and onset of three-dimensional island growth, with consequent surface roughness. We should note, however, that in 1983, when the Bell team began their studies, self-organised InGaAs three-dimensional dot growth had still not been reported—there was therefore no role model to illuminate the two-dimensional/three-dimensional landscape. Perhaps unsurprisingly, then, Bean et al. looked at the situation from a slightly different viewpoint, that of atom surface mobility. They examined *thin* layers of SiGe as a function of growth temperature and showed that, at any specified temperature, smooth, 2D growth would occur for values of Ge fraction x up to a critical value x_c and found that x_c increased from 0.1 at $T_s = 750 °C$

to 1.0 at $T_s = 550\,°C$, an observation explicable on the understanding that at low temperatures atom surface mobility is too low to allow island formation—atoms stay more or less where they land on the substrate plateau, and a uniform beam flux produces a uniformly flat film. Of course, if T_s is low enough, atom mobility may be so restricted that impinging atoms are unable even to reach appropriate lattice sites, resulting in the formation of an amorphous film. In the work reported here, however, Rutherford backscattering revealed good crystallinity, consistent with pseudomorphic (i.e. strained-layer) growth.

In a paper published a few months later (Bean et al. 1984a) they reported measurement of the critical thickness h_c for these samples as a function of Ge content x and found them to be 'significantly larger than observed in earlier studies or predicted by theory'. This suggested a need to reconsider the theoretical models used to explain critical thickness and they proceeded to do just that (People and Bean 1985), making a new assumption concerning the energy balance which determined the generation of dislocations. Their model assumed that the substrate was free of threading dislocations and that 'the critical layer thickness [is] obtained by assuming interfacial misfit dislocations will be generated when the areal strain energy density of the film exceeds the self-energy of an isolated screw dislocation at a distance h from the free surface.' As can be seen from Figure 8.16, they obtained remarkably good agreement between this theory (with no adjustable parameters) and their earlier experimental data, in marked contrast with the predictions of previous theories. The implication was that the high quality of MBE growth on carefully prepared substrates allowed, for the first time, a reliable comparison to be made. It may have had little direct relevance to potential commercial applications but there can be no doubting its scientific significance.

Thinking of potential commercial interest brings us neatly to the next development, that of modulation doping (see Section 6.6). People et al. (1984) reported the first observation of a two-dimensional hole gas in a SiGe/Si strained-layer heterostructure. That it should be a hole gas rather than an electron gas was a consequence of the band alignment associated with this material system. It is of type I but with only a tiny conduction-band offset; nearly all the band-gap difference appears in the valence band. The structure grown by People et al. consisted of $Si_{0.8}Ge_{0.2}/Si$, for which the valence-band offset was $\Delta E_v \sim 100\,meV$. The Si layer was doped with B to a level of $10^{24}\,m^{-3}$, while the SiGe layer was nominally undoped. Samples were prepared both with and without a 10 nm set-back to move the ionised acceptors further away from the 2DHG. Temperature-dependent Hall mobilities for these two structures are compared in Figure 8.17 with a uniformly doped $Si_{0.8}Ge_{0.2}$ epilayer, showing the behaviour anticipated by comparison with the AlGaAs/GaAs 2DEG structures discussed in Chapter 6 (see Figure 6.26). The sample with offset shows a mobility tending to level off at low temperature, indicative of the absence of ionised impurity scattering. The maximum value of $0.33\,m^2\,V^{-1}s^{-1}$ was comparable with the best hole mobilities measured on Si/SiO_2 inversion layers. Low-temperature Schubnikov–de Haas measurements confirmed the existence of a 2DHG and yielded a value of free-carrier density $p_s \sim 3.5 \times 10^{15}\,m^{-2}$. Attempts to generate a 2DEG were unsuccessful as a consequence of the very small conduction-band offset.

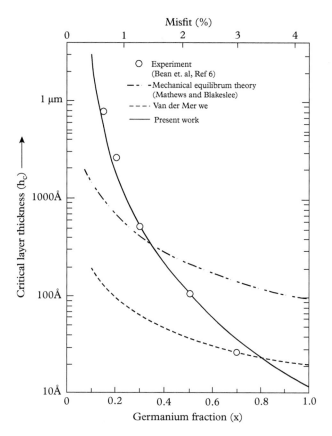

Figure 8.16 *Comparison between experimental data and theoretical predictions concerning the critical thickness h_c for dislocation generation in strained-layer $Si_{1-x}Ge_x/Si$ structures. Open circles represent the data, the solid line the theory presented by People and Bean, while the earlier theories of Van Der Merwe and by Mathews and Blakeslee are represented by the dashed and dot–dashed lines, respectively. (From People and Bean 1984.) Reproduced with permission from People, R and Bean, J C (1985) Appl Phys Lett 47, 322. Copyright 1985, AIP Publishing LLC.*

The fact that the room temperature electron mobility in lightly doped Si is roughly three times greater than that of holes suggested that it might well be advantageous to find a method of growing a 2DEG structure, which, in turn, demanded a heterojunction characterised by a much larger CB offset. For this important step we have to shift our attention back across the Atlantic to the AEG group. No sooner had People et al. reported their 2DHG than Jorke and Herzog (1985) described the first SiGe 2DEG—and

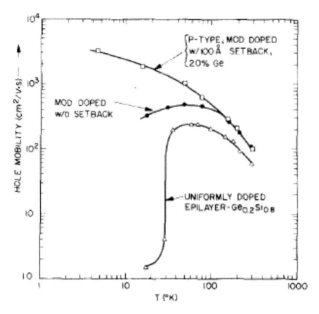

Figure 8.17 *Hole Hall mobilities measured on $Si_{0.8}Ge_{0.2}/Si$*
modulation-doped structures, with and without dopant
set-back, as a function of temperature. Results measured on
a uniformly doped sample of $Si_{0.8}Ge_{0.2}$ are shown for
comparison, illustrating the sharp drop in mobility due
to ionised impurity scattering. (From People et al. 1984.)
Reproduced with permission from People, R, Bean, J C, Lang,
D V, Sergent, A M, Stormer, H L, Wecht, K W, Lynch, R T
and Baldwin, K (1984) Appl Phys Let 45, 1231. Copyright
1985, AIP Publishing LLC.

it came with something of a surprise. Though they presented clear evidence of mobility enhancement, it was not immediately obvious how the necessary CB offset had arisen. The structure they employed consisted of a $Si_{0.55}Ge_{0.45}/Si$ superlattice, grown not directly on a Si substrate but on a thick $Si_{0.75}Ge_{0.25}$ buffer layer, itself grown on Si. The buffer thickness was considerably in excess of the critical thickness and was therefore relaxed, its lattice constant lying between those of the SL constituents. Crucially (as it turned out) this meant that, while the SiGe alloy was, as usual, under compressive stress, the Si was under tensile stress, this being the essential distinction between this and the earlier structures. It came to be known as a symmetrically strained structure and it was essential that both the Si and SiGe components of the SL should be thinner than the appropriate critical thickness, so as to retain pseudomorphic growth. The question arising concerned the nature of the CB offset—given that ΔE_c was negligibly small when the Si was unstrained, how, exactly, did the strain in the Si layers produce a significant offset?

The key (Abstreiter et al. 1985) came from their observation that mobility enhancement occurred only when the Sb dopant atoms were located within the SiGe layers, implying that the 2DEG lay within the Si component. In other words, the Si CB was lower in energy than that of the SiGe. As proposed in this paper, the tensile stress in the Si split the conduction-band states so that the twofold degenerate level was pushed downwards *below* the fourfold degenerate level in the $Si_{0.55}Ge_{0.45}$. Thus, electrons in the SiGe layers spilled over into the Si layers and the 2DEG was located on the Si side of the interface. Because the band gap of the SiGe layer was smaller than that of Si, this implied a type II heterostructure. Abstreiter et al. supported this idea with measurements of Raman spectra (to confirm the various strains) and Shubnikov–de Haas and cyclotron resonance measurements to confirm the nature of the 2DEG. Further confirmation of the model was later provided by People and Bean (1986) who reported calculations of band gap and band offsets as a function of material composition and uniaxial strain. (We provide a brief summary of all this in Box 8.3.)

Box 8.3 Band offsets in Si/SiGe strained-layer heterojunctions

---

The band gaps of silicon and germanium are 1.12 eV and 0.667 eV, respectively, (at room temperature) and their lattice parameters are 0.5431 nm and 0.5657 nm, respectively. Both elements have valence-band maxima at the Brillouin zone centre, while their conduction-band minima lie near the X point (Si) and at the L point (Ge). In the form of an unstrained alloy $Si_{1-x}Ge_x$, the gap varies according to the solid lines in Figure 8.18. As can be seen, the X-point

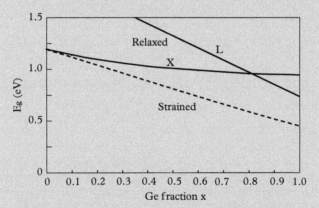

Figure 8.18 *Band gaps of Si/Ge alloys as a function of Ge content x. The solid lines represent the $\Gamma - L$ and $\Gamma - X$ gaps for unstrained alloys, the dashed curve represents the case of compressively strained alloys grown as thin ($h < h_c$) films on Si substrates.*

continued

Box 8.3 *continued*

band gap appropriate to Si decreases slowly (and bows slightly) with the addition of Ge, while the corresponding L-point gap decreases more rapidly, resulting in a crossover at approximately 82% Ge. If, on the other hand, the alloy is grown as a pseudomorphic thin film on a (100) silicon substrate with thickness less than the critical thickness h_c, the gap is considerably reduced and varies as shown by the dashed curve. We should first examine how this comes about.

Because of the 4.2% difference in lattice parameter, the Si/Ge alloy is under biaxial compressive strain in the plane of the film, while the interatomic distance is increased normal to the plane. Formally, this can be expressed in terms of a hydrostatic compression, together with a uniaxial tension along [001], and these strains produce both a shift and a splitting of the band-edge degeneracies. For example, the Si-like CB X valleys are sixfold degenerate (three distinct [100]-type axes, together with spin) and the hydrostatic term shifts these bodily downwards, while the uniaxial tension splits them into lower fourfold and upper twofold degenerate states. The overall effect is therefore to leave the fourfold X states lower than the original unstrained CB minimum. The unstrained VB maximum is characterised by heavy-hole, light-hole and split-off states and both hydrostatic and uniaxial strain components shift the heavy-hole state upwards. Thus, the band gap of the alloy $(E(X_4)-E_{hh})$ is reduced approximately in proportion to the Ge content x, as shown in Figure 8.18. While the uniaxial strain does not split the Ge-like L CB states (all three [111]-type axes make the same angle with the [001] direction), the hydrostatic term shifts them to higher energies so the band gap remains Si-like over the whole range of x-values. To a fair approximation, the difference in gap between the alloy and Si is given by

$$\Delta E_g(x) = 0.72x \, eV \tag{B8.6}$$

(For a quantitative account of the above the reader should refer to the excellent review article by Brunner (2002)).

Turning attention to the band offset between the unstrained Si substrate and the strained Si/Ge film, we have to appreciate that knowing the relevant band gaps tells us nothing about the band *alignments*—these are determined by the interaction between Si and Ge atoms at the interface. They have been calculated (see People and Bean 1986), confirming the observation made in the text that very nearly all the offset appears at the valence-band step. Though the alignment is just of type I, the CB step is very small. The VB step is therefore given roughly by equation (B8.6).

Let us now look at the case of the alloy $Si_{0.5}Ge_{0.5}$ grown on (a) a Si substrate and (b) on a relaxed $Si_{0.75}Ge_{0.25}$ substrate. In case (a) we can use equation (B8.6) to estimate the VB step as 0.36 eV, as shown in Figure 8.19(a). That is straightforward enough. In case (b), which corresponds approximately to the 2DEG structure grown by Jorge and Herzog (1985), we shall simply adopt a plausibility argument to explain the resulting band structure. The important point about this structure is that both the Si and the $Si_{0.5}Ge_{0.5}$ films are strained, the Si being under tensile and the alloy under compressional strain. Note, however, that the magnitudes of these strains are only half that of the compressional strain appropriate to case (a). Consider the Si film first: the hydrostatic component will shift the CB upwards in energy, while the axial component will lift the degeneracy, though in the opposite sense to that considered above. The doubly degenerate CB state will therefore be lowest but, because of the lower overall strain and the opposing tendencies of the two components, the net result will be only a small change in band gap. The effect on the VB will be to shift the light-hole band upwards to a modest extent so we can anticipate a slightly reduced Si band gap, the calculated value being 1.05 eV (see Figure 8.19(b)). On the other hand, the alloy band gap can be estimated from Figure 8.18 by assuming the strain effect will be only half that appropriate to such an alloy grown on a Si substrate. The result is $E_g \sim 0.9$ eV, and the type II band alignment is shown in Figure 8.19(b), which explains why the 2DEG is situated on the Si side of the heterojunction.

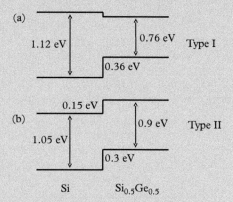

Figure 8.19 *Approximate band alignments for two examples of Si/Si$_{0.5}$Ge$_{0.5}$ heterojunctions. In case (a) the Si/Ge alloy is grown as a pseudomorphic thin film on a Si substrate, so the alloy is under biaxial compressional strain. In case (b) both the alloy and the Si are grown as thin films on a relaxed Si$_{0.75}$Ge$_{0.25}$ substrate, implying that the Si is under tensile and the alloy is under compressional strain.*

We might offer our own comment on the choice of structure for this first 2DEG. It would have been relatively straightforward to grow the SL on a bulk SiGe substrate—such crystals certainly existed. The argument in favour of using a relaxed buffer layer on a bulk Si substrate was based on the hope that SiGe MBE was destined to play a serious role in the future of Si technology (all this work had been performed, of course, in industrial laboratories). Because there was no likelihood of microelectronics ever being based on SiGe substrates, there was no point in demonstrating a 2DEG which relied on such a component—if MBE was to be taken seriously it must demonstrate compatibility with established technology. But, indeed, there was an obvious drawback—the fact that the buffer layer was relaxed implied that it contained misfit dislocations which were more than likely to propagate through the active layers and degrade their electronic performance. Could anything be done about it? Similar situations had been known before—compare the case of GaAsP light emitters, grown on GaAs substrates. The solution to that dilemma was to grade the epilayer and so it was in the case of the SiGe buffer layer. As an example of how this could influence the 2DEG electron mobility we quote the paper by Schaffler et al. (1992) (noting in passing that AEG Telefunken had, in the interim, become Daimler-Benz AG). From the low-temperature mobility of 0.23 $m^2 V^{-1}s^{-1}$ measured on the original sample, use of a carefully graded 3 μm buffer resulted in a value as high as 17.3 $m^2 V^{-1} s^{-1}$, very nearly two orders of magnitude improvement. Of course, the mobility enhancement at room temperature was far smaller—Schaffler et al. quote a room temperature mobility of roughly 0.18 $m^2 V^{-1}s^{-1}$ in their best sample, while (see Sze 1985) a typical Si MOSFET might show an electron mobility of about 0.08 $m^2 V^{-1} s^{-1}$. Later work demonstrated the possibility of achieving even higher 2DEG mobilities—for example, Ismail et al. (1994), using a carefully graded buffer layer obtained values as high as $\mu_e = 40 m^2 V^{-1} s^{-1}$. They claimed that their success lay in minimising the density of dislocations at the Si/SiGe interface. Similar attention to detail also led to much enhanced 2DHG mobilities. Xie et al. (1993), using a graded SiGe buffer layer and a thin Ge channel measured low-temperature hole mobilities as high as 5.5 $m^2 V^{-1}s^{-1}$. In a later review, Whall and Parker (2000) of Warwick University came to the conclusion that the most probable limitation to even higher mobilities lay in the realm of interface roughness and material inhomogeneity.

In conclusion, it is worth noting that both p-channel and n-channel high electron mobility transistors (HEMTs) were reported following almost immediately on the demonstration of these 2D gas structures (Pearsall et al. 1985 and Daembkes et al. 1985) but we shall leave further discussion until Chapter 9. The reader looking for a more detailed account of Si/SiGe superlattices, 2DEGs and 2DHGs will find an excellent account in Kasper and Schaffler (1991). We shall now (briefly) turn our attention to the heterojunction bipolar transistor (HBT).

The story of the HBT is a fascinating one and involved the unusual spectacle of Si technology chasing that based on III-V compounds, rather than vice versa. The concept goes back to a patent taken out by William Shockley in 1948, soon after the first observation of transistor action in 1947, together with important developments proposed by Herb Kroemer in the 1950s. The basic idea depends on the use of a narrow-gap base layer, sandwiched between wider-gap emitter and collector regions. A glance at Figure 8.19(a) will make clear the fact that the step in the valence band between an n-type Si emitter and a p-type SiGe base will readily allow electron flow into the base but will strongly oppose hole flow from base to emitter. This allows both emitter and base to be heavily doped, which minimises base resistance, while favouring minority carrier injection. A further advantage, suggested by Kroemer, depends on the use of a graded base layer to produce an electric field within the base, thus sweeping electrons rapidly to the collector, both these features favouring high frequency operation. With regard to materials, Shockley presumably had in mind the use of Si and Ge but, the necessary technology not being available, the HBT's first appearance was to depend on the more malleable III-V technology appropriate to the AlGaAs/GaAs material system. Such devices were demonstrated in the 1970s, some ten years before the first successful Si/SiGe device, and rapidly found application in microwave circuitry, values of cut-off frequency f_T reaching the prodigious heights of 300 GHz. It was a while before SiGe could catch up.

The first Si/SiGe HBT was reported by IBM in 1987 but this was rapidly followed by three others, one from Japan, one from Bell and one from Linkoping University in Denmark. All four groups published their findings in 1988 (Patton et al. 1988; Tatsumi et al. 1988; Temkin et al. 1988; Xu et al. 1988) in a flurry of competitive outpouring comparable with that last seen in 1962 when a similar number of papers had described the first observation of laser action in p–n junction diodes (see e.g. Orton 2004, Chapter 5). From our viewpoint, it was perhaps remarkable that all four had used MBE to grow their devices, n-Si/p-SiGe/nSi heterostructures. Clearly, the expertise in strained-layer epitaxy built up over the previous decade had persuaded all four protagonists that MBE was the best way to go but sadly that proved true only in the research environment. When it came to commercialisation, the natural preference of the silicon community for the chemically based CVD process led to the development of a low-temperature version which carried the day. MBE, in this instance could claim no more than the inspiration for all that followed—but, again, details must wait until Chapter 9.

Our next topic is concerned with an interesting modification to the properties of Si/SiGe heterostructures, resulting from the incorporation of small amounts of carbon. The lattice constant of diamond ($a_0 = 0.3567$ nm) is approximately 50% smaller than that of Si, which suggests that it may not be easy to form random alloys. Indeed, experimental studies tend to suggest that, while Si and Ge are miscible in all proportions, no more than 5% C can be incorporated before crystal quality is seriously degraded. There is a strong possibility of precipitating the compound SiC, particularly at higher growth temperatures. However, the large difference in lattice parameter between C and SiGe implies that even small amounts will have a significant effect in modifying the degree of strain in SiGe layers grown on Si and, what is more, C and Ge produce opposite effects. The possibility exists, therefore, of growing strain-free epilayers in the system $Si_{1-x-y}Ge_xC_y$. As a broad generalisation, we can observe that 5% C might be expected to balance the effect of about 50% Ge.

The first reported growth of $Si_{1-x}C_x$ alloys in 1990 made use of a modified CVD process but this was soon followed by extensive studies from IBM using solid source MBE (Eberl et al. 1992a, 1992b; Iyer et al. 1992; Tsang et al. 1992). They grew both $Si_{1-x}C_x$ and $Si_{1-x-y}Ge_xC_y$ alloys and showed that, as predicted, 1% C compensated the strain associated with approximately 10% Ge. There existed a fairly narrow growth temperature window between 400 °C and 550 °C—below 400 °C films were amorphous, while above 550 °C the compound SiC was produced and crystal quality was poor. Samples grown within the window showed no indication of compound SiC, while Raman measurements of lattice vibrational spectra confirmed that C was incorporated in substitutional lattice sites. Osten et al. (1994) introduced as much as 2% C into SiGe films on Si substrates and obtained SiGe films which were under *tensile* stress, without the need for a graded buffer layer. Lanzerotti et al. (1996b) described the use of C-doped SiGe in an HBT and stumbled upon a fascinating, and quite unexpected interaction between C and the B atoms used to dope the base p-type. It became clear that the introduction of the order of 10^{26} C atoms per cubic metre strongly inhibited out-diffusion of B from the base, which allowed the use of a very thin, highly doped base layer, thus further benefitting high frequency operation. They also noted (Lanzerotti et al. 1996a) that, though C tended to increase the alloy band gap, nevertheless, a SiGeC film was less strained than a simple SiGe film with the same band gap. While these authors grew their samples by CVD, numerous further studies made use of MBE with a view to understanding the mechanism of diffusion inhibition. For example, Osten et al. (1999) pointed out that C doping inhibited B diffusion which, otherwise, occurred as a result of a variety of device processing steps, such as ion-implantation and annealing while Rucker et al. (1999) showed evidence to support a model involving C–Si co-diffusion, which removed Si point defects from the B-doped region.

The interest in C doping has continued right up to the present day. We shall mention just two examples of recent work to illustrate progress. Both papers are concerned with a concept originating in the early 1990s, aimed at improving hole mobility in a PMOSFET. Following the outline presented in Box 8.3, it should be clear that, if a thin layer of Si is grown on a relaxed SiGe buffer, it will be under tensile strain and, in consequence, the light-hole valence band will be uppermost and, the hole effective mass being reduced, this will favour an increase in mobility. Such a device was demonstrated in 1993 by Nayak et al. (1993) from the University of California and they reported a hole mobility some 50% greater than measured on a standard PMOSFET. It follows that a similar concept can be envisaged by replacing the SiGe film with a similarly relaxed $Si_{1-x}C_x$ film, and several groups have pursued the idea. An important feature in growing such structures is the balance between the degree of strain introduced and the quality of the resulting material—there is little virtue in promoting the light hole if the resulting material is full of dislocations and point defects. Ishihara et al. (2007) therefore approached this dilemma by choosing to grow (by gas source MBE) the Si-C film in the form of a series of step-graded layers—increasing the C content in each step. They showed that the uppermost layer was not only more relaxed than when using a single layer but also contained far fewer defects. More recently, Arimoto et al. (2013) also used gas source MBE to study the growth of single $Si_{1-x}C_x$ films, emphasised the importance of growth temperature—they found that a very narrow window at 550 °C existed, within which relaxed material could be grown. AFM measurements then showed excellently smooth surfaces, there being no sign of the cross-hatch pattern, characteristic of relaxed SiGe layers. It therefore appears that there is still hope for practical realisation of high-mobility PMOSFETs based on $Si/Si_{1-x}C_x$ structures. We might also comment that, even though MBE may never be preferred in the silicon commercial world, it is still very much alive in the parallel world of silicon research.

Finally, we note that the SiGe alloy system is well suited to the self-assembled, Stranski–Krastanow type of quantum dot formation such as described earlier for InAs/GaAs, in Section 6.5, and for PbSe/PbEuTe in Section 8.2. The first observation of Ge dots followed closely on the heels of the discovery of InAs dots in the middle of the 1980s. It is interesting to recognise that, whereas the development of low-dimensional structures in the III-V compounds was initially based on the lattice-matched AlAs/GaAs pairing, Si/Ge epitaxy was obliged to cope immediately with a considerable lattice mismatch. While AlAs and GaAs had a matter of ten years in which to get used to living together, before their offspring InAs introduced strain into the epitaxial relationship, strain was inherent in Si/Ge LDS from the very beginning. We have already referred to the Bell studies of critical layer thickness (People and Bean 1985) and their concern with the nature of island growth when the overall thickness of the Ge layer exceeded this limit. While it was widely recognised that strain relief might imply the generation of misfit dislocations, which could degrade the crystal quality of the resulting quantum dots, an alternative model had been put forward (in the case of InAs/GaAs dots) by Berger et al. (1988) and which involved a rippling of the growing surface so as to allow lattice-matching of the dot without the need for dislocations. Such an arrangement would clearly be of benefit to the optical and electrical properties of dots so formed. Interestingly, just such a structure was observed in the Si/Ge system by Eaglesham and Cerullo (1990). They grew Ge dots at 500 °C on Si buffer layers on Si substrates by solid source MBE and studied their microstructure by TEM. Dots were approximately 40 nm in diameter, spaced roughly 200 nm apart (a density of 2.5×10^{13} m^{-2}) and showed no evidence for the presence of dislocations. They pointed out, though, that this mechanism of dislocation-free dot formation applied only to such low densities—any significant increase in density would lead to the more common dislocation-mediated strain relaxation. They also demonstrated the onset of dislocations in thicker Ge layers, in which dot diameters were as large as 0.14 μm. We should probably add the comment that island formation in MBE depends critically on growth temperature. As we have remarked previously, low temperatures imply low diffusion coefficients for surface atoms, while high temperatures imply diffusive intermixing of Si and Ge atoms, thus modifying dot constitution. Effusive loss from the surface is also significant, which implies that growth rate may also be relevant.

Apart from the considerable interest in dot formation as a means of studying crystal growth mechanisms, it was clear that there might be important commercial applications of Ge dots, on account of their obvious compatibility with Si integrated circuit technology. In particular, there was hope that high densities of dots could form the basis for efficient electroluminescence (LEDs) or sensitive IR photodetectors. Both the near-IR communications wavelengths in the region of 1.55 μm and the longer thermal detector wavelength windows at 3–5 μm and 8–12 μm were targeted. As an example (selected at random from a vast literature!), we cite the 'invited paper' by Wang et al. from the University of California (Wang et al. 2007), which gives a helpful account of progress towards these ambitious aims, but it might be well, first, to emphasise one or two important criteria.

With regard to possible LEDs, the hope was that radiative recombination within a dot might be significantly more probable than that in bulk Ge (or in Ge quantum wells), the inherent difficulty being the indirect band gap, with its implied need for phonon co-operation. An initial thought in this direction was the suggestion that close confinement of free carriers in a small Ge dot could enhance the radiative rate in much the same manner as observed for N-doped GaP (high precision in spatial coordinate implying a wide spread in momentum, thereby allowing a pseudo-direct optical transition). This was considered theoretically by Takagahara and Takeda (1992), who showed that it would be necessary to grow dots with dimensions as small as 2 nm, whereas typical practical values lie in the range 5–10 nm high by 50–100 nm diameter. What is more, it would be necessary to employ high dot areal densities and thick multilayers to obtain the light intensities essential for practical application. Another hope lay in the confinement of phonon modes as a possible means for modifying radiative and non-radiative recombination rates. Finally, we might make the obvious point that a high density of dislocations within the dots could only be detrimental to radiative efficiency.

Photodetectors operating in the region of 1.55 μm wavelength would be dependent on interband absorption between holes confined in the Ge valence-band potential well and free electrons in the Si conduction band. Again, the spectre of the indirect gap threatens the efficiency of photon absorption and implies the need for a thickness of absorbing material much greater than a micron. Detection at longer wavelengths implies the use of intraband transitions (holes being ejected from confined states into the VB continuum) so the nature of the band gap is no longer relevant. Whether Ge dot technology can truly rival that of MCT in thermal imaging applications still remains to be seen but much effort has been expended in attempts to find out. Let us now look briefly at the details, as outlined by Wang et al.

They described the growth of self-assembled Ge dots on (100) Si by MBE and, in particular, demonstrated the importance of substrate temperature. In samples grown at 550 °C AFM images showed the presence of two distinct sets of dots, large multifaceted domes with diameters close to 70 nm and heights of about 15 nm, together with smaller, square-based pyramids. The ratio R of pyramids to domes was a strong function of growth temperature, showing a sharp minimum at $T_S = 600$ °C, when R ~ 5%. This was obviously the condition which resulted in the optimum uniformity in dot size, at which point the areal density was approximately 10^{13} m^{-2}, a relatively modest figure. To improve both density and uniformity, the authors made use of a patterned substrate, obtained by the use of a 'diblock copolymer', on top of a thin SiO$_2$ film. This latter involved spinning on a mixture of polystyrene and polymethylmethacrylate (PS-PMMA) and annealing at 170 °C to separate the two components. UV light was used to cross-link the PS, and the PMMA was then removed in acetic acid, allowing localised removal of the SiO$_2$ to expose the Si beneath. Ge was then grown in the holes to produce a high density (~ 10^{15} m^{-2}) of dots with diameter close to 30 nm and height of 5 nm. Finally, the SiO$_2$ mask material was etched away, to leave the dot array, which could then be capped with a layer of Si, thus allowing the possibility for growing dot multilayers.

At this point we interject a brief comment concerning a quite different approach to obtaining high densities of small dots. This was reported in 1999 by a group from Switzerland (Leifeld et al. 1999) who grew self-organised Ge dots on (100) Si substrates on which had been deposited a small fraction of a monolayer of C.* Compared to similar deposition on bare Si surfaces, the resulting dots, when grown at temperatures between 450 °C and 550 °C were found to be smaller (~20 nm diameter) and of much greater density (~ 10^{15} m^{-2}) but suffering from considerably greater irregularity. The authors

---

* The reader will be well aware of the irony—MBE growers had spent sleepless nights devising methods of removing C from substrate surfaces—now we see yet another reason for welcoming it back!

offered an explanation, consistent with RHEED and STM measurements, based on a surface arrangement of six C atoms positioned between two Si dimers.

Retracing our steps to the Wang et al. paper, the authors did report multilayer growth of Stranski–Krastanov dots and showed the vertical correlation of dot formation which we described earlier for InAs and PbSe dots but they discovered an aspect of importance for practical application, namely that there was a limit to the number of layers which could be grown before the onset of dislocation generation. They grew at 540 °C, using nominal thicknesses of 1.5 nm for Ge and 20 nm for Si. Up to 35 periods could be grown free of dislocations but a 50 period sample showed high densities of dislocations associated with the 25th layer. They also found a clear correlation with the intensity of photoluminescence. Figure 8.20 shows PL spectra from a set of multilayer Ge dot samples, the number of layers being indicated at the left of each spectrum. The sharp lines at photon energies above 1.0 eV originate from Si and are well known. The broader lines at about 0.95 eV are from Ge quantum wells, while the single broad line at 0.75 eV is from Ge dots. The precise photon energy was shown to vary slightly according to the height of the dot—that is, according to the degree of quantum confinement. As can be seen in Figure 8.20, the intensity of the Ge dot emission increases with the number of layers but collapses at 50 layers on account of the dislocation count.

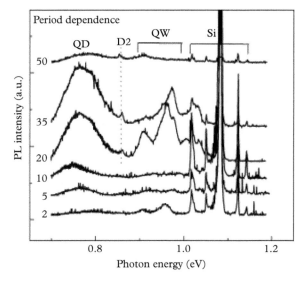

Figure 8.20 *PL spectra, measured at 4 K, from a series of Ge quantum dot multilayer samples. The number of layers is shown at the left. The sharp lines above 1.0 eV come from Si, and the broader lines near 0.95 eV are from Ge quantum wells, while the broad line at 0.75 eV is from Ge dots. Note that, for the 50 layer sample, the PL efficiency is much reduced by a high density of dislocations. (From Wang et al. 2007, courtesy IEEE).*

Following these PL measurements, Wang et al. demonstrated their ability to make LEDs emitting over the wavelength range 1.3 –1.6 μm by inserting an undoped ten-layer Ge dot structure between n⁺ and p⁺ Si contacts but they were obliged to admit that the quantum efficiency was probably too low to excite much interest amongst the practitioners of fibre-optic communications. They also demonstrated a p-i-n Ge dot photodiode with a response over the range 1.3–1.5 μm, with peak efficiency of 8%, comparable with results from earlier work on a strained-layer superlattice diode. In this mode, the absorbed photons generate holes in confined states which must then be thermally emitted into the Ge VB in order to contribute to the photocurrent. Biasing the diode increases sensitivity by way of the quantum confined Stark effect.

To complete the gamut of devices, they also described both p⁺-p-p⁺ and n⁺-n-n⁺ long-wavelength photodetectors. The former device showed sensitivity in the 3–5 μm band and operated by the excitation of holes in the Ge dots between confined hole states and the Ge VB continuum. It offered an important advantage over similar quantum well devices in that selection rules for dots allow a response to normal incidence light, which simplifies the design considerably. The n⁺-n-n⁺ device depended on the use of delta-doping in the Ge dots to provide confined electron states. It showed sensitivity across both the 3–5 μm and 8–12 μm bands. Further to this, an interesting paper was published recently by a Russian group from Novosibirsk (Yakimov et al. 2013), describing a similar long-wavelength detector based on ten layers of Ge dots grown by MBE on a $Si_{0.6}Ge_{0.4}$ pseudo-substrate. The dots were small, 10–15 nm in diameter, 1.0–1.5 nm in height and with a density $\sim 3 \times 10^{15}$ m⁻² and the layers were separated by 35 nm of $Si_{0.6}Ge_{0.4}$. The dot structure was contained between p⁺ (B-doped) $Si_{0.6}Ge_{0.4}$ contact layers and the dots themselves were doped p-type by inserting a delta-doping layer of B within each of the spacer layers, designed so as to supply approximately three holes to each dot. The detector showed background-limited performance when cooled to 100 K and could be tuned (by varying the bias voltage) to respond in either the 3–5 μm or the 8–12 μm band. Such performance, allied to silicon-compatibility, certainly promises to be of real commercial interest.

By way of a postscript to this brief account of SiGe strained-layer nanostructures, we might advise the reader interested in gaining a deeper understanding to consult the excellent reviews by Brunner (2002) and by Berbezier et al. (2002).

REFERENCES

Abbink, H C, Broudy, R M and McCarthy, G P (1968) Surface processes in the growth of silicon on (111) silicon in ultrahigh vacuum. J Appl Phys 39, 4673.
Abstreiter, G, Brugger, H, Wolf, T, Jorke, H and Herzog, H J (1985) Strain-induced two-dimensional electron gas in selectively doped Si/Si_xGe_{1-x} superlattices. Phys Rev Lett 54, 2441.
Akimoto, K, Miyajima, T and Mori, Y (1989) Electroluminescence in an oxygen-doped ZnSe pn junction grown by molecular beam epitaxy. Jap J Appl Phys 28, L531.

Arimoto, K, Furakawa, H, Yamanaka, J, Yamamoto, C, Nakagawa, K, Usami, N, Sawano, K and Shiraki, Y (2013) Formation of compressively strained $Si/Si_{1-x}C_x/Si(100)$ heterostructures using gas-source molecular beam epitaxy. J Cryst Growth 362, 276.

Aven, M and Prener J S (eds) (1967) 'Physics and Chemistry of II-VI Compounds', North Holland, Amsterdam.

Bean, J C (1978) Arbitrary doping profiles produced by Sb-doped Si MBE. Appl Phys Lett 33, 654.

Bean, J C (1981) Silicon molecular beam epitaxy. J Vac Sci Technol 18, 769.

Bean, J C (1987) Silicon molecular beam epitaxy: 1984–1986. J Cryst Growth 81, 411.

Bean, J C, Feldman, L C, Fiory, A T, Nakahara, S and Robinson, I K (1984a) Ge_xSi_{1-x}/Si strained-layer superlattice grown by molecular beam epitaxy. J Vac Sci Technol A 2, 436.

Bean, J C and Sadowski, E A (1982) Silicon MBE apparatus for uniform high-rate deposition on standard format wafers. J Vac Sci Technol 20, 137.

Bean, J C, Sheng, T T, Feldman, L C, Fiory, A T and Lynch, R T (1984b) Pseudomorphic growth of Ge_xSi_{1-x} on silicon by molecular beam epitaxy. Appl Phys Lett 44, 102.

Becker, G E and Bean, J C (1977) Acceptor dopants in silicon molecular-beam epitaxy. J Appl Phys 48, 3395.

Benz, R G II, Huang, P C, Stock, S R and Summers, C J (1988) Molecular beam epitaxial growth and structural characterization of ZnS on (001) GaAs. J Cryst Growth 86, 303.

Benz, R G, Wagner, B K, Rajavael, D and Summers, C J (1991) Chemical beam epitaxy of CdTe, HgTe, and HgCdTe. J Cryst Growth 111, 725.

Berbezier, I, Ronda, A and Portavoce, A (2002) SiGe nanostructures: new insights into growth processes. J Phys: Condens Matter 14, 8283.

Berger, P R, Chang, K, Bhattacharya, P, Singh, J and Bajaj, K K (1988) Role of strain and growth conditions on the growth front profile of $In_xGa_{1-x}As$ on GaAs during the pseudomorphic growth regime. Appl Phys Lett 53, 684.

Bicknell, R N, Giles, N C and Schetzina, J F (1986) p-type CdTe epilayers grown by photoassisted molecular beam epitaxy. Appl Phys Lett 49, 1735.

Boukerche, M, Wijewarnasuriya, P S, Sivananthan, S, Sou, I K, Kim, Y J, Mahavadi, K K and Faurie, J P (1988) The doping of mercury cadmium telluride grown by molecular-beam epitaxy. J Vac Sci Technol A 6, 2830.

Brunner, K (2002) Si/Ge nanostructures. Rep Prog Phys 65, 27.

Butler, J F, Calawa, A R and Rediker, R H (1964) 'PbSe Diode Laser' in 'Electron Devices Meeting, 1964 International', Vol 10, p 36.

Capper, P (1991) A review of impurity behavior in bulk and epitaxial $Hg_{1-x}Cd_xTe$. J Vac Sci Technol B 9, 1667.

Chang, J H, Takai, T, Koo, B H, Song, J S, Handa, T and Yao, T (2001) Aluminum-doped n-type ZnTe layers grown by molecular-beam epitaxy. Appl Phys Lett 79, 785.

Chen, Y, Bagnall, D M, Koh, H, Park, K, Hiraga, H, Zhu, Z and Yao, T (1998) Plasma assisted molecular beam epitaxy of ZnO on c-plane sapphire: growth and characterization. J Appl Phys 84, 3912.

Chen, Y, Bagnall, D M, Zhu, Z, Sekiuchi, T, Park, K, Hiraga, K, Yao, T, Koyama, S, Shen, M Y and Goto, T (1997) Growth of ZnO single crystal thin films on c-plane (0 0 0 1) sapphire by plasma enhanced molecular beam epitaxy. J Cryst Growth 181, 165.

Cho, A Y and Arthur, J R (1975) Molecular beam epitaxy. Prog Sol St Chem 10, 157.

Chu, S, Lim, J H, Mandalapu, L J, Yang, Z, Li, L and Liu, J L (2008a) Sb-doped p-ZnO/Ga-doped n-ZnO homojunction ultraviolet light emitting diodes. Appl Phys Lett 92, 152103.

Chu, S, Olmedo, M, Yang, Z, Kong, J and Liu, J (2008b) Electrically pumped ultraviolet ZnO diode lasers on Si. Appl Phys Lett 93, 181106.

Cook, J W, Eason, D B, Vaudo, R P and Schetzina, J F (1992) Molecular-beam epitaxy of ZnS using an elemental S source. J Vac Sci Technol B 10, 901.

Cullis, A G and Booker, G R (1971) The epitaxial growth of silicon and germanium films on (111) silicon surfaces using UHV sublimation and evaporation techniques. J Cryst Growth 9, 132.

Daembkes, H, Herzog, H J, Jorke, H, Kibbel, H and Kasper, E (1985) 'Fabrication and Properties of n-Channel SiGe/Si Modulation Doped Field-Effect Transistors Grown by MBE' in 'Electron Devices Meeting, 1985 International', Vol 31, p 768.

Dai, N, Brown, F, Barsic, P, Khodaparast, G A, Doezema, R E, Johnson, M B, Chung, K, Goldammer, K J and Santos, M B (1998) Observation of excitonic transitions in InSb quantum wells. Appl Phys Lett 73, 1101.

Dean, P J, Stutius, W, Neumark, G F, Fitzpatrick, B J and Bhargava, R N (1983) Ionization energy of the shallow nitrogen acceptor in zinc selenide. Phys Rev B 27, 2419.

Dean, P J, Williams, G M and Blackmore (1984) Novel type of optical transition observed in MBE grown CdTe. J Phys D: Appl Phys 17, 2291.

Dziawa, P, Sadowski, J, Dluzewski, P, Lusakowska, E, Domukhovski, V, Taliashvili, B, Wojiechowski, T, Baczewski, L T, Bukala, M, Galicka, M, Bukzko, R, Kacman, P and Story, T (2010) Defect free PbTe nanowires grown by molecular beam epitaxy on GaAs(111)B substrates. Cryst Growth Des 10, 109.

Eaglesham, D J and Cerullo, M (1990) Dislocation-free Stranski-Krastanow growth of Ge on Si(100). Phys Rev Lett 64, 1943.

Eason, D B, Yu, Z, Hughes, W C, Roland, W H, Boney, C, Cook, J W, Schetzina, J F, Cantwell, G and Harsch, W C (1995) High-brightness blue and green light-emitting diodes. Appl Phys Lett 66, 115.

Eberl, K, Iyer, S S, Tsang, J C, Goorsky, M S and Legoues, F K (1992a) The growth and characterization of $Si_{1-y}C_y$ alloys on Si(001) substrate. J Vac Sci Technol B 10, 934.

Eberl, K, Iyer, S S, Zollner, S, Tsang, J C and Legoues, F K (1992b) Growth and strain compensation effects in the ternary $Si_{1-x-y}Ge_xCy$ alloy system. Appl Phys Lett 60, 3033.

Farrow, R F C (1985) MBE growth of CdTe, $Hg_{1-x}Cd_xTe$, and multilayer structures: achievements, problems, and prospects. J Vac Sci Technol A 3, 60.

Farrow, R F C, Jones, G R, Williams, G M, Sullivan, P W, Boyle, W J O and Wotherspoon, J T M (1979) The vaporisation of $Hg_{1-x}Cd_xTe$ crystals-a case of gross incongruency. J Phys D: Appl Phys 12, L117.

Farrow R F C, Jones, G R, Williams, G M and Young, I M (1981) Molecular beam epitaxial growth of high structural perfection, heteroepitaxial CdTe films on InSb (001). Appl Phys Lett 39, 954.

Faurie, J P and Million, A (1981a) J Cryst Growth 54, 577.

Faurie, J P and Million, A (1981b) Molecular beam epitaxy of II–VI compounds: $Cd_xHg_{1-x}Te$. J Cryst Growth 54, 582.

Faurie, J P and Million A (1982) $Cd_xHg_{1-x}Te$ n-type layers grown by molecular beam epitaxy. Appl Phys Lett 41, 264.

Faurie, J P, Million, A, Boch, R and Tissot, J L (1983) Latest developments in the growth of $Cd_xHg_{1-x}Te$ and CdTe–HgTe superlattices by molecular beam epitaxy. J Vac Sci Technol A 1, 1593.

Faurie, J P, Million, A and Piaguet, J (1982) CdTe-HgTe multilayers grown by molecular beam epitaxy. Appl Phys Lett 41, 713.

Faurie, J P, Reno, J, Sivananthan, S, Sou, I K, Chu, X, Boukerche, M and Wijewarnasuriya, P S (1986) Molecular beam epitaxial growth and characterization of HgCdTe, HgZnTe, and HgMnTe on GaAs(100). J Vac Sci Technol, A 4, 2067.

Faurie, J P, Sporken, R, Sivananthan, S and Lange, M D (1991) New development on the control of homoepitaxial and heteroepitaxial growth of CdTe and HgCdTe by MBE. J Cryst Growth 111, 698.

Foxon, C T, Joyce, B A, Farrow, R F C and Griffiths, R M (1974) The identification of species evolved in the evaporation of III-V compounds. J Phys D: Appl Phys 7, 2422.

Gaines, J M, Petruzzello, J and Greenberg, B (1993) Structural properties of ZnSe films grown by migration enhanced epitaxy. J Appl Phys 73, 2835.

Goltsos, W, Nakahara, J, Nurmikko, A V and Partin, D L (1985) Photoluminescence in PbTe-PbEuTeSe multiquantum wells. Appl Phys Lett 46, 1173.

Guldner, Y, Bastard, G, Vieren, J P, Voos, M, Faurie, J P and Million, A (1983) Magneto-optical investigations of a novel superlattice: HgTe-CdTe. Phys Rev Lett 51, 907.

Haase, M A, Cheng, H, DePuydt, J M and Potts, J E (1990) Characterization of p-type ZnSe. J Appl Phys 67, 448.

Haase, M A, Qiu, J, DePuydt, J M and Cheng, H (1991) Blue-green laser diodes. Appl Phys Lett 59, 1272.

Han, J and Gunshor, R L (1997) 'MBE Growth and Electrical Properties of Wide Bandgap ZnSe-based II–VI Semiconductors' in 'Semiconductors and Semimetals', Vol 44 (ed R K Willardson, A C Beer and E R Weber), Academic Press, New York, p 1.

Han, J, Ringle, M D, Fan, Y and Gunshore, R L (1994) D (donor) X center behavior for holes implied from observation of metastable acceptor states. Appl Phys Lett 65, 3230.

Haseyama, K, Sugiura, O, Haga, R, Sugahara, S and Matsumura, M (1991) Newly designed Hg cell for molecular beam epitaxy growth of CdHgTe. Jap J Appl Phys 30, L202.

Holloway, H and Jesion, G (1982) Lead strontium sulfide and lead calcium sulfide, two new alloy semiconductors. Phys Rev B 26, 5617.

Holloway, H, Logothetis, E M and Wilkes, E (1970) Epitaxial growth of lead tin telluride. J Appl Phys 41, 3543.

Holloway, H and Walpole, J N (1980) 'MBE Techniques for IV-VI Optoelectronic Devices' in 'Molecular Beam Epitaxy' (ed B R Pamplin), Pergamon Press, Oxford, p 49.

Horikoshi, Y (1985) 'Semiconductor Lasers with Wavelengths Exceeding 2 μm' in 'Semiconductors and Semimetals', Vol 22C (ed R K Willardson and A C Beer), Academic Press, New York, p 93.

Ichino, K, Kotani, A, Tanaka, H and Kawai, T (2010) ZnS:N and ZnS:N, Ag grown by molecular beam epitaxy. Phys Stat Sol (c) 7, 1504.

Ichino, K, Matsuki, Y, Lee, S T, Nishikawa, T, Kitagawa, M and Kobayashi, H (2004) Li-doped p-type ZnS grown by molecular beam epitaxy. Phys Stat Sol (c) 1, 710.

Ichino, K, Yoshida, H, Kawai, T, Matsumoto, H and Kobayashi, H (2008) J Korean Phys Soc 53, 2939.

Iida, S, Yatabe, T and Kinto, H (1989) P-type conduction in ZnS grown by vapor phase epitaxy. Jap J Appl Phys 28, L535.

Ishibashi, A (1996) II–VI blue-green light emitters. J Cryst Growth 159, 555.

Ishihara, H, Murano, M, Yamada, A and Konagai, M (2007) Growth of strain-relaxed $Si_{1-y}C_y$ films with compositionally graded buffer layers by gas source molecular beam epitaxy. Jap J Appl Phys 46, 1600.

Ishizaka, A, Nakagawa, K and Shiraki, Y (1982) 'Low-Temperature Surface Cleaning of Silicon and Its Application to Silicon MBE' in 'Collected Papers of the 2nd International Symposium on Molecular Beam Epitaxy and Related Clean Surface Techniques, 27–30, August 1982, Tokyo', The Japan Society of Applied Physics, Tokyo, p 183.

Ismail, K, LeGouse, F K, Saenger, K L, Arafa, M, Chu, J O, Mooney, P M and Meyerson, B S (1994) Identification of a mobility-limiting scattering mechanism in modulation-doped Si/SiGe heterostructures. Phys Rev Lett 73, 3447.

Iyer, S S, Eberl, K, Powell, A R and Ek, B A (1992) SiCGe ternary alloys—extending Si-based heterostructures. Microelectron Eng 19, 351.

Janotti, A and Van de Walle, C G (2009) Fundamentals of zinc oxide as a semiconductor. Rep Prog Phys 72, 126501.

Jeon, H, Ding, J, Patterson, W, Nurmikko, A V, Xie, W, Grillo, D C, Kobayashi, M and Gunshor, R L (1991) Blue-green injection laser diodes in (Zn,Cd)Se/ZnSe quantum wells. Appl Phys Lett 59, 3619.

Jobst, B, Hommel, D, Lunz, U, Gerhard, T and Landwehr, G (1996) E0 band-gap energy and lattice constant of ternary $Zn_{1-x}Mg_xSe$ as functions of composition. Appl Phys Lett 69, 97.

Johnson, M A L, Fujita, S, Rowland, W H, Hughes, W C, Cook, J W and Schetzina, J F (1996) MBE growth and properties of ZnO on sapphire and SiC substrates. J Electronic Matls 25, 855.

Jona, F (1966) Study of the early stages of the epitaxy of silicon on silicon. Appl Phys Lett 9, 235.

Jorke, H and Herzog, H J (1985) in 'Proceedings of the 1st International Symposium on Silicon Molecular Beam Epitaxy' (ed J C Bean), Electrochemical Society, Pennington, New Jersey, p 352.

Jung, H D, Song, C D, Wang, S Q, Arai, K, Wu, Y H, Zhu, Z, Yao, T and Katayma-Yoshima, H (1997) Carrier concentration enhancement of p-type ZnSe and ZnS by codoping with active nitrogen and tellurium by using a δ-doping technique. Appl Phys Lett 70, 1143.

Kanie, H, Araki, H, Ishizaka, K, Ohta, H and Murakami, S (1994) Epitaxial growth of ZnS on GaP by molecular beam deposition. J Cryst Growth 138, 145.

Kasap, S and Capper, P (eds) (2007) 'Springer Handbook of Electronic and Photonic Materials', Springer Science & Business Media, New York.

Kasper, E and Schäffler, F (1991) 'Group-IV Compounds' in 'Semiconductors and Semimetals', Vol 33 (ed TP Pearsall, R K Willardson and A C Beer), Academic Press, New York, p 223.

Kasper, E and Wörner, K (1985) High speed integrated circuit using silicon molecular beam epitaxy (Si-MBE). J Electrochem Soc 132, 2481.

Katayama, Y, Shiraki, Y, Kobayashi, K L I, Komatsubara, K F and Hashimoto, N (1979) An MOS field-effect transistor fabricated on a molecular-beam epitaxial silicon layer. Appl Phys Lett 34, 740.

Kitagawa, F, Mishima, T and Takahashi, K (1980) Molecular beam epitaxial growth of ZnTe and ZnSe. J Electrochem Soc 127, 937.

Kittel, C (1967) 'Introduction to Solid State Physics', Wiley, New York.

Klingshirn, C L, Meyer, B K, Waag, A, Hoffman, A and Geurts, J (2010) 'Zinc Oxide from Fundamental Principles towards Novel Applications', Springer Series in Materials Science 120, Springer-Verlag, Berlin.

Konig, U, Kibbel, H and Kaspar, E (1979) Si–MBE: growth and Sb doping. J Vac Sci Technol 16, 985.

Kröger, F A and Vink, H J (1954) The origin of the fluorescence in self-activated ZnS, CdS, and ZnO. J Chem Phys 22, 250.

Kruse, P W (1981) 'The Emergence of $Hg_{1-x}Cd_xTe$ as a Modern Infrared Sensitive Material' in 'Semiconductors and Semimetals', Vol 18 (ed R K Willardson and A C Beer), Academic Press, New York, p 1.

Lagally, M G (Ed) (1990) 'Kinetics of ordering and growth at surfaces', NATO ASI Series B: Physics Vol. 239, Plenum Press, New York.

Lanzerotti, L D, Amour, A St, Liu, C W, Sturm, J C, Watanabe, J K and Theodore, ND (1996a) $Si/Si/_{1-xy}/Ge_xC_y/Si$ heterojunction bipolar transistors. IEEE Electron Device Letters 17, 334.

Lanzerotti, L, Sturm, J C, Stach, E, Hull, R, Buyuklimanli, T and Magee, C (1996b) 'Suppression of Boron Outdiffusion in SiGe HBTs by Carbon Incorporation' in 'Electron Devices Meeting, 1996. IEDM '96., International', p 249.

Larach, S, Shrader, R E and Stocker, C F (1957) Anomalous variation of band gap with composition in zinc sulfo-and seleno-tellurides. Phys Rev 108, 587.

Lawson, W D, Nielsen, S, Putley, E H, and Young, A S (1959) Preparation and properties of HgTe and mixed crystals of HgTe-CdTe. J Phys Chem Solids 9, 325.

Leifeld, O, Hartmann, R, Müller, E, Kaxiras, E, Kern, K and Grützmacher, D (1999) Self-organized growth of Ge quantum dots on Si(001) substrates induced by sub-monolayer C coverages. Nanotechnology 10, 122.

Look, DC (2005) Electrical and optical properties of p-type ZnO. Semicond Sci Technol 20, S55.

Look, D C, Reynolds, D C, Litton, C W, Jones, R L, Eason, D B and Cantwell, G (2002) Characterization of homoepitaxial p-type ZnO grown by molecular beam epitaxy. Appl Phys Lett 81, 1830.

Lovett, D R (1977) 'Semimetals and Narrow Bandgap Semiconductors', Pion Ltd, London.

Mandalapu, L J, Yang, Z, Chu, S and Liu, J L (2008) Ultraviolet emission from Sb-doped p-type ZnO based heterojunction light-emitting diodes. Appl Phys Lett 92, 122101.

Mandalapu, L J, Yang, Z, Xiu, F X, Zhao, D T and Liu, J L (2006) Homojunction photodiodes based on Sb-doped p-type ZnO for ultraviolet detection. Appl Phys Lett 88, 092103.

Merz, J L, Kukimoto, H, Nassau, K and Shiever, J W (1972) Optical Properties of Substitutional Donors in ZnSe. Phys Rev B 6, 545.

Micklethwaite, W F H (1981) 'The Crystal Growth of Cadmium Mercury Telluride' in 'Semiconductors and Semimetals', Vol 18 (ed R K Willardson and A C Beer), Academic Press, New York, p 47.

Nakahara, K, Akasaka, S, Yuji, H, Tamura, K, Fujii, T, Nishimoto, Y, Takamizu, D, Sasaki, A, Tanabe, T, Takasu, H, Amaike, H, Onuma, T, Chichibu, S F, Tsukazaki, A, Ohtomo, A and Kawasaki, M (2010) Nitrogen doped $Mg_xZn_{1-x}O/ZnO$ single heterostructure ultraviolet light-emitting diodes on ZnO substrates. Appl Phys Lett 97, 013501.

Nayak, D K, Woo, J C S, Park, J S, Wang, K L and MacWilliams, K P (1993) High-mobility p-channel metal-oxide-semiconductor field-effect transistor on strained Si. Appl Phys Lett 62, 2853.

Neumark, G F (1988) Site dependence of donor properties in ZnSe and validity of effective-mass theory. Phys Rev B 37, 4778.

Oh, D C, Chang, J H, Takai, T, Song, J S, Gudo, K, Park, Y K, Shindo, K and Yao, T (2003) Electrical properties of heavily Al-doped ZnSe grown by molecular beam epitaxy. J Cryst Growth 251, 607.

Ohtomo, A, Tamura, K, Saikusa, K, Takahashi, K, Makino, T, Segawa, Y, Koinuma, H and Kawasaki, M (1999) Single crystalline ZnO films grown on lattice-matched $ScAlMgO_4(0001)$ substrates. Appl Phys Lett 75, 2635.

Okuyama, H, Morinaga, Y and Akimoto, K (1993) High-temperature blue lasing in photopumped ZnSSe-ZnMgSSe double heterostructures. J Cryst Growth 127, 335.

Okuyama, H, Nakano, K, Miyajima, T and Akimoto, K (1991) Epitaxial growth of ZnMgSSe on GaAs substrate by molecular beam epitaxy. Jap J Appl Phys 30, L1620.

Orton, J W (2004) 'The Story of Semiconductors', Oxford University Press, Oxford.

Osten, H J, Bugiel, E and Zaumseil, P (1994) Growth of an inverse tetragonal distorted SiGe layer on Si(001) by adding small amounts of carbon. Appl Phys Lett 64, 3440.

Osten, H J, Knoll, D, Heinemann, B and Schley, P (1999) Increasing process margin in SiGe heterojunction bipolar technology by adding carbon. IEEE Trans Elect Devices 46, 1910.

Ota, Y (1977) Si molecular beam epitaxy (n on n^+) with wide range doping control. J Electrochem Soc 124, 1795.

Ota, Y (1979) n-Type doping techniques in silicon molecular beam epitaxy by simultaneous arsenic ion implantation and by antimony evaporation. J Electrochem Soc 126, 1761.

Park, R M, Troffer, M B, Rouleau, C M, DePuydt, J M and Haase, M A (1990) p-type ZnSe by nitrogen atom beam doping during molecular beam epitaxial growth. Appl Phys Lett 57, 2127.

Partin, D L (1984) Single quantum well lead-europium-selenide-telluride diode lasers. Appl Phys Lett 45, 487.

Partin, D L (1991) 'Molecular-Beam Epitaxy of IV-VI Compound Heterojunctions and Super-lattices' in 'Semiconductors and Semimetals', Vol 33 (ed R K Willardson and A C Beer), Academic Press, New York, p 311.

Partin, D L, Thrush, C M, Simko, S J and Gaarenstroom, S W (1989) Bismuth-and thallium-doped lead telluride grown by molecular-beam epitaxy. J Appl Phys 66, 6115.

Patton, G L, Iyer, S S, Delarge, S L, Tiwari, S and Stork, J M C (1988) Silicon-germanium-base heterojunction bipolar transistors by molecular beam epitaxy. IEEE Electron Device Lett 9, 165.

Pearsall, T P, Bean, J C, People, R, and Fiory, A T (1985) in 'Proceedings of the 1st International Symposium on Silicon MBE' (ed J C Bean), Electro Chem Soc, New Jersey, p 400.

People, R and Bean, J C (1985) Calculation of critical layer thickness versus lattice mismatch for Ge_xSi_{1-x}/Si strained-layer heterostructures. Appl Phys Lett 47, 322.

People, R and Bean J C (1986) Band alignments of coherently strained Ge_xSi_{1-x}/Si heterostructures on <001> Ge_ySi_{1-y} substrates. Appl Phys Lett 48, 538.

People, R, Bean, J C, Lang, D V, Sergent, A M, Störmer, H L, Wecht, K W, Lynch, R T and Baldwin, K (1984) Modulation doping in Ge_xSi_{1-x}/Si strained layer heterostructures. Appl Phys Lett 45, 1231.

Pinczolits, M, Springholz, G and Bauer, G (1998) Direct formation of self-assembled quantum dots under tensile strain by heteroepitaxy of PbSe on PbTe (111). Appl Phys Lett 73, 250.

Qiu, J, DePuydt, J M, Cheng, H and Haase, M A (1991) Heavily doped p-ZnSe:N grown by molecular beam epitaxy. Appl Phys Lett 59, 2992.

Rucker, H, Heinemann, B, Bolze, D, Kurps, R, Kruger, D, Lippert, G and Osten, H J (1999) The impact of supersaturated carbon on transient enhanced diffusion. Appl Phys Lett 74, 3377.

Saitoh, S, Ishiwara, H, Asano, T and Furukawa, S (1981) Single crystalline silicide formation. Jap J Appl Phys 20, 1649.

Sasaki, T, Tomono, M and Oda, N (1995) Crystallinity improvement of HgCdTe on GaAs grown by molecular beam epitaxy. J Cryst Growth 150, 785.

Schaffler, F, Tobben, D, Herzog, H-J, Abstreiter, G and Hollander, B (1992) High-electron-mobility Si/SiGe heterostructures: influence of the relaxed SiGe buffer layer. Semicond Sci Technol 7, 260.

Segawa, Y, Ohtomo, A, Kawasaki, M, Koinuma, H, Tang, Z K, Yu, P and Wong, G K L (1997) Growth of ZnO thin film by laser MBE: lasing of exciton at room temperature. Phys Stat Sol (b) 202, 669.

Seong, M J, Alawadhi, H, Miotkowski, I, Ramdas, A K and Miotkowska, S (1999) The anomalous variation of band gap with alloy composition: cation vs anion substitution in ZnTe. Sol State Commun 112, 329.

Shiraki, Y (1982) 'Recent Japanese Developments in Si MBE' in 'Collected Papers of the 2nd International Symposium on Molecular Beam Epitaxy and Related Clean Surface Techniques, 27–30, August 1982, Tokyo', The Japan Society of Applied Physics, Tokyo, p 179.

Simma, M, Bauer, G and Springholz, G (2012) Temperature dependent band offsets in PbSe/PbEuSe quantum well heterostructures. Appl Phys Lett 101, 172106.

Simma, M, Fromherz, T, Bauer, G and Springholz, G (2009) Type I/type II band alignment transition in strained PbSe/PbEuSeTe multiquantum wells. Appl Phys Lett 95, 212103.

Sivananthan, S, Chu, X, Reno, J and Faurie, J P (1986) Relation between crystallographic orientation and the condensation coefficients of Hg, Cd, and Te during molecular-beam-epitaxial growth of $Hg_{1-x}Cd_xTe$ and CdTe. J Appl Phys 60, 1359.

Smith, D L and Pickhardt, V Y (1975) Molecular beam epitaxy of II-VI compounds. J Appl Phys 46, 2366.

Smith, D L and Pickardt, V Y (1976) Low carrier concentration (PbSn)Te by molecular beam epitaxy. J Electron Matls 5, 247.

Smith, D L and Pickardt, V Y (1978) Impurity dopant incorporation and diffusion during molecular beam epitaxial growth of IV–VI semiconductors. J Electrochem Soc 125, 2042.

Smith, R A (1959) 'Semiconductors', Cambridge University Press, Cambridge.

Sporken, R, Sivananthan, S, Mahavadi, K K, Monfroy, G, Boukerche, M and Faurie, J P (1989) Molecular beam epitaxial growth of CdTe and HgCdTe on Si (100). Appl Phys Lett 55, 1879.

Springholz, G and Bauer, G (2013) 'IV-VI Semiconductors', in 'Semiconductor Quantum Structures: Growth and Structuring. Landolt-Bornstein Group III Vol 34A' (ed C. F. Klingshirn), Springer, New York, p 415.

Sugiura, H (1980) Silicon molecular beam epitaxy with antimony ion doping. J Appl Phys 51, 2630.

Svob, L, Thiandoume, C, Lusson, A, Bouanani, M, Marfaing, Y and Gorochov, O (2000) p-type doping with N and Li acceptors of ZnS grown by metalorganic vapor phase epitaxy. Appl Phys Lett 76, 1695.

Sze, S M (1985) 'Semiconductor Devices: Physics and Technology', Wiley, New York.

Taike, A, Migita, M and Yamamoto, H (1990) p-type conductivity control of ZnSe highly doped with nitrogen by metalorganic molecular beam epitaxy. Appl Phys Lett 56, 1989.

Takagahara, T and Takeda, K (1992) Theory of the quantum confinement effect on excitons in quantum dots of indirect-gap materials. Phys Rev 46, 15578.

Takamizu, D, Nishimoto, Y, Akasaka, S, Yuji, H, Tamura, K, Nakahara, K, Onuma, T, Tanabe, T, Takasu, H, Kawasaki, M and Chichibu, S F (2008) Direct correlation between the internal quantum efficiency and photoluminescence lifetime in undoped ZnO epilayers grown on Zn-polar ZnO substrates by plasma-assisted molecular beam epitaxy. J Appl Phys 103, 063502.

Tao, I W, Jurkovic, M and Wang, W I (1994) Doping of ZnTe by molecular beam epitaxy. Appl Phys Lett 64, 1848.

Tatsumi, T, Hirayama, H and Azarki, N (1988) $Si/Ge_{0.3}Si_{0.7}/Si$ heterojunction bipolar transistor made with Si molecular beam epitaxy. Appl Phys Lett 52, 895.

Taylor, P J, Dhar, N K, Harris, E, Swaminathan, V, Chen, Y and Jesser, W A (2009) Analysis of dislocation density in $Pb_{(1-x)}Sn_xSe$ grown on ZnTe/Si by MBE. J Electron Matls 38, 2343.

Temkin, H, Bean, J C, Antreasyan, A, and Leibenguth, R (1988) Ge_xSi_{1-x} strained-layer heterostructure bipolar transistors. Appl Phys Lett 52, 1089.

Theuerer, H C (1961) Epitaxial silicon films by the hydrogen reduction of SiCl4. J Electrochem Soc 108, 649.

Tong, W, Wagner, B K, Tran, T K, Ogle, W, Park, W and Summers, C J (1996) Kinetics of chemical beam epitaxy for high quality ZnS film growth. J Cryst Growth 164, 202.

Tran, T K, Park, W, Tong, W, Kyi, M M, Wagner, B K and Summers, C J (1997) Photoluminescence properties of ZnS epilayers. J Appl Phys 81, 2803.

Tsang, J C, Eberl, K, Zollner, S and Iyer, S S (1992) Raman spectroscopy of $C_y Si_{1-y}$ alloys grown by molecular beam epitaxy. Appl Phys Lett 61, 961.

Tsukazaki, A, Ohtomo, A, Onuma, T, Ohtani, M, Makino, T, Sumiya, M, Ohtani, K, Chichibu, S F, Fuke, S, Segawa, Y, Ohno, H, Koinuma, H and Kawasaki, M (2004) Repeated temperature modulation epitaxy for p-type doping and light-emitting diode based on ZnO. Nat Mater 4, 42.

Van de Walle, C G (1997) 'Doping of Wide-Band-Gap II-VI Compounds–Theory' in 'Semiconductors and Semimetals', Vol 44 (ed R K Willardson, A C Beer and E R Weber), Academic Press, New York, p 121.

Volobuev, V V, Stetsenko, A N, Mateychenko, P V, Zubarev, E N, Samburskaya, T, Dziawa, P, Reszka, A, Story, T and Sipatov, A Yu (2011) Bi catalyzed VLS growth of PbTe (0 0 1) nanowires. J Cryst Growth 318, 1105.

Walukiewicz, W, Shan, W, Yu, K M, Ager, J A, Haller, E E, Miotkowski, I, Seong, M J, Alawadhi, H and Ramdas, A K (2000) Interaction of localized electronic states with the conduction band: band anticrossing in II-VI semiconductor ternaries. Phys Rev Lett 85, 1552.

Wang, K L, Cha, D, Liu, J and Chen, C (2007) Ge/Si self-assembled quantum dots and their optoelectronic device applications. Proc IEEE 95, 1866.

Werkhoven, C, Fitzpatrick, B J, Herko, S P, Bhargava, R N and Dean, P J (1981) High-purity ZnSe grown by liquid phase epitaxy. Appl Phys Lett 38, 540.

Whall, T E and Parker, E CH (2000) Si/SiGe/Si pMOS Performance-alloy scattering and other considerations. Thin Solid Films 369, 297.

Widmer, H (1964) Epitaxial growth of Si on Si in ultra high vacuum. Appl Phys Lett 5, 108.

Williams, F E (1960) Theory of the energy levels of donor-acceptor pairs. J Phys Chem Solids 12, 265.

Williams, G M, Whitehouse, C R, Chew, N G, Blackmore, G W and Cullis, A G (1985) An MBE route towards CdTe/InSb superlattices. J Vac Sci Technol B 3, 704.

Wroge, M L, Peterman, D J, Morris, B J, Leopold, D J, Broerman, J G and Feldman, B J (1988) Controlled p-type impurity doping of $Hg_{1-x}Cd_x Te$ during growth by molecular-beam epitaxy. J Vac Sci Technol A 6, 2826.

Xie, Y H, Monroe, D, Fitzgerald, E A, Silverman, P J, Thiel, F A and Watson, G P (1993) Very high mobility two-dimensional hole gas in $Si/Ge_x Si_{1-x}/Ge$ structures grown by molecular beam epitaxy. Appl Phys Lett 63, 2263.

Xu, D -X, Shen, G -D, Willander, M, Ni, W -X and Hansson, G V (1988) n-Si/p-$Si_{1-x}Ge_x$/n-Si double-heterojunction bipolar transistors. Appl Phys Lett 52, 2239.

Yakimov, A, Kirienko, V, Armbrister, V and Dvurechenskii, A (2013) Broadband Ge/SiGe quantum dot photodetector on pseudosubstrate. Nanoscale Res Lett 8, 217.

Yanka, R W, Harris, K A, Mohnkern, L M and Myers, T H (1991) A novel technique for the MBE growth of twin-free HgCdTe. J Cryst Growth 111, 715.

Yao, T, Amano, S, Makita, Y and Maekawa, S (1976) Molecular beam epitaxy of ZnTe single crystal thin films. Jap J Appl Phys 15, 1001.

Yao, T, Fujimoto, M, Uesugi, K and Kamiyama, S (1989) 'MBE and ALE of II-VI Compounds: Growth Processes and Lattice Strain in Heteroepitaxy' in 'Growth and Optical Properties of Wide-Gap II-VI Low-Dimensional Semiconductors', NATO ASI Series, Vol 200 (ed T C McGill, C M Sotomayer-Torres and W Gebhardt), Springer, New York, p 209.

Yao, T, Miyoshi, Y, Makita, Y and Maekawa, S (1977) Growth rate and sticking coefficient of ZnSe and ZnTe grown by molecular beam epitaxy. Jap J Appl Phys 16, 369.

Yao, T and Okada, Y (1986) Phosphorus acceptor levels in ZnSe grown by molecular beam epitaxy. Jap J Appl Phys 25, 821.

Yao, T, Tenada, H and Funaki, M (1986) Reflection high-energy electron diffraction oscillations during molecular beam epitaxial growth of ZnSe on (001)GaAs. Jap J Appl Phys 25, L952.

Yao, T, Zhu, Z, Uesugi, K, Kamiyama, S, and Fujimoto, M (1990) Molecular beam epitaxy/atomic layer epitaxy growth processes of wide-band-gap II–VI compounds: characterization of surface stoichiometry by reflection high-energy electron diffraction. J Vac Sci Technol A 8, 997.

Yokoyama, M, Kashiro, K and Ohta, S (1986) High quality zinc sulfide epitaxial layers grown on (100) silicon by molecular beam epitaxy. Appl Phys Lett 49, 411.

Yokoyama, M and Ohta, S (1986) Growth of crystalline zinc sulfide films on a (111)-oriented silicon by molecular-beam epitaxy. J Appl Phys, 59, 3919.

Yoneta, M, Ohishi, M, Saito, H and Hamasaki, T (1993) Low temperature molecular beam epitaxial growth of ZnS/GaAs(001) by using elemental sulfur source. J Cryst Growth 127, 314.

Yoshikawa, A, Nomura, H, Yamaga, S and Kasai, H (1989) Controlled conductivity in iodine-doped ZnSe films grown by metalorganic vapor-phase epitaxy. J Appl Phys 65, 1223.

Zhang, Q, Liu, X, DiNezza, M J, Fan, J, Ding, D, Furdyna, J K and Zhang, Y -H (2012) Influence of Te/Zn flux ratio on aluminium doped ZnTe grown by MBE on GaSb substrates. Phys Stat Sol (c) 9, 1724.

Zhao, K, Ye, L, Tamargo, M C and Shen, A (2012) Plasma-assisted MBE growth of ZnO on GaAs substrate with a ZnSe buffer layer. Phys Stat Sol (c) 9, 1809.

Zhu, Z, Mori, H, Kawashima, T and Yao, T (1992a) Planar doping of p-type ZnSe layers with lithium grown by molecular beam epitaxy. J Cryst Growth 117, 400.

Zhu, Z, Mori, H and Yao, T (1992b) Extremely low resistivity, high electron concentration ZnSe grown by planar-doping method. Appl Phys Lett 61, 2811.

Zogg, H (2012) Dislocation reduction by glide in epitaxial IV–VI layers on Si substrates. J Electron Matls 41, 1931.

Zogg, H and Hüppi, M (1985) Growth of high quality epitaxial PbSe onto Si using a (Ca,Ba)F$_2$ buffer layer. Appl Phys Lett 47, 133.

Zogg, H, Maissen, C, Masek, J, Hoshino, T, Blunier, S and Tiwari, A N (1991) Photovoltaic infrared sensor arrays in monolithic lead chalcogenides on silicon . Semicond Sci Technol 6, C36.

9

Electronic Devices

9.1 Introduction

There has been much discussion over the years as to whether MBE should be seen as a growth technique suitable for commercial exploitation or merely(!) a clever method of doing good material science. From our present vantage point it is clear that the answer has to be an unequivocal 'both'—there being more than adequate evidence of commercial devices grown by MBE in machines designed for that specific purpose (see Chapter 3, Section 3.7). In this chapter and in Chapter 10 we shall look in more detail at some of the major contributions made by MBE to the commercial development of a variety of electronic and optical devices, recognising that such contributions come in two guises, in the research stage and in the production stage.

It was not, of course, always clear that MBE was capable of development into a production method, and, in Chapter 2, we pointed out the importance of Al Cho's pioneering work to demonstrate the MBE growth of various device structures which performed at least as well as those grown by VPE and LPE. During the period 1972–1975 Cho and co-workers reported the MBE growth of GaAs mixer diodes, varactor diodes, IMPATT diodes, laser diodes, optical waveguides and FETs (see Cho and Arthur 1975). They also showed how it would be possible to fabricate optical waveguides and demonstrated an MBE-based planar technology. Such work can, perhaps, be seen in the context of a need to prove MBE's capabilities to a, sometimes, sceptical management, and it was not necessarily pursued into production. Nevertheless, we have documentary evidence to remind us that our own laboratory (PRL) was actively supplying MBE-grown GaAs layers to the Mullard factory in Hazel Grove (near Manchester) for the commercial production of microwave mixer diodes as early as 1982. Could this be the very first example of a commercial outlet for MBE material? We can only speculate. On the other hand, there can be no denying the importance of MBE growth in the commercial development of the high electron mobility transistor (HEMT), high-power microwave transistors and several other electronic devices so we shall concentrate attention on these in this chapter, while taking up the equally important question of semiconductor lasers and photodetectors in Chapter 10.

9.2 High Electron Mobility Transistors (HEMTs)

The early days of research into III-V compounds saw considerable argument concerning the virtue of GaAs as a medium for high-speed electronic devices. Because of its small conduction-band effective mass it could boast an electron mobility roughly six times greater than that of its, by then, well-established Si rival and this was seen as a major justification for investment in III-V technology during the 1960s. As solid state electronic devices moved into the megahertz and even gigahertz frequency ranges, silicon's relatively slow response began to look increasingly vulnerable to this thrusting newcomer—until, alas, it became clear that gallium arsenide's high electron mobility was compromised by an extremely short minority-carrier lifetime—carrier trapping by defects in even the best available GaAs dashed all hope for the development of a viable high-speed bipolar transistor. However, not all hope was lost. The alternative majority carrier device in the shape of the Si MOSFET had emerged from Bell Labs in the early 1960s, thus opening the way to the development of the GaAs MESFET (metal–semiconductor field effect transistor) which effectively cocked a snook at the short minority-carrier lifetime by the simple ruse of avoiding the use of minority carriers. In terms of technological convenience, Si had the advantage of its unique oxide which offered the essential low density of interface states but GaAs could claim the counter advantage of a sixfold increase in electron mobility. As the ultimate device response depended on the transit time of electrons under the gate electrode, this appeared to give GaAs an unassailable advantage—though not quite so much as at first seemed likely. The saturation drift velocity also plays a major role and this yields a somewhat smaller ratio in favour of GaAs. However, this is no place to become embroiled in a discussion of the finer details of FET design—we merely note that it was felt well worthwhile to develop GaAs MESFETs both for logic circuitry and for microwave applications, considerable success being achieved during the 1970s (for more details see Orton 2004, Chapter 5).

Much of the early work on MESFETs was based on vapour-phase epitaxial growth, a thin film of n-type material (t ~ 1 μm, n ~ 10^{23} m^{-3}) being deposited on a semi-insulating GaAs substrate. The gate took the form of an evaporated metal film, acting as a Schottky barrier, to modulate the electron current flowing between source and drain. An important improvement was introduced in 1974 by the Japanese company NEC in the shape of a thin, undoped buffer layer which served to isolate the conducting channel from the (relatively poor quality) substrate (see Figure 9.1). The fact that the thicknesses of buffer layer and channel were both of order 1 μm or less made it possible to grow them by MBE, and several groups took the opportunity do so. Cho et al. (1977) compared FETs grown directly on the substrate with those grown on MBE buffer layers and found much improved performance from the latter. Low-noise FETs showed noise figures as low as 1.9 dB at 6 GHz. Bandy et al. (1979) at Varian used heavily doped source and drain contact layers to enhance FET performance and measured noise figures of 1.5 dB at 8 GHz, comparable with those from the best VPE devices. Omori et al. (1981) used Si doping of the active layer (in contrast to earlier reports which relied on Sn doping) and measured noise figures of 1.2 dB at 4 GHz and 2 dB at 12 GHz. Feng et al. (1982) of Hughes obtained slightly better performance (1.5 dB at 12 GHz). Su et al. (1982)

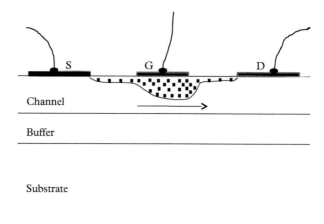

Figure 9.1 *Schematic diagram of a GaAs MESFET illustrating the use of an undoped buffer layer to isolate the channel from the SI substrate. The channel is typically 1 μm thick and doped n-type to a density of approximately 10^{23} m^{-3}. The gate is an evaporated metal Schottky barrier.*

(Illinois) studied the effect of buffer-layer thickness and showed that optimum FET performance demanded buffer layers with thicknesses greater than 1 μm.

From these examples, we see that MBE was certainly capable of producing state-of-the-art MESFET structures but further progress became submerged in the rush of papers reporting the use of a two-dimensional electron gas (2DEG) as the active layer in a field effect device, the so-called high electron mobility transistor, or HEMT. The innovative demonstration of modulation doping by Dingle et al. (1978) at Bell Labs led to numerous demonstrations of enhanced electron mobility (most particularly at low temperatures), as we discussed in Chapter 6 (Section 6.6), and it was immediately clear that this might lead to significant improvement in FET device characteristics. A surprising feature of these developments, however, was the fact that they came from sources other than those of the original inventors—Bell authors were not amongst the list of those claiming enhanced FET performance, even though they had been involved in research into the GaAs MESFET. In fact, the first HEMT report came from Fujitsu in Japan (Mimura et al. 1980a), and it was they who gave it its original name (various others emerged, as we shall see in a moment!). It was based on the MBE growth of the structure shown in Figure 9.2. The 2DEG was provided by Si doping $\left(n = 6.6 \times 10^{23} \text{ m}^{-3}\right)$ of the Al$_{0.32}$Ga$_{0.68}$As layer, its density being estimated as 7×10^5 m^{-2}. (Interestingly, they did not use an undoped spacer layer in the AlGaAs to increase the ionised donor–electron separation, even though this had been demonstrated in the Bell work, but it may well have been a deliberate choice to maximise the sheet-carrier density in the 2DEG.) The electron mobility was measured as 0.62 m^2 V^{-1} s^{-1} at room temperature, and 3.25 m^2 V^{-1} s^{-1} at 77 K. At 77 K the transconductance was roughly three times that of a comparable MESFET, though the advantage at room temperature was rather small. We should emphasise, too, that this was a large area structure, so it was by no

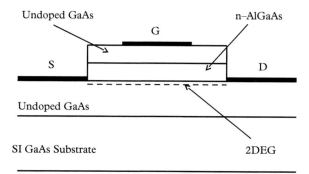

Figure 9.2 *Outline of the first HEMT, described by Mimura et al. (1980a). The AlGaAs layer was doped with Si to a level of* 6.6×10^{23} *m^{-3}. The 2DEG was formed in the undoped GaAs layer close to the interface. Complete pinch-off of the drain current was obtained.*

means a high-speed transistor. However, it was soon followed by a second brief publication describing a 2 μm-gate-length device with a cut-off frequency of 8.2 GHz, some 20% greater than an equivalent MESFET (Mimura et al. 1980b). At almost the same time, Delagebeaudeuf et al. (1980) at Thomson CSF independently reported the use of a 2DEG as the conducting channel in a FET device. The race was now on to realise HEMT devices with truly impressive microwave performance. (As an aid to the understanding of HEMT performance, we offer a simplified account in Box 9.1.)

Box 9.1 Theory of the microwave HEMT

To assist with the understanding of HEMT performance, it may be useful to present here a simplified model of its operation. (Note: this material is taken directly from Chapter 6 of Orton (2004), Box 6.3.) Needless to say, there are numerous detailed modifications which have to be included in a proper engineering model but the account we give here at least makes clear the general relationship between the various parameters used to describe it.

We concentrate attention on the region of the device under the gate. This has length (in the direction from source to drain) L and width W. It contains a sheet electron density n_s (m^{-2}), which is controlled by the gate voltage V_G. We treat the AlGaAs barrier layer as an insulator of thickness d, in which case the gate electrode and the channel charge $q_s = n_s eLW$ act as a parallel plate capacitor C_G, given by

$$C_G = \varepsilon LW/d \qquad (B9.1)$$

where ε is the dielectric constant of the AlGaAs layer ($\varepsilon \sim 10\varepsilon_0$). This gives us a simple relationship between n_s and V_G:

$$q_s = C_G V_G$$

or

$$n_s = (\varepsilon LW/d)V_G/eLW$$
$$= \varepsilon V_G/ed \qquad (B9.2)$$

continued

Box 9.1 *continued*

For a short-gate-length device which operates under saturated drift velocity conditions, we can write the source–drain current as

$$I_{SD} = en_s v_s W$$
$$= \varepsilon V_G v_s W/d \qquad \text{(B9.3)}$$

This gives us the important relationship for the transconductance of the device g_m:

$$g_m = dI_{SD}/dV_G$$
$$= \varepsilon v_s W/d \qquad \text{(B9.4)}$$

Finally, we need an expression for the limiting speed of the device, which depends on the transit time τ for electrons to pass under the gate:

$$\tau = L/v_s \qquad \text{(B9.5)}$$

The effective gain of the transistor will drop to unity at a frequency f_T given by $2\pi f_T \tau = 1$. Thus,

$$f_T = 1/2\pi\tau$$
$$= v_s/2\pi L \qquad \text{(B9.6)}$$

which, using (B9.1) and (B9.4), can also be written in the form

$$f_T = g_m/2\pi C_G \qquad \text{(B9.7)}$$

Equation (B9.6) shows that, for high operating frequencies, it is necessary to maximise the saturation drift velocity v_s and minimise the gate length L. The former is, of course, a pure material parameter which is larger in InGaAs than in pure GaAs, and the latter is a matter for the technologist who has obliged with values of L down to 0.1 μm, no mean achievement.

It may be useful at this point to insert appropriate parameter values into some of these equations to see how they relate to practical device results. Taking $v_s = 2 \times 10^5$ ms^{-1} and L = 0.1 μm gives us the value $f_T = 300$ GHz (to be compared with the experimental value of about 200 GHz) or a switching time of $\tau = 0.5$ ps. Typical transconductance values can be obtained from (B9.4), using an AlGaAs thickness d = 20 nm, as $(g_m/W) = 1000$ Sm^{-1} (usually written as 1000 mS mm^{-1}), in other words, 1 S mm^{-1} of gate width (such values have actually been achieved). The saturation value of source–drain current is found from (B9.3), using values of $n_s = 2 \times 10^{16}$ m^{-2} and $v_s = 2 \times 10^5$ ms^{-1}, as $I_{sat} = 600$ W amps—in other words, a gate width of 100 μm would give $I_{sat} = 60$ mA, an output power level of roughly 100 mW (100 W mm^{-1}). Note that the gate capacitance C_G takes values of about 1 pF mm^{-1}, which implies actual capacitance of order 0.1 pF or less, demanding extreme care in minimising stray capacitances.

Postscript: we should emphasise again that the above modelling applies only to short-gate devices where the electric field in the channel is large enough to ensure electrons move with saturation drift velocity. In the case of long gates and correspondingly lower velocities, both g_m and f_T will be significantly smaller, so care is necessary when making comparisons between devices with differing geometries.

Lester Eastman's group at Cornell University and Hadis Morkoc's group at the University of Illinois, both characterised by well-established MBE facilities applied to MESFET development, were quick to join the fray, using appropriate layers of AlGaAs and GaAs. Initial efforts were concerned to demonstrate promising DC characteristics, with emphasis on the considerable improvement achieved by cooling the devices to 77 K. However, genuine high-frequency transistors were soon fabricated, the French group describing what they preferred to call a 'TEGFET (two-dimensional electron gas FET)' with a gate length of 0.8 μm, showing a noise figure of 2.3 dB and an associated gain of 10.3 dB at 10 GHz (Laviron et al. 1981). They did employ an undoped spacer layer, 6 nm thick. Meanwhile the Fujitsu workers showed their preference not only for the name 'HEMT' but also for its application in fast-switching logic circuitry (Mimura et al. 1981). They measured room-temperature switching delays as short as 56.5 ps on devices with a 1.7 μm gate length. Not to be outdone, the French group also demonstrated logic circuitry with TEGFET propagation delay times as short as 19.1 ps (Tung et al. 1982). Judaprawira et al. (1981) at Cornell and Drummond et al. (1982) at the University of Illinois introduced a further structural modification in the shape of an $Al_{0.3}Ga_{0.7}As$ buffer layer between the GaAs buffer and the modulation-doped GaAs active layer, a modification which resulted in improved transconductance values. Performance depended on growth temperature for this buffer layer, 700 °C being optimum—contrasting with the more usual 600 °C used for GaAs growth. It is perhaps worth noting their measured room-temperature electron mobility of 0.80 $m^2 V^{-1} s^{-1}$, which compares with values of about 0.4 $m^2 V^{-1} s^{-1}$ typical of conventional MESFET structures.

We should also comment on the need for device structures somewhat more sophisticated than the rather crude mesa used initially by the Fujitsu group (Figure 9.2). In particular, high-frequency performance was limited by resistance inherent in source and drain contacts and it was vital that these should be much improved by including a heavily doped GaAs contact layer. This also led to the use of a recessed gate electrode and other subtleties such as had already been introduced in MESFET technology but we shall not go into detail—it is the MBE aspects of these devices which concerns us here. As an example of the way in which the AlGaAs/GaAs HEMT developed, we refer to a short review written by the Fujitsu group in 1985 (Mimura et al. 1985), where they considered both high-frequency analogue devices and high-speed logic circuits. Current gain cut-off frequency, f_T (the frequency at which transistor gain falls to unity), was shown to exceed that of a corresponding MESFET by a factor of between 1.5 and 2.0 for devices with gate length between 0.25 and 0.5 μm, while noise figures measured at frequencies of 12 and 18 GHz were also consistently better. Both enhancement mode and depletion mode devices could be made on the same substrate, and various examples of logic circuit had been reported, showing propagation delays of 12–18 ps at 77 K. (Values at room temperature were typically about 20–25 ps.) Preliminary attempts at making large-scale integrated circuits showed considerable promise, though there was a significant growth problem, in the shape of the 'oval defect', a surface irregularity characteristic of MBE growth. They pointed out that it would be necessary to reduce such defects to a minimum if LSI technology was to progress satisfactorily, and, as most MBE growers are well aware, that has subsequently happened. Two major sources of such defects were recognised, one associated with the Group III cell and one with residual oxide on the substrate. So-called gallium spitting occurred as a result of Ga droplets forming round the (cold) lip of the Ga cell, then dropping back into the hot liquid—it was effectively eliminated by using a 'hot-lip' cell. Substrate oxide was eliminated by better surface preparation—with the use of modern-day 'oven-ready' substrates, oval defects are now a thing of the past.

We now return to the fascinating question of why Bell were not included in this initial scramble to establish the claims of the HEMT as an important element in both fast logic circuits and in microwave integrated circuits. (In discussing this, we gratefully acknowledge Art Gossard's helpful comments.) The first point to make clear is that the Bell team did actually patent the idea of the 'modulation-doped field effect transistor' or MODFET—at a stroke it already rejoiced in three different names! Second, they did, indeed, report on the gating of modulation-doped quantum wells at a 1980 conference. However, no publication appeared as follow-up for the simple reason that their transistor characteristics were no better than those available from the best MESFETs. Third, and, perhaps, of even greater long-term significance, was the fact that Takashi Mimura came to appreciate the future possibilities of the device during a shared taxi ride with Ray Dingle, from the Kawasaki airport to the Fujitsu laboratory. On such stray occurrences does the long-term future of modern consumer electronics depend!

It would be seriously misleading, though, to suggest that the Bell interest in HEMT devices had come to a premature end—they were to be very much involved in its future development and, in 1982–1983, Bell workers were responsible for demonstrating the use of InGaAs as the channel material, on account of its higher electron mobility and velocity. The alloy $In_{0.53}Ga_{0.47}As$, lattice-matched to InP was recognised as a special case of the commercially important alloy system InGaAsP, which was developed for fibre-optic communication lasers during the late 1970s to the early 1980s. This activity came to be dominated by MOVPE growth, though MBE growth of the heterostructure $In_{0.53}Ga_{0.47}As/In_{0.52}Al_{0.48}As$ on InP was demonstrated by Cheng et al. (1982) (a collaboration between Bell Labs and the University of Illinois cemented by Al Cho's dual affiliation). They reported the formation of a 2DEG in the InGaAs by doping the InAlAs n-type with Si at a level of $\sim 10^{23}$ m^{-3} and measured room-temperature electron-sheet densities of $1.3 - 2.5 \times 10^{16}$ m^{-2} and mobilities as high as 0.89 m^2 V^{-1} s^{-1}. The Bell team went on to demonstrate the use of this combination in HEMT devices in a series of papers published during 1982–1983—we refer only to the final one, Pearsall et al. (1983), which describes a 2 μm-gate-length device. As they point out, the 2DEG, in this case, occurs as an accumulation layer close to the interface, rather than as the inversion layer characteristic of AlGaAs/GaAs structures (see Figure 9.3) and this led to difficulties with parallel conduction in the InGaAs—whereas undoped GaAs grown by MBE turns out to be lightly p-type, InGaAs tends to be n-type at a level of $\sim 1 \times 10^{22}$ m^{-3}.

Interesting though this approach to HEMT fabrication may have been, there were drawbacks. In the first instance, the Fe-doped semi-insulating substrates were smaller and of inferior quality to standard SI GaAs substrates—they were also more expensive, an argument to be repeated over many future years of HEMT development! What was more, preparation of a clean, smooth surface also required a degree of ingenuity

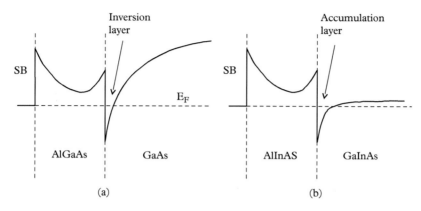

Figure 9.3 *Comparison between the conduction bands of (a) the AlGaAs/GaAs and (b) AlInAs/GaInAs HEMT devices, showing the inversion and accumulation layers, respectively. The Schottky barrier gate electrode is indicated as 'SB'.*

(Cheng et al. 1981b). The polished surface was first etched, then allowed to oxidise before the substrate was introduced into the MBE machine. In contrast to the case of GaAs, InP substrates could not easily be heat-cleaned on account of the low congruent evaporation temperature (360 °C), together with the undesirability of using a P_4 stabilising beam in the MBE chamber. Instead, substrates were heated to 500 °C under an As_4 background pressure of ~1.5 × 10⁻⁶ Torr, while monitoring the surface with Auger and RHEED. A second problem concerned uniformity of lattice match over the substrate surface—use of two standard cells for Ga and In resulted in a quite unacceptable variation of mole fraction of In by roughly 0.03 over a distance of 2.5 cm. Substrate rotation, not yet being available in their MBE machine, the uniformity problem was solved by the use of a coaxial In-Ga effusion cell, together with careful choice of aperture ratio and cell temperature. Final lattice match was maintained to within 5 × 10⁻⁴ over a lateral dimension of 2.5 cm. Note, however, the small size of such substrates—the commercial future would demand diameters of three inches or more, when substrate rotation would be essential. Indeed, the Bell workers very soon reported the growth of $Ga_{0.47}In_{0.53}As$ using a rotating sample holder, which produced material of even greater uniformity (Cheng et al. 1981a).

All this was typical of Bell Labs' innovative expertise and yet another example was concerned with the question of complementary-pair transistors. In the field of silicon integrated circuitry, the use of such pairs had become widely accepted on account of their low-power consumption—power only being dissipated during the act of switching. However, it was necessary to have available a pair of matching transistors and the development of complementary pairs of GaAs FETs was complicated by the considerable difference between electron and hole mobilities. In fact, hole mobility in GaAs differs very little from that in p-type Si and this makes for slow switching speeds in GaAs complementary circuits. In an attempt to remove this particular dilemma Kiehl and Gossard (1984) employed a combination of an n-GaAs MESFET with a p-HEMT, though it must be said that this only realised significantly faster switching speed when cooled to 77 K. Nevertheless, later work certainly justified the idea and today's IEEE 'International Electron Device Meetings' (IEDM) are apparently dominated by research into III-V-based CMOS (Rick Kiehl, private communication). Dwelling for a further moment on Bell innovations, we might make mention of a paper by Stormer et al. (1991) in which they describe the ultimate in short-gate structures by use of their so-called 'cleaved edge overgrowth' technique (this they also applied to the MBE growth of quantum wires, which we met in Chapter 6—see Section 6.5 and Figure 6.21). Briefly reiterating, this consists of growing an AlGaAs/GaAs quantum well, removing the sample from the MBE chamber, cleaving along a (110) plane and then growing a second well at right angles to the first. The result is shown in Figure 9.4, where it can be seen how one of the wells serves as gate while the other serves as channel. By such a stratagem, the Bell team were able to demonstrate a HEMT structure with a 20 nm gate length—minute even by the standards of electron-beam technology, which was rapidly being introduced at this juncture. However, ingenious though it certainly was, we are not aware that the idea has ever been taken up commercially.

Returning to our main theme, the practical demand for large, relatively inexpensive substrates steered research back to GaAs-based structures, while the use of InGaAs active layers to achieve greater CB offset and correspondingly greater sheet-carrier densities led to the development of the so-called 'pseudomorphic-HEMT' or P-HEMT. A 'Physical Review' article from Sandia (Osbourn 1983) discussed the

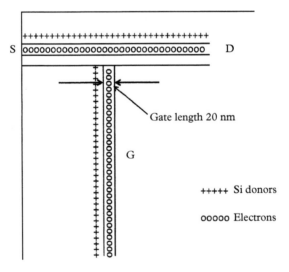

Figure 9.4 *Schematic cross-section of an ultra-short gate HEMT formed by cleaved-edge overgrowth. S and D represent the source and drain contacts to the channel, while the other quantum well acts as the gate, G. The well width (20 nm) defines the gate length. (Following Stormer et al. 1991.)*

general usefulness of InGaAs strained-layer superlattices and we might make a very brief detour to recall their application in the development of strained-layer quantum well lasers. It was in the year 1986 that Alf Adams of Surrey University made his noteworthy proposal for the use of pseudomorphic strained wells to raise the valence-band degeneracy, thereby deploying the light-hole band and effecting a considerable reduction in laser threshold current (Adams 1986). This innovation has attracted universal application in the field of InGaAsP lasers for optical fibre communications but does not feature in our discussion of quantum well lasers in Chapter 10 because most of this work relies on MOVPE growth. However, it certainly features here in our discussion of HEMTs, much of the work being based on MBE. At the same time, we should also draw attention to another important (undesirable!) material property which came to light during the 1980s and which has particular relevance to HEMT development, that of the DX centre in AlGaAs.

It soon became clear that HEMT performance could be improved by increasing the sheet-carrier density in the 2DEG, and one approach to this required both heavy doping in the supply layer and an increase in conduction-band offset at the interface. In the case of the lattice-matched AlGaAs/GaAs system, it demanded an increase in the Al content of the supply layer but, unfortunately, this led to the DX problem. Silicon-doped AlGaAs tended to show a density of deep trapping levels (DX centres) which increased with increasing doping level and whose depth increased with Al content, leading to undesirable effects, including persistent photoconduction and the collapse of the source–drain current–voltage characteristics in HEMTs. Limiting these effects to an acceptable level implied a limit to x in $Al_xGa_{1-x}As$ of approximately x < 0.25 and this, in turn, implied a limit on the CB offset of approximately 0.2 eV. This could, of course, be increased by substituting $In_yGa_{1-y}As$ for the GaAs channel material. The band gap of pseudomorphic $In_yGa_{1-y}As$ on GaAs is given in Figure 12 of the review by Schaff et al. (1991) and, assuming a 67:33 CB:VB split of the band-gap difference, we can write the CB offset for small values of y as

$$\Delta E_{CB} = 0.835x + 0.556y \; (eV)s \tag{9.1}$$

from which we can estimate an offset of approximately 0.28 eV for a pseudomorphic structure of $Al_{0.20}Ga_{0.80}As/In_{0.20}Ga_{0.80}As$. Compared with this, however, the offset for lattice-matched $In_{0.53}Ga_{0.47}As/In_{0.52}Al_{0.48}As$ on InP, about 0.5 eV, is considerably larger. It would be interesting to see whether this might result in corresponding differences in sheet-carrier density.

The situation of HEMT development is well summarised by Schaff et al. (1991), from which we can deduce that the difference was rather small. The Bell workers report values of $n_S \sim 1 - 2 \times 10^{16} \, m^{-2}$ for their InP-based devices, while later work at Hughes realised even larger values—for example, Mishra et al. (1988b) measured $n_s = 3.4 \times 10^{16} \, m^{-2}$. On the other hand, somewhat similar values $\left(n_S = 2.5 \times 10^{16} \, m^{-2}\right)$ were observed in pseudomorphic devices with 15% In content (see Schaff et al. 1991, Figure 24). What was clear, however, was the superior microwave performance of pseudomorphic transistors compared to the corresponding AlGaAs/GaAs devices and, with the development of sub-micron gate-length technology (courtesy of electron-beam lithography), cut-off frequency f_T had reached the impressive figure of 150 GHz, there being a definite dependence of f_T on sheet-carrier density. Lattice-matched In-GaAs/InAlAs devices also showed f_T values greater than 100 GHz, running more or less neck and neck with pseudomorphic devices. Within the space of a single decade, the HEMT had gone from being a gleam in Ray Dingle's eye to a fully fledged high-speed transistor with application as a low-noise amplifier, a high-power output device and a high-speed logic element. Also significant, the number of research groups working to perfect it had increased dramatically, most of them growing their device structures by MBE. Companies included Fujitsu, Bell, Thomson, CNET, IBM, Sandia, GE, and Hughes—universities included Illinois, Cornell and Minnesota.

This was not, of course, the end of the story. Now that it was clear that an increase in In content of the channel material was a 'good thing', efforts were made to increase it further. The group at Hughes had reported several advances in short-gate-length AlInAs/GaInAs/InP devices (see e.g. Mishra et al. 1988b, in which the authors describe a 0.1 μm gate microwave transistor with $f_T = 170$ GHz) but then took the concept of the P-HEMT a step further by increasing the In content to 63% while retaining the lattice-matched $Al_{0.48}In_{0.52}As$ supply layer (Mishra et al. 1988a). In terms of the inherent strain, this was a bit like growing $In_{0.1}Ga_{0.9}As$ on a GaAs buffer layer but it yielded a CB offset of roughly 0.55 eV. In device terms, this produced a transistor with an $f_T > 200$ GHz, an impressive performance in itself but of practical import in view of certain military requirements for devices to work at 94 GHz. The question remained; could this be the limit to the amount of In in the channel layer? The more In there was, the greater the mismatch and the thinner must be the pseudomorphic InGaAs layer. It was not clear whether the process could be pushed yet further until a careful study was made by Tacano et al. (1991) of the Electrotechnical Laboratory in Ibaraki. They demonstrated convincingly that it was possible to grow pseudomorphic $In_xGa_{1-x}As$ films on an AlInAs layer lattice-matched to InP with x varying all the way from 0 to 1 but that the maximum room-temperature electron mobility of 1.5 m^2 V^{-1} s^{-1} occurred for x ~ 0.8. In doing this, they introduced an important modification to the shuttering procedure for the Group III sources to ensure a constant flux from the instant of opening the shutter (in other words, to avoid shutter transient effects). This involved setting the cell temperature several degrees below its 'running' value before opening the shutter, then increasing it according to a square-root function so that it reached the operating temperature at the moment the shutter was opened. The resulting temperature overshoot was thereby confined to less than 1%—compared with almost 10% in the case of the 'standard' method. (We should perhaps point out that, with modern cells, shutter transients are no longer of significance but they were seen as a serious problem in the 1980s and 1990s.) One year later, the Hughes workers (Nguyen et al. 1992) reported on a 65 nm gate HEMT device with 80% In in the channel and boasting a current gain cut-off frequency of $f_T > 300$ GHz. The sheet-carrier density was 2.9×10^{16} m^{-2}, and the electron mobility just over 1.0 m^2 V^{-1} s^{-1}.

This represented truly exciting progress but, on the other hand, one must never overlook the commercial disadvantage of relying on InP substrates and there were those who saw this to be of overriding importance. Such a viewpoint led to the rival HEMT structure known as 'metamorphic'. (The 'Shorter Oxford English Dictionary' defines 'metamorphosis' as 'the action or process of changing in form, shape or substance; esp. transformation by supernatural means'.) We suspect that MBE growers would probably prefer to think their application of scientific principle had more than a little to do with the matter! However, philological disputation apart, the principle concerned was one involving the growth on a GaAs substrate of a compositionally graded $In_xAl_{1-x}As$ layer, x varying from 0 to 0.5, so as to provide a platform for growing an $In_{0.5}Ga_{0.5}As/In_{0.5}Al_{0.5}As$ heterostructure. The MBE growth of such a structure was demonstrated by Harmand et al. (1989) at the Matsushita laboratory, achieving a room-temperature electron mobility of $1.05~m^2\,V^{-1}\,s^{-1}$ and they followed this with the first report of a working HEMT (Inoue et al. 1991). It was far from being an ideal device, lacking full saturation in its source–drain characteristics, but nevertheless demonstrated the possibility of employing an $In_{0.5}Ga_{0.5}As$ channel on a GaAs substrate. A much improved transistor (with gate length of 130 nm) came from the Walter Schottky Institute in Munich in 1996 (Chertouk et al. 1996), characterised by a value of $f_T = 160GHz$ and the high-breakdown voltages essential for a power device. To illustrate the complexity demanded of the MBE growth, we show a schematic of the relevant structure in Figure 9.5. Having grown the GaAs buffer at 540 °C, they proceeded to grow the graded layer at 420 °C, then the actual HEMT at 470 °C—as we have seen before, it was necessary to grow In-containing material at significantly lower temperatures because of its higher vapour pressure. This particular M-HEMT introduced an interesting subtlety in the shape of a 'composite' channel. Adjacent to the $In_{0.32}Al_{0.68}As$ supply layer were 12 nm

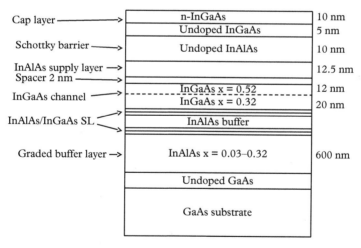

Figure 9.5 *Outline of the M-HEMT structure grown by Chertouk et al. (1996) to produce transistors with composite $In_xGa_{1-x}As$ channels, where x takes values of 0.52 and 0.32. It was grown on a GaAs substrate and employed a graded layer of InAlAs to effect a lattice-match with the device layers.*

of InGaAs with 52% In content, a pseudomorphic layer with reduced band gap, and 20 nm with 32% In. In operation, electrons near the source are concentrated in the narrow gap region but as they accelerate towards the drain contact they become hot enough to shift into the wider gap region, thus favouring higher breakdown voltage. The net result (gate length for gate length) was a transistor closely comparable in performance to the best available on an InP substrate. Clearly, the battle to win heads and wallets was still very much on! Let's look briefly at how it developed.

An example of a low-noise, MBE-grown P-HEMT was described by Lee et al. (1995) from Korea who were particularly concerned with the design of the T-shaped gate electrode and its lithographic realisation. Their devices were grown on three-inch GaAs substrates. The channel was formed from a 12 nm layer of undoped $In_{0.15}Ga_{0.85}As$ grown on a buffer of an AlAs/GaAs superlattice and 600 nm of GaAs, the supply layer was Si-planar-doped $Al_{0.24}Ga_{0.76}As$, and the 2DEG density and mobility were 2.1×10^{16} m^{-2} and 0.61 m^2 V^{-1} s^{-1}, respectively. All this was relatively ordinary but they measured excellent low-noise characteristics of 0.31 dB at 12 GHz, and 0.45 dB at 18 GHz, which were of particular interest for radio astronomy and satellite communication systems. Duran et al. (1996) from Zurich also measured low-noise properties on lattice-matched InGaAs/InAlAs/InP HEMTs, emphasising the successful use of reactive ion etching of the gate recess. Devices with gate lengths of 0.2 μm showed $f_T = 110$GHz, with noise figures of 0.5 dB at 10 GHz, and 0.75 dB at 25 GHz. An interesting development was later described by Chen et al. (2002), who were from Taiwan and who were aiming their P-HEMTs specifically at an application in the Japanese 'personal handy-phone handset' market. Important requirements were for high-power output at 1.9 GHz, with maximum efficiency, and for an enhancement mode device (rather than a depletion mode device, requiring a two-polarity supply rail). The channel material consisted of $In_{0.2}Ga_{0.8}As$, with the two supply layers of $Al_{0.3}Ga_{0.7}As$, planar-doped with Si. We may also note that this *consumer* application was based on the use of three-inch GaAs substrates. An interesting example of a GaAs-based M-HEMT was described by Ouchi et al. (2002) from Hitachi. The versatility of MBE is well illustrated by their employing a composite channel, consisting of 20 nm $In_{0.5}Ga_{0.5}As$, 5 nm graded InAsP and 5 nm InP, the supply layer being Si-doped $In_{0.5}Al_{0.5}As$. Because of the phosphorus content, they preferred gas source, rather than solid source MBE. Electron density and mobility as high as 3.5×10^{16} m^{-2} and 0.96 m^2 V^{-1} s^{-1} were obtained. Ono et al. (2004) from Sony made an interesting comparison between GaAs-based M-HEMTs and InP-based lattice-matched HEMTs, both of which demonstrated closely similar f_T values for devices with 53% In content in the channels. Values of the order of 150 GHz were obtained in both cases. Somewhat surprisingly, crystalline defects which penetrate the channel in M-HEMTs appear not to influence electron velocity. They did, however, notice an influence on electron–hole recombination via these defects which they thought might affect long-term reliability of M-HEMT devices. Finally, we draw attention to a dramatic development in M-HEMT technology, in the shape of a 200 GHz monolithic integrated power amplifier described by Kallfass et al. (2009). MBE growth took place on four-inch GaAs substrates, using a graded InAlGaAs buffer and a composite channel consisting of $In_{0.65}Ga_{0.35}As$ and $In_{0.53}Ga_{0.47}As$ layers. Devices with gate lengths of

100 nm showed high output power up to 200 GHz, which compared with 270 GHz for the best InP lattice-matched HEMTs. As we remarked immediately prior to this paragraph, the race between the InP and the GaAs-based devices was well and truly joined and, though the former appeared to be keeping its performance nose in front, the latter could probably challenge it closely enough to be preferred commercially. We leave the matter there, while turning our attention to another usurper.

It was clear that a major driving force behind the development of P-HEMTs and M-HEMTs was the increase of the In content of the channel material. Both low-field mobility and saturation drift velocity were thus increased and the obvious logical conclusion lay in using pure InAs. This was all very well but implied ever increasing strain in the channel, with consequently greater probability for dislocation incorporation. It therefore led to a rethink concerning the optimum choice for the adjacent barrier material, GaSb ($a_0 = 0.6096$ nm) and AlSb ($a_0 = 0.6136$ nm) being relatively close in lattice constant to InAs ($a_0 = 0.6059$ nm). However, AlSb, with a band gap of almost 1.6 eV, also offered the possibility of a large conduction-band offset ($\Delta E_c = 1.35$ eV). As we discussed in Chapter 7 (Section 7.4), an intriguing feature of the InAs/AlSb heterojunction is the possibility of two distinct interface types (illustrated in Figure 7.9), one being AlAs-like, the other InSb-like. Tuttle et al. (1990) demonstrated that it was possible to control which of these actually occurred during MBE growth by suitable choice of shutter sequence and incorporation of 'soak' periods. Soaking in Sb flux produced the InSb-like interface, soaking in Al the AlAs-like interface. They also showed that the nature of the bottom interface was crucial to the electronic properties of AlSb-InAs-AlSb quantum wells. The use of an AlAs bottom interface resulted in high electron-sheet density but very low electron mobility, an observation they interpreted as indicating the presence of a high density of As_{Al} anti-site defects which acted as donors. Their model's veracity was supported by the fact that post-growth soaking in an As flux led to a significant increase in carrier density. They also demonstrated that remote anti-site defects in the AlSb barriers could provide modulation doping of the well material. Leaving aside the subtleties of heterojunction interface structure, from the viewpoint of the device engineer the significant fact was that, using InSb interfaces, it was possible to achieve sheet-carrier densities of 4×10^{16} m^{-2} allied with electron mobilities of slightly more than 2 m^2 V^{-1} s^{-1}. Of even more significance was the InAs electron-drift velocity of 4×10^5 ms^{-1}. These should be compared with typical values for GaAs-based devices of 2×10^{16} m^{-2}, 0.8 m^2 V^{-1} s^{-1} and 2×10^5 ms^{-1}. What was more, these structures were grown (by MBE) on GaAs substrates, while the reduced band gap also promised significantly reduced operating voltages. The question remaining was the familiar one of whether future promise could overcome the prejudice (apparently inevitably) associated with a well-established technology.

Early work to demonstrate practical HEMT performance is typified by that from the University of California (Werking et al. 1992) employing the structure shown in Figure 9.6. The active layer took the form of a 12.5 nm InAs quantum well, the supply layer consisting of a 40 nm AlSb barrier delta-doped with two sheets of Te donor atoms, while the structure was capped with GaSb to protect the AlSb from oxidation. (Note that the Si donor commonly used in arsenide semiconductors is amphoteric in

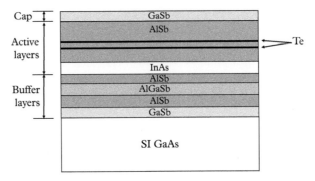

Figure 9.6 *An example of an early InAs/AlSb HEMT structure using Te delta-doping to achieve electron sheet densities of* $3.8 \times 10^{16}\,\mathrm{m^{-2}}$ *and mobilities of* $2.1\,m^2\,V^{-1}s^{-1}$. *The various buffer layers were chosen to accommodate the large lattice mismatch between the GaAs substrate and the active layers. (Following Werking et al. 1992.)*

the antimonides—Te is considerably more reliable.) A series of buffer layers was grown to accommodate the mismatch between the GaAs substrate and the active material. Dislocation densities were in the region of $10^{11}\,\mathrm{m^{-2}}$ but in-plane transport properties were good. Sheet-carrier density was measured as $3.8 \times 10^{16}\,\mathrm{m^{-2}}$, and electron mobility, $2.1\,\mathrm{m^2\,V^{-1}\,s^{-1}}$. Transistors with gate lengths of $2\,\mu\mathrm{m}$ showed transconductances greater than 400 mS/mm. There was evidence of carrier multiplication in the InAs well at source–drain voltages greater than about 0.7 V, a possible drawback to using a channel material with such a small band gap. However, the current–voltage characteristics confirmed that such devices did, indeed, function at low-input-power levels.

Moving on a little, it is convenient to refer to the review of antimonide devices prepared by Bennett et al. (2005) of the Naval Research Laboratory in Washington. By this time, many different groups were involved in attempts to improve and commercialise AlSb/InAs HEMTs, most of the work being based on MBE growth. The low-power capability of these HEMTs was confirmed, typical results on 60 nm-gate-length devices showing values of f_T as high as 90 GHz at drain voltages of 0.1 V, and 160 GHz at 0.4 V, significantly greater than available from GaAs or InP-based devices at such voltages. Transconductance was measured as 1.1 S/mm at $V_{SD} = 0.35$ V. Channel characteristics of these devices were $n_s = 1.6 \times 10^{16}\,\mathrm{m^{-2}}$ and $\mu = 2.13\,\mathrm{m^2\,V^{-1}\,s^{-1}}$, in the case of n_s much lower than the values referred to above in order to reduce the value of threshold current to one more acceptable for low-power circuits. (Real life can be relied upon to throw up problems unforeseen in one's initial enthusiasm for exciting new material properties!) Device characteristics again showed evidence of impact ionisation in the channel material, and various modifications had been reported in attempts to minimise it. The concept of dual channel devices was mentioned earlier in connection with the development of the metamorphic HEMT, the channel containing parallel regions of lower and higher band gap. At low electron velocity, charge was confined to the narrow gap material, while, as electrons became hot, they could transfer to the higher band-gap region, thus minimising the possibility of impact ionisation. Boos et al. (1998) achieved this ingeniously by growing a channel of the form of two InAs quantum wells separated by a thin barrier of AlSb, thus obtaining 10 nm InAs/3 nm AlSb/4.2 nm InAs. The narrower well performed the role of wider-energy gap material by virtue of its greater confinement energy. The 100 nm-gate-length HEMTs using this design achieved the impressive f_T value of 250 GHz. Lin et al. (2005) used the more conventional method of combining InAs with InAsP to form the composite channel and, in both cases, the resulting HEMTs exhibited noticeably increased breakdown voltages.

While high-frequency microwave transistors certainly demand high f_T, low-noise figures are also of major importance, and progress in this direction was reported, too. Best results at 4 GHz and 10 GHz were 0.3 dB and 0.5 dB, respectively, while a value of 5.5 dB was achieved at 100 GHz. Though noise performance was certainly less impressive than that reported for the best GaAs or InP-based HEMTs, this nevertheless represented excellent progress, allowing for there being considerable scope for further improvement in material quality. In particular, dislocation densities were still in the range 10^{12}–10^{13} m^{-2} and there was evidence for a deep level in the AlSb barrier layers. The low-power consumption of these AlSb/InAs HEMTs made them candidates for space applications, where available power is severely limited and where tolerance to radiation effects is also at a premium. Measurements of radiation hardness by Weaver et al. (2005) showed excellent promise, the radiation-induced decrease in drain current being two orders of magnitude less than comparable results for AlGaAs/GaAs HEMTs. Finally, we should note yet another potential advance (Ashley et al. 1995) in the shape of InSb channel HEMTs, pioneered at QinetiQ (a company grown out of the RSRE government laboratory in Malvern, UK). InSb has the highest electron mobility of any known semiconductor, and quantum wells can be made using InAlSb barriers. Utilising these structures to make HEMTs was not without its difficulties but preliminary device performance reported by Ashley et al. (2004) appeared suitably encouraging. A device with a 200 nm gate length achieved an f_T value of 150 GHz at $V_{DS} = 0.5$ V. Later work (Ashley et al. 2006) utilised gate lengths of 85 nm to achieve cut-off frequencies as high as 340 GHz.

Clearly antimonide HEMTs were staking a strong claim to be taken seriously by systems engineers worldwide and this was emphasised by reports of low-noise, low-power microwave amplifiers. Hacker et al. (2004) claimed to have built the first antimonide MMIC (monolithic microwave integrated circuit) in 2004 using MBE to grow the three-stage AlSb/InAs structure. The heterostructures were characterised by $n_s = 3.7 \times 10^{16}$ m^{-2} and $\mu = 1.9$ m^2 V^{-1} s^{-1}, the 0.25 μm-gate HEMTs showing transconductances of 2.0 S/mm at $V_{SD} = 0.4$ V, and $f_T = 160$ GHz. The amplifier operated at 35 GHz with overall gain of 22 dB and an associated noise figure of 2.1 dB. DC power dissipation was 4.5 mW, less than 10% of the corresponding figure for an equivalent InGaAs/AlGaAs low-noise HEMT amplifier. Riemer et al. (2006) described a single-stage amplifier which operated at 90 GHz with a 5.6 dB gain and a noise figure of 2.5 dB while consuming only 2.0 mW of DC power. Ma et al. (2006) also used MBE to fabricate a three-stage AlSb/InAs HEMT wide-band, low-noise amplifier providing a 30 dB gain over a bandwidth of 0.3 –11 GHz and an associated noise figure of 2.6 dB. The DC power dissipated was only 7.5 mW, and the maximum f_T value for an individual stage (gate length 0.1 μm) was 270 GHz.

However, perhaps the most significant development associated with the antimonides came to the fore in 2008, when a consortium involving QinetiQ and Intel (Radosavljevic et al. 2008) reported the first high-performance, low-power, p-channel FET, thus opening the way to complementary-pair logic circuitry. Here might well be a serious rival to the long-established Si CMOS logic element, with the possibility of markedly superior switching speed. Up to this moment, the relentlessly increasing speed of Si ICs had been achieved by the simple (!) expedient of decreasing gate length but it had been clear for some time that this strategy must eventually run into a fundamental limit. Si, it had to be said, was a distinctly modest semiconductor in terms of its carrier mobilities and drift velocities, and, as we have already pointed out, the electron mobility in InSb far outstripped that of Si. If, therefore, a p-channel antimonide FET could be produced with even a moderately improved switching speed over its Si rival, the possibility of a CMOS replacement with an increment in speed determined by *material* properties might take the microprocessor world by storm! What was more, it might, at the same time, offer an entrée for MBE into the most prestigious device market of the twenty-first century! Just how realistic was the reality?

The structure of the p-HEMT transistor, grown by solid source MBE on (100) SI GaAs, is shown in Figure 9.7. It used a 5 nm InSb quantum well as active channel, sandwiched between $Al_xIn_{1-x}Sb$ barrier layers, the upper one being delta-doped with Be acceptors. Because the AlSb lattice parameter is some 5.6% smaller than that of InSb, the channel was under biaxial compressive strain, which could be modulated between 1.0–2.0% by varying x between 0.15 and 0.35. This had the vital effect of raising the degeneracy of the light- and heavy-hole bands and reducing the heavy-hole effective mass, the measured hole mobility being 0.123 $m^2\,V^{-1}\,s^{-1}$ (combined with $p_s = 1 \times 10^{16}\,m^{-2}$) for a device with x = 0.35. This was nearly an order of magnitude greater than the hole mobility available to the strained Si layers currently employed in Si p-MOS devices and resulted in significantly improved switching speed. Values of effective hole velocity extracted from analysis of 40 nm-gate-length InSb and Si device performances suggested that the former was more than double the latter. In practical device terms, the antimonide HEMTs offered ten times lower power dissipation for same-speed operation, or double the speed for same-power dissipation. The transconductance and cut-off frequency of the antimonide device with x = 0.35 were 0.51 S/mm and 140 GHz, respectively, at a supply voltage of 0.5 V, the highest values ever measured for a p-channel FET. For completeness, it is only fair to note that the Naval Research Labs group had earlier studied mobility enhancement in p-InGaSb quantum wells (Bennett et al. 2007) and had measured hole mobilities as high as 0.15 $m^2\,V^{-1}\,s^{-1}$. They later showed, in a detailed study of InGaSb channels (Nainani et al. 2012), that it was possible to achieve high hole mobilities in InGaSb channels together with sheet-hole densities as high as $7 \times 10^{16}\,m^{-2}$, providing sheet conductance at least three times greater than the best available in uniaxially strained Si. They also confirmed the reduction in hole effective mass in strained

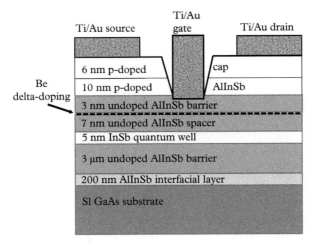

Figure 9.7 *The structure of the compressively strained InSb quantum well p-type QW FET demonstrated by Radosavljevic et al. (2008). The Al composition of the AlInSb barriers was varied between 0.15 and 0.35 to vary the degree of strain. Devices with 1.9% strain showed the highest transconductances and cut-off frequencies measured on any III-V p-type FET.*

QW layers by performing cyclotron resonance measurements. So, the idea may not have been new but the QinetiQ/Intel work certainly made the world of CMOS technology sit up and take notice.

However, before we leave ourselves open to accusations of inadequately justified euphoria, it would be well to point out that making a high-speed p-type FET represents only the first step towards the ultimate challenge of achieving practical CMOS circuits. The next stage, that of combining n- and p-type FETs on the same substrate, presents an even greater challenge. It is here that the picture becomes somewhat blurred as an inevitable consequence of commercial sensitivity and we make no pretence of being privy to the secrets of Intel et al. In any case, we are attempting to write history, rather than to speculate on possible future development. What we can say is that interest is certainly at fever pitch and that there are no less than 25 groups around the world active in bringing the future to pass (Brian Bennett, private communication). Perhaps, to offer an indication of the esoteric nature of the chase, we might simply refer to just one of the techniques being explored, the 'two-step layer transfer' technique described by an interesting group of Far-Eastern scientists (perhaps temporarily?) resident in California and New Mexico (Nah et al. 2012). In a word, two thin films of material were grown on separate substrates, one of InGaSb, the other of InAs; these were then etched to form strips about 0.5 μm wide, removed from their substrates by selective etching and mounted on a third substrate, this time of Si/SiO$_2$. FETs were then formed as shown in Figure 9.8 (note the use of ZrO$_2$ gate isolation to minimise gate-leakage current).

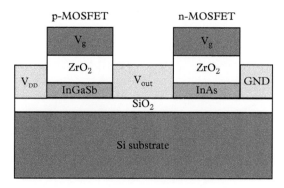

Figure 9.8 *Cross-section of the III-V CMOS inverter structure described by Nah et al. (2012). This was 'built' by growing separate InGaSb and InAs layers, cutting them into strips and mounting them on the Si/SiO₂ substrate. A 10 nm-thick ZrO₂ was added as gate dielectric.*

One's mind boggles at the effrontery of such innovative technology but one might also be forgiven for asking what all this has to do with MBE—the answer is, of course, that the original layers were grown by MBE!

Reference, above, to the use of a Si substrate reminds one of the controversy concerning GaAs versus InP substrates. Just as GaAs is more readily available in larger sizes and at lower cost than InP, so is Si preferable to GaAs for the very same reasons. What is more, the Si IC community is obviously strongly committed to Si on the grounds of familiarity, and this raises the question of whether antimonide HEMTs might be grown directly on Si, rather than on GaAs. Evidence pointing the way is provided by successful growth of AlGaSb/InAs HEMTs on Si. Two examples of such studies have been reported from the Far East. In 2007, a Japanese/Chinese consortium (Lin et al. 2007) measured electron mobilities in AlGaSb/InAs quantum wells of 2.73 $m^2\,V^{-1}\,s^{-1}$ as a result of a careful choice of buffer layers. The active layers were grown by MBE but the buffer layers of $Si_{0.1}Ge_{0.9}$, $Si_{0.05}Ge_{0.95}$, Ge and GaAs were grown by MOVPE. The total buffer-layer thickness in this work amounted to some 1.6 μm, which can be seen as a significant drawback (as can the use of two very different growth techniques) but in 2009 Ko et al. (2009) from Korea achieved success with a very much thinner buffer layer, the whole structure being grown by MBE. This employed an unusual technique for preventing dislocations from propagating into the AlSb/InAs active region by use of an AlGaSb buffer containing AlSb and InSb quantum dots. They measured defect densities as low as 10^{10} m^{-2} and an electron mobility of 1.6 $m^2\,V^{-1}\,s^{-1}$, at a sheet density of 2.5 × 10^{16} m^{-2}. Apparently, sufficient attention to buffer-layer design can overcome the disadvantages of serious lattice mismatch and also avoid the generation of anti-phase domains. Clearly, the future for III-V-compound, high-frequency, low-noise, low-power HEMTs holds much promise, and that promise carries with it an intriguing future for

MBE growth—we have to leave it there and move on to consider the contrasting subject of high-power HEMTs based on Group III nitride materials.

The Group III nitride semiconductors came into prominence during the early 1990s, largely as a result of their impact on the visible LED and laser market, and, while this work was dominated by MOVPE growth, MBE can be seen to have made a significant contribution to the field of nitride HEMTs (Kaun et al. 2013; Wong et al. 2013). The combination of large band gaps, high saturated electron velocities $\left(\sim2.5 \times 10^5 \, ms^{-1}\right)$ and high-breakdown electric fields $\left(\sim3.3 \times 10^8 \, Vm^{-1}\right)$ makes the nitrides well suited to application as power transistors, and MBE-grown devices continue to impact on this field up to the present day. For example, RF power densities of 40 W mm^{-1} at 4 GHz, 30 W mm^{-1} at 8 GHz, 15 W mm^{-1} at 30 GHz, and 10 W mm^{-1} at 40 GHz are typical of reported results at the time of writing. Current gain cut-off frequencies f_T greater than 100 GHz are also well established. A further feature of interest for space applications is their much improved stability against radiation damage (Chen et al. 2013).

The first AlGaN/GaN HEMTs, grown by MOVPE, were reported in 1993 by Asif Kahn et al. (1993). A 2DEG was formed at the interface between 600 nm GaN and 100 nm Al$_{0.14}$Ga$_{0.86}$N films, grown on a sapphire (0001) substrate with an AlN buffer. A room-temperature sheet-carrier density of $1.15 \times 10^{17} \, m^{-2}$ and a mobility of $0.0563 \, m^2 \, V^{-1} \, s^{-1}$ were measured. A year later (Asif Khan et al. 1994), devices with 0.25 μm gates showed current density of 60 mA/mm, transconductance of 27 mS/mm and $f_T = 11$ GHz (compared with today's typical values of 400 mA/mm, 100 mS/mm and 60 GHz). The first MBE-grown devices originated from the Morkoc group at the University of Illinois (Ozgur et al. 1995). They had very similar values for sheet-carrier density, electron mobility and transconductance on 3 μm gate geometries, consistent with the earlier MOCVD devices. It was clear that the large values of n_s and of saturation drift velocity would favour the use of nitride HEMT structures in high-frequency, high-power applications. An interesting feature of these early results was the fact that no attempt was made to dope the AlGaN supply layers—large values of n_s appeared as if by magic, a subject taken up by Asbeck et al. (1997) who emphasised the importance of piezoelectric effects in nitride heterostructures. Because the nitride films used to make HEMTs grow with (0001) hexagonal faces and because there are significant differences in lattice parameter between the three compounds InN, GaN and AlN, one must expect significant polarisation of the structures, resulting in sheet-charge densities at each heterojunction which are compensated by corresponding sheets of free electrons. Indeed, there is no need to employ doping as a means of producing the desired 2DEG (see Box 9.2 for a brief account of the background theory). Asbeck et al. showed that, as would be expected from the increasing interface strain, the value of n_s increased approximately linearly with the aluminium fraction in the AlGaN layer. An MBE-grown sample with 35% Al yielded a value of $n_s = 1.8 \times 10^{17} \, m^{-2}$, something like five times greater even than values measured on InGaAs P-HEMT and M-HEMT devices. They also showed that results obtained on MBE- and MOCVD-grown structures were entirely consistent.

Box 9.2 Polarisation doping of nitride HEMTs

An interesting feature of nitride HEMT devices is the fact that high levels of sheet electron density can be obtained without the need for deliberate doping of the supply layer. This arises as a result of the polarisation charge, which is associated with the pseudomorphic growth of the thin AlGaN barrier layer on a relatively thick GaN layer. The in-plane lattice parameter of wurtzite GaN ($a_0 = 0.3189\,\text{nm}$) exceeds that of AlN ($a_0 = 0.3112\,\text{nm}$) by approximately 2.5%; this implies a step change in strain at the interface, and this change increases roughly linearly with the fraction of Al in the AlGaN. Because the wurtzite structure is electrically polarised in the [0001] direction, this gives rise to a donor-like positive sheet charge on the AlGaN side of the interface and this attracts a corresponding sheet of electrons which constitutes the 2DEG. Detailed calculations of n_s using the known elastic and piezoelectric constants of the two materials confirm the veracity of this thesis, as shown, for example, by Ambacher et al. (1999).

In the following simplified account we shall use the approach adopted by Asbeck et al. (1997), who considered a 50 nm thick barrier layer on top of a 3 μm GaN layer. The tensile strain in the AlGaN in the plane of the heterojunction results in a compressive strain along the z axis (the [0001] direction) and this produces a piezoelectric polarisation P_z given by

$$P_z = 2d_{31}\left(c_{11} + c_{12} - 2c^2_{13}/c_{33}\right)\varepsilon_{xx} \tag{B9.8}$$

where d_{31} is the piezoelectric strain coefficient for AlGaN, c_{ij} denotes the elastic stiffness coefficient for the different materials, and ε_{xx} is the strain in the x direction (assumed equal to that in the y direction). The resulting fixed sheet-charge density is given by

$$Q_{pz} = eN_{pz} = P_z \tag{B9.9}$$

If it were not for the complicating fact that the gate electrode is close to the interface, this sheet charge would be exactly compensated by an equal two-dimensional electron charge (i.e. $n_s = N_{pz}$) but, in practice, one must allow for the pinning of the Fermi level at the gate Schottky barrier and this reduces n_s somewhat. Asbeck et al. derive the equation

$$n_s = N_{pz} - C_b(\phi_s - \Delta E_c + E_{Fn})/e \tag{B9.10}$$

where C_b is the geometrical capacitance of the AlGaN layer, ϕ_s is the depth of the Fermi level at the AlGaN surface, ΔE_c is the conduction-band offset, and E_{Fn} is the height of the Fermi level above the CB edge in the GaN at the interface. This is (hopefully) made clearer by a glance at Figure 9.9.

Without wishing to delve further into mathematical complexity, we should finally note that there is also a contribution to n_s from the fact that the GaN and the AlGaN are individually spontaneously polarised to a slightly different degree (further details can be found in Ambacher et al. 1999). Thus, the total fixed sheet charge is given by

$$\begin{aligned} Q &= P(\text{top}) - P(\text{bottom}) \\ &= \{P_{sp}\,(\text{top}) + P_{pz}\,(\text{top})\} - \{P_{sp}\,(\text{bottom}) + P_{pz}\,(\text{bottom})\} \end{aligned} \tag{B9.11}$$

and the overall effect depends on the relative magnitudes and directions of these various components.

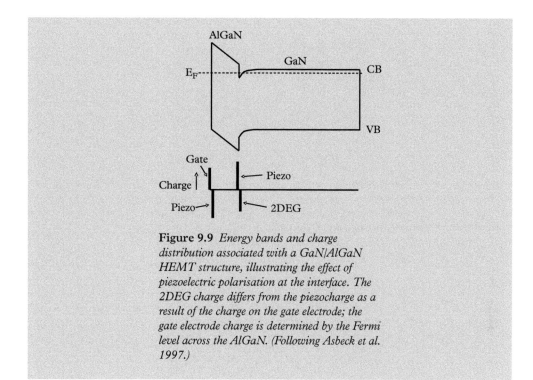

Figure 9.9 *Energy bands and charge distribution associated with a GaN/AlGaN HEMT structure, illustrating the effect of piezoelectric polarisation at the interface. The 2DEG charge differs from the piezocharge as a result of the charge on the gate electrode; the gate electrode charge is determined by the Fermi level across the AlGaN. (Following Asbeck et al. 1997.)*

Once it was appreciated that such polarisation effects could dominate the behaviour of nitride heterostructures, there was a flood of papers describing similar results and, because interface quality was obviously of paramount importance, many of these made use of MBE. For example, Dimitrov et al. (1999) described the use of plasma-assisted MBE to grow nominally undoped HEMTs with sheet-carrier densities of more than 10^{17} m^{-2}, mobilities in the range 0.10–0.12 m^2 V^{-1} s^{-1}, maximum source–drain currents of 800–850 mA/mm, and transconductances up to 250 mS/mm. An interesting feature was their ability to control the polarity of the films by including (or not) an AlN nucleation layer. The two structures are shown schematically in Figure 9.10, the N-face sample being designated 'inverted', and the Ga-face, 'normal'. Note that in the former the 2DEG, as measured by C–V profiling, is formed at the upper AlGaN/GaN heterojunction, not at the lower one, as is the case for the Ga-face structure. This results from the opposite polarisation. As explained by Ambacher et al. (1999), in this case the lower interface is characterised by a *negative* fixed-charge sheet which would attract holes, rather than electrons—only the upper heterojunction has the required *positive* sheet charge, and the explanation only becomes clear when one considers the spontaneous as well as the piezoelectric polarisation (see Box 9.2). In growing (at 800 °C) the GaN buffer layer for the N-face structure, a background free electron density of 5×10^{22} m^{-3} was obtained, which served as a parallel conduction path so it was necessary to compensate with a carefully

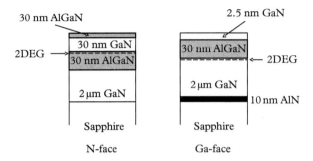

Figure 9.10 *The two undoped structures grown using PAMBE by Dimitrov et al. (1999). The polarisation could be controlled by inserting a 10 nm AlN buffer layer, with the N-face surface being referred to as 'inverted', and the Ga-face as 'normal'. Note that the position of the 2DEG is different in the two structures.*

controlled Mg-acceptor doping, the net free-carrier density being thus reduced below $1 \times 10^{20} \, \text{m}^{-3}$. However, such counter-doping was not required when growing on the AlN nucleation layer. Finally, we note that the N-face structure showed evidence of three-dimensional island growth—AFM imaging showing an rms surface roughness of 5 nm. The corresponding parameter for the Ga-face structure was 0.6 nm. Murphy et al. (1999) reported the PAMBE growth (incorporating an AlN nucleation layer) of 0.25 μm gate transistors with $f_T = 50 \, \text{GHz}$, output power = 1.88 W/mm at 4 GHz and power-added efficiency of 34%, showing obvious promise as future high-power amplifiers.

Further progress, however, would be dependent on the choice of substrate. Sapphire had been used for most nitride growth up to this point because of its ready availability but it was clear that it was far from ideal for high-power devices on account of its relatively poor thermal conductivity. SiC was very much better in this regard but was more expensive and this led to a surge of activity to explore the alternative of MBE growth of nitride films on Si substrates. Just as GaAs substrates were preferred for commercial development of P-HEMTs and M-HEMTs, Si would be favoured for nitride devices and in this case there was little competition from lattice-matched substrates, bulk GaN being available in only small quantities. We shall look first at progress with SiC substrates and then concentrate on what promises to represent the long-term future—Si-based power HEMTs.

Perhaps, to set the scene, we might note the high-power AlGaN/GaN HEMT grown by PAMBE on a SiC substrate, as described by Palacios et al. (2005). A 160 nm-gate-length device generated 8.6 W/mm at 40 GHz with 32% power-added efficiency. Alternatively, Corrion et al. (2006) obtained 13.7 W/mm output at 4 GHz, with 55% efficiency. Kaun et al. (2013) summarise later developments, transistor characteristics demonstrating very clearly the thermal advantage offered by the SiC substrate over similar structures grown on sapphire. A typical Ga-face structure on 4H–SiC consists of a 100 nm AlN nucleation layer, 1 µm (undoped) GaN, 3 nm AlN, 20 nm AlGaN (polarisation doping layer), and is passivated with SiN_x, the 2DEG being formed at the AlN/GaN interface. In addition to determining Ga-face growth, the nucleation layer serves as a barrier to Si and O dopant atoms propagating from the substrate into the GaN layer, while also reducing the lattice mismatch somewhat. Nevertheless, the GaN grows by the Volmer–Weber three-dimensional mode and threading dislocations propagate through the 2DEG.

Two problem areas demanded attention before PAMBE could be seen as a viable approach to GaN HEMT production, those of improving surface morphology and of eliminating parallel conduction. The former was seen as particularly serious in the context of growth on large area substrates, where modest variations in growth temperature over the surface were difficult to avoid. It had been established (see Section 7.2) that good surface morphology depended on growing at a precisely controlled III:V flux ratio, on the boundary between Ga drop formation and an intermediate regime characterised by incomplete Ga wetting and this, as shown in Figure 9.11 for typical temperatures of 700 °C–720 °C, was very sensitive to precise substrate temperature. Under continuous growth, it would be virtually impossible to obtain acceptable surface quality across even a two-inch wafer so some modified growth technique was essential. The solution (described by Poblenz et al. 2005 from the University of California, Santa Barbara) involved pulse-modulating the N and Ga beams so as to divide the growth into alternating periods of growth with droplet formation and Ga desorption (not dissimilar to the modulation techniques described in Chapter 5, Section 5.2). During the first period, of length $t_1 = 180$ s, N, together with *excess* Ga, was supplied so as to produce droplets (which were known to be undesirable for surface morphology); then, during the second period ($t_2 = 120$ s), both beams were shuttered off to allow the excess Ga to desorb, leaving a surface with a wetting layer but without droplets, ready for the next growth pulse. The essential requirement was to ensure that, over the range of temperatures present on the substrate, the excess Ga flux was everywhere sufficient to push growth into the droplet regime.

The problem of parallel conduction in the thick GaN film was also a serious one, and various strategies were adopted to avoid it. Initially attempts were made to neutralise the n-type doping with compensating acceptors such as Be, Mg or C and, while this could be made to work successfully, it required very careful control and was clearly not a technique to recommend itself to anyone looking to establish a commercial process. It was clearly important to understand the mechanism involved and the Santa Barbara group again came up with the answer (Corrion et al. 2006). They first of all demonstrated a very strong correlation between parallel conduction and the Al:N flux ratio employed in growing the AlN nucleation layer—in this instance it was essential to use nitrogen-rich conditions. By studying the Si concentration in AlN/GaN test structures, they then showed that Si from the SiC substrate dissolved in liquid Al in Al-rich conditions and was transported to the AlN/GaN interface, where it doped the GaN buffer layer. The

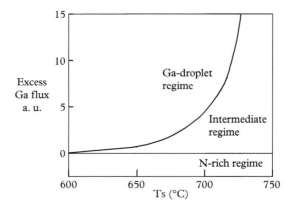

Figure 9.11 *Growth diagram showing the relationship between excess Ga flux and substrate temperature for MBE growth of GaN. Three distinct regimes can be distinguished, characterised by Ga-droplet formation at one extreme, N-rich growth at the opposite extreme and an intermediate regime in which growth is Ga-rich but not sufficiently to produce Ga droplets. To obtain smooth surfaces, growth should be maintained close to the upper demarcation line (see Figure 7.3).*

overall secret of making high-power HEMTs by PAMBE apparently lay in growing the nucleation layer under N-rich conditions and then switching to pulsed mode for the GaN buffer. We have already quoted examples of the transistor characteristics which resulted.

Clearly, excellent progress had been made in developing high-power, high-frequency HEMTs but there were still limitations associated with Ga-rich PAMBE, apparently related to high dislocation densities. More recently, therefore, effort has gone into the alternatives of N-rich PAMBE and NH_3-MBE. While not wishing to introduce national bias into a totally unbiased scientific discussion, we cannot help noticing that French scientists showed a marked preference for NH_3-MBE as a means for producing nitride HEMTs, the French company Picogiga apparently employing it in production. They also made a major contribution to the study of the use of silicon substrates. As early as 1999, the group in CNRS, Valbonne, reported on NH_3-MBE growth of GaN on Si (111) (Semond et al. 1999), showing that low temperature luminescence spectra were dominated by donor-bound-exciton recombination and that AlGaN/GaN quantum well luminescence compared closely with that from similar samples grown on sapphire substrates. RHEED oscillations indicated two-dimensional growth of both GaN and AlGaN alloys. Two-inch Si (111) wafers were outgassed at 600 °C, then rapid-thermally annealed at 950 °C to remove the native oxide before the growth of a thin AlN buffer layer. This was achieved by depositing 1–3 MLs of Al at ~400 °C, then exposing the surface to NH_3 at 850 °C. A 1.5 μm layer of GaN was then grown at 800 °C (notice that this is almost 100 °C hotter than the temperature commonly used for PAMBE growth of GaN). They followed this in 2001 with an account of the growth of AlGaN/GaN HEMT structures (Semond et al. 2001). As they pointed out, a major difficulty with GaN growth on Si is a marked tendency for cracking during cool-down—the thermal expansion coefficient of GaN $\left(\alpha = 5.59 \times 10^{-6} \, \text{K}^{-1}\right)$ being more than double that of Si $\left(\alpha = 2.59 \times 10^{-6} \, \text{K}^{-1}\right)$. To minimise this tendency they grew the following structure: 40 nm AlN nucleation layer, 250 nm GaN, 250 nm AlN, 2.5 μm undoped GaN buffer, 25 nm $Al_{0.31}Ga_{0.69}N$ barrier and 1 nm undoped GaN cap. This allowed the growth of crack-free GaN buffer layers with dislocation densities ~$5 \times 10^{13} \, \text{m}^{-2}$, sheet-carrier density = $1.25 \times 10^{17} \, \text{m}^{-2}$, and electron mobility = $0.148 \, \text{m}^2 \, \text{V}^{-1} \, \text{s}^{-1}$. HEMT devices with 0.3 μm gates yielded power densities of the order of 2 W/mm at 10 GHz (comparing very well with earlier device results on Si substrates). More recently, Renneson et al. (2013) optimised $Al_{0.29}Ga_{0.71}N$/GaN HEMT structures, grown by NH_3-MBE on Si (111) substrates, and measured a power density of 1.5 W/mm at 40 GHz, further enhancing the promise of this approach.

While Si (111) was appropriate to the growth of free-standing microwave HEMTs, the alternative application in fast-switching ICs demanded either (001) or (110) substrates in the interest of compatibility with well-established Si technology. The Valbonne group duly responded. In 2005 they reported HEMTs grown on Si (001) (Joblot et al. 2005) and in 2008 on Si (110) (Cordier et al. 2008). The familiar difficulty associated with anti-site domains was minimised by using substrates misoriented by 5° towards [110] and by the use of a high-temperature (~1150 °C) thermal anneal prior to growth, which was known to stabilise double atomic steps on Si (001) surfaces. The sheet-carrier density and electron mobility obtained on Si (001) were relatively modest ($4.2 \times 10^{16} \, \text{m}^{-2}$ and $0.073 \, \text{m}^2 \, \text{V}^{-1} \, \text{s}^{-1}$) and HEMT performance was less good than obtained on Si (111), probably on account of high densities of dislocations. Somewhat improved performance was reported a year later (Joblot et al. 2006) but results on Si (110) were comparable with those obtained on Si (111), suggesting that this might represent the way ahead. On the other hand, the same group also explored the use of alternative substrates in the shape of Si-on-polySiC (Cordier et al. 2006), GaN-on-sapphire (Cordier et al. 2007) and free-standing GaN (Fontserè et al. 2013) which, perhaps, gives the impression that the way ahead is, as yet, far from clear.

The paper by Fontserè et al. is concerned with another important aspect of high-power HEMTs, namely their ability to function at high ambient temperatures, and reports measurements up to 300 °C on identical device structures grown on three different substrates: Si (111), MOCVD GaN-on-sapphire and free-standing GaN. They dismiss the alternative of SiC as being 'prohibitively expensive'! Thermal stabilities were found to improve in the order sapphire, silicon and GaN, presumably due to increasing thermal conductivity but it is hard to believe that free-standing GaN substrates can be regarded as serious contenders for other than a few very specific applications. We can only repeat our impression that the long-term future of nitride HEMTs is still unclear.

Finally, we shall comment briefly on recent work reported from other sources. An interesting collaboration between the University of Illinois and Kyungpook National

University, Republic of Korea, was concerned with the use of PAMBE-SAG (selective area growth) to grow AlGaN/GaN HEMTs on sapphire/MOCVD GaN substrates (Pang et al. 2014). By connecting as many as 13 units in parallel, they obtained a total output power of ~20 W. A group from the University of Notre Dame, Indiana, produced p-channel heterostructure FETs by MBE growth on sapphire, achieving hole sheet densities as high as 5×10^{17} m^{-2}, very significantly higher than ever reported for n-channel devices (Li et al. 2013). Finally, we note again that Kaun et al. (2013) suggest there may be lasting advantage in using high-temperature N-rich PAMBE or NH$_3$-MBE rather than the more popular Ga-rich PAMBE. They point out that the latter suffers from two disadvantages: the fact that dislocations are decorated with excess Ga, thereby generating leakage paths. and the need for very precise control of temperature and the Ga:N flux ratio, as we discussed above. By using N-rich PAMBE at temperatures close to 800 °C and Ga:N flux ratios of approximately 0.8, it is possible to obtain smooth surfaces, the higher temperature favouring adatom surface mobility. Experimental sheet-carrier density and electron mobility leave something to be desired but HEMT performance (on SiC substrates) looks distinctly promising. We have already discussed the virtues of NH$_3$-MBE (where N:Ga ratios tend to be much larger, typically ~1000, and temperature control is far less critical) in connection with the French contribution. Their success has, indeed, been tangible and there seems little doubt that MBE-grown nitride HEMTs show considerable promise but, as we implied earlier, it still remains to be seen what the future will bring.

9.3   Heterojunction Bipolar Transistors (HBTs)

In discussing the MBE contribution to transistor development, we should not forget that the bipolar device came first (the point-contact Ge transistor made its dramatic debut just before Christmas 1947), and dominated the solid state electronics scene during its early years. It was only with the discovery of the remarkably low interface state density of the Si/SiO$_2$ interface in 1958 that a viable MOS transistor could be realised and it was not until the 1960s that such devices were introduced into integrated circuits. Indeed, it was not until the development of CMOS transistor pairs towards the end of the 1960s that FET-type transistors took over the dominant role in IC design, and one reason for this was the struggle to make bipolar devices which could switch at the desired speed. The source of this difficulty lay in the need for a narrow base region to minimise the transit time between emitter and collector, which, in turn, demanded heavy doping of the base in order to achieve acceptably low base resistance (the RC time constant representing a further limitation to switching speed). However, a highly doped base meant, that for an n-p-n device, emitter-base current tended to be dominated by the injection of holes from the base into the emitter, when what was required for efficient transistor action was the emission of electrons from emitter into base. It began to take on the aspect of an insoluble impasse but, in fact, the solution had been available ever since 1948(!)—the concept of an HBT having been included in a patent filed by William Shockley during his period of post-point-contact transistor introspection, which produced the junction

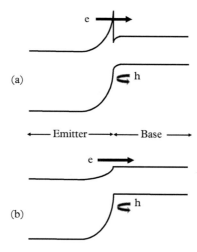

Figure 9.12 *Energy level diagram to illustrate the emitter-base junction of a typical n-p-n HBT. In (a) is shown the conduction-band spike and notch which occur at an abrupt junction, while (b) shows the effect of grading the emitter layer close to the junction. (Following Malik et al. 1983 and Kroemer 1982.)*

transistor (see Kroemer 1982). The idea was simple enough (as clever ideas so often are)—it was merely necessary to provide a barrier to hole injection by making the emitter from a material with a band gap somewhat larger than that of the base, as is made clear in Figure 9.12. If the step in the valence band could be chosen to be of order of a few times kT, hole injection would be minimised, no matter how heavy the p-type doping in the base region. The problem was, of course, that the necessary technology only became available with the development of compound semiconductors and, in particular, that of the lattice-matched AlGaAs/GaAs system, over 20 years later.

The first examples of HBTs using AlGaAs/GaAs emitter/base combination appeared during the early 1970s (see Kroemer 1982), being grown by LPE. Though these showed promise in demonstrating the principle of operation, it soon became clear that LPE lacked the necessary degree of control to reproduce narrow base layers on any kind of commercial scale, and attention turned to various other approaches, such as the use of a GaP emitter combined with Si base. Indeed, such a combination was grown by MBE in 1982, though performance was disappointing. (The problems of growing III-V compounds on Si have been referred to already.) However, as pointed out by Kroemer, the introduction of MBE and MOCVD growth techniques, with much improved control over layer thickness, down to the monolayer scale, changed the situation dramatically with respect to III-V materials and led to a resurgence of interest in the HBT during the

1980s. AlGaAs/GaAs HBT phototransistors were demonstrated in 1979 (by MOCVD) and 1982 (by MBE). The need to integrate appropriate drive circuitry with a GaAs laser also stimulated the use of AlGaAs/GaAs HBTs in the early 1980s. To quote from Kroemer's 1982 review article, this began to look like 'an idea whose time has come'. He went on to analyse the function of the HBT in some detail and discussed various possible innovations such as the grading of the junction region to eliminate the spike in the conduction band (see Figure 9.12) and possible alternative material combinations such as InP/GaInAsP and InAs/GaInSb. Readers seeking further understanding of HBT performance would be well advised to read Kroemer's review paper. He discusses, for example, the pros and cons of compositional grading—abrupt junctions result in CB spikes which reduce emitter efficiency, while, on the other hand, injecting electrons at high velocity ($v \sim 10^6$ m/s), offering the possibility of ballistic transport through the base. Here we see an interesting trade-off between speed and current gain. He also comments on the disadvantage associated with recombination of electrons trapped in the CB notch and points out that this can be removed (without altering the spike) by introducing a plane of acceptors at the heterojunction.

The possibilities offered by MBE growth were clearly demonstrated in 1982 by McLevige et al. (1982), who discussed the application of AlGaAs/GaAs HBTs to logic circuitry. As they pointed out, for I^2L (integrated injection) logic, the emitter should be located beneath the base, which meant growing the GaAs base on top of the AlGaAs emitter and this presented a problem for MBE, growth conditions having to be controlled extremely carefully in order to obtain an adequate interface quality (Morkoc et al. 1982). Rather pedestrian HBTs with current gain of the order of 100 were demonstrated for both GaAs-on-top and AlGaAs-on-top geometries (base width ~50 nm), though these were rapidly improved upon by the same Illinois group (Su et al. 1983). In this later work they grew double heterostructures, both emitter and collector being formed from $Al_{0.5}Ga_{0.5}As$. The use of a large Al content improved emitter efficiency and current gain—values as large as 1650 being measured on devices with base width of 100 nm—but also resulted in undesirable turn-on voltages, which, in turn, meant excess power dissipation. The symmetrical heterostructure was introduced with the aim of balancing out asymmetrical contact voltages, which it did to a degree, reducing the turn-on voltage from about 0.8 V to 0.2 V. These devices were grown on Si-doped GaAs substrates, employing a GaAs buffer doped with Si at 8×10^{23} m^{-3}. The 0.5 μm collector layer was doped with Si at 5×10^{23} m^{-3}, followed by a GaAs base doped with Be at 1×10^{24} m^{-3}, and a 0.5 μm emitter doped with Si at 2×10^{23} m^{-3}. The $Al_xGa_{1-x}As$ layers on both sides of the base were graded from $x = 0.5$ to $x = 0.08$ over a distance of 15 nm to minimise band-edge spikes. At this point, MBE growth had achieved performance comparable with the best LPE-grown devices and showed obvious promise for application to low-power, high-speed logic circuits.

A similar approach was adopted by Ankri et al. (1983) of Cornell University in making a phototransistor (a possible alternative to the avalanche photodiode, but with built-in gain). They employed an $Al_{0.25}Ga_{0.75}As$ emitter, together with a double base, part $Al_{0.12}Ga_{0.88}As$ and part GaAs, the total thickness being 130 nm. Measured response times of 250 ps suggested that electrons injected over the emitter–base barrier were swept out in a time small compared to the recombination time in the base. Maximum current gain for a device with a base width of 75 nm was 300. An interesting paper about the growth of AlGaAs/GaAs structures was published by Ito and Ishibashi (1985) of NTT. They were concerned with the commercial importance of using In-free substrate mounting to avoid In-contamination and to improve wafer throughput. Recognising that current gain must depend on non-radiative recombination within the base layer, which could be strongly influenced by variation in substrate temperature, they examined the variation of current gain across a radiatively heated HBT wafer. Growth temperature was 600 °C, uniform to better than 30 °C across a two-inch GaAs wafer, and current gains were found to be constant within about 10% from centre to edge, measured values being typically greater than 80.

Another early HBT innovation was described by Malik et al. (1983) of Bell Labs, who used MBE to grow an AlInAs/GaInAs/InP structure with a base width of 250 nm (on a semi-insulating (SI) Fe-doped InP substrate). This combination of materials is significant, of course, because it is compatible with lasers for fibre-optic communications based on the InGaAsP/InP system. They demonstrated the expected improvement in current gain resulting from grading the emitter–base interface to remove the CB spike, measuring values of 280 for the graded case, compared with 140 for the ungraded. An interesting, but somewhat ephemeral, variation on this particular theme was the MBE growth of AlGaAs/GaAs HBTs on InP substrates by the NTT group (Ito and Ishibashi 1987). They demonstrated functional devices but the maximum current gains achieved were no greater than 5, leaving the field to the more conventional AlGaInAs material system, which was taken up by several different laboratories.

As an example, we quote recent MBE work at the Fraunhofer Institute in Freiburg (Driad et al. 2011, 2012), which has been concerned with the effect of grading both composition and doping of the base layer. In the first paper the authors used a lattice-mismatched InGaAs/GaAs strained-layer superlattice base which was digitally graded to induce an electric field of ~6.6 $MV\,m^{-1}$ (as they point out, accurate grading is much easier to achieve in this manner, compared to smoothly varying composition). This was compared with results from devices using a uniform $In_{0.53}Ga_{0.47}As$ base layer and showed an 18% improvement in current gain (though the maximum measured gain was only 58 for a 50 nm-base-width device). Their second paper was concerned with comparing four different device structures, the base consisting of (1) a lattice-matched, constant composition, uniformly doped InGaAs layer, (2) a compositionally graded InGaAs layer, (3) a digitally graded strained-layer superlattice and (4) a lattice-matched InGaAs layer with graded doping (from $2 \times 10^{25}\,m^{-3}$ at the collector junction to $7 \times 10^{25}\,m^{-3}$ at the emitter junction). The last three cases were designed to produce an electric field accelerating electrons across the base, to enhance current gain and response speed. Phosphorus for the InP emitter was provided from a valved cracker cell, as was the arsenic. Conventional solid sources were used to supply the Group III elements. Si was used as n-type dopant, C as p-type, the latter being supplied from a CBr_4 source. Growth temperature for the various base layers was chosen to be in the range 430 °C–450 °C, for which the hole density was shown to be close to that of the acceptors (as measured by SIMS). All base widths were 30 nm. Current gain was, in all cases, found to be improved by use of a graded base, the largest improvement being 42% for linear composition grading. This type of grading was then used in the design of a double heterostructure HBT which produced current gains as high as 95 and values of $f_T = 220\,\text{GHz}$, emphasising the promise of such devices for application to high-frequency transistors and high-speed digital circuitry. It would seem that solid source MBE was at least the equal of MOVPE as an appropriate growth technology.

The international nature of HBT research is illustrated by a recent paper from China (Beijing and Shanghai) (Teng et al. 2013), as it describes the use of gas source MBE to grow InP/InGaAs/InP structures, the As and P beams being obtained by thermal cracking of arsine and phosphine, respectively. A CBr$_4$ source was used to achieve p-type doping of the 65 nm graded base layer, at a level of $3 \times 10^{25}\,\text{m}^{-3}$, the measured hole mobility $\mu_H = 6.63 \times 10^{-3}\,\text{m}^2\,\text{V}^{-1}\,\text{s}^{-1}$ agreeing well with values reported by the Freiburg group. The current gain of a large area device was 60, consistent with the somewhat larger base width than that used by the German group.

As we saw already, in the case of HEMT development, the ever-widening range of appropriate material systems has encouraged a similar diversification in HBT research. In addition to the 'classic' AlGaAs/GaAs and the InP/InGaAs systems, we should take note of no less than four others: InGaP/GaAs, Si/SiGe, AlGaN/GaN and what might be called 'Sb-based' materials, and, historically, this is the order in which they appeared. Herb Kroemer (1983) had proposed the use of the lattice-matched In$_{0.49}$Ga$_{0.51}$P/GaAs combination on the basis of its advantageous band line-up, most of the offset being predicted to appear in the valence band, rather than, as with the AlGaAs/GaAs combination, in the conduction band. So it was appropriate that the first experimental study should emerge from his Santa Barbara laboratory (Mondry and Kroemer 1985). The InGaP emitter was grown by solid source MBE, P being supplied from a novel GaP cell, fitted with a baffle to remove unwanted Ga flux. Acceptor doping was with Be at a level of order $10^{25}\,\text{m}^{-3}$, while Si was used as donor in the emitter $\left(5 \times 10^{23}\,\text{m}^{-3}\right)$ and collector. In spite of the anticipated smaller CB spike, they emphasised the importance of grading the emitter close to its junction with the base. Their HBT devices were of large area so no high-frequency data were measured. Maximum current gain from devices with 150 nm base widths was limited to 30 by device burn-out at high current densities and there was evidence for space–charge recombination at the emitter–base junction, so this work could be seen as no more than a preliminary material study; but it, nevertheless, suggested the possibility of future success. However, progress was slow. Seven years later Abernathy et al. (1992) reported the growth of similar structures by MOMBE, using C as base dopant rather than Be, on the premise that C had the advantage of better thermal stability. Their principal success lay in obtaining a much improved ideality factor for the emitter–base junction, thus resulting in current gain holding up well at low current densities. However, the maximum gain was only 20, so, again, this represented little more than an exercise in material control. Significant device progress *was* achieved at the University of Manchester by Joe and Missous (2005), who again used solid source MBE with a GaP cell to grow their InGaP emitter layers. They used Be as p-type dopant and added the refinement of compositionally grading the base by using InGaAs with up to 10% In at the collector junction. HBTs with base widths of 100 nm showed significantly better performance than had been obtained previously, the current gain for a uniform base structure being 163, while for an InGaAs-graded base they measured gains as high as 397 and $f_T = 16.3$ GHz, much improved but still well below the performance available with AlGaAs/GaAs devices. In a much later paper Narang and Shukla (2013), who presented an extensive theoretical analysis of the performance of InGaP/GaAs HBTs, referred to current data on IQE devices which showed current gains of only 100–120, so one is driven to the overall conclusion that the InGaP/GaAs system may continue to show promise for some time yet.

Moving on to the Si/SiGe system, we examine the work of the Daimler Benz group in Ulm, Gruhle et al. (1992), who used MBE to grow Si/SiGe HBTs with current gains as high as 550 at room temperature (increasing to 13 000 at 77 K). They pointed out that straightforward Si bipolar transistors had reached a limiting transit frequency of 50 GHz, imposed by the compromise between, on the one hand, a narrow and, on the other, a low-resistivity base layer. The only way forward seemed to require a SiGe base which would allow much heavier base doping. They grew the complete structure without interruption, the emitter and collector being doped with Sb and the base with B. The important parameter was the Ge content of the base which lay in the range 21–28%, significantly greater than that employed in earlier work. Their initial attempts led to transistors with values of $f_T = 42$ GHz, which was then a record for MBE-grown devices, though showing little improvement over current bipolar transistor performance. However, a year later (Gruhle et al. 1993) they reported on HBTs with $f_T = 91$ GHz, an achievement resulting largely from a reduction in base thickness down to 20 nm. As the authors pointed out, MBE growth at temperatures in the region of 500 °C minimises diffusion of both dopant and Ge atoms and allows excellent control of base width and grading. It also avoids relaxation of the strained SiGe alloy material, as this relaxation becomes of increasing importance as the Ge fraction is increased. Yet further improvements were reported by the same group in 1994 (Schuppen et al. 1994). Reduction of the base width to 7 nm and an increase in Ge content to upwards of 40% resulted in current gains approaching 200, and f_T values as high as 116 GHz. It was clear that MBE was capable of producing Si/SiGe devices with optimum performance but it is not at all clear that this particular technology has ever been taken up commercially.

With regard to the application of the Group III nitrides to HBT development, we note that they were relatively late to the starting gate. As we discussed earlier, the possibility of doping GaN p-type with Mg only became apparent during the early 1990s, when Isamu Akasaki and Shuji Nakamura in Japan demonstrated their ability to grow (by MOVPE) p–n junctions for highly efficient LEDs and laser diodes. The first attempts to apply this technology to HBTs therefore had to wait until the second half of the 1990s and, even then met with a major difficulty. The Mg acceptor is characterised by a rather large ionisation energy, $E_A \sim 250$ meV, which implies low doping efficiency, the free hole density being of the order of 1% of the acceptor density. This puts a severe restriction on the base conductivity, with corresponding limitations on current gain and base transit time. Early attempts to make AlGaN/GaN HBTs by MBE (Ren et al. 2000; McCarthy et al. 2001—which rejoice, between them, in no less than 30 authors!) clearly suffered from this, room-temperature current gains being of the order of 5 (though increasing to 10 when devices were operated at 300 °C). Both groups compared devices grown by plasma-assisted-MBE and MOVPE on sapphire substrates and found favour with MBE in so far as it produced sharper Mg profiles in the base region. McCarthy et al. also showed that dislocations caused serious diode leakage currents, which are of considerable significance for materials that must be grown on lattice-mismatched substrates. Later work by Raman et al. (2012) used ammonia-MBE to grow the AlGaN/GaN structures and this resulted in much reduced leakage currents and improved device performance, though maximum current gains were still only 15 at room temperature. The reason for these improvements was probably associated with the change from the Ga-rich growth conditions in the case of plasma-MBE to N-rich for ammonia-MBE. The former was likely to produce Ga-filled dislocations which would provide leakage paths through the junction regions, while this was unlikely to be the case in ammonia-MBE. While nitride HBTs have potential for high-temperature operation and high-breakdown voltages, and therefore high-power applications, the difficulties involved make it unlikely that we shall see them in production for some little time yet. However, we should not dismiss them without reference to an interesting alternative studied by a Japanese group at Kyoto University (Miyake et al. 2012, 2013). Their approach involved combining a SiC collector and base with an MBE-grown GaN emitter, the structure being grown on a SiC substrate. It certainly has the advantage of allowing the base to be doped at levels as high as 5×10^{25} m^{-3}, though still being faced with the problem of a lattice-mismatched emitter–base junction. Studies of the junction I-V behaviour suggested significant leakage due to tunnelling via interface traps which could be reduced by reducing the base doping level—not a desirable procedure—or by the introduction of a thin AlN layer, which is characterised by a smaller mismatch. Finally, they obtained much improved performance by employing an AlGaN emitter in the shape of an AlN/GaN short-period superlattice but, even so, the maximum current gain was only 13. No RF data were presented—which probably sums up the present state of development of these devices.

Finally we should make mention of the application of antimony-based materials to HBT development. As we saw earlier, there is considerable interest in HEMT devices on account of the high electron mobilities characteristic of Sb-based semiconductors but one must not overlook possible HBT applications, too, as summarised by Yeh et al. (2013). Perhaps the first comment to be made concerns the wide range of material combinations—quoting from their abstract, 'Sb-based semiconductors incorporating heterostructures of InP, InAs, AlSb, InSb, GaSb, InGaAs, InGaSb, GaAsSb and InGaAsSb can be used for high speed, low power applications'—but we shall concentrate on just two of these: InP/GaAsSb and InP/InGaAsSb. We are already familiar with the virtues of lattice-matched InP/In$_{0.53}$Ga$_{0.47}$As HBTs, which have found application to 40 GB s^{-1} optical communications and whose frequency responses have reached almost to the terahertz band. Indeed, one may well ask 'why the need for any alternative?' The answer concerns the problem of the step in the CB at the base/collector junction in InP/InGaAs/InP double heterostructures. To minimise the current blocking associated with this, it is necessary to incorporate a graded composition layer which complicates growth procedures, and such apparently minor modifications can become of major significance when facing the challenge of large-scale commercial application.

The advantage offered by using Sb-based materials lies in the type II band alignment of the InP/GaAsSb heterojunction and its controlled changeover effected by the incorporation of In to form InGaAsSb (see Figure 9.13). The alloy $In_{0.37}Ga_{0.63}As_{0.89}Sb_{0.11}$, which is approximately lattice-matched to InP, has a band gap of 0.67 eV while showing a negligibly small step in the CB. This makes it ideal as base material to be combined with an InP emitter and, what is more, it is then possible to use $In_{0.53}Ga_{0.47}As$ as collector. Quoting again from Yeh et al., we also note that 'the epilayers are typically grown on Fe-doped semi-insulating (100) InP substrates in a solid source MBE system, equipped with arsenic, phosphorus and antimony valve cracker cells. Silicon and beryllium are utilised as the n-type and p-type dopants, respectively.' Further advantages of this structure include a reduced turn-on voltage (0.35 V; cf. 0.51 V for the conventional InGaAs device), high electron velocity, long recombination lifetime and low base resistivity. It is also possible, by small variations in composition of the base material, to use strain to improve device characteristics still further. HBTs with 42 nm base widths have shown values of f_T as high as 238 GHz (some 7% greater than identical devices using InGaAs bases).

Finally, even greater advantage can be gleaned from the use of an $In_{0.52}Al_{0.48}As$ emitter, combined with an InGaAsSb base. The wider gap of this emitter (1.53 eV compared with 1.35 eV of InP) allows greater Sb content in the base, while still maintaining a type I emitter–base junction and type II base–collector junction. HBTs with 44 nm base width have demonstrated turn-on voltages of 0.38 V, current gains as high as 130 and f_T of 260 GHz. The potential for terahertz operation is clear.

In summary, one may simply record the obvious fact that HBTs have come a long way since the days when Si bipolar devices were labouring to reach operating frequencies

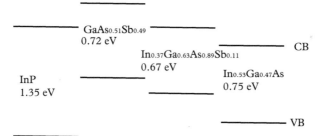

Figure 9.13 *Some important band line-ups for materials lattice-matched to InP and used in HBT design. The GaAsSb alloy shows a type II band line-up which is valuable in improving base/collector characteristics in InP/GaAsSb/InP double heterostructures. Incorporation of In can be used to achieve an accurate alignment of conduction bands while maintaining an effective barrier to hole injection into the emitter layer. The use of an InGaAs collector results in a built-in field which sweeps electrons to the collector contact.*

of 100 MHz as required for FM radio. Equally clear is the major contribution made to this impressive progress by the finely controlled growth of complex material structures successfully achieved by the several versions of molecular beam epitaxy.

9.4 Hall Effect Devices

Perhaps the best known example of a 'commercial' application for the Hall effect is its use as a resistance standard. We discussed the quantum Hall effect in Chapter 6 (Section 6.6). It was discovered in 1980 in the 2D electron gas occurring at a Si/SiO_2 interface and taken up with remarkable intensity in AlGaAs/GaAs 2DEGs during the early 1980s. The experimental observation can be summed up in the statement that a plot of Hall resistance R_H (i.e. Hall voltage divided by sample current) against magnetic field B showed a sequence of plateaux at precisely defined values, given by

$$R_i = h/ie^2 \tag{9.2}$$

where h is Planck's constant, e is the electron charge, and i is a so-called filling factor which defines the individual plateaux (i = 1, 2, 3, 4, ...). Given that h and e were known to considerable accuracy, it soon became clear that equation (9.2) defined a precise value of resistance which could be used as an international standard—assuming, of course, that the quantum Hall effect itself was independent of material inconsistencies and experimental uncertainties.

As explained in the elementary account of the effect given in Orton (2004), Chapter 6, it is much easier to observe it in these latter structures because of the smaller electron effective mass in GaAs compared with Si. Two criteria must be satisfied. First, it is essential that the electrons complete several cyclotron orbits before being scattered, a condition which is most conveniently expressed as

$$\mu_e B \gg 1 \tag{9.3}$$

where μ_e is the electron mobility, and B is the magnetic field, normal to the plane of the 2DEG. The smaller the electron effective mass, and therefore, the larger the mobility, the smaller the magnetic field required for observing the Hall plateau. The second criterion was based on the need for the energies of adjacent Landau levels to be spaced much wider apart than kT so that thermal transfer between them was minimised and, as the cyclotron frequency $\omega_c = eB/m_e$, the smaller effective mass in GaAs implied that this second criterion could be satisfied at higher temperatures (or at lower magnetic fields at the same temperature).

Of greater interest to the crystal grower was the question (which we discussed briefly in Chapter 6) of the desired material quality—in particular, interface quality—and sheet-carrier density. As it turned out, the ease of observing well-defined Hall plateaux in AlGaAs/GaAs 2DEGs made relatively little demand on crystal quality. Indeed, it soon became clear that samples which showed the best plateaux were those characterised by a significant lack of perfection, observation of the effect demanding inhomogeneously broadened Landau levels with localised states in the wings of the energy distribution. Though most of the samples were grown by MBE, it was actually necessary, in some cases, to deliberately downgrade the quality in order to optimise their suitability as resistance standard samples, a rare experience for most materials scientists! There is one condition that we do need to consider, however, and that is the relation between magnetic field and sheet-carrier density n_s. The relationship has the form

$$R_i = B_i/en_s \qquad (9.4)$$

So, in order to reach the $i = 1$ plateau, for which $R_i \sim 26\,k\Omega$, we require that $B/n_s = 4.2 \times 10^{-15}\,Tm^2$. Thus, if $n_s = 1 \times 10^{15}\,m^{-2}$, we require $B = 4.2\,T$. (Because of silicon's larger effective mass, the corresponding figure for the Si/SiO_2 MOSFET would be 15 T.) Then, in order to satisfy equation (9.3), we require an electron mobility of at least $1\,m^2\,V^{-1}\,s^{-1}$. One may ask how this compared with readily available samples and the answer was 'very satisfactorily'. There was no difficulty in meeting such a specification at a temperature of 4 K. It was only when meeting the challenge of the fractional quantum Hall effect that MBE growers were obliged to put their metaphorical backs into it! But that need not concern us here.

In practice, those concerned with establishing a practical resistance standard chose to work with the $i = 2$ or $i = 4$ plateau ($R \sim 12.9\,k\Omega$ or $R \sim 6.4\,k\Omega$), which could be reached at lower fields and which could more readily be compared with the wire-wound resistors previously employed as standards. The first challenge was to investigate the degree of reproducibility between different AlGaAs/GaAs samples and the degree of generality (e.g. how did these compare with Si/SiO_2, for instance?). The answer to the first question appears to be that the uncertainty is no more than a few parts in 10^{11} (and a few parts in 10^{12} might well be possible!), while the answer to the second question can be expressed as a few parts in 10^{10}. The reader interested in following up this particular topic might like to consult the paper by Janssen et al. (2013) as that is where these remarkable statistics were gleaned. It is primarily concerned with the study of the quantum Hall effect in graphene, and one may reasonably suppose that MBE growth of that exciting new material will soon be contributing to further progress. (However, we must continue to remind ourselves that we are concerned, here, with history!)

Our final topic in this brief look at the application of the Hall effect concerns its use in measuring magnetic fields in down-to-earth situations such as applications in brushless electric motors. The first commercial application of Hall effect sensors dates back to the mid 1950s (Wang et al. 1997) and (we admit to having been unaware of it) these devices have been widely used in such equipment as video cassette recorders, floppy disc drives, CD-ROMs and DVD players. In 1995 about 800 million InSb thin film Hall elements were sold commercially (Shibasaki 1997), the total world market being well over a billion, and there is a growing demand for cheap, reliable sensors in the automotive industry (Behet et al. 1998). To our certain knowledge, there is interest in China, Japan and Belgium—there may well be several other centres. But what has this to do with MBE? Many such sensors are grown by MBE in the form of InAs epilayers grown on GaAs substrates and we are assured that this technology forms the basis for today's mass production technology (I. Shibasaki, private communication).

The criterion for a good sensor material is that it should have low free-carrier density and high mobility. It should also generate a Hall voltage which remains sensibly constant over a wide temperature range—particularly important in the context of an automobile engine compartment. Early sensors consisted of InSb bulk crystals which had high electron mobility but, because of the small energy gap $\left(E_g = 0.18\,\text{eV}\right)$, resulting in high intrinsic carrier density, they could not satisfy the last of these criteria. InAs $\left(E_g = 0.36\,\text{eV}\right)$ bulk material proved to be no more suitable, so many sensors depended on Si or GaAs, in spite of their relatively poor carrier mobilities. It was the development of epitaxial InAs growth on GaAs substrates which led to the present commercial success of Hall sensors, MBE growth being described both from China (Wang et al. 1997) and Japan (Shibasaki 1997). As we are well aware, InAs growth on GaAs results initially in quantum dot formation but, if growth is continued to a depth of about a micron, the 2D growth mode is recovered and adequate quality films result. Si doping at a level of about $1 \times 10^{23}\,\text{m}^{-3}$ represented a suitable compromise in providing electron mobility of the order of 1.0–$2.0\,\text{m}^2\,\text{V}^{-1}\,\text{s}^{-1}$, which remained fairly constant over the temperature range of 100–$400\,\text{K}$ (Wang et al. 1997). Even better results were obtained by Shibasaki (1997) using a 15 nm InAs quantum well, within AlGaAsSb barrier layers which were lattice-matched to InAs. These barriers appeared to absorb many of the defects arising from the mismatch with the GaAs substrate and led to significantly improved electron mobility—$\mu_e \sim 2$–$3\,\text{m}^2\,\text{V}^{-1}\,\text{s}^{-1}$—and correspondingly improved sensitivity. The workers at IMEC in Belgium (Behet et al. 1998) followed a similar path, growing 15 nm InAs wells within AlGaSb barriers on SI GaAs substrates. They employed nucleation layers of AlAs and AlSb, followed by buffer layers of $Al_{0.2}Ga_{0.8}Sb$ ($1.1\,\mu$m) and GaSb ($0.5\,\mu$m) and then a short-period superlattice of GaSb and AlSb before the quantum well structure. Room-temperature electron mobility was measured to be $2.76\,\text{m}^2\,\text{V}^{-1}\,\text{s}^{-1}$, and magnetic sensitivity was similar to that of InSb elements—roughly four times better than a comparable GaAs sample.

That concludes our account of the MBE contribution to electronic device development—which has certainly been of considerable significance. We now move on to a similar discussion of the even greater impact that MBE has had on optical devices.

··

REFERENCES

Abernathy, C R, Ren, F, Wisk, P W, Pearton, S J and Esagui, R (1992) Improved performance of carbon-doped GaAs base heterojunction bipolar transistors through the use of InGaP. Appl Phys Lett 61, 1092.

Adams, A R (1986) Band-structure engineering for low-threshold high-efficiency semiconductor lasers. Electron Lett 22, 249.

Ambacher, O, Smart, J, Shealy, J R, Weimann, N G, Chu, K, Murphy, M, Schaff, QW J, Eastman, L F, Dimitrov, R, Wittmer, L, Stutzmann, M, Rieger, W and Hilsenbeck, J (1999)

Two-dimensional electron gases induced by spontaneous and piezoelectric polarization charges in N-and Ga-face AlGaN/GaN heterostructures. J Appl Phys 85, 3222.

Ankri, D, Schaff, W J, Barnard, J, Lunardi, L and Eastman, L F (1983) High-speed GaAs heterojunction bipolar phototransistor grown by molecular beam epitaxy. Electron Lett 19, 278.

Asbeck, P M, Yu, E T, Lau, S S, Sullivan, G J, Van Hove, J and Redwing, J (1997) Piezoelectric charge densities in AlGaN/GaN HFETs. Electron Lett 33, 1230.

Ashley, T, Barnes, A R, Buckle, L, Datta, S, Dean, A B, Emery, M T, Fearn, M, Hayes, D G, Hilton, K P, Jefferies, R, Martin, T, Nash, K J, Phillips, T J, Tang, W A, Wilding, P J, and Chau, R (2004) 'Novel InSb-based Quantum Well Transistors for Ultra-high Speed, Low Power Logic Applications' in 'Proceedings of the 7th International Conference on Solid-State and Integrated Circuits Technology, 2004', vol. 3, pp. 2253–2256, 18–21 Oct. 2004, doi: 10.1109/ICSICT.2004.1435293.

Ashley, T, Buckle, K L, Emeny, M T, Fearn, M, Hayes, D G, Hilton, K P, Jefferies, R, Martin, T, Phillips, T J, Powell, J, Tang, A W H, Wallis, D and Wilding, P J (2006) 'Indium Antimonide Based Quantum Well FETs for Ultra-High Frequency, Low Power Dissipation Circuits' in 'Proceedings of the 1st European Microwave Integrated Circuits Conference, 2006', p 29.

Ashley, T, Dean, A B, Elliott, C T, Pryce, G J, Johnson, A D and Willis, H (1995) Uncooled high-speed InSb field-effect transistors. Appl Phys Lett 66, 481.

Asif Khan, M, Bhattarai, A, Kuznia, J N and Olsen, D T (1993) High electron mobility transistor based on a GaN-$Al_xGa_{1-x}N$ heterojunction. Appl Phys Lett 63, 1214.

Asif Khan, M, Kuznia, J N, Olsen D T, Schaff, W J, Burm, J W and Shur, M S (1994) Microwave performance of a 0.25 μm gate AlGaN/GaN heterostructure field effect transistor. Appl Phys Lett 65, 1121.

Bandy, S G, Collins, D M and Nishimoto, C K (1979) Low-noise microwave f.e.t.s fabricated by molecular-beam epitaxy. Electron Lett 15, 218.

Behet, M, Das, J, De Boeck, J and Borghs, G (1998) InAs/(Al,Ga)Sb quantum well structures for magnetic sensors. IEEE Trans Magnetics 34, 1300.

Bennett, B R, Ancona, M G, Boos, J B and Shanabrook, B V (2007) Mobility enhancement in strained p-InGaSb quantum wells. Appl Phys Lett 91, 042104.

Bennett, B R, Magno, R, Boos, J B, Kruppa, W and Ancona, M G (2005) Antimonide-based compound semiconductors for electronic devices: a review. Sol St Electron 49, 1875.

Boos, J B, Yang, M J, Bennett, B R, Park, D, Kruppa, W and Yang, C H (1998) 0.1 μm AlSb/InAs HEMTs with InAs subchannel. Electron Lett 34, 1525.

Chen, S -H, Chang, L, Chang, E Y and Chang, C -Y (2002) Low-voltage-operation high-power-density AlGaAs/InGaAs enhancement-mode pseudomorphic high-electron-mobility transistor for personal handy-phone handset application. Jap J Appl Phys 41, L20.

Chen, J, Puzyrev, Y S, Zhang, C X, Zhang, E X, McCurdy, M W, Fleetwood, D M, Schrimpf, R D, Pantelides, S T, Kaun, S W, Kyle, E C H and Speck, J S (2013) Proton-induced dehydrogenation of defects in AlGaN/GaN HEMTs. IEEE Trans Nuc Sci 60, 4080.

Cheng, K Y, Cho, A Y, Drummond, T J and Morkoc, H (1982) Electron mobilities in modulation doped $Ga_{0.47}In_{0.53}As/Al_{0.48}In_{0.52}$ As heterojunctions grown by molecular beam epitaxy. Appl Phys Lett 40, 147.

Cheng, K Y, Cho, A Y and Wagner, W R (1981a) Molecular-beam epitaxial growth of uniform $Ga_{0.47}In_{0.53}As$ with a rotating sample holder. Appl Phys Lett 39, 607.

Cheng, K Y, Cho, A Y, Wagner, W R and Bonner, W A (1981b) Molecular beam epitaxial growth of uniform $In_{0.53}Ga_{0.47}As$ on InP with a coaxial In-Ga oven. J Appl Phys 52, 1015.

Chertouk, M, Heiss, H, Xu, D, Kraus, S, Klein, W, Bohm, G, Trankle, G and Weimann, G (1996) Metamorphic InAlAs/InGaAs HEMTs on GaAs substrates with a novel composite channels design. IEEE Elcctron Dev Lett 17, 273.

Cho, A Y and Arthur, J R (1975) Molecular beam epitaxy. Prog Sol St Chem 10, 157.

Cho, A Y, DiLorenzo, J V, Hewitt, B S, Niehaus, W C, Schlosser, W O and Radice, C (1977) Low-noise and high-power GaAs microwave field-effect transistors prepared by molecular beam epitaxy. J Appl Phys 48, 346.

Cordier, Y, Azize, M, Baron, N, Chenot, S, Tottereau, O and Massies, J (2007) AlGaN/GaN HEMTs regrown by MBE on epi-ready semi-insulating GaN-on-sapphire with inhibited interface contamination. J Cryst Growth 309, 1.

Cordier, Y, Chenot, S, Laugt, M, Tottereau, O, Joblot, S, Semond, F, Massies, J, Di Cioccio, L and Moriceau, H (2006) Growth by molecular beam epitaxy of AlGaN/GaN high electron mobility transistors on Si-on-polySiC. Superlattice Microst, 40, 359.

Cordier, Y, Moreno, J-C, Baron, N, Frayssinet, E, Chenot, S, Damilano, B and Semond, F (2008) IEEE Electron Dev Lett 29, 1187.

Corrion, A, Poblenz, C, Waltereit, P, Palacios, T, Rajan, S, Mishra, U K and Speck, J S (2006) Review of recent developments in growth of AlGaN/GaN high-electron mobility transistors on 4H-SiC by plasma-assisted molecular beam epitaxy. IEICE Trans Electron, 89 906.

Delagebeaudeuf, D, Delescluse, P, Etienne, P, Laviron, M, Chaplart, J and Linh, N T (1980) Two-dimensional electron gas MESFET structure. Electron Lett 16, 667.

Dimitrov, R, Mitchell, A, Wittmer, L, Ambacher, O, Stutzmann, M, Hilsenbeck, J and Rieger, W (1999) Comparison of N-face and Ga-face AlGaN/GaN-based high electron mobility transistors grown by plasma-induced molecular beam epitaxy. Jap J Appl Phys 38, 4962.

Dingle, R, Störmer, H L, Gossard, A C and Wiegmann, W (1978) Electron mobilities in modulation-doped semiconductor heterojunction superlattices. Appl Phys Lett 33, 665.

Driad, R, Aidam, R and Yang, Q (2012) Influence of built-in drift fields on the performance of InP-based HBTs grown by solid-source MBE. IEEE Trans Elect Devices 59, 1915.

Driad, R, Aidam, R, Yang, Q, Maier, M, Gullich, H, Schlechtweg, M and Ambacher, O (2011) InP-based heterojunction bipolar transistors with InGaAs/GaAs strained-layer-superlattice. Appl Phys Lett 98, 043503.

Drummond, T J, Kopp, W, Thorne, R E, Fischer, R and Morkoc, H (1982) Influence of $Al_xGa_{1-x}As$ buffer layers on the performance of modulation-doped field-effect transistors. Appl Phys Lett 40, 879.

Duran, H C, Klepser, B-U H and Bachtold, W (1996) Low-noise properties of dry gate recess etched InP HEMTs. IEEE Electron Dev Lett 17, 482.

Feng, M, Eu, V K, D'Haenens, I J and Braunstein, M (1982) Low-noise GaAs field-effect transistor made by molecular beam epitaxy. Appl Phys Lett 41, 633.

Fontserè, A, Pérez-Tomás, A, Placidi, M, Baron, N, Chenot, S, Moreno, J C and Cordier, Y (2013) Bulk temperature impact on the AlGaN/GaN HEMT forward current on Si, sapphire and free-standing GaN. ECS Sol St Lett 2, P4.

Gruhle, A, Kibbel, H, Erben, U and Kasper, E (1993) 91 GHz SiGe HBTs grown by MBE. Electron Lett 29, 415.

Gruhle, A, Kibbel, H, Konig, U, Erben, U and Kasper, E (1992) MBE-grown Si/SiGe HBTs with high beta, f/sub T/, and f/sub max. IEEE Electron Dev Lett 13, 206.

Hacker, J B, Bergman, J, Nagy, G, Sullivan, G, Kadow, C, Lin, H-K, Gossard, A C and Rodwell, M (2004) An ultra-low power InAs/AlSb HEMT Ka-band low-noise amplifier. IEEE Microw Wirel Compon Lett 14, 156.

Harmand, J C, Matsuno, T and Inoue, K (1989) Lattice-mismatched growth and transport properties of InAlAs/InGaAs heterostructures on GaAs substrates. Jap J Appl Phys, 28, L1101.

Inoue, K, Harmand, J C and Matsuno, T (1991) High-quality $In_xGa_{1-x}As$/InAlAs modulation-doped heterostructures grown lattice-mismatched on GaAs substrates. J Cryst Growth 111, 313.

Ito, H and Ishibashi, T (1985) MBE grown AlGaAs/GaAs HBTs with direct-radiation substrate heating. Jap J Appl Phys 24, 1567.

Ito, H and Ishibashi, T (1987) MBE-grown AlGaAs/GaAs HBTs on InP substrate. Electron Lett 23, 394.

Janssen, T J B M, Tzalenchuk, A, Lara-Avila, S, Kubatkin, S and Fal'ko, V I (2013) Quantum resistance metrology using graphene. Rep Prog Phys 76, 104501.

Joblot, S, Cordier, Y, Semond, F, Chenot, S, Vennéguès, P, Tottereau, O, Lorenzini, P and Massies, J (2006) AlGaN/GaN HEMTs grown on silicon (001) substrates by molecular beam epitaxy. Superlattice Microst, 40, 295.

Joblot, S, Semond, F, Cordier, Y, Lorenzini, P and Massies, J (2005) High-electron-mobility AlGaN/GaN heterostructures grown on Si(001) by molecular-beam epitaxy. Appl Phys Lett 87, 133505.

Joe, J H and Missous, M (2005) High-performance InGaP/GaAs HBTs with compositionally graded bases grown by solid-source MBE. IEEE Trans Electron Dev 52, 1693.

Judaprawira, S, Wang, W I, Chao, P C, Wood, C E C, Woodward, D W and Eastman L F (1981) Modulation-doped MBE GaAs/n-AlxGa1–xAs MESFETs. IEEE Electron Dev Lett 2, 14.

Kallfass, I, Pahl, P, Massler, H, Leuther, A, Tessmann, A, Koch, S and Zwick, T (2009) A 200 GHz monolithic integrated power amplifier in metamorphic HEMT technology. IEEE Microw Wirel Compon Lett 19, 410.

Kaun, S W, Wong, M H, Mishra, U K and Speck, J S (2013) Molecular beam epitaxy for high-performance Ga-face GaN electron devices. Semicond Sci Technol 28, 074001.

Kiehl, R A and Gossard, A C (1984) Complementary p-MODFET and n-HB MESFET (Al,Ga)As transistors. IEEE Electron Dev Lett 5, 521.

Ko, K -M, Seo, J -H, Kim, D -E, Lee, S -T, Noh, Y -K, Kim, M -D and Oh, J -E (2009) The growth of a low defect InAs HEMT structure on Si by using an AlGaSb buffer layer containing InSb quantum dots for dislocation termination. Nanotechnology 20, 225201.

Kroemer, H (1982) Heterostructure bipolar transistors and integrated circuits. Proc IEEE 70, 13.

Kroemer, H (1983) Heterostructure bipolar transistors: what should we build? J Vac Sci Technol B 1, 126.

Laviron, M, Delagebeaudeuf, D, Delescluse, P, Chaplart, J and Linh, N T (1981) Low-noise two-dimensional electron gas FET. Electron Lett 17, 536.

Lee, J -H, Yoon, H -S, Park, C -S and Park, H -M (1995) Ultra low noise characteristics of AlGaAs/InGaAs/GaAs pseudomorphic HEMT's with wide head T-shaped gate. IEEE Electron Dev Lett 16, 271.

Li, G, Wang, R, Song, B, Verma, J, Cao, Y, Ganguly, S, Verma, A, Guo, J, Xing, H G, and Jena, D (2013) Polarization-induced GaN-on-insulator E/D mode p-channel heterostructure FETs. IEEE Electron Dev Lett 34, 852.

Lin, H K, Kadow, C, Bae, J U, Rodwell, M J W, Gossard, A C and Brar, B (2005) Design and characteristics of strained InAs/InAlAs composite-channel heterostructure field-effect transistors. J Appl Phys 97, 024505.

Lin, Y C, Yamaguchi, H, Chang, E Y, Hsieh, Y C, Ueki, M, Hirayama, Y and Chang, C Y (2007) Growth of very-high-mobility AlGaSb/InAs high-electron-mobility transistor structure on si substrate for high speed electronic applications. Appl Phys Lett 90, 023509.

Ma, B Y, Bergman, J, Chen, P, Hacker, J B, Sullivan, G, Nagy, G and Brar, B (2006) InAs/AlSb HEMT and its application to ultra-low-power wideband high-gain low-noise amplifiers. IEEE Trans Microw Theory Tech 54, 4448.

Malik, R J, Hayes, J R, Capasso, F, Alavi, K and Cho, A Y (1983) High-gain Al0.48In0.52As/Ga0.53As vertical n-p-n heterojunction bipolar transistors grown by molecular-beam epitaxy. IEEE Electron Dev Lett 4, 383.

McCarthy, L S, Smorchkova, I P, Xing, H, Fini, P, Limb, J, Pulfrey, D L, Speck, J S, Rodwell, M J W, DenBaars, S P and Mishra, U K (2001) GaN HBT: toward an RF device. IEEE Trans Electron Dev 48, 543.

McLevige, W V, Yuan, H T, Duncan, W M, Frensley, W R, Doerbeck, H, Morkoc, H and Drummond, T J (1982) GaAs/AlGaAs heterojunction bipolar transistors for integrated circuit applications. IEEE Electron Dev Lett 3, 43.

Mimura, T, Hiyamizu, S, Fujii, T and Nanbu, K (1980a) A new field-effect transistor with selectively doped GaAs/n-Al$_x$Ga$_{1-x}$As heterojunctions. Jap J Appl Phys 19, L225.

Mimura, T, Hiyamizu, S, Hashimoto, H and Fukuta, M (1980b) WA-B5 high-electron mobility transistors with selectively doped GaAs/n-AlGaAs heterojunctions. IEEE Trans Electron Dev 27, 2197.

Mimura, T, Joshin, K, Hiyamizu, S, Hikosaka, K and Abe, M (1981) High electron mobility transistor logic. Jap J Appl Phys 20, L598.

Mimura, T, Nishiuchi, K, Abe, M, Shibatomi, A and Kobayashi, M (1985) Status and trends of hemt technology. Superlattice Microst 1, 369.

Mishra, U K, Brown, A S and Rosenblaum, S E (1988a) 'DC and RF Performance of 0.1 mu m Gate Length Al/sub 0.48/In/sub 0.52/As-Ga/sub 0.38/In/sub 0.62/As Pseudomorphic HEMTs' in 'Electron Devices Meeting, 1988. IEDM '88. Technical Digest., International', p 180.

Mishra, U K, Brown, A S, Rosenbaum, S E, Hooper, C E, Pierce, M W, Delaney, M J, Vaughn, S and White, K (1988b) Microwave performance of AlInAs-GaInAs HEMTs with 0.2- and 0.1-mu m gate length. IEEE Electron Dev Lett 9, 647.

Miyake, H, Aman, K, Kimoto, T and Suda, J (2012) Growth, electrical characterization, and electroluminescence of GaN/SiC heterojunction diodes and bipolar transistors fabricated on SiC off-axis substrates. Jap J Appl Phys 52, 124102.

Miyake, H, Kimoto, T and Suda, J (2013) AlGaN/SiC heterojunction bipolar transistors featuring AlN/GaN short-period superlattice emitter. IEEE Trans Electron Dev 60, 2768.

Mondry, M J and Kroemer, H (1985) Heterojunction bipolar transistor using a (Ga,In)P emitter on a GaAs base, grown by molecular beam epitaxy. IEEE Electron Dev Lett 6, 175.

Morkoc, H, Drummond, T J and Fischer, R (1982) Interfacial properties of (Al,Ga)As/GaAs structures: effect of substrate temperature during growth by molecular beam epitaxy. J Appl Phys 53, 1030.

Murphy, M J, Chu, K, Wu, H, Yeo, W, Schaff, W J, Ambacher, O, Eastman, L F, Eustis, T J, Silcox, J, Dimitrov, R and Stutzmann, M (1999) High-frequency AlGaN/GaN polarization-induced high electron mobility transistors grown by plasma-assisted molecular-beam epitaxy. Appl Phys Lett 75, 3653.

Nah, J, Fang, H, Wang, C, Takei, K, Lee, M H, Plis, E, Krishna, S and Javey A (2012) III–V complementary metal–oxide–semiconductor electronics on silicon substrates. Nano Lett 12, 3592.

Nainani, A, Bennett, B R, Boos, J B, Ancona, M G and Saraswat, C (2012) Enhancing hole mobility in III-V semiconductors. J Appl Phys 111, 103706.

Narang, V B and Shukla, S R (2013) Quantitative analysis of dc characteristics of Ga$_{0.5}$In$_{0.5}$P/GaAs heterojunction bipolar devices. J Vac Sci Technol B 31, 061201.

Nguyen, L D, Brown, A S, Thompson, M A, Jelloian, L M, Larson, L E and Matloubian, M (1992) 650-AA self-aligned-gate pseudomorphic Al/sub 0.48/In/sub 0.52/As/Ga/sub 0.2/In/sub 0.8/As high electron mobility transistors. IEEE Electron Dev Lett 13, 143.

Omori, M, Drummond, T J and Morkoc, H (1981) Low-noise GaAs field-effect transistors prepared by molecular beam epitaxy. Appl Phys Lett 39, 566.

Ono, H, Taniguchi, S and Suzuki, T -K (2004) Indium content dependence of electron velocity and impact ionization in InAlAs/InGaAs metamorphic HEMTs. Jap J Appl Phys 43, 2259.

Orton, J W (2004) 'The Story of Semiconductors', Oxford University Press, Oxford.

Osbourn, G C (1983) $In_xGa_{1-x}As - In_yGa_{1-y}As$ strained-layer superlattices: a proposal for useful, new electronic materials. Phys Rev 27, 5126.

Ouchi, K, Mishima, T, Kudo, M and Ohta, H (2002) Gas-source molecular beam epitaxy growth of metamorphic $InP/In_{0.5}Al_{0.5}As/In_{0.5}Ga_{0.5}As/InAsP$ high-electron-mobility structures on GaAs substrates. Jap J Appl Phys 41, 1004.

Ozgur, A, Kim, W, Fan, Z, Botchkarev, A, Salvador, A, Mohammad, S N, Sverdlov, B and Morkoc, H (1995) High transconductance normally-off GaN MODFETs. Electron Lett 31, 1389.

Palacios, T, Chakraborty, A, Rajan, S, Poblenz, C, Keller, S, DenBaars, S P, Speck, J S and Mishra, U K (2005) High-power AlGaN/GaN HEMTs for Ka-band applications. IEEE Electron Dev Lett 26, 781.

Pang, L, Krein, P, Kim, K-W, Lee, J-H and Kim, K (2014) High-current AlGaN/GaN high electron mobility transistors achieved by selective-area growth via plasma-assisted molecular beam epitaxy. Phys Stat Sol (a) 211, 180.

Pearsall, T P, Hendel, R, O'Connor, P, Alavi, K and Cho, A Y (1983) Selectively-doped Al0.48In0.52As/Ga0.47In0.53As heterostructure field effect transistor. IEEE Electron Dev Lett 4, 5.

Poblenz, C, Waltereit, P and Speck, J S (2005) Uniformity and control of surface morphology during growth of GaN by molecular beam epitaxy. J Vac Sci Technol B 23, 1379.

Radosavljevic, M, Ashley, T, Andreev, A, Coomber, S D, Dewey, G, Emeny, M T, Fearn, M, Hayes, D G, Hilton, K P, Hudait, M K, Jefferies, R, Martin, T, Pillarisetty, R, Rachmady, W, Rakshit, T, Smith, S J, Uren, M J, Wallis, D J, Wilding, P J and Chau, R (2008) 'High-Performance 40nm Gate Length InSb P-Channel Compressively Strained Quantum Well Field Effect Transistors for Low Power (V(CC) = 0.5 V) Logic Applications' in 'Proceedings of International Electron Devices Meeting', p. 727.

Raman, A, Hurni, C A, Speck, S and Mishra, U K (2012) AlGaN/GaN heterojunction bipolar transistors by ammonia molecular beam epitaxy. Phys Stat Sol (a) 209, 216.

Ren, F, Han, J, Hickman, R, Van Hove, J M, Chow, P P, Klaasen, J J, LaRoche, J R, Jung, K B, Cho, H, Cao, X A, Donovan, S M, Kopf, R F, Wilson, R G, Baca, A G, Shul, R J, Zhang, L, Willison, C G, Abernathy, C R and Pearton, S J (2000) GaN/AlGaN HBT fabrication. Sol St Electron 44, 239.

Renneson, S, Lecourt, F, Defrance, N, Chmielowska, M, Chenot, S, Lesecq, M, Hoel, V, Okada, E, Cordier, Y and De Jaeger, J -C (2013) Optimization of $Al_{0.29}Ga_{0.71}N$/GaN high electron mobility heterostructures for high-power/frequency performances. IEEE Trans Electron Devices 60, 3105.

Riemer, P J, Buhrow, R, Hacker, J B, Bergman, J, Brar, B, Gilbert, B K and Daniel, E S (2006) Low-power W-band CPWG InAs/AlSb HEMT low-noise amplifier. IEEE Microw Wirel Compon Lett 16, 40.

Schaff, W J, Tasker, P J, Foisy, M C and Eastman, L F (1991) 'Strained Layer Superlattices: Materials Science and Technology' in 'Semiconductors and Semimetals', Vol 33 (ed R K Willardson and A C Bccr), Academic Press, New York, p 73.

Schuppen, A, Gruhle, A, Kibbel, H Erben, U and Konig, U (1994) SiGe-HBTs with high f_T at moderate current densities. Electron Lett 30, 1187.

Semond, F, Damilano, B, Vezian, S, Grandjean, N, Leroux, M and Massies, J (1999) GaN grown on Si(111) substrate: from two-dimensional growth to quantum well assessment. Appl Phys Lett 75, 82.

Semond, F, Lorenzini, P, Grandjean, N and Massies, J (2001) High-electron-mobility Al-GaN/GaN heterostructures grown on Si(111) by molecular-beam epitaxy. Appl Phys Lett 78, 335.

Shibasaki, I (1997) Mass production of InAs Hall elements by MBE. J Cryst Growth 175, 13.

Stormer, H L, Baldwin, K W, Pfeiffer, L N and West, K W (1991) GaAs field-effect transistor with an atomically precise ultrashort gate. Appl Phys Lett 59, 1111.

Su, S L, Tejayadi, O, Drummond, T J, Fischer, R and Morkoc, H (1983) Double heterojunction AlxGa1-xAs/GaAs bipolar transistors (DHBJT's) by MBE with a current gain of 1650. IEEE Electron Dev Lett 4 130.

Su, S L, Thorne, R E, Fischer, R, Lyons, W G and Morkoc, H (1982) Influence of buffer thickness on the performance of GaAs field effect transistors prepared by molecular beam epitaxy. J Vac Sci Technol 21, 961.

Tacano, M, Sugiyama, Y and Takeuchi, Y (1991) Critical-layer thickness of a pseudomorphic $In_{0.8}Ga_{0.2}As$ heterostructure grown on InP. Appl Phys Lett 58, 2420.

Teng, T, Xu, A, Ai, L, Sun, H and Qi, M (2013) InP/InGaAs/InP DHBT structures with high carbon-doped base grown by gas source molecular beam epitaxy. J Cryst Growth 378, 618.

Tung, P N, Delagebeaudeuf, D, Laviron, M, Delescluse, P, Chaplart, J and Linh, N T (1982) High-speed low-power DCFL using planar two-dimensional electron gas FET technology. Electron Lett 18, 109.

Tuttle, G, Kroemer, H and English, J H (1990) Effects of interface layer sequencing on the transport properties of InAs/AlSb quantum wells: evidence for antisite donors at the InAs/AlSb interface. J Appl Phys 67, 3032.

Wang, H M, Zeng, Y P, Fan, T W, Zhou, H W, Pan, D, Dong, J R and Kong, M Y (1997) Characteristics of InAs epilayers for Hall effect devices grown on GaAs substrates by MBE. J Cryst Growth 179, 658.

Weaver, B D, Boos, J B, Papanicolau, N A, Bennett, B R, Park, D and Bass, R (2005) High radiation tolerance of InAs/AlSb high-electron-mobility transistors. Appl Phys Lett 87, 173501.

Werking, J D, Bolognesi, R, Chang, L -D, Nguyen, C, Hu, E L and Kroemer, H (1992) High-transconductance InAs/AlSb heterojunction field-effect transistors with delta-doped AlSb upper barriers. IEEE Electron Dev Lett 13, 164.

Wong, M H, Keller, S, Dasgupta, N S, Denninghoff, J, Kolluri, S, Brown, D F, Lu, J, Fichtenbaum, N A, Ahmadi, E, Singisetti, U, Chini, A, Rajan, S, DenBaars, S P, Speck, J S and Mishra, U K (2013) N-polar GaN epitaxy and high electron mobility transistors. Semicond Sci Technol 28, 074009.

Yeh, N -T, Chiu, P -C, Chyi, J -I, Ren, F and Pearton, S J (2013) Sb-based semiconductors for low power electronics. J Mat Chem C 1, 4616.

10

Optical Devices

10.1 GaAs Lasers

Development of the GaAs laser must rank as one of the highlights of semiconductor device innovation. It represented the optical equivalent of the transistor, providing, as it did, the basis for semiconductor optoelectronics. Just as the invention of the germanium transistor in 1947 led to the silicon integrated circuit and a dramatic invasion of semiconductor electronics into the very fabric of modern society, the invention of the GaAs laser in 1962 provided the basis for a similar, and arguably equally important, contribution from photonics. Numerous other semiconductor lasers have, of course, been demonstrated but it is clear that the GaAs version represents the prototype on which they all depend. We therefore describe its development in some detail (bearing in mind that it should strictly be referred to as the AlGaInAs laser) before moving on to consider these alternatives in later sections. First of all, however, we should explain yet another aspect of our presentational philosophy—lasers obviously grew out of initial work on LEDs but we nevertheless have not included a section on LEDs themselves for the simple reason that MBE has played only a very minor role in LED development. On the other hand, perhaps because laser structures are intrinsically more complicated than those employed in LED manufacture, MBE has played a much more significant part in the success of the semiconductor laser. It is also worth noting that, conversely, the GaAs laser played an important part in the development of MBE as a technology for growing very-high-quality semiconductor materials and, ultimately, semiconductor heterostructures with monolayer precision.

In order to appreciate the MBE contribution it may be helpful to begin with a brief outline of GaAs laser development over the period from its inception in 1962 as an inevitably crude demonstration of an existence theorem to the sophistication of the modern quantum dot laser with its threshold current density some four orders of magnitude lower than measured on the early pioneering devices. The first proposal of the concept of a p–n junction injection laser was probably that put forward by von Neumann as early as 1953 (but unpublished—see Bardeen 1963), the first publication being that of Basov et al. (1961) from the P. N. Lebedev Institute in Moscow, while the conditions for obtaining an inverted carrier population were clarified by Bernard and Duraffourg (1961) from the CNET laboratory at Issy-les-Moulineaux, in France. Efforts were concentrated on seeking stimulated emission from a p–n homojunction structure in 1962, once it became evident that radiative recombination in a GaAs p–n diode could be close to 100% efficient at low temperatures, an essential ingredient being the addition of a Fabry–Perot optical cavity. No less than four groups in the USA reported laser action from p–n diodes in September of that year (Hall et al. 1962 (GE Schenectady); Holonyak and Bevacqua 1962 (GE Syracuse); Nathan et al. 1962 (IBM); Quist et al. 1962 (Lincoln Labs)). With the exception of Holonyak, who used GaAsP, all the authors employed bulk samples of n-GaAs, Zn-diffused to form the p–n junction. All these devices operated at 77 K with (short pulsed) threshold current densities of 10^4 A/cm^2 or greater—room-temperature operation appeared to be out of the question. There was to be a lengthy period of attritional, nose-to-the-grindstone research before the commercially significant CW room-temperature operation could be achieved.

The two keys to success were, of course, those of material quality and the introduction of heterostructures, both of which required epitaxial growth— initially, LPE and, later, MBE and MOVPE. The concept of the double heterostructure, confining the recombining free carriers close to the junction, was proposed from both Russia (Alferov and Kazarinov 1963; see Alferov 2001) and from the USA (Kroemer 1963). The further use of heterostructures, forming an optical waveguide to concentrate radiation close to the recombination region, proposed independently from England (Thompson and Kirkby 1973) and from the USA (Hayashi 1972; see Casey et al. 1974), had to wait another ten years, and the first commercial application, that to the CD player, even longer— Sony and Philips finally marketed it in 1980. From the materials viewpoint, the crucial advance came with the realisation that it was essential for successful heterojunction development to find lattice-matched compounds and that this could be achieved within the AlGaAs system. In 1967 the application of LPE growth to AlGaAs at both IBM in Yorktown Heights and the Ioffe Institute in Leningrad (Alferov et al. 1967; Ruprecht et al. 1967) made possible the required defect-free heterostructures and opened the way to the long-sought room-temperature, CW operation, which was finally demonstrated in 1970 (Alferov et al. 1970; Hayashi et al. 1970). This success came from the use of DH structures and brought the threshold current density down from being greater than 10^4 A/cm^2 to being less than 10^3 A/cm^2. Further progress depended on the introduction of quantum well active regions in the early 1980s and of quantum dot active regions in the late 1990s. This is well summarised by the data plotted in Figure 10.1, taken from Alferov's Nobel Prize lecture, published in 'Reviews of Modern Physics' (Alferov 2001). Readers interested in a more detailed account of this little gem of semiconductor history may find it in Alferov's review, the interesting article by Depuis (1987) or Chapter 5 of Orton (2004).

While LPE certainly provided the basis for the initial achievement of RT CW operation, the introduction of quantum wells demanded the introduction of more sophisticated growth methods, exemplified by MOVPE and MBE, and these two techniques fought a lengthy battle for domination of the commercial laser world. Our purpose here is to describe the contribution made by MBE rather than to adjudicate between the rival technologies so we shall say nothing more concerning victories or defeats but simply catalogue the progress made by MBE in furthering the technical development of this vitally important photonic device.

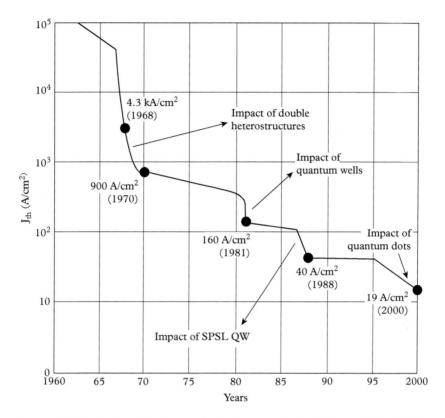

Figure 10.1 *Evolution of GaAs laser threshold current density $\mathcal{J}_{th}$ from 1962 to 2000, showing the effect of various technological innovations. (From Alferov 2001, courtesy American Physical Society.)*

MBE growth of a DH GaAs laser was first reported by Cho and Casey (1974) at Bell Labs. It can be seen as an important factor in their ambition to demonstrate the practical usefulness of the technique and as justification for the considerable financial investment required. Bell were seriously interested in developing efficient laser sources for the pioneering experiments in fibre-optic communication and, as it was yet to be realised that the quartz fibres used showed minimum loss at a wavelength of 1.55 μm and zero dispersion at 1.3 μm, a GaAs laser operating at 0.88 μm was seen as optimum choice for the signal source. As far as MBE was concerned, the immediate task was to achieve device performance comparable with that offered by LPE growth, and this first sally by Cho and Casey certainly showed promise. They grew a simple DH structure, using Sn and Mg as n- and p-type dopants and, though threshold currents $\left(J_{th} = 3.5 \times 10^4 \text{A/cm}^2\right)$ were of an order greater than those of the current LPE devices, a 2 h anneal at 750 °C brought them down to within a factor of 1.4 $\left(J_{th} = 4.0 \times 10^3 \text{A/cm}^2\right)$, the improvement resulting from a reduction of non-radiative centres somewhere within the structure. As we commented above, material quality was as vital to success as was the use of high-quality heterostructures, and MBE was clearly facing a challenge. Would it be possible to meet the demand for GaAs with adequately long non-radiative lifetimes? MBE was still very much in its infancy and there was scope for improvement, much of which was, as we shall see, a direct consequence of these attempts to grow more efficient lasers.

An interesting next step saw the application of MBE to the hybrid LPE/MBE growth of a distributed feedback (DFB) laser which included optical-confinement layers of $Al_{0.3}Ga_{0.7}As$ (Casey et al. 1975). The DFB structure included an ion-milled corrugation, forming a Bragg grating which had the virtue of selecting a single longitudinal optical mode, highly desirable in the context of fibre communications. However, there was a problem in introducing this to a simple DH laser, in so far as the milling process degraded the recombination efficiency of the active layer, so Casey et al. were concerned to separate it from the active layer by including it within the uppermost separate confinement layer. They grew the basic structure by LPE, then formed the grating but were unable to complete it because of the tendency for melt-back in LPE growth. MBE, of course, did not suffer from this difficulty and was therefore used to complete the structure, in the shape of a p-type $Al_{0.3}Ga_{0.7}As$ layer. The result was a DFB laser which operated at room temperature, a significant improvement over earlier devices which required cooling to 77 K. Noteworthy in the MBE context was the use of Be as p-type dopant. As we saw in Chapter 4 (Section 4.5), Be had the advantage over Mg with regard to its reactivity and diffusivity and came to be widely adopted by MBE growers worldwide but it is probably not well recognised that its first application was in the context of laser development.

Those concerned with advancing the application of MBE were never likely to be satisfied with the minor role of growing one additional layer on top of an LPE structure—their target had to be the growth of complete structures with performance at least as good as that offered by other growth methods. An important step in this direction came just one year later when Cho et al. (1976) reported CW operation at room temperature of a simple DH laser grown entirely by MBE. As they saw it, it was important to minimise the deleterious effects of carbon and oxygen, particularly with regard to the AlGaAs layers, so they concentrated on reducing the influence of hydrocarbons and water vapour. The principal modification concerned the introduction of a liquid nitrogen cryopanel to act as a getter, a feature now taken for granted, of course, in modern MBE machines. They also replaced the graphite source cells with pyrolytic boron nitride. In terms of the laser structure, they introduced an interesting change by grading the interfaces between GaAs and AlGaAs, on the basis that there is a small, but finite, difference in lattice parameter between them, and an abrupt interface was therefore likely to introduce defects which might induce non-radiative recombination. Here was a novel technique made possible by the fine control inherent in the MBE growth process. As dopants, they used Sn and Mn, noting that the Sn profile was far from sharp. As we now know, this results from surface segregation and eventually led to the rejection of Sn as n-type dopant in favour of the better-behaved Si. It is interesting to note that they still did not incorporate Be for p-doping—in fact, the full benefit of Be was only clarified in 1977 when Marc Ilegems published his detailed study of its properties (Ilegems 1977).

The year 1978 saw the beginning of a remarkable sequence of 'Applied Physics Letters' from another of the Bell Labs workers, Won Tsang, describing a series of MBE-grown DH lasers which, ultimately, achieved performance levels at least as good as and, in some cases, better than those attainable with LPE or MOVPE. He began by studying the reasons why MBE lasers performed less well than their rivals and traced this to non-radiative recombination at the interfaces between GaAs and AlGaAs layers and which resulted from the presence of trapping centres in the AlGaAs (Tsang 1978). It therefore became even more important to minimise the densities of residual gases, such as water vapour, CO_2 and O_2 in the MBE system and Tsang therefore introduced a load-lock system for exchanging samples. This, of course, meant that the growth chamber could be maintained under UHV for a long period of time, and the quality of both vacuum and laser performance improved in consequence. DH GaAs lasers showed threshold current densities at least as low as those measured on comparable LPE samples (Tsang 1979). (Following source refurbishment, three runs were sufficient to achieve stable performance again (Tsang et al. 1980).) At this time, Tsang had changed to using Be as p-type dopant but was still using Sn as n-type dopant and noted the importance of timing the switching of the Sn beam so as to place the p–n junction correctly within the GaAs active layer (hardly a plus point for MBE!—but the later introduction of Si as n-type dopant would change all that). Theory and experiment agreed that, as the active layer decreased in thickness, threshold current density should follow suit (though eventually reaching a minimum value)—Tsang (1980a) showed that MBE-grown samples fell in with this trend and measured values as low as 800 A/cm^2. He also showed (Tsang et al. 1980) that laser threshold current decreased with increasing substrate temperature in the range 450 °C–650 °C, while this correlated with improvement in PL efficiency of the GaAs active and AlGaAs confinement layers. Gradually, understanding of the links between growth parameters and laser performance was emerging.

The year 1980 also saw a significant extension to Tsang's repertoire in so far as he grew a series of lasers with AlGaAs active layers, reducing the lasing wavelength from 0.89 μm to 0.72 μm (Tsang 1980b) and this happened to be of particular significance in the context of the commercial launch of the CD player in the same year. It was clear that a reduction in wavelength would allow more music to be stored on each disc and, after a great deal of soul-searching, Philips and Sony had decided to standardise on a laser wavelength of 780 nm. This was all very well in principle but it put their technologists in a difficult position—no one had succeeded in making a reliable 780 nm laser at that time and the Philips player was initially launched with an 840 nm device. The difficulty lay in the growth of high-quality AlGaAs—while, previously, it may have been adequate for confining layers, it was just not good enough for use as the active layer, and something had to be done about it. Tsang was clearly a world expert in this particular specialisation and his demonstration of efficient short-wavelength devices grown by MBE raised important commercial questions. It certainly appeared to stimulate the Japanese company Rohm to choose MBE as a vehicle for the production of CD lasers, as we discuss in a moment. Indeed, yet another letter from Tsang (1981a) reported that DH lasers could be grown with high yield and good reproducibility using unusually high growth rates (as large as 11.5 μm/h) without compromising threshold current density—he quoted average values as low as 700 A/cm^2 for GaAs lasers and later (Tsang 1984) measured operating lifetimes of more than 10^6 h. Here was proof positive that MBE must be considered seriously as a possible technology for laser production and, while Bell, as a communication systems company, were obviously interested, they would (a) choose to outsource and (b) concentrate on 1.3–1.55 μm wavelength devices for long-distance transmission.

There can be little doubt, however, that Tsang's work on AlGaAs lasers was of more than passing interest to the Japanese company Rohm in the context of the CD laser market (Tanaka and Mushiage 1991)—not only did they opt for MBE but they actually produced something like half the CD lasers used. This was an important demonstration of the large-scale commercial possibilities for MBE growth as, prior to this, no one had thought it a serious possibility. It is interesting to examine Rohm's thinking as described in their paper (published, we should note, several years after the decision to take the MBE plunge). Initially, they saw two main problems, those of reproducibility and of designing a suitable stripe structure compatible with MBE growth. They also emphasised the importance of easy substrate mounting, large effusion cells (capable of 2400 h operation), precise temperature control and shutter reliability. Their plan was to run the MBE machines continuously, under computer control for 100 days, then reload the cells, pump down, bake for three days and restart growth. Using the PL efficiency of a DH structure as guide, they found that, after two days of growth, the system was in good order to grow successful lasers—single transverse mode, self-aligned AlGaAs DH structures. Formation of 4 μm stripes required an etching step so the lasers had to be taken out of the MBE machine, then returned for a second growth step, the use of a vacuum load lock being of paramount importance. The final structure consisted essentially of four layers—n-Al$_{0.6}$Ga$_{0.4}$As (Si: 5×10^{23} m^{-3}), an undoped Al$_{0.15}$Ga$_{0.85}$As active layer, p-Al$_{0.6}$Ga$_{0.4}$As (Be: 5×10^{24} m^{-3}), and a p$^+$-GaAs contact layer (Be: 5×10^{25} m^{-3}). (Note the use of Si and Be as dopants, the preferred choice of most MBE users from this time onwards.) The active layer was 0.07 μm in thickness, the current threshold density was 3 kA/cm^2 and the output power was 3 mW at a 780 nm wavelength. (For comparison, Tsang measured values of about 1 kA/cm^2 on a broad-area laser at the same wavelength.)

Tsang's work may already have been a major stimulant to the first commercial use of MBE but he was far from finished. Up to this point he had concentrated on simple DH laser structures but, as we saw earlier, the advent of optical confinement in addition to carrier confinement was to offer ultimate laser performance and there was every reason to believe that MBE could supply the necessary structures. Thus, later in 1981, (Tsang 1981b) he published a letter describing the growth of a graded-index, separate confinement (GRIN-SCH) laser with a threshold current density of 500 A/cm^2 and superior output-beam characteristics. This made use of MBE's ability to grow carefully controlled graded layers, in this instance by varying the temperature of the Ga cell when growing AlGaAs. Thus, the band gap of the laser structure was graded as shown in Figure 10.2, the graded region serving to form an optical waveguide (the refractive index of the AlGaAs decreasing as the band gap increased), while the step change confined recombining carriers in a very narrow (typically about 50 nm) central region. It is

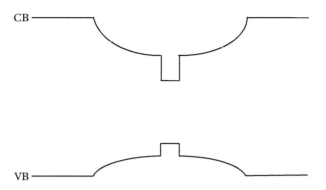

Figure 10.2 *Schematic band diagram of the GRIN-SCH laser structure introduced by Tsang (1981b). The graded regions serve to form an optical waveguide, confining the light beam close to the central recombination region.*

certainly not our purpose to attempt any detailed account of laser design principles but we can simply point out that the index profile has a direct influence on the quality of the emitted light beam. In particular, the far-field pattern can be chosen to match that demanded by specific optical waveguides. These GRIN-SCH lasers with parabolic, graded indices produced Gaussian beams with narrow beam divergence between 20° and 30°, normal to the junction plane. What was more, by careful analysis of the various loss parameters within these lasers, Tsang was able to adjust doping levels and active layer thickness so as to minimise threshold current density and achieved values as small as 160 A/cm^2 (Tsang 1982). Over a period of just four years, improvements in MBE technology (largely stimulated by the quest for better lasers—allied with Tsang's tenacity) had fostered improvements in material quality and device design that reduced threshold current densities by more than an order of magnitude. At the same time, the excellent control inherent in MBE growth played a vital role in making this possible.

As an interesting postscript to this account, we may, perhaps, be excused for recalling some results from our own MBE studies of AlGaAs/GaAs heterostructures, made in the context of GaAs laser development. Non-radiative recombination at the AlGaAs–GaAs interface had frequently been cited as responsible for poor laser performance, and oxygen was posited as being the most likely culprit. It was well known that Al had a propensity for combining with oxygen and there was evidence to suggest that O formed a deep centre within the AlGaAs band gap. Plausible though this certainly was, we felt it desirable to attempt quantitative verification by measuring the PL efficiency of a lightly n-doped Al$_{0.2}$Ga$_{0.8}$As layers as a function of the oxygen concentration, as measured by SIMS (secondary ion mass spectrometry; Foxon et al. 1985). The result was a very clear correlation between the two, as shown in Figure 10.3. On reloading the MBE system, oxygen concentrations as high as 8×10^{24} m^{-3} were measured, falling gradually as the system cleaned up to a few times 10^{22} m^{-3}. Over the same range, PL efficiency improved by three orders of magnitude. In order to relate this specifically to the behaviour of DH

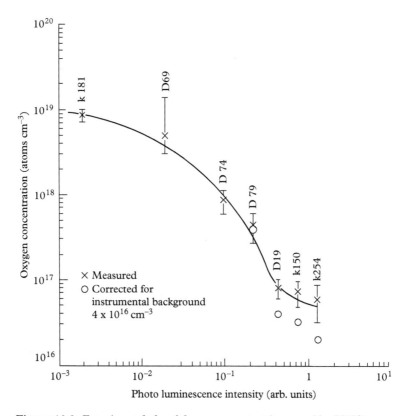

Figure 10.3 *Experimental plot of the oxygen content (measured by SIMS) vs the PL efficiency of AlGaAs films grown by MBE after reloading the machine. (From Foxon et al. 1985.) Reproduced with permission from Foxon, C T, Clegg, J B, Woodbridge, K, Hilton, D, Dawson, P and Blood, P (1985) J Vac Sci Technol B 3, 703. Copyright 1985, AIP Publishing LLC.*

structures, a further series of measurements was made to quantify the recombination velocity characteristic of recombination at the AlGaAs–GaAs interface as a function of various pre- and post- layers grown near the DH structure (Dawson and Woodbridge 1984). The DH structures were composed of $Al_{0.3}Ga_{0.7}As$ cladding and GaAs active regions, the PL decay time of the GaAs luminescence being measured for samples with and without pre-layer. The latter took the form of a single GaAs layer or of a short-period superlattice within the lower cladding layer, while in one instance the SL was grown *after* the DH structure. The results showed that lifetimes increased with system clean-up after reloading, in accord with the above PL efficiency measurements, and that the presence of a pre-layer could increase recombination lifetimes by as much as a factor of 30, the effect being significant even when the pre-layer was 100 nm distant from the DH structure. On the other hand, a SL after the DH structure had no effect, favouring a model in which the pre-layer traps impurities which might otherwise accumulate close to the central GaAs layer. We might note, in passing, that similar SL pre-layers have also been successful in improving electron mobility in high-quality 2DEG structures (see e.g. English et al. 1987).

10.2   Quantum Well and Quantum Dot Lasers

So much for DH lasers—at much the same time as Tsang's work was published, it had become clear that there were worthwhile advantages to be had from the use of quantum wells, wires or dots on account of the sharpening of the density of states in conduction and valence bands. As we discussed in Chapter 6 (Section 6.5), not only could one anticipate a reduction in the threshold current but also in its temperature dependence (Arakawa and Sakaki 1982). Temperature dependence of laser performance was of particular significance in respect of optical communications, where sources could experience very wide variation in ambient temperature but it was of general interest for lasers designed to operate at high power levels, where there was the possibility of thermal runaway, so the prospect of low, or even zero dependence (in the case of quantum dots) on temperature had considerable appeal. The commonly used (empirical) formula for the threshold current density J was

$$J(T) = J_0 \exp\{T/T_0\} \tag{10.1}$$

so a large value of the parameter T_0 was to be preferred. Typical values for DH lasers were in the region of $T_0 = 150 - 200\,\mathrm{K}$, while measurements on quantum well devices demonstrated values as large as 400 K (though see our later comments on this aspect). Nor should we overlook the use of quantum wells as a means for reducing the operating wavelength of a laser. Simply by growing narrow wells it would be possible to obtain visible emission without the need to include aluminium in the active regions, that is, in the wells. It would be interesting to see how real devices matched up with such hopeful prognostications.

The first report of laser action in a quantum well structure (van der Ziel et al. 1975) came very soon after the first study of the optical properties of quantum wells (Dingle et al. 1974) and inevitably (!) it came from Bell Labs. The authors used optical pumping so the sample consisted of a simple AlGaAs/GaAs MQW structure cleaved to form a laser cavity. The wells were 8 nm wide and the emission wavelength was shifted to an energy some 27 meV above that of the GaAs buffer layer. Equivalent threshold current density at T = 15 K was $3.6 \times 10^4 \,\mathrm{A/cm^2}$, greater than that for emission from the buffer layer, so it would appear that the MQW was of rather modest quality (though their pump photon energy of 1.67 eV was too small for it to pump the barrier material, as would be the case for a diode laser)—here was another example, perhaps, of establishing an existence theorem, rather than attempting to make a practical device.

Won Tsang, on the other hand, took up the practical challenge, not only growing diode laser structures but modifying their design so as to optimise performance (Tsang

1981c). As we pointed out above, the threshold current density for a MQW laser was expected to be lower than that of a corresponding DH laser, on account of the different density of band states and, though first attempts showed no such improvement, Tsang was eventually successful in bringing down the values for MQW devices well below those of DH structures. Key features concerned the adjustment of the height and thickness of the interwell barriers so as to allow carrier injection into all the wells. A schematic diagram of the structure is shown in Figure 10.4, the cladding layers consisting of $Al_{0.4}Ga_{0.6}As$, while the optimum barrier composition was found to be $Al_{0.2}Ga_{0.8}As$. A very low threshold current density (250 A/cm^2) was achieved with well widths of 12 nm and barrier widths of 3.5 nm, a factor of 2 lower than his best DH lasers. Note, however, that these were relatively wide wells and narrow barriers.

MBE-grown QW laser threshold currents continued to fall over the next few years, as other contestants entered the fray. For example, in 1985 workers at the Fujitsu laboratory in Kawasaki, Japan, reported on a GRIN-SCH MQW laser with J_{th} = 175 A/cm^2, an important feature being the use of a short-period superlattice (SPSL) pre-layer to improve the quality of the active material (Fujii et al. 1985). Then, in 1987 Chen et al. (1987) of the University of California grew SQW GRIN-SCH lasers with J_{th} as low as 93 A/cm^2, while the coup de grace was delivered from the Ioffe Institute in Leningrad by Alferov et al. (1988) who described a SQW GRIN-SCH device with J_{th} as low as 52 A/cm^2 (later improved to 40 A/cm^2; see Alferov 2001). This latter device represented something of a triumph for MBE growth, employing, as it did, SPSLs to simulate the graded optical confining regions. A schematic diagram of this structure is shown in Figure 10.5 to illustrate the complexity made possible by computer control of an MBE system.

Another prediction of QW laser performance concerned its control of emission wavelength and this was convincingly demonstrated by work in our own laboratory

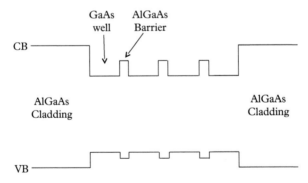

Figure 10.4 *Band diagram of a MQW sample as used by Tsang (1981c) in an attempt to optimise the threshold current density of a GaAs laser. The confining material was $Al_{0.4}Ga_{0.6}As$, while the barriers were $Al_{0.2}Ga_{0.8}As$. The GaAs wells were 12 nm, and the barriers 3.5 nm in width.*

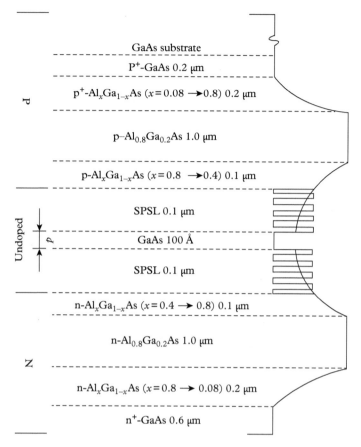

Figure 10.5 *Quantum well heterostructure laser using variable short-period superlattices to simulate the graded-index, separate confinement region. Such a structure provides an excellent example of the capability of a computer-controlled MBE growth machine. (From Alferov 2001, courtesy American Physical Society.)*

(Woodbridge et al. 1984). MQW lasers were grown, in a 'home-built' diffusion-pumped MBE system, with GaAs well widths varying between 5.5 and 1.3 nm. As anticipated from well-established understanding of carrier confinement energies, emission wavelength varied monotonically with well width, the shortest value achieved being 704 nm in the red part of the visible spectrum. Thus, it was possible to generate visible light while still employing GaAs in the active region of the device—there being no need to include Al. However, it was noteworthy that, in all samples, measured laser wavelengths exceeded by about 20 nm values calculated for appropriate values of well width, while absorption measurements showed band edge energies in good agreement with calculation. Even shorter wavelength devices were grown by MBE, when Al was included in the wells, Saku et al. (1985) achieving an emission wavelength of 650 nm from a laser with wells containing 35% Al.

In addition to the discrepancy in wavelength referred to above, it was also noted (see Blood 1988) that laser threshold currents were, in general greater than those calculated on the basis of a two-dimensional density of states (DOS) and that they also rose rapidly with decreasing wavelength. What is more, the calculated increase in T_0 was rarely observed in practice. In other words, QW lasers did not behave as originally predicted. It would be taking us well outside our brief to enter into any detailed discussion—the interested reader should consult Peter Blood's paper—but we may simply note that broadening of the DOS function by, amongst other things, local fluctuations in well width and the effect of gain saturation due to the flat (energy-independent) DOS could account for the disappointing values of J_{th}, while carrier recombination within the barrier material could probably account for the poor T_0 values. Rapid rise in J_{th} with reduction in wavelength was, at least in part, due to the quality of the Al-containing layers in the laser structures. Even when Al is not included within the well, reduction of well width inevitably leads to penetration of the carrier wave functions into the barriers and non-radiative recombination therein may, for example, result from the presence of oxygen impurity atoms.

Following our departure from PRL in 1991, we took the opportunity to delve a little deeper into this question by studying the PL efficiency and minority-carrier lifetime in AlGaAs/GaAs MQW structures grown over a range of substrate temperatures, using either As_2 or As_4 beams. Our first conclusion (Cheng et al. 1993) was that lifetimes were sensibly independent of growth temperature when using As_2, while showing a sharp peak at $T_S = 675\,°C$ when using As_4. This, incidentally, correlated well with several reports that J_{th} showed a minimum value at this temperature when QW lasers were grown with As_4. Following up on the idea that both lifetime and J_{th} depend on the quality of the AlGaAs barriers, we showed that measured lifetimes were at least a factor of 5 shorter than radiative lifetimes expected for our samples (of carefully controlled doping level) and interpreted the measured values in terms of a detailed Shockley–Read model of recombination within the barriers, close to the barrier/well interface (Orton et al. 1994). This, combined with our earlier correlation between PL efficiency and oxygen content in AlGaAs films, provided convincing evidence for the importance of oxygen in limiting laser performance.

As we saw above, by the end of the 1980s, the quantum well laser had shown evidence of fulfilling its destiny, threshold current densities being as low as 40 A/cm², but LDS was not yet finished—quantum dots were still to be incorporated in laser structures, with the hope of even lower values of J_{th} and even larger values of the temperature dependence parameter T_0 (see equation (10.1)). In Chapter 6 (Section 6.5) we discussed a variety of methods for growing quantum dots by MBE but device engineers overwhelmingly chose to use self-organised arrays of dots as the only practical method of realising quantum dot lasers and, not surprisingly, initial attempts were concerned with the InAs/GaAs material system (see Box 10.1). What is more, it would appear that by far the majority of the work was based on MBE growth—indeed, this is one area where MBE can probably claim dominance. The first report of lasing action in InGaAs dots (Kirstaedter et al. 1994) came from a 13-strong consortium of Russian and German physicists (in recognition of the newly available collaboration made possible by the end of communist rule in the Soviet Union). Their lasers consisted of conventional DH structures grown by solid source MBE at 600 °C in which the active region (grown at 460 °C–490 °C) took the form of a single layer of InGaAs dots encapsulated in GaAs. Short-period AlGaAs superlattices were also incorporated to improve growth morphology. Dot formation was confirmed by the change of RHEED patterns from streaky to spotty. Lasing action at wavelengths close to 1.3 μm was observed with promisingly low values of $J_{th} = 120A/cm^2$ at 77 K (though rising sharply to 950 A/cm² at room temperature, suggesting thermal excitation of carriers into the GaAs barrier layer) and a value of $T_0 = 350\,K$ over the range $T = 50 - 120\,K$.

Box 10.1 Self-organised quantum dots

The discovery of self-organised quantum dots which form as a result of strain relaxation during the epitaxial growth of InGaAs on (100) GaAs surfaces has proved to be of tremendous importance not only from the viewpoint of crystal growth and semiconductor physical understanding but also from that of optical device technology. We discussed the early development of these dots in Chapter 6 but, in relation to their device application, it seems appropriate to summarise some of their properties here as helpful background to our account of quantum dot lasers.

Whereas the mathematical description of quantum well energy states (see Box 6.4) can proceed in a straightforward fashion, based on a simple and reasonably accurate model of the physical structure, the same can definitely *not* be said of self-organised quantum dots. Not only the precise shape but also the composition of such dots is far from accurately known. Several groups have described their dots as having pyramidal form, whereas others prefer to think of them being lenticular, while the evidence of electron microscopy favours less well-defined geometries. What most researchers agree on is their having a significant degree of anisotropy, typically with lateral dimensions of 30–50 nm while being only 15–25 nm in height. This implies a high degree of difficulty (impossibility?) in the theoretical modelling of their energy states—while it may be straightforward to derive analytical expressions for spherical dots (Bimberg et al. 1999), the same is certainly not true for such anisotropic and ill-defined *real* shapes. This, alone, means that the confined energy levels must be regarded as unknowns to be estimated from experimental data, rather than as precisely defined quantities to be *compared* with experiment, as is well-established practice in quantum well physics. We should note that confinement energy is probably defined largely by the smallest of the dot dimensions, that is, the height, so it would be helpful to the experimenter if there were some simple method of controlling it and this may be part of the reason why growers have adopted various modified procedures, such as 'alternating epitaxy' in which In and Ga are supplied separately in sub-monolayer quantities. It also emphasises the merit of the multilayer growth technique pioneered by Egorov et al. (1996) to build several dots, one atop the other, thus obtaining

continued

Box 10.1 *continued*

dots with heights much greater than their lateral dimensions (though characterised by even less well-defined shapes).

In attempts to control the emission wavelength of QDs it is obviously important to know their indium content, as this has more effect on photon energy than dot size or shape. The band gap of InAs at room temperature is 0.36 eV (corresponding to a wavelength of 3.4 μm) while for the alloy $In_{1-x}Ga_xAs$ the band gap is given by

$$E_g = 0.36 + 1.064x \text{ eV} \tag{B10.1}$$

It may seem a simple matter to determine the In content of dots by controlling beam fluxes but there are complications associated with the tendency of In to diffuse, surface-segregate and re-evaporate. This implies a non-uniform distribution of In atoms within a dot (which makes the meaningful calculation of confined energy states even less likely!) and a tendency for In to diffuse into the GaAs capping layer, thus blurring its outline. Needless to say, this has serious implications for the choice of substrate temperature, most workers preferring to grow their dots at temperatures in the range 500 °C–530 °C (though temperatures as low as 450 °C have also been employed (Ledentsov et al. 1994; Wang et al. 1994), while growing the rest of the structure at 610 °C–620 °C. We should also bear in mind the observation of Saito et al. (1998) to the effect that growth of the GaAs capping layer also results in strain within the dot which can increase the emission energy by something approaching 300 meV. Finally, in the case of high injection levels such as occur in lasers, emission may arise from excited confined states. For example, it appears that luminescence may result from the transition between E_1 and HH_2 states which, unlike the case of a quantum well, is allowed in the case of a dot. Huffaker and Deppe (1998) observed several emission lines from $In_{0.5}Ga_{0.5}As$ dots, separated by roughly 70 meV, emission shifting to shorter wavelengths as the injection level was increased. This so-called band-filling effect can result in laser action being similarly shifted to shorter wavelength, as compared with PL emission from the same samples, an effect that is hardly welcomed by those concerned to develop QD lasers operating at 1.3 μm or longer wavelengths. If we assume x = 0.5 and make reasonable estimates of confinement energies, it is easy to show that emission associated with ground-state levels would occur at a wavelength close to 1.3 μm and this is consistent with many experimental observations but we must also acknowledge that others show significant differences. One can only conclude that estimating emission wavelengths is not a very precise science!

Yet another parameter of importance for optical devices is that of dot uniformity. Predictions of low threshold current and temperature insensitivity are based on the concept of a delta-function density of states but these are negated by a distribution of dot dimensions. Several workers have measured such distributions directly using AFM, while a simple measure of PL line width also serves to quantify the effect. Mirin et al. (1995) provided a good example of AFM measurements, showing that their dots were characterised by a height distribution of 23.7 ± 3.6 nm, a major axis of length 53.7 ± 8.2 nm and a minor axis of 36.4 ± 6.8 nm. PL line widths were about 28 meV, roughly consistent with the above size distributions and showing little variation with temperature, as would be expected for inhomogeneous broadening. As we noted in Chapter 6, line widths measured on *single* dots are less than 0.1 meV.

With regard to radiative efficiency, it is important that the dots are free from dislocations, which may serve as non-radiative recombination sinks, and this implies the need to keep them separate from one another. There is clear evidence that when dots coalesce dislocations form, even though the original dots may be of high structural quality. A related parameter is that of areal density. If we take a typical lateral dimension to be 50 nm, it is easy to derive a maximum value for this of 4×10^{14} m^{-2}, while, unsurprisingly, densities measured by atomic force microscopy tend to lie slightly lower. Again, while it is possible to grow stacked layers of dots to produce a quasi-three-dimensional layer of material, most laser samples have utilised single sheets, this being a logical approach to achieving low threshold currents. Needless to say, high output power would demand thicker active regions and therefore stacked layers (cf. multi-quantum wells).

In summary, we can do no more than note that there are many factors which come into play in determining the performance of a QD laser and it is far from easy to predict the quantitative relationship between this and growth parameters.

This early result was particularly favourable in so far as it demonstrated a degree of uniformity in dot sizes adequate for achieving low threshold currents. It was also significant in demonstrating laser action at 1.3 μm wavelength, of obvious interest for fibre-optic communications (1.3 μm being favoured for short-haul local area networks), the appeal being based on the use of a GaAs rather than an InP substrate. But it also revealed a significant problem with regard to thermal loss of carriers which would need correction if useful room-temperature devices were to be realised. Effort would also be required towards achieving lower threshold current, higher output power and, finally, an issue emerged with regard to the high-frequency modulation required in optical communication systems. The question of structural perfection of quantum dots was also of critical importance with regard to radiative efficiency—once carriers were captured into a defect-free dot, they might be expected to recombine radiatively—however, the presence of a dislocation, inevitably close by, could be catastrophic. These topics have been discussed by Ledentsov et al. (2000) and by Bhattacharya et el (2004) and a particularly useful summary of dot properties and applications has been provided by Mowbray and Skolnick (2005)—we shall look at them in turn but first it is necessary to say something of the relevant growth techniques.

Control of dot dimensions and structural quality was clearly an essential aspect of growth technology and led to the widely accepted use of sub-monolayer epitaxy (sometimes referred to as 'alternating epitaxy'). Wang et al. (1994) explored the growth of fractional monolayer films of InAs on GaAs substrates at 450 °C and found that such layers were characterised by very high luminescence efficiency. Ledentsov et al. (1994) extended this approach to the growth of quantum dots and observed optically pumped laser action at a wavelength of about 1 μm. Mirin et al. (1995) grew similar structures at 515 °C, supplying first In, then As, then Ga, then As again, with a 3 s pause between the Group III and Group V elements. They also reported very high PL efficiency at room temperature ($\lambda = 1.3\,\mu$m) and later (Mirin et al. 1996) reported laser action at a wavelength of 1.02 μm. The principle of the method has already been discussed in Chapter 5—the central idea is that the Group III elements diffuse much further in the absence of a Group V beam and this clearly aids the formation of well-resolved (hopefully defect-free) dots.

Returning to our discussion of laser development, the observation of thermal loss of carriers from confined states in InGaAs quantum dots was confirmed by several workers (see Mowbray and Skolnick 2005). While relatively large values of T_0 were observed at low temperatures, J_{th} became rapidly more temperature sensitive in the region of room temperature (where it really mattered!). Various mechanisms were considered, including carrier redistribution over non-lasing confined states (more probable for holes because the larger hole effective mass results in more closely spaced states), loss of carriers into the GaAs cladding layer and non-radiative recombination via Auger processes. The first of these mechanisms could be tested by doping the dots strongly p-type, thus filling all confined hole states, and both Shchekin et al. (2002) and Fathpour et al. (2004) obtained high values of T_0 by the use of modulation doping. It also helped, apparently, with the capture of free carriers from barrier states into the dot ground state. However, at temperatures above about 60 °C, there was evidence of a strong temperature dependence, suggesting the involvement of other loss processes, and p-type doping was not the only method of achieving good temperature stability. For example, Tokranov et al. (2003) obtained a value of $T_0 = 380$ K for temperatures up to 55 °C by thermal annealing of dots after capping with 6 nm of GaAs. Yet another approach involved growing a cap material of larger band gap, such as AlGaAs, thus increasing the thermal energy required for emission into barrier states, and an ingenious method described by Bhattacharya et al. (2004) involved the use of tunnelling injection of carriers directly into the ground-state level of the dot via a narrow barrier grown within a quantum well (see Figure 10.6). By arranging for the rate of injection to be equal to the recombination rate,

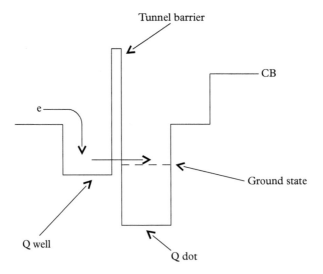

Figure 10.6 *Conduction-band energy level diagram of a quantum dot within a quantum well, showing the use of a narrow tunnelling barrier to inject electrons directly into the ground state of the dot. (Following Bhattacharya et al. 2004.)*

it was possible to maintain a population of 'cold' carriers in the ground state, thereby avoiding the problem of carrier loss inherent in a system under thermal equilibrium at the lattice temperature. Though there was still evidence for severe carrier loss at ambient temperatures greater than about 50 °C, values of $T_0 \sim 350-400\,\text{K}$ at room temperature were typical for QD lasers employing one or other of these technologies. The ideal value of $T_0 = \infty$ remained out of reach but these practical results represented a worthwhile improvement on those typically available for either simple DH or QW devices.

Threshold current densities (for room-temperature operation) also showed progressive improvement. Kamath et al. (1996) reported a 1 μm laser (operating, interestingly, on the HH_2 hole level, rather than the HH_1 ground state) with $J_{th} = 650\,\text{A/cm}^2$, Huffaker et al. (1998) a 10 mW, 1.3 μm laser with $J_{th} = 270\,\text{A/cm}^2$, Ledentsov et al. (1996) reached 63 A/cm², while Liu et al. (1999) measured a value as low as 26 A/cm², significantly less than the best QW laser available at the time. This latter device employed a single layer of InAs dots located within a 10 nm quantum well (referred to as a DWELL structure) in order to improve the capture of injected carriers (the lack of luminescence from the well itself suggesting a capture time much less than the well recombination time). At much the same time, Park et al. (1999) reported $J_{th} = 25\,\text{A/cm}^2$ for a 1.3 μm laser based on a single layer of $In_{0.5}Ga_{0.5}As$ dots and in 2001 Klopf et al. (2001) added to the sophistication of the DH laser by making a DBR version which operated in a single longitudinal mode at 1.3 μm wavelength. At this point, the relatively rapid reduction came to an end but 2004 saw further minor improvement in the guise of a 1.3 μm DWELL device with high-reflection coatings showing $J_{th} = 17\,\text{A/cm}^2$ (Sellers et al. 2004). Clearly the QD laser was approaching its limit in respect of threshold current but there remained the questions of output power and modulation frequency.

The approach to high-power dot lasers required the incorporation of high densities of dots within the plane and stacking of dot planes so as to increase the number of recombination centres. As an example, we might quote the work of Kovsh et al. (1999) from the St Petersburg Ioffe Institute. Dense arrays of InAs dots were produced by a 'seeding' process in which a plane of InAlAs dots is first deposited, followed by planes of standard InAs dots. The InAlAs dots form at significantly higher density $(\sim 1.5 \times 10^{15}\,\text{m}^{-2})$ and the subsequent InAs dots (of which three layers were deposited, separated by $Al_{0.15}Ga_{0.85}As$ barriers) grow in vertically coupled fashion at the same density. This composite active region was grown within a standard AlGaAs GRINCH structure from which 100 μm-wide ridge lasers were fabricated. These devices operated CW at room temperature with a wavelength of 870 nm and output powers up to approximately 3.5 W, before failing due to catastrophic mirror damage. The same group (Zhukov et al. 1999) also reported similar devices emitting at the important wavelength of 947 nm appropriate to pumping of Er-doped fibre-optic amplifiers. The active region consisting of ten layers of InAs dots was grown by the usual process of sub-monolayer epitaxy, and the resulting 100 μm-ridge lasers were driven CW at room temperature at output power levels up to 3.9 W.

Finally, we look briefly at the problem of direct high-frequency modulation of QD lasers, a matter of considerable importance if they were to challenge for the role of 1.3 μm fibre-optic sources. This was discussed by Bhattarcharya et al. (2004) who emphasised the importance of the thermal distribution of carriers over the various confined dot states and those associated with the GaAs cladding material. They pointed out that, whereas at low temperatures (80 K) it was possible to reach modulation bandwidths of 30 GHz, at room temperature this appeared to be limited to no more than 7.5 GHz. The problem lay with the fact that electron relaxation by way of phonon processes is suppressed in quantum dots because the separation of confined electron states is greater than the optical phonon energy of ~36 meV within the dot material—the so-called phonon bottleneck. The solution, according to them, was to use tunnelling injection of electrons as shown in Figure 10.6. As we pointed out earlier, this allows the electron distribution within the dot to remain 'cold' and this not only improves threshold current performance but also permits faster modulation. Modulation bandwidths in excess of 20 GHz were measured at room temperature on a QD device emitting at 1.0 μm wavelength.

Given that preliminary measurements of operating lifetime on QD lasers were looking hopeful, we can say, in summary, that by the year 2005, all aspects of such lasers, including operation at wavelengths as long as 1.5 μm, appeared very promising. It remained yet to be seen how well such promise would be fulfilled in practice. Could the QD laser, based on GaAs substrates, offer a serious challenge to the incumbent InP-based long-wavelength device which had supplied the fibre-optic communications industry since its burgeoning in the early 1980s? The answer comes in three parts—first, there was the need for greater understanding of laser performance, second, we note that (at the time of writing) there were two successful industrial companies making and selling InGaAs QD lasers based on the conventional AlGaAs/GaAs/InGaAs/GaAs/AlGaAs type of structure and, third, there has been a strong move towards a successful development of QD lasers on Si substrates. Let us look briefly at each of these.

The first point to make concerns the structural quality of the dot layers. Structural defects result in non-radiative recombination with consequent increase in threshold current, a problem addressed by the Sheffield group in England (Liu et al. 2004a, 2006a, b). They demonstrated that threading dislocation densities could be drastically reduced by careful attention to the growth of the GaAs capping layers. While they grew the InGaAs material and the first few nanometres of the GaAs layers at 510 °C, the rest of the GaAs was grown at 580°C. (They coined the acronym HGTSL for such structures (high-growth-temperature spacer layer), thus adding conspicuously to a long line of hard-to-remember sets of upper case characters which bedevil the lot of the struggling reviewer!) TEM studies clearly showed such procedures to be beneficial, particularly for multilayer structures, leading to a reduction of laser threshold current density from ~500 A/cm^2 to ~20 A/cm^2. The second major issue was with temperature stability of laser performance—the question of achieving an acceptably high value of the parameter T_0. As we saw above, the use of p-type doping of dots showed promise for improved stability at room temperature but there was considerable doubt concerning the mechanism(s) involved, a question addressed by several groups (Marko et al. 2005; Jin et al. 2006; Sandall et al. 2006; Badcock et al. 2007). It was general experience that threshold currents varied in the manner shown in Figure 10.7 for p-doped and undoped dot samples. The effect of doping was to *increase* J_{th} over the temperature range 100 K–300 K, accompanied by a *negative* value of T_0 just below room temperature, while above room

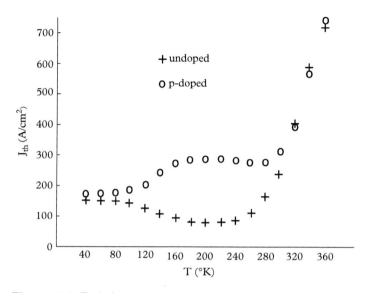

Figure 10.7 *Typical experimental data showing the temperature dependence of QD laser threshold current density $\mathcal{J}_{th}$ for modulation p-doped samples, compared with undoped samples. Note that p-doped lasers show an infinite value of T_0 just below room temperature but that it decreases rapidly above RT.*

temperature J_{th} increased rather rapidly for both doped and undoped samples. Modulation p-type doping has at least two effects on the recombination process—in the first instance, it tends to fill all the confined hole states (as already mentioned above) and thus prevents holes being thermally excited from the ground state into excited states but, second, it causes the dots to be positively charged, thereby increasing the confinement energies of electrons (i.e. making it more difficult for them to escape). This also tends to increase the occupancy of the electron ground states and militates against thermal distribution of electrons between the various dots. Marko et al. (2005) also claimed that p-doping leads to increased Auger recombination (though this has been disputed by other authors), which could account for the increase of J_{th} at temperatures below room temperature. Alternative models include the effect of photon coupling between dots (as an alternative to thermal coupling; Jin et al. 2006; Badcock et al. 2007). We make no attempt to resolve the interpretational problem here, merely commenting that it is clearly far from trivial. Perhaps of greater significance is the fact that J_{th} increases rapidly above room temperature in both doped and undoped samples, making it difficult to achieve high-temperature operation. Indeed, most workers report values of T_0 on the order of 100 K for temperatures above room temperature. Nevertheless, there have certainly been reports of lasers operating up to 100 °C (Ishida et al. 2007; Takada et al. 2011), and these reports encourage hope for the future application of such lasers to the fibre communications business.

The two commercial companies referred to, one European, 'Innolume', and one Japanese, 'QDL Laser, Inc.', grow all their structures by MBE, thus exploiting years of research into MBE-grown devices. Innolume is based in Dortmund and was set up in 2003 by a group of scientists emanating originally from the Ioffe Institute in St Petersburg. In 2006 they bought out the only other company producing QD lasers, Zia Laser, in New Mexico, though, coincidentally, QDL was launched in the same year as an offshoot of Fujitsu research. Examples of their products include DFB lasers with output wavelengths in the range 1.0 μm to 1.3 μm and modulation capabilities up to 10 GB/s, single-mode lasers producing powers of up to 600 mW, multimode devices capable of reaching close to 10 W output, and devices operating at temperatures as high as 200 °C. An interesting development from QDL is a green laser employing frequency doubling from a 1.064 μm QD laser. Epitaxial growth is on 3-inch GaAs substrates, a multiwafer facility being a vital component of commercial success.

The need for efficient and reliable optoelectronic devices integrated on Si substrates has been appreciated for many years. As circuit speeds increase, it becomes ever more desirable to employ optical signal transfer within and between silicon chips and this requires a technology for integrating high-quality lasers under conditions compatible with CMOS processing. The first successful QD laser on a silicon substrate dates from 2005, when Mi et al. (2005) reported MBE growth of suitable structures on an (001)-oriented Si slice, misoriented 4° towards the [111] direction, a well-established technique for avoiding anti-phase domains. Pulsed threshold current density was high ($\sim 1.5\,\text{kA/cm}^2$) but outputs up to 50 mW were achieved at temperatures up to 95 °C. The T_0 value was 244 K over the temperature range 5 °C–95 °C. Laser structures were grown on a 2 μm GaAs buffer layer, deposited by MOVPE and, though this contained a relatively high dislocation density, the devices could be seen as (marginally!) practical light sources. More recent progress was reviewed by Lee et al. (2013) with specific reference to 1.3 μm QD lasers grown by solid source MBE. The quality of the GaAs buffer layer was found to depend strongly on growth temperature and growth rate, the first 30 nm being grown at 400 °C. This proved helpful in lowering threshold current but, even then, the best value of J_{th} (pulsed) was 725 A/cm^2. The issue of the large misfit between Si and GaAs was clearly still a problem, as illustrated by the fact that similar methods applied to the growth of QD lasers on Ge substrates resulted in much improved devices. Threshold current densities as low as 55 A/cm^2 were measured on 1.3 μm lasers, some 13 times lower than equivalent devices grown on Si. Finally, values of $J_{th} \sim 64\,\text{A/cm}^2$ were measured on devices grown on Si/Ge substrates, a result perhaps pointing the way to really useful light sources on Si. Nevertheless, we should not overlook an alternative approach described by Tanabe et al. (2013) and in which QD lasers were grown on a GaAs substrate, wafer-bonded to the silicon slice and the substrate then etched away. The process is not only compatible with CMOS technology but it is also appropriate to substrates accurately aligned to the (001) plane, as employed in standard IC technology. While the lasers showed good thermal characteristics above room temperature (attributed to the use of p-type doping of the dots), threshold current densities were high—it could not, therefore, be seriously claimed that the technology had reached a state of readiness for immediate commercialisation. Once again, one was left with the all-too-common statement that 'this result is an encouraging demonstration—with application as a *promising* light source' [our italics]. (One can't help wondering how many times such hopes have been expressed in the context of silicon light sources during the past four decades but MBE-grown QD lasers do certainly show promise!)

10.3 Alternative Materials

The two previous sections have covered the development of GaAs-based lasers in some detail with the intention of bringing out some of the growth issues inherent in producing commercially viable devices. We now move on to take a much briefer look at a range of alternative laser materials which have been grown by MBE techniques. In essence, the reason for developing these alternatives is concerned with the requirement for emitters at different wavelengths, for example, the growth of mercury cadmium telluride (MCT) and lead tin chalcogenides for far-infrared devices and GaInNAs for lasers emitting in the important region of 1.5–1.6 μm, a region relevant to long-distance optic fibre communication systems. We shall begin by considering this last example as it follows naturally from the discussion of near-infrared lasers in the previous section.

GaInNAs is an example of the so-called 'dilute nitrides' which we discussed in Chapter 7 (Section 7.5). It was developed from the simpler system GaNAs once it was recognised that addition of N to GaAs had the somewhat surprising effect of *reducing* the band gap, rather than increasing it (Weyers et al. 1992). Kondow et al. (1996) first proposed it as a suitable material for lasers operating in the 1.3–1.6 μm wavelength range, later expanding on their initial suggestion with experimental proof of room-temperature laser action at ~1.2 μm, based on MBE growth of the appropriate alloy (Kondow et al. 1997). As they pointed out, adding In to GaAs increased the lattice constant while adding N decreased it; so, by controlling the ratio of In to N appropriately, it would be possible to grow films lattice-matched to a GaAs substrate and, as both had the effect of reducing the band gap, the resulting alloy gap might be adjusted to match the requirement for optical-fibre communication sources (see Box 10.2). Up to this time, of course, these demands had been met by InP/InGaAsP lasers but these suffered from certain infelicities such as expensive substrates, difficulty in controlling the Group V element ratio and poor thermal behaviour—it was generally necessary to employ thermoelectric cooling to maintain acceptable wavelength and threshold current performance. This latter aspect was becoming ever more significant as communication traffic requirements grew, a demand which could only be satisfied by the use of wavelength division multiplexing, involving a sequence of lasers operating at different, closely spaced wavelengths. On the other hand, GaAs substrates were cheaper, of better quality and available with larger diameters, while the larger conduction-band offset characteristic of the GaAs/GaInNAs heterojunction would offer better temperature stability from devices made in the new material. The argument concerning substrate cheapness may sound a trifle artificial until one recognises the urgent need for fibre communication to be extended right into subscribers' homes (the famous 'last mile')—a laser in every home means the cheapest possible laser and it should certainly not be in need of its own cooling system! However, in case this argument sounds overwhelmingly to favour GaInNAs, we should pause to reflect that there just might be difficulties in growing it, too! Let us look at the reality.

Box 10.2 GaInNAs

The alloy GaInNAs is of particular interest for near-infrared lasers because it can be lattice-matched to GaAs substrates while possessing a band gap which is reduced below that of GaAs (1.43 eV) by the addition of *both* In and N. In this box we provide a few numerical values for the various relevant parameters as an aid to understanding. First, we look at the question of achieving the appropriate lattice match. Tables 10.1 and 10.2 show values for the covalent radii of the elements involved, showing just how small the N atom is in comparison with the other elements, together with appropriate lattice parameters and band gaps for the binary compounds. GaN, of course, usually crystallises in hexagonal (wurtzite) form but it is also reasonably well known in cubic (zinc blende) modification (see Orton and Foxon 1998).

From these figures and assuming Vegard's Law, it is easy to derive the result that the compound $Ga_{1-x}In_xN_yAs_{1-y}$ will be lattice-matched to GaAs provided that $(x/y) = 2.73$ and this is in good agreement with the data presented in Figure 10.8 (following that given by Kondow et al. (1996)).

We next require knowledge of the way in which the band gaps of the two alloys GaN_yAs_{1-y} and $In_xGa_{1-x}As$ depend on the parameters x and y. The latter is well established (see Casey and Panish 1974) and can be written as

$$E_g = 1.43 - 1.064x \text{ (eV)} \tag{B10.2}$$

while the former can be obtained from the experimental data of Uesugi et al. (1999) (see Figure 7.12). To first approximation, this can similarly be written as

$$E_g = 1.43 - 13y \text{ (eV)} \tag{B10.3}$$

If we now make the (outrageous?) assumption that the two contributions to band-gap reduction in $Ga_{1-x}In_xN_yAs_{1-y}$ can simply be added together, while the condition $(x/y) = 2.73$ is satisfied, we arrive at the result that

$$E_g = 1.43 - 16y \text{ (eV)} \tag{B10.4}$$

Table 10.1 *Covalent radii of the elements, in angstroms.*

Ga	In	As	N
1.26	1.44	1.18	0.70

Table 10.2 *Lattice parameters (Å) and band gaps (eV) of the binary ZB compounds.*

	GaAs	GaN	InAs
Lattice parameter	5.65	4.53	6.06
Band gap	1.43	3.23	0.36

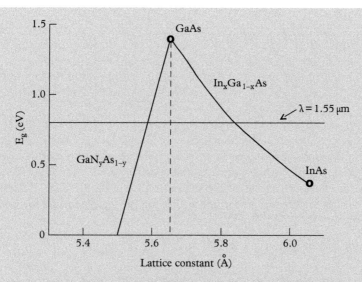

Figure 10.8 *Plot of band gap vs lattice parameter for the alloys InGaAs and GaNAs, showing the rapid decrease of gap as N is substituted for As in GaAs. By adjusting the relative amounts of In and N, it is clearly possible to produce a quaternary alloy which is lattice-matched to GaAs.*

To obtain a band gap of $0.80\,\text{eV}$ ($\lambda = 1.55\,\mu\text{m}$) therefore requires that the alloy should contain 4% N and 11% In. By far the larger contribution to the reduction is provided by the N so our assumption, though it may be outrageous, is unlikely to lead to serious error. A similar calculation suggests that $E_g = 0.95\,\text{eV}$ ($\lambda = 1.3\mu\text{m}$) would require 3% N and 8% In.

Before the reader takes such simple results too seriously, we should make clear that practical lasers are made in the form of quantum well structures so estimating the emission wavelength requires one to know the electron and hole confinement energies, which depend on well width. Clearly the band gap of the well material must be less than the laser photon energy. If we allow 100 meV for electron and hole confinement, this implies the need for a material band gap of 0.7 eV for emission at $\lambda = 1.55\,\mu\text{m}$ and this demands a composition of roughly 4.5% N and 12% In. Our purpose here is simply to help with understanding, rather than to lay down design rules so we shall leave the matter at that. It should be clear, however, that the crystal grower is faced with the challenge of incorporating up to 5% N in the alloy films and, though this may sound relatively trivial, it proved far from easy in practice, as we discuss in the main text. Finally, this account is oversimplified in so far as it neglects the influence of antimony in modifying alloy properties and the fact that practical lasers employ non-lattice-matched structures with considerably greater In content, strain adding yet another variable to the already complicated mix.

Perhaps the first point to emphasise is the fact that thermodynamic arguments predict the occurrence of a miscibility gap in the GaInNAs system and, though MBE growth takes place under kinetic control, we should not be surprised to find difficulty in incorporating significant quantities of nitrogen. Indeed, it was the common experience of all growers that no more than a few percent of nitrogen could be incorporated while maintaining adequate crystal quality. However, as discussed in Box 10.2, this would nevertheless prove adequate to achieve laser action at wavelengths out to 1.6 μm. The issues involved in MBE growth have been clearly presented by Harris (2002). A growth-temperature window exists between about 420 °C and 450 °C, above which the desired two-dimensional deposition breaks up into 3D island growth with phase segregation. Ga, In and As can be supplied from conventional cells, while, as we discussed in Chapter 7, the preferred method of supplying nitrogen is from a plasma source. It seems probable that only atomic nitrogen is incorporated and this constitutes no more than a few percent of the output but, according to Harris, its sticking coefficient is close to unity. In fact, GaInNAs differs from all other III-Vs with two Group V components in so far as it can be grown under strongly As-rich conditions, in which case the N:As ratio is determined solely by the growth rate, it being necessary only to control the In:Ga ratio, and this offers a major advantage as seen from the commercial viewpoint in terms of yield and reproducibility.

A further development in MBE growth emerged when Yang et al. (1999) studied the effect of adding an Sb beam to the four already present. The result was an improvement in surface morphology at the higher N and In contents necessary to reach 1.5 μm wavelength and it appeared that Sb was acting as a surfactant in much the same manner as had been observed for other large atoms such as lead—in the early years of MBE development, Pb had been seen to improve the structural quality of GaAs films even though there was no evidence for its incorporation in the grown film. However, later work by the Harris group demonstrated that Sb could also be incorporated in the film at levels up to 7% and this correlated with the nature of the Sb beam—Yang et al. used an Sb_4 beam from a conventional solid source, while Harris et al. used Sb_2 from a cracker. It was far from clear exactly what was the mechanism of the Sb contribution but it appeared not only to improve morphology but also to aid incorporation of N. This seemed strange in view of the alleged unity sticking coefficient for N—though one might speculate that Sb was acting as a catalyst for the breakdown of excited N_2 molecules to provide an increased supply of atomic nitrogen, we certainly have no evidence to support such an idea. No matter, the introduction of Sb certainly offered considerable assistance in the struggle to achieve the desired 3–5% of N incorporation.

Further complications could be seen as the result of deep level defects which became ever more pervasive as the amounts of N and In were increased towards the desired 1.5 μm wavelength target. PL efficiency deteriorated rapidly as more and more N was incorporated. Post-growth annealing had the effect of improving PL efficiency but also produced an undesirable blue shift in photon energy. Again, it was difficult to ascertain the origin of such recombination centres but Harris suggested that N could bond with Ga in two different configurations, one of which may have been responsible for encouraging non-radiative recombination. In this context, a somewhat bizarre effect was reported very recently by Kondo and Ishikawa (2012) (having 'invented' GaInNAs in 1995, they were still very much involved!). They chose to grow the alloy by gas source MBE but using a plasma N source and conventional Al, Ga and In solid sources. Their study of defects concerned the Al cell used to grow the laser confining layers and revealed (using quadrupole mass spectroscopy) that Al atoms were incorporated in the alloy film when the N source was activated, even when the Al cell shutter was closed. What was more, the Al appeared to be taking with it impurity atoms of O and C. The authors developed a gas-phase scattering model to explain their observations but, in more practical vein, they discovered that the effect could be suppressed by reducing the Al cell temperature from its working value of 1020 °C to an idling value of 750 °C. Such are the vagaries of crystal growth!

But enough of crystal growth problems—how successful have been the various efforts to develop viable near-infrared laser devices? A convenient starting point is the report of the first room-temperature operation of a GaInNAs laser emitting at 1.52 μm, from the University of Wurzburg (Fischer et al. 2000). This device was grown by solid source MBE, using a plasma source for N. It consisted of two $Ga_{0.62}In_{0.38}N_{0.05}As_{0.95}$ QWs within GaAs barriers, together with $Al_{0.34}Ga_{0.66}As$ cladding layers. Threshold current density was of the order of 40 kA/cm^2 so it was operated in pulsed mode only. The authors remarked that, compared with similar devices operating at shorter wavelengths, J_{th} increased rapidly as the wavelength became longer, probably as a result of degraded material quality in these highly strained layers. The possibility for a GaAs-based 1.5 μm laser may have been demonstrated but it was clear that much needed to be done before such devices could be thought of as practically viable, and material quality was obviously at the centre of this particular challenge.

As we have already hinted, the addition of Sb to the QW material was to play a major role in improving the situation. Its surfactant function could be seen as reducing the surface diffusion length of the Group III atoms, particularly In, thereby minimising the tendency of InGaAs and InGaN to segregate and form individual islands on the growth surface, a tendency implied by their different crystal structures (ZB and WZ, respectively). Thus, the use of Sb allowed films to be grown at somewhat higher temperatures than would otherwise be the case and this improved material quality. The fact that a few percent of Sb was also incorporated in the alloy probably helped in reducing the band gap but of even greater significance was the use of a relatively large fraction of In. We show in Box 10.2 that lattice-matching of the alloy $Ga_{1-x}In_xN_yAs_{1-y}$ to GaAs requires that (x/y) = 2.73 so 5% N implies roughly 14% In though, in practice, the quantities used have been typically 3% N and 30% In. Of course, the excess In does reduce the alloy band gap but at the expense of introducing considerable compressive strain. However, as argued by Kondow et al. (1996), this is beneficial in increasing the valence-band offset between GaInNAs and GaAs—yet another factor which must be taken into account in designing an optimum material strategy! On the other hand, it was eventually found necessary, in the interest of reducing defect densities, to reduce this strain by employing GaInN barriers rather than GaAs. These various ramifications go to show just how complicated was the task of achieving acceptable laser performance at these long wavelengths. Nevertheless, considerable progress was, indeed, made.

By 2007, threshold current densities had come down appreciably. Bank et al. (2007) showed how PL efficiency increased with the As:Group III ratio, saturating at ratios greater than ten, and demonstrated a peak as a function of Sb flux, suggesting that the Group III surface diffusion should not be overly inhibited, to the point where atoms could no longer find optimal lattice sites. They adopted the philosophy of keeping the nitrogen plasma conditions constant while controlling N content by varying growth rate. They also took the precaution of removing N ions from the beam by the use of DC-biased deflector plates. Growth temperature was seen as the most critical parameter, PL efficiency peaking at $T_S = 440\,^{\circ}\text{C}$, and the temperature window was found to be rather small ($\sim 20\,^{\circ}\text{C}$), though still large enough to allow reproducible material quality with the use of careful temperature monitoring. Post-growth annealing of samples was found to be necessary, and a detailed study by Bae et al. (2007) showed the importance of relatively long annealing times at moderate temperatures, typically 1 h at $700\,^{\circ}\text{C}$ (rather than the earlier method of a very short anneal at $780\,^{\circ}\text{C}$). This suggested that non-radiative recombination still represented a serious problem, as borne out by the work of Ferguson et al. (2009) in a long-range collaboration between the universities of Cardiff and Stanford. They confirmed that Shockley–Read recombination was the dominant process, the barrier material making the major contribution, presumably due to its having a greater defect density than the well material. The origin of the recombination centres remained to be fully established—the work of Kondow et al., referred to above, might be seen as an example of such an investigation. However, as a result of this careful attention to detail, they were able to make lasers reproducibly, operating CW at room temperature at 1.55 μm wavelength, with threshold current densities below 400 A/cm^2 and producing output powers greater than 100 mW. Values of $T_0 \sim 70\,\text{K}$ were measured at room temperature which, though not spectacular, were still significantly greater than those appropriate to the incumbent InGaAsP devices. Though some improvement in device design, to yield improved temperature stability, would be desirable, it was now clear that the GaInNAsSb system was capable of mounting a serious challenge to InGaAsP in this commercially important area. What is more, it seemed likely that any such development would be based on MBE growth, the necessary low-growth temperatures appearing to rule out the MOVPE alternative.

Let us now take a very brief look at some other materials employed in making laser diodes. The first of these is GaSb, which has been applied to the growth of lasers in the wavelength range of 2–4 μm, appropriate for trace-gas sensing. An account of the early work on antimonide lasers can be found in Section 9.6 of the work by Orton (2004); from this it is clear that technologists enjoyed considerable success but struggled to achieve room-temperature operation. Looking at more recent work however, we see that this particular hurdle has finally been overcome. Vizbaras et al. (2011), from the Walter Schottky Institute, Technical University of Munchen, describe MBE growth of QW lasers on GaSb substrates, employing GaInAsSb wells and GaSb barriers. The cladding material was AlGaAsSb. These devices operated CW at room temperature, emitting radiation at wavelengths of 2.5 and 2.7 μm with output powers up to 40 mW, threshold current densities as low as 70 A/cm^2 and temperature performance characterised by values of $T_0 \sim 42\text{--}65\,\text{K}$. This clearly represents an important step forward and gives reason to believe that this important wavelength range may soon be adequately covered by a corresponding range of practical laser sources.

While GaSb lasers appear somewhat limited in wavelength, this is certainly not true of their principal rivals, the IV-VI compounds, and MBE growth of IV-VI laser structures has been a busy area of activity, right up to the present. As we saw in Chapter 8 (Section 8.2), their optical and electronic properties have been well reviewed by Springholz and Bauer (2013) and these authors also provide an excellent account of mid-infrared laser development. Emission wavelengths range from about 3 μm to 8 μm, covering the whole of the 3–5 μm atmospheric window, which is of interest for IR astronomy. Perhaps inevitably, IV-VI lasers followed the better known III-V compound devices from simple homojunction structures to separate confinement double heterostructures and, finally to quantum well devices. In Chapter 8 we discussed the development of suitable material combinations offering narrow-band-gap active material and wider gap-barrier and optical-confinement layers. Thus, active materials were based on the lead salts, PbS, PbSe and PbTe, (sometimes combined with Sn) while wider gaps were obtained by the addition of Eu or Sr. Substrates were usually either PbTe or BaF_2. An example of early development (Feit et al. 1996a, b) took the form of an electrically pumped, buried heterostructure, a separate confinement MQW laser based on PbSnTe quantum wells, lattice-matched PbEuSeTe barriers, and electrical and optical-confinement layers. Effusion cells contained PbTe, $Pb_{0.7}Sn_{0.3}Te$, Eu, PbSe and Te, p and n-type dopants were Tl_2Te and Bi_2Te_3, substrates were PbTe and substrate temperatures were in the 300 °C–350 °C range. These lasers represented a major effort to reach room-temperature operation (highly desirable for practical application in gas analysis) and were nearly successful, achieving CW operation at temperatures up to 223 K. This, at least, made possible the use of a thermoelectric cooler, rather than the much less convenient liquid nitrogen cryostat. Threshold current density varied from 100 A/cm² at 20 K to a little under 100 kA/cm² at 200 K which illustrates the difficulty inherent in attempts to obtain room-temperature operation at these mid-infrared wavelengths. A few years later, Schießl and Rohr (1999) achieved pulsed operation of an MBE-grown 5 μm PbSrSe/PbSe/PbSrSe laser up to 60 °C. Further progress has been based on the development of surface-emitting laser structures, which we shall discuss in the next section.

Turning now to shorter wavelengths, the reader may be reminded of the glorious failure of the II-VI green laser struggle, which we outlined in Chapter 8 (Section 8.1). The long-term problem of p-type doping in ZnSe was finally solved by the use of the N acceptor in MBE growth and then green laser emission was achieved (under pulsed excitation at 77 K) at the 3M Company laboratory in St Paul, Minnesota, in 1991 (Haase et al. 1991). All that was then required was to extend this undoubted success to CW operation at room temperature and, when this was achieved (see Ishibashi 1996), to prove the long-term stability of the resulting lasers. Alas, it was at this final hurdle that disaster befell a decade of earnest endeavour—not only lasers but LEDs were found to die premature deaths as a result of persistent material defects, and this misfortune was compounded by the considerably greater success of the rival AlGaInN system. Shuji Nakamura, at the Nichia Chemical Company on the island of Shikoku in southern Japan, first reported his blue-violet laser in 1995 on Japanese television, then in the 'Japanese Journal of Applied Physics' (Nakamura et al. 1996). Operating lifetimes were soon shown to be more than acceptable to the device community and the nitride blue laser has become a byword for innovative technology in the guise of the Blu-ray video recording system. It was based on InGaN QWs with GaN barriers and was realised, of course, by the use of MOVPE growth!

Following this remarkable success, it is hardly surprising that the commercial development of nitride semiconductors has been dominated by MOVPE material but we should not be misled into ignoring the contribution of MBE altogether. The first point to emphasise is the struggle to achieve room-temperature CW operation experienced by the Nichia group—it took something like 18 months of dedicated application to produce long-lived, CW, RT diodes, the single most significant detail being a large decrease in threshold voltage (from 34 V to 5.5 V; see Orton and Foxon 1998). Given the obvious difficulties, it is not surprising that MBE growers suffered similar frustration. The first report of MBE-grown InGaN lasers came from the Sharp Laboratories of Europe (Hooper et al. 2004), based on the use of ammonia as a N source. These, like their MOVPE contemporaries, were grown on sapphire substrates and suffered from threshold voltages comparable to the early Nichia devices but with the additional disadvantage of much greater threshold current densities. They later brought these values down to $V_{th} \sim 7$ V and $J_{th} \sim 3.6$ kA/cm^2 (Tan et al. 2008) and recorded a RT CW operating lifetime of 42 h. These compared with 'MOVPE values' of the order of 3 V, 2 kA/cm^2 and >3000 h. As the Sharp workers concluded, 'These results demonstrate the feasibility of MBE for high-quality growth of InGaN laser diodes' but there was obviously a long way to go before MBE could hope to rival MOVPE commercially and this has remained true to the present day. The Polish consortium (Skierbiszewski et al. 2005) described MBE growth of laser structures on GaN substrates by plasma-assisted MBE but recorded values of $J_{th} \sim 12$ kA/cm^2 under pulsed operation. This was later improved (Skierbiszewski et al. 2011) to a CW threshold current density of 5.5 kA/cm^2 and voltage of 5.7 V, still well short of the best MOVPE values, though more promising. They also reported on optically and electrically pumped lasers at longer wavelengths (410–455 nm) (Skierbiszewski et al. 2012), concluding that 'the long-term prospects for PAMBE in this area appear to be as good as for MOVPE.' While this may just be possible and the study of MBE-grown laser structures will no doubt reveal much worthwhile data on material properties, our feeling is that the likelihood of MBE proving a serious commercial rival to MOVPE lacks credibility. We shall leave the matter there.

10.4   Vertical Cavity and Quantum Cascade Lasers

All the lasers discussed so far have been of the edge-emitting variety, taking the form of a mini-brick of typical dimension $500 \times 10 \times 2\ \mu m^3$, light emerging from the cleaved ends of the brick which serve to form a Fabry–Perot optical cavity. The reflectivity of the end mirrors (uncoated) is typically R = 0.3, an observation which obviously implies a mirror loss of approximately 70% and this must be made up by gain in the active material.

Assuming this to be uniform along the light path between the mirrors, the required gain is proportional to the reciprocal of the cavity length L. It is easy to show that

$$R \sim 1 - LG \qquad (10.2)$$

where G is the so-called 'modal gain', the net gain, allowing for the fact that the light wave extends across the width of the optical waveguide, whereas the gain region is only as wide as the active region (which may be no more than a quantum well). This is taken care of by introducing a 'filling factor' Γ so that

$$G = \Gamma g \qquad (10.3)$$

where g is the local gain. Typically $\Gamma \sim 0.01$ so it becomes clear that the local gain must be of order $g \sim 1 \times 10^5 \mathrm{m}^{-1}$ to satisfy equation (10.2).

For many applications this conventional structure is far from ideal, and during the 1980s much effort was devoted to developing an alternative laser which emitted light normal to the sample surface, the so-called vertical cavity surface-emitting laser (or VC-SEL). This was originally proposed in 1977 by Kenichi Iga of the Tokyo Institute of Technology (see Iga 2000). Obvious advantages included compatibility with planar electronic circuitry, the possibility of using lithographic techniques to define large arrays and (marginally less obvious?) the fact that the VCSEL has a circular far-field pattern, compatible with coupling into an optic fibre and, because of the short cavity length, it emits in essentially a single longitudinal mode. A further advantage comes with the much reduced area—a device of radius 1 μm compared with a conventional $500 \times 10 \ \mu\mathrm{m}^2$ laser structure offers a reduction of approximately 1000 times in terms of its threshold current. Assuming a threshold current density of 100 A/cm^2 suggests a threshold current of order 3 μA, small enough to allow output to be generated by signal currents in integrated circuits. However, there was, of course, a disadvantage. This is inherent in the use of epitaxy to grow the necessary structures and limits the cavity length to no more than a few microns, thus implying a relatively low output power. Then, looking again at the above equations, one can anticipate that the filling factor Γ might be close to unity but the small value of L nevertheless implies the need for a much greater reflectivity from the F–P mirrors. Putting $\Gamma = 1$, $g = 10^5 \mathrm{m}^{-1}$ and $L = 1 \mu\mathrm{m}$ implies $R \sim 0.9$. However, this is not the end of the story. If, in order to minimise the active volume in the interest of achieving low threshold current, one chooses a quantum well as the active region, this means that $L \sim 10 \, \mathrm{nm}$ and this then demands $R \sim 0.999$ (i.e. 99.9%), a very different requirement indeed. While early attempts to make practical VCSELs used metal mirrors (Soda et al. 1979), the only hope for obtaining such near-perfect reflectors resided in the use of Bragg mirrors, and the requirement for a structure lattice-matched to a convenient substrate suggested the use of AlAs/GaAs multilayer Bragg stacks grown on a GaAs substrate. This, indeed, proved to be the way forward.

For example, an early use of MBE to fabricate a GaAs VCSEL (Iga et al. 1986), using one metal and one AlGaAs/air mirror, resulted in a room-temperature threshold current density of order 100 kA/cm^2, while, as we shall see in a moment, the use of

Bragg mirrors in an InGaAs/GaAs QW laser brought this down below 1 kA/cm^2. The first CW room-temperature VCSEL (Koyama et al. 1989) used dielectric mirrors rather than Bragg stacks but this involved ring electrodes and current spreading. The threshold current density was approximately 10 kA/cm^2 and, in any case, it was grown by MOVPE! Of some significance, however, was an early industrial interest, represented by the work of Ibaraki et al. (1989) of Sanyo Electric—they reported the LPE growth of a room temperature, CW device with $J_{th} \sim 30$ kA/cm^2, using a combination of an epitaxial Bragg mirror and a dielectric mirror.

Our own interest is better served by the first MBE-grown CW room-temperature devices. The first of these (Lee et al. 1989) came from Bell (Holmdel and Red Bank) and used a single In$_{0.2}$Ga$_{0.8}$As quantum well and Bragg mirrors. Threshold currents as small as 1.1 mA were obtained but J_{th} was still rather high (~ 15 kA/cm^2). The real break-through came from Larry Coldren's group at the University of California, Santa Barbara (Geels et al. 1990). Their VCSEL was grown in all its complexity (see Figure 10.9) by solid source MBE and involved over 900 distinct layers! The active region was a single, 8 nm, strained In$_{0.2}$Ga$_{0.8}$As quantum well with 10 nm GaAs barriers, while the n-type AlAs/GaAs Bragg mirror consisted of 28.5 pairs, and the p-type mirror of 23 pairs of quarter-wavelength thickness (approximately 80 nm). The drive current flowed through both mirrors so it was necessary to grade the interfaces in order to reduce the effect of the large number of CB and VB steps and this was achieved by using a graded short-period AlAs/GaAs superlattice. The p-side of the laser was etched by reactive ion etching to

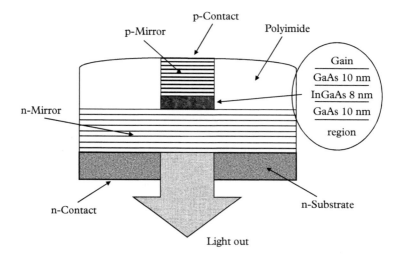

Figure 10.9 *Schematic diagram of the VCSEL structure used by Geels et al. (1990) to achieve room-temperature CW operation of an InGaAs/GaAs QW VCSEL. The mirrors were AlAs/GaAs Bragg stacks and the laser cross-section was defined by reactive ion etching to form a mesa which was passivated by a film of polyimide. Note that the GaAs substrate is transparent to the light from the InGaAs QW.*

form a square cross-section, and the pillar thus formed was passivated by a film of polyimide. Threshold current for a $12 \times 12 \, \mu m^2$ sample was approximately 1 mA, threshold current densities being 760 A/cm^2, while larger, 75 µm diameter devices achieved values as low as 600 A/cm^2 with output powers of 25 mW. The applied voltage at threshold was 4.0 V, somewhat greater than the AlAs band gap but dramatically less than test structures in which the superlattice grading was omitted. The output wavelength was 963 nm, well below the effective band gap of the Bragg mirrors.

This development clearly set the AlGaInAs VCSEL on the road to practical application and it is interesting to follow its development. Selecting references (all based on MBE growth) more or less at random, we note that small area devices were capable of threshold current densities below 1 mA (Geels and Coldren 1990), power output could exceed 100 mW (Peters et al. 1993), the operating wavelength was reduced to 850 nm, using GaAs QWs and AlAs/AlGaAs mirror stacks (Jager et al. 1997), again, power output reached 350 mW (Grabherr et al. 1998) and potential commercialisation was described by Jäger and Riedl (2011). A particularly important development came in 1994 with the adoption of the oxide aperture control, first described by Huffaker et al. (1994). This involved the introduction of a single layer of AlAs (or high Al-content AlGaAs), which could be partially oxidised by a 'wet oxidation' process to define a current-confining aperture, as shown in Figure 10.10. This was subsequently widely adopted. The commercial aspect of these developments is of particular interest from the growth viewpoint and is worth looking at in a little more detail.

The idea that the GaAs VCSEL might soon be of commercial importance was discussed as early as 1991 by Jewell et al. (1991) and some of the inherent challenges made clear. The use of 20–30-period Bragg mirrors implies something like 1000 shutter operations (placing a premium on reliable shutter operation) and a total device thickness

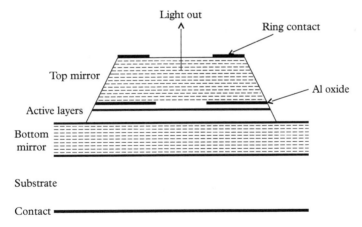

Figure 10.10 *Schematic diagram of a typical VCSEL structure, showing the manner in which an oxidised AlAs layer can be used to define the current flow through the active region of the laser.*

amounting to 5–6 μm, equivalent to a day's MBE growth. This would be acceptable provided yields were of the order of 80% or more and an important aspect of this concerns uniformity of device characteristics over a 3- or 4-inch wafer. A device footprint of 200 μm × 200 μm suggests that roughly 90 000 (150 000) devices could be obtained from a 3- or 4-inch wafer and that these should lie within a wavelength range of 1% (9 nm in 890 nm), implying a similar limit on the reproducibility of layer thickness. (Note that the laser gain versus wavelength curve has a width typically of 1% and it is essential that the active-region cavity resonance lies close to the centre of this.) The use of substrate rotation in a well-designed MBE chamber allows such a degree of uniformity but there is also a question of absolute accuracy in meeting a design wavelength. Growth rate may be calibrated to this accuracy using RHEED oscillations but this calibration is usually done with a small substrate without rotation and care is necessary to ensure that the growth temperature in the two cases is closely similar. Another aspect of growth temperature centres on the degree of re-evaporation of the Group III atoms. Jewell et al. chose to grow at temperatures below those at which this effect became significant $-T_S = 580\,°C$ for AlAs and GaAs, $T_S = 530\,°C$ for InGaAs. They also pointed out the need to minimise shutter transients resulting from the change in cell temperature when shutters are opened and closed. In a structure involving 1000 shutter operations, such fluctuations in beam flux can amount to serious errors in layer thickness, so the researchers set the appropriate cells well back behind their shutters.

Having noted the observations by Jewell et al., one cannot help being fascinated to come across the paper by Jäger and Riedl (2011), published just 20 years later, which describes the realisation of these very concepts. The authors started research into VCSEL technology when at the University of Ulm but this later paper was published from Philips Technologie in Ulm. It is concerned with high-volume applications of 850 nm VCSELs and demonstrates that MBE growth can achieve wavelength uniformity over 3- or 4-inch slices of ±1.8 nm, which represents ±0.2%, and reproducibility over 100 wafers (or roughly three million devices) within 8 nm. Threshold currents were between 300 and 400 μA, operating voltage approximately 2 V and power output 1 mW at 2 mA drive current. Device yields were generally greater than 80%. One final point concerns the use of carbon as the preferred method of p-type doping. Be had been widely used for a long time but these authors claim some clear advantage for C doping.

Broadening our outlook a little, we should note that the VCSEL concept has been applied to a range of applications and therefore wavelengths, and an excellent summary of such developments is available in the review written by Iga (2000). We should note, in particular, that much effort has been devoted to attempts to make lasers for the 1.3 μm and 1.55 μm bands, based on the InGaAsP/InP material system. In most cases, the mirrors used consist of dielectric multilayers because the refractive-index differences available within this material system are too small to allow effective Bragg stacks to be employed. Needless to say, because of the essential P component, these structures were generally grown by MOVPE—not by MBE. However, we should not leave this topic without mention of the possibility of using GaAs substrates and AlAs/GaAs Bragg mirrors in conjunction with GaInNAs active regions or, alternatively, InGaAs quantum dot active regions. An example of a GaInNAs VCSEL grown by MBE is reported by Yang et al. (2005), using an active layer of $In_{0.35}Ga_{0.65}N_{0.02}As_{0.98}$. They achieved 1 mW of CW power at room temperature in a device operating at 1.3 μm wavelength with a threshold current density of 1.25 kA/cm^2. An example of a QD VC-SEL is described by Yu et al. (2006). These authors obtained 0.33 mW CW operation at room temperature at a wavelength close to 1.3 μm. The threshold current density in this case was 1.5 kA/cm^2. It may be of interest that nearly all the authors of these papers come from Taiwan, a sign of the ever-increasing oriental contribution to modern day optoelectronics.

One of the most important applications of the VCSEL concept concerns mid-infrared IV-VI lasers which were introduced in the previous section. An early example of a PbTe VCSEL is that reported by Schwarzl et al. (2000) from the University of Linz. It was based on the use of PbEuTe Bragg mirrors, the alternating layers containing 1% and 6% of Eu and consisting of 18 and 24 periods for the top and bottom mirrors. Such a small difference in composition between adjacent quarter-wave layers is made possible by the large refractive indices of the lead chalcogenides. On the other hand, the addition of Eu results in severely degraded carrier mobilities and consequent high resistivities which are highly undesirable when the laser drive current must be transmitted via the mirrors. The above values probably represent an acceptable compromise, though, in this particular case, the laser was optically pumped (with 1.064 μm radiation from a pulsed NdYVO$_4$ laser). The VCSEL emitted at a wavelength of 6 μm at 2 K. The same group also demonstrated optically pumped VCSEL laser action at a wavelength of 4.8 μm from PbTe quantum wells (Springholz et al. 2000). These were grown by MBE on (111) BaF$_2$ substrates (their insulating nature being of no detriment for optically pumped devices), using $Pb_{0.95}Eu_{0.05}Te$ barriers and $EuTe/Pb_{0.95}Eu_{0.05}Te$ Bragg mirror stacks. Because of the large difference in refractive index between EuTe and PbTe, only three periods were needed to obtain reflectivities greater than 99.5%, thus effecting a considerable simplification in the growth process. As a result of quantum confinement and the fact that the energy gap of the lead salts *increases* with increasing temperature, the emission wavelength (measured between 45 K and 85 K) varied between 4.86 μm and 4.81 μm. They also reported laser emission from similar VCSEL structures in which the active region consisted of PbSe self-organised quantum dots (Springholz et al. 2001). The wavelength varied between 3.9 μm and 4.2 μm at temperatures up to 90 K. Attempts to reach room-temperature operation were eventually successful, with pulsed devices operating up to 65 °C, as summarised in the review by Springholz and Bauer (2013).

All this work was concerned with optical pumping and this brings us to yet another chapter in the semiconductor laser saga: the development of the vertical external cavity surface-emitting laser, or VECSEL. This took place during the 1990s and appears to have been motivated by the need for high output power combined with high-quality far-field pattern (see e.g. Kuznetsov 2010). The use of one external mirror, as shown schematically in Figure 10.11, infers that the use of electrical pumping via the mirrors is no longer possible—the VECSEL is usually pumped optically with a high-power, multi-mode, poor-beam-quality laser source. As Kuznetsov points out, the VECSEL can be

Light out

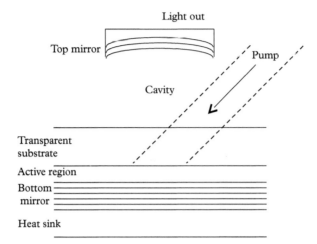

Top mirror

Pump

Cavity

Transparent
substrate

Active region

Bottom
mirror

Heat sink

Figure 10.11 *Schematic of a typical optically pumped
VECSEL laser. The use of a transparent substrate allows the
MBE-grown active region and lower mirror to be inverted
onto a metal heat sink. The external cavity is formed by
mounting a concave dielectric mirror a distance of perhaps
1 mm above the substrate, allowing the pump beam easy
access to the active region.*

thought of as a simple method of converting this crude pump power into a sophisticated,
narrow, diffraction-limited beam with circular far-field pattern.

As we pointed out above, because of its short resonant cavity, the VCSEL can be
relied upon to lase in a single longitudinal mode. However, there are still difficulties in
respect of transverse modes and these become more significant as the laser diameter
is increased. Thus the combination of high-power and single-mode output proved extremely difficult to achieve, a serious drawback, for example, when faced with the need
to launch light signals into a single-mode optical fibre. One method of avoiding this is
to employ an external cavity to control the transverse modes and, by adjusting the cavity length and introducing a variable intracavity aperture, it proved possible, at least,
to come within touching distance of this particular holy grail. An early example is provided by Wilson et al. (1993), who used MBE to grow an electrically pumped VECSEL
producing 2 mW of CW power at 980 nm in a single mode. The device consisted of
three $In_{0.2}Ga_{0.8}As/GaAs$ QWs sandwiched between a pair of $Al_{0.9}Ga_{0.1}As/GaAs$ Bragg
mirrors, the top mirror having a reflectivity of only about 86% to allow coupling with
the external cavity. Light was output through the external mirror of reflectivity 96%.
An example of an optically pumped MBE-grown GaAs VECSEL, emitting at 860 nm,
was reported by Sandusky and Brueck (1996). The top mirror in their structure was
designed as an anti-reflection stack so as to enhance the coupling to the external cavity
and this led to the observation of many longitudinal modes separated by 7.4×10^{-12} m (in

agreement with the anticipated value of $\lambda^2/2L$ for the case of $L = 50$ mm, the measured length of the external cavity).

Unsurprisingly, the VECSEL idea found favour with those interested in other material systems, and the dilute nitrides provide an interesting example. The group at Tampere University of Technology, Finland, reported plasma-assisted MBE-grown lasers producing several watts of CW power at room temperature at wavelengths close to 1.2 μm and demonstrated the generation of red/orange visible radiation by means of intracavity frequency doubling (Konttinen et al. 2007; Korpijärvi et al. 2009). The structures were grown on GaAs substrates and employed 12 GaInNAs 7 nm QWs as active elements, each well being provided with a GaNAs strain-compensation layer. A 30-section AlAs/GaAs Bragg stack served as bottom mirror, light being coupled into the external cavity through an $Al_{0.37}Ga_{0.63}As$ window. Pump power was supplied from an 808 nm diode laser focussed to a spot of about 180 μm in diameter. Chips were bonded to diamond heat spreaders and the whole mounted on a water-cooled copper heat sink to cope with the high thermal challenge.

Finally, we look briefly at VECSELs designed for mid-infrared wavelengths. First we note that Manz et al. (2009) at the Fraunhoffer Institute in Freiburg obtained upwards of 3 W of power at wavelengths in the range 2.0–2.3 μm from optically pumped GaInAsSb/AlGaAsSb VECSEL devices grown by solid source MBE, a 980 nm fibre-coupled diode laser providing the pump power. They discovered some interesting challenges during the growth of the appropriate structures, grown on GaSb substrates. These consisted of a GaSb/AlAsSb Bragg mirror (the AlAsSb being lattice-matched), an active region of strained GaInAsSb QWs with lattice-matched AlGaAsSb barriers (with 30% Al) and a window layer consisting of lattice-matched AlGaAsSb (with 85% Al). These individual sections were each grown without the need for growth interruption but, because of the very different conditions required for the various AlGaAsSb layers, it was necessary to introduce pauses between the sections (in order to reset the Group III cell temperatures) and this resulted in an accumulation of In at each growth interrupt. Apparently, In atoms were creeping past the In cell shutter and being 'dragged' to the substrate by the heavy Sb_2 molecules (the reader may recall a similar phenomenon to which we referred in Section 10.3 in which N atoms appeared to be dragging Al atoms which escaped from their cell, even when the shutter was closed; see Kondow and Ishikawa 2012).

Finally, there has been some well-directed attention to the MBE growth of IV-VI VECSEL devices in both Austria and Switzerland, with two distinct motivations. The first objective was that of producing room-temperature laser sources covering the wavelength range 3–5 μm, and the second, that of fulfilling the frequently occurring desire to grow optical devices on silicon substrates. At the University of Linz, Eibelhuber et al. (2010) achieved a rather chilly room-temperature (2 °C), single-mode, CW operation of an optically pumped PbSe/PbSrSe QW VECSEL at a 4.3 μm wavelength. The refractive-index difference between the BaF_2 substrate and the PbSrSe was sufficient to provide adequate optical feedback so the device consisted simply of a disc of QW material with diameter of the order of 10–50 μm (and often referred to as a 'micro-disc laser'), pumped by a Yb:YAG laser. At the Technical University in Zurich, Rahim et al. (2009b) grew an active region consisting of a single PbSe layer on a BaF_2 substrate, followed by a $PbEuTe/BaF_2$ Bragg mirror (only 2.5 pairs were necessary) which was mounted on an aluminium heatsink. A similar $PbEuTe/BaF_2$ external concave mirror completed the structure, which was pumped by a 1.55 μm laser, and laser action was observed at a wavelength of 4.5 μm. Pulsed powers of 6 mW were obtained at temperatures as high as 27 °C. Clearly, the goal of practical room-temperature sources was rapidly approaching. The Zurich group also reported both PbTe and PbSe VEC-SELs on Si (111) substrates, covering the wavelength range from 3.3 μm to 5.6 μm (Rahim et al. 2009a; Khiar et al. 2010; Fill et al. 2011). Successful growth on Si depended on the use of a 2 nm-thick CaF_2 buffer layer. Pulsed powers of 1 W and 17 mW CW were obtained at temperatures close to 100 K. The use of a moveable external concave mirror allowed 5% tuning of the output wavelength with a piezoelectric movement.

We now turn our attention to another exciting development in laser technology, that of the quantum cascade laser. This was yet another product of the ever-fecund Bell Laboratories and emerged in 1994 in a report in the journal 'Science' (Faist et al. 1994). The concept was originally suggested as early as 1971 by Kazarinov and Suris (1971) at the Ioffe Institute but many technological difficulties had to be overcome before this first practical demonstration could be realised. It is of particular interest in so far as, unlike all the lasers we have considered previously, it is a unipolar device, depending on only one type of carrier. What is more, the emission wavelength is determined not by a semiconductor band gap but by intraband states within a quantum well and can therefore be selected by a simple choice of well width. In practice, lasers have been reported over the whole of the mid-infrared and out into the terahertz region ($\lambda = 3$ μm–500 μm), a remarkable extension of the available electromagnetic spectrum. The principle of its operation is illustrated (in simplified fashion) in Figure 10.12. Electrons (in this case) are injected from a doped region into Level 3 in a suitable quantum well, from which they drop into Level 2 by the emission of an infrared photon. Level 2 is kept effectively empty by rapid phonon emission into Level 1 so it is possible to maintain an inverted population difference between Levels 2 and 3. Electrons can then be injected from Level 1 into Level 3 in an adjacent well and the process repeated, in practice up to as many as 50 or 100 times. Thus, each electron flowing through the structure stimulates the emission of up to 50 or 100 photons, compared to the single photon resulting from a recombination process in a conventional semiconductor laser, and this immediately promises the possibility of high output power. Given that the injection and QW system are typically only a matter of 50 nm thick, such a structure requires no more than a few microns of epitaxial growth—though the details of its design are a good deal more complicated than this simplified account suggests, some of the complications being discussed in Box 10.3. Practical structures involve short-period superlattices and a multiplicity of wells, implying a very large number of carefully defined epilayers and, once again, computer-controlled MBE rose spectacularly to the challenge (Al Cho's name being on the original paper for good reason).

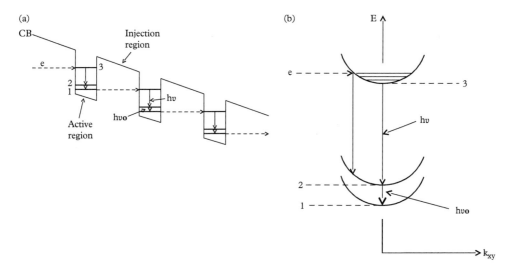

Figure 10.12 *(a) Simplified conduction-band structure to illustrate the operation of a quantum cascade laser. Electrons are injected into Level 3 in the quantum well, drop to Level 2 with emission of a photon hν and then to Level 1 with the emission of an optic phonon hν$_0$. The electron is then injected from Level 1 into Level 3 in the next well and the process repeated. (b) Plot of energy bands in k-space, showing how all three bands curve in the same sense, which implies a sharply defined laser-gain spectrum.*

Box 10.3 Quantum cascade laser—band structure engineering

The simplified structure of a quantum cascade laser presented in Figure 10.12(a) ignores a number of important complications which we try to illuminate in this box. The arrangement of energy levels with $(E_3 - E_2) \gg (E_2 - E_1)$ was originally chosen so that $(E_2 - E_1) \sim h\nu_0$, the optic phonon energy at the Γ point of the Brillouin zone. This ensures a strong coupling with the phonon mode and therefore a rapid depopulation of level 2. In the III-V compounds employed $h\nu_0 \sim 35$ meV, whereas, for a laser operating at $\lambda \sim 4\,\mu$m, $(E_3 - E_2) \sim 300$ meV, nearly ten times greater. The question then arose: could one choose a quantum well containing three confined levels in which these particular differences occur? And the answer was very definitely 'no'. Some more complex structure was needed, such as that offered by a system of coupled wells. In the original paper three wells were chosen, each with a confined level that could be individually tailored to meet the above requirement. The first well had to be narrow to lift level 3 to the required height, while the other two wells differed only slightly in width so that $(E_2 - E_1) \sim h\nu_0$. It should also be emphasised that in order for transitions to occur between the three levels, the barriers between them had to be thin enough to allow the appropriate degree of coupling. It also means that any calculation of the energies must include this coupling so life for the laser designer is rendered correspondingly more difficult. It would take us far beyond our modest aims to attempt a more detailed account—we merely hope that this level

continued

Box 10.3 *continued*

of presentation serves to explain the complexity of the band structure engineering required and the corresponding demands on crystal growth skills.

So much for the active regions—what should be said concerning the injection regions? It is clear from Figure 10.12(a) that the material between adjacent active layers should function rather like a semiconductor conduction band to transport electrons from the left-hand output level to the right-hand input and that there should be an accurate energy match at this injection point. Given that the applied field imposes a slope on the energy diagram, it is important that the intermediate material should be designed so as to counter this slope and give an approximately flat band and this implies the use of a suitably graded superlattice where the band takes the form of a SL mini-band. Once again, we make no attempt to delve deeper into design details—suffice it to say that, once designed, it places another challenge to the crystal grower, who must work hand-in-glove with the designer to ensure that the resulting design can actually be grown!

Finally, we take the opportunity, with reference to Figure 10.12(b), to emphasise an important feature of the cascade laser, namely its inherently narrow line width, with corresponding reduction in temperature sensitivity. In the conventional interband laser transition, the upper and lower energy levels curve in opposite directions in k-space and thermally induced distributions in band occupation lead inevitably to a broadened gain spectrum. In the case of the cascade laser, where the levels E_3 and E_2 curve in the same direction, this effect is minimised and one can anticipate a much less temperature-sensitive threshold current—that is, a larger value of the parameter T_0. Another significant advantage concerns the development of terahertz lasers—the emission energy in this case lies in the range 2 meV to 40 meV (600–30 μm) which is of the order of thermal energies in the temperature range 25 K to 500 K and this makes it practically impossible to make interband lasers for this wavelength range. The remarkable success achieved with quantum cascade devices certainly owes a great deal to the above feature.

In the original example, the authors chose to base their laser design on the material system $Al_{0.48}In_{0.52}As/Ga_{0.47}In_{0.53}As$, lattice-matched to an InP substrate. Outer layers of AlInAs served to provide optical confinement, the injection regions were 'digitally graded' AlGaInAs superlattices and the active regions were GaInAs/AlInAs coupled quantum wells (as explained in Box 10.3). Overall, 25 identical sections were grown to complete the structure. Stripe lasers were made by mesa etching and FP mirrors formed by cleaving. The emission wavelength was $\lambda = 4.26$ μm and, at a temperature of 10 K, 8.5 mW of pulsed power were obtained with threshold current density of 11 kA/cm². The device operated up to a maximum temperature of ~90K with $J_{th} \sim 14$ kA/cm². Clearly, as explained in Box 10.3, J_{th} did not vary rapidly with increasing temperature.

This was, without doubt, a splendid achievement, both in terms of complex design and of finely tuned crystal growth but it was, of course, only the first shot in the war to develop lasers covering the whole of the mid-infrared spectrum from 3 μm to 15 μm and of the terahertz region from 30 μm to 600 μm, in the struggle to reach the highly desirable vantage point of room-temperature operation. Needless to say, we can do no more than sketch an outline of the battles which followed—they were certainly numerous.

Two years later (Faist et al. 1996) came a report of room-temperature operation of a 5.2 μm laser producing 100 mW of power under pulsed conditions. Then, in 1997 Sirtori et al. (1997) achieved 15 mW of power at 8.5 μm with a threshold current density of 8 kA/cm^2. Both devices employed MBE-grown AlInAs/GaInAs quantum wells and superlattices on InP substrates but included a cap layer of InP, grown by MOVPE, which served to improve heat sinking. The latter showed good temperature performance, characterised by a value of $T_0 = 107$ K. Several of the original Bell team then returned to Europe and continued their QCL activity in their new locations, with Sirtori at Thomson CSF, and Fait at the University of Neuchatel in Switzerland. In 1998, they reported the first QCL in the AlGaAs/GaAs system, again grown by MBE (Sirtori et al. 1998). Because the refractive index of GaAs is larger than that of the QW/SL active region, it could not be used for optical confinement and this led to a different approach, in the shape of a plasmon-enhanced waveguide (using the refractive index of a heavily doped layer of AlGaAs). This 9.4 μm laser operated only at temperatures up to 140 K, the value of J_{th} being 7.2 kA/cm^2 at 77 K. Then, in the interest of single-mode operation, Hofstetter et al. (1999, 2001) developed DFB lasers at wavelengths of 10 μm and 5 μm, respectively. Once again, the structure was MBE-grown AlInAs/GaInAs/InP but with the addition of a grating on top. Both operated at or close to room temperature with improved threshold currents, and the latter device generated up to 1 W of power. This success was, at least partly, due to the use of four wells in the active region to provide a four-level system, rather than the usual three, in order to speed the emptying of the lower laser level. Faist et al. (2001) described a QCL using optical transitions between bound and continuum states which produced 700 mW of power at 30 °C, and 90 mW at 150 °C. The characteristic temperature T_0 was a very satisfactory 190 K. Beck et al. (2002) reported CW operation at room temperature from an InP/AlInAs/GaInAs/InP structure grown by the MBE/MOVPE combination. The power output was 17 mW, the wavelength was 9.1 μm, $J_{th} = 4$ kA/cm^2, and $T_0 = 171$ K. This again used a four-level system and a buried stripe geometry to aid heat sinking. Clearly these devices were beginning to reach the kind of performance required by those interested in applying them to pollution monitoring, medical diagnostics and even astronomy. One desirable feature remaining to be developed was the ability to tune the output over a significant band of wavelengths but this problem was also soon to be solved by the use of an external cavity which included a diffraction grating. Maulin et al. (2004) achieved a 15% tuning range from a 10 μm laser, and Hugi et al. (2009) also used an external cavity to tune a QCL from 7.6 μm to 11.4 μm. This latter device was remarkable in including no less than five *different* cascades within the one structure. It also produced an output as high as 1 W at room temperature. Without resort to an external cavity, Bismuto et al. (2010) employed electrical tuning on an 8.5 μm QCL producing 450 mW of CW power at room temperature. Finally, we note that it was also possible to move to *shorter* wavelengths by choosing a QW barrier material providing a larger conduction-band offset. This was demonstrated by the group at the University of Sheffield, Revin et al. (2007), who used the InGaAs/AlAsSb/InP system to make a 3.5 μm laser working at room temperature. And let us not forget that all these devices depended on MBE for their vital parts!

Hopefully, that gives a flavour of QCL development covering the mid-infrared. However, it still leaves the important region of terahertz radiation (30–500 μm) unaccounted for. This had been but little explored due to the lack of suitable sources. However, as explained in Box 10.3, QCL devices have a significant advantage in this spectral region on account of their much sharper joint density of states, due to the fact that the bands have approximately the same curvature in k-space (see Figure 10.12(b)). From the first demonstration of quantum cascade laser action in 1994, it was clear that this was a device which could have tremendous significance for the terahertz region but the first working laser did not materialise until the year 2002 (Köhler et al. 2002). (The fact that the output transition energy was less than the LO phonon energy imposed severe restrictions on the design of the active region.) A consortium of researchers based in Italy (Pisa and Turin) and England (Cambridge), supported by a combination of European Union and Japanese (Toshiba) funding produced a prototype laser emitting at 4.4 THz. The single-mode power output was 2 mW at temperatures up to 50 K, and threshold current density was as low as a few hundred amperes per square centimetre. The material system was AlGaAs/GaAs and the structure was grown, of course, by solid source MBE. It represented a very encouraging start and stimulated a surge of interest—the rapid progress during the next ten years reflecting the urgency with which the opening up of this spectral range was widely regarded. But first, we shall take up the question of the challenge presented to the MBE grower, a challenge admirably discussed by Beere et al. (2005).

Given the inherently tiny size of the laser photon energy in this case, the twin questions of uniformity and absolute accuracy in growth rates and doping levels take on an even greater significance. The laser they describe comprised no less than 104 repeats of the basic injection/active region which is made up of a series of short-period super-lattices, the whole adding up to approximately 1450 different layers, beating even the Geels et al. VCSEL hands down in complexity. The resulting total thickness of 12 μm presented a challenge in itself but the greater difficulties concerned absolute thickness control and uniformity both across the substrate and in time over a 12 h run. To meet design criteria, it was necessary to maintain an accuracy of ±1% in the GaAs growth rate, which relied on minimal drift in flux from the Ga cell (demands on AlAs growth rate being much less stringent due to AlAs amounting to only 3% of the structure). In the first instance, it was essential to calibrate the Group III fluxes accurately and this was done not by RHEED oscillations but by optical thickness oscillations, a technique first described by SpringThorpe and Majeed (1990). The accuracy is claimed to be better than 1%. Flux drift was checked by measuring growth rates before and after the growth of several complete structures—the results suggested a drift of only –0.6% for the Ga cell and –0.8% for the Al cell over a 12 h run. More significant was a –1.3% (–2.9%) variation of GaAs (AlAs) growth from centre to periphery of a 2-inch wafer (though a variation of no more than 1% across a 4-inch wafer would be typical for an MBE machine designed for commercial use). Additionally, the Si doping level was checked by growing Hall effect samples before and after device growth, though a variation of ±10% seemed acceptable in practice and could be achieved satisfactorily, the variation across the 2-inch wafer being no more than 5%. Finally, as a test of overall accuracy, four nominally identical samples were processed and their laser frequencies found to lie within the range 4.40 ± 0.07 THz, consistent with a design frequency of 4.35 THz. In this respect, at least, 1% accuracy appears to be a very practical possibility. An interesting observation was the dependence of device properties on the cleanliness of the MBE system. It is well established that low temperature electron mobilities in two-dimensional electron gases (2DEGs) improve considerably as a function of run number prior to a system download and, though relying on the limited evidence of these four samples, this also appears true of QCL performance—their threshold current densities correlated well with the MBE chamber quality (as measured by 2DEG mobility).

It was clear that the widespread use of QCLs in the terahertz region would need some improvement in performance, in particular, increased power and a range of wavelengths, but, more significantly, operation at higher temperatures. Room temperature would be a 'must' for many applications, such as the detection of illegal substances at airport security checks, though the compromise of 250 K, which would allow thermoelectric cooling might be acceptable in many situations. There was also a specific requirement to develop devices at longer wavelengths (~1 THz) because explosives and drugs, for example, show absorption lines in the region of 1–2 THz, while clothing becomes more transparent. Unfortunately, this latter wish was in direct conflict with the former—the smaller sub-band separation made for greater difficulty in achieving an inverted population, while free-carrier losses increase as the square of the wavelength. Progress was nonetheless made, the situation in 2007 being well summarised by Williams (2007). The design of the QCL active region was improved by changing to a bound-to-continuum laser transition and by giving more attention to the resonant extraction of electrons from the laser ground state, while improved waveguides reduced losses. Output powers increased to the 200 mW region for 4.5 THz devices, and maximum operating temperatures reached 160 K. CW powers of 100 mW had also been reported. An interesting approach to the generation of terahertz radiation was described by Belkin et al. (2007) based on 'intra-cavity difference frequency generation'. Whereas the use of an external cavity providing a non-linear dielectric medium has been demonstrated to produce terahertz difference frequencies, this is nevertheless a clumsy and complex technique—the advantage of the method proposed here was the use of the internal cavity of a dual wavelength QCL device, generating radiation at both 7.6 μm and 8.7 μm and a difference wavelength of 60 μm (5 THz). Sadly, they only succeeded in producing 60 nW of power—some considerable improvement would be necessary before such an approach could be taken seriously.

Looking a little closer to the present time, we see that, though further progress has certainly been made, the ultimate target of room-temperature operation still appears to be some distance away. The highest maximum temperature so far reported (by yet another international consortium from Canada, the USA, Germany and China—Fathololoumi et al. 2012) is a fraction below 200 K, with a laser emitting at λ ~ 3 THz. The crucial factors in their design were concerned with optimising the oscillator strength of the bound-to-continuum laser transition, the resonant phonon depopulation of the lower level and the tunnelling probability for the injection of electrons into the next stage. It is worth emphasising that such optimisations involve considerable computation of the properties of the various superlattices involved and, while MBE is concerned only to implement the design, once finalised, we must nevertheless admire the computational skills required.

Finally, it is worth noting that, even though available lasers are still far from ideal, they are certainly being used in practical spectroscopic applications. An example is provided by the work of the Leeds University team (Dean et al. 2012), who grew by MBE a series of QCL devices emitting in the range 2.7–4.0 THz and used them to detect the highly explosive compound pentacrythritol.

10.5   Photodetectors

By far the most significant application of semiconductor photodetectors is to infrared imaging in the interest of, for example, night vision, pollution control, medical imaging, astronomy and environmental monitoring. This has been an important area of research ever since the development of PbS, PbSe and PbTe during the Second World War, the early photocells being made with polycrystalline films deposited on quartz or glass (see Orton 2004, Chapter 9). Bulk single crystals of the lead salts made their bow in the early 1950s, followed by those of InSb. Both photoconductive and photovoltaic detectors were demonstrated in all these materials but they suffered from the obvious limitation of providing response only at specific wavelengths determined by the relevant band gaps, thereby lacking flexibility. There was a particular need for detectors sensitive in the 8–14 μm atmospheric window and this saw the introduction of the ternary compounds HgCdTe (MCT) and PbSnTe. The advantages offered by these compounds were band gaps which could be adjusted over the whole range of night-vision wavelengths simply by controlling the Hg:Cd or Pb:Sn ratios. Bulk single crystals were grown by the Bridgman or Czochralski processes but it soon came to be recognised that the future lay with epitaxy. In both cases, suitable substrates were available in the shape of CdTe and PbTe and both LPE and VPE growth methods were developed. Gradually, however, MCT was preferred over PbSnTe, and MBE over LPE and VPE and, by 1991, no less than 20 laboratories around the world had invested in MBE technology for MCT growth. The many difficulties associated with MCT growth have been outlined in Chapter 8 (Section 8.1) so we shall concentrate here on the specifically device-oriented challenges, with emphasis on the need for staring arrays associated with CCD (or similar) electronic readout. However, before delving into detail, it would be well to take note of yet another recent trend—the application of quantum structures to infrared detector fabrication. Quantum wells and quantum dots offer energy levels whose energy separations can readily be adjusted to match the requirements of the various IR bands, and MBE has played a major role in their development, as we saw above in relation to laser applications. A recent summary of this activity, 'Progress in Infrared Photodetectors Since 2000', has been presented by Downs and Vandervelde (2013), and the reader is recommended to consult it for further detail. For convenience, we might, nevertheless, list their designations of the various important wavebands—near-infrared (NIR; 0.7–1.0 μm), shortwave infrared (SWIR; 1–3 μm), mid-wave infrared (MWIR; 3–5 μm), longwave infrared (LWIR; 8–12 μm) and very-longwave infrared (VLWIR; 12 μm onwards), each of these having its own application areas and preferred device structures.

While acknowledging the fact that MCT appears capable of achieving the best detectivities, which make it first choice for military night-vision systems, it is probably worth noting that the much wider range of applications for which these photodiodes are now appropriate suggests there may be an argument for alternative technologies. An interesting example concerns the use of MBE-grown InSb to fabricate a diode sensitive in the range 2–7 μm and which functions at room temperature (Ueno et al. 2013). The concept is one of monitoring the presence (or absence) of human beings in an isolated building. Such a detector, requiring no cooling, can be left on at all times to control the use of electric power and thus effect considerable energy savings. The structure employed is shown in Figure 10.13—the InSb layers are grown on a semi-insulating GaAs substrate, and a 20 nm-thick AlInSb barrier layer is incorporated to minimise carrier leakage.

Perhaps of much wider application is the use of IV-VI compounds, which more closely rival MCT in performance, in particular, the fabrication of staring arrays on silicon substrates, a process which has been vigorously pursued by the Swiss group Zogg et al. (1991, 1994, 2002, 2008). As they point out, the IV-VI compounds may not quite equal MCT in performance but do offer a number of practical advantages. For instance,

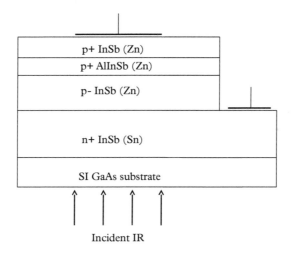

Incident IR

Figure 10.13 *Schematic structure of a room-temperature InSb IR photodiode capable of detecting human beings in an isolated building. Radiation is incident through the SI GaAs substrate, whose surface is roughened to minimise reflection. Current leakage is minimised by the introduction of an* $Al_{0.18}In_{0.82}Sb$ *barrier layer (Following Ueno et al. 2013.)*

material for LWIR devices is easier to grow with uniform wavelength response over the large area appropriate to a staring array because the compound $Pb_{1-x}Sn_xSe$ is five times less sensitive to compositional variation than the rival, $Cd_{1-x}Hg_xTe$. This arises on account of the CdTe band gap's being very much larger than that of SnSe (which is a semi-metal). It is also relatively easy to grow structurally acceptable PbSnSe films on silicon substrates because of their softness—its being standard practice to employ a CaF_2 or BaF_2 buffer layer. Though the extremely high permitivities of the IV-VI materials are often cited as limiting the frequency response of photodiodes, it is also true that they shield free carriers from the fields due to charged defects, making these materials fault tolerant. Finally, their optical absorption coefficients are significantly greater than those of MCT, allowing the use of thinner layers.

As an example of the possibilities for infra-red focal plane arrays (IRFPAs), we might refer to the work of Zogg et al. (2002) as it purports to be the first report of a (IV-VI) IRFPA on a silicon substrate containing electronic readout circuits. It contained a fairly modest array of 96×128 photodiodes for the 3–5 μm window but was sufficient to demonstrate an IR camera working at a temperature of about 100 K. The PbTe layers and a 3 nm CaF_2 buffer layer were grown by solid source MBE at temperatures in the range 250 °C–400 °C, the substrate being a (111) Si wafer, it being important to use this particular orientation because it allows dislocation glide under thermal stress, even down to cryogenic temperatures—only thus can the diode material retain its structural quality under operational conditions. Measured dislocation densities were within the range 10^{11}–10^{12} m^{-2}. Doping levels for optimum sensitivity lay in the low 10^{23} m^{-3} range, which is significantly higher than that appropriate to MCT devices and therefore considerably easier to control to any desired accuracy. Photodiodes were formed by depositing Pb Schottky diodes on the 3–4 μm-thick PbTe absorbing films. Image quality may have been somewhat less than ideal but it is worth emphasising that the growth temperatures employed were low enough to avoid damaging the silicon circuitry, an attribute essential to any commercially viable process.

Promising as such chalcogenide technologies were, there seems to be little doubt that international preference has centred on MCT so we should perhaps concentrate our attention thereon. The ultimate aim has been to develop large-area, multipixel staring arrays with readout circuitry built into the substrate so we need to consider the problems of MBE growth on appropriate substrates, suitable doping to form photodiodes, the achievement of the necessary uniformity over the relevant area and, finally, that of the resultant image quality. We have already discussed the general problems associated with MBE growth of MCT in Chapter 8 (Section 8.1) so in this chapter we shall concentrate on the specific question of realising viable staring arrays on substrates carrying the necessary readout circuitry and, in so doing, bringing the subject of MCT up to date.

Suitable substrates have been discussed by Garland and Sporken (2010). The choice covers $Cd_{0.96}Zn_{0.04}Te$ (CZT), GaAs, Ge, Si and InSb, only the first being accurately lattice-matched to MCT, though InSb comes relatively close (mismatch only 0.04%). However, because of its narrow band gap, InSb is only suitable for LWIR and VLWIR devices. Furthermore, there are many other factors to be considered, such as thermal expansion match, thermal conductivity, size, cost, uniformity, strength and hardness.

Substrate orientation can also be of crucial importance. Yet another criterion for MBE growth is that of substrate surface quality. Because MBE proceeds at relatively low temperatures (~180 °C) (a considerable advantage, of course, with regard to interdiffusion of dopants or substrate atoms) it is important that the surface should be free of structural defects, there being clear evidence that MBE-grown MCT films replicate any defects present on the substrate surface. It is possible, however, to employ a suitable buffer layer to block the deleterious effects of defects, one such being a HgTe/CdTe superlattice (and there may well be others but they tend to be proprietary!). There is a long history of MCT films being grown on CZT substrates and, while MCT detectors grown on CZT can certainly be combined, via In bump technology, with silicon readout circuitry, it is worth bearing in mind that CZT is expensive, brittle and available only in small sizes. Nevertheless, this approach to FPAs appears to be in serious practical use so we begin by outlining some of the issues and describing a typical application, before moving on to consider Si, GaAs and InSb.

The first issue of note concerns the CZT orientation for optimum film quality. It was recognised at an early stage that the Hg sticking coefficient is a strong function of substrate orientation, while the use of low-index surfaces such as (100) or (111) generally leads to micro-twinning. Modern practice is to grow on (211)B which gives MCT films of adequate structural quality (dislocation densities being less than $8 \times 10^8 \, \text{m}^{-2}$) while minimising the necessary Hg vapour pressure in the MBE chamber. Problems, nonetheless, abound. The low thermal conductivity of CZT can result in non-uniform temperature distribution across the growing layer which, in turn leads to non-uniform composition in the MCT active material, a serious problem for LWIR devices. Similarly, the Zn content of substrates may vary by as much as ±1% between centre and circumference, causing a corresponding variation in threading dislocation density. Surface quality has also been found to vary significantly between different samples of CZT, resulting in poor control of dislocation density in the MCT, with consequent variation in minority-carrier lifetime and, therefore, device performance. Surface roughness is yet another source of defective film growth, while impurities in the substrate can diffuse into the growing MCT layer, leading to leakage currents which degrade photodiode performance. As already mentioned, the use of appropriate buffer layers can mitigate some of these difficulties but the limitation in size of CZT substrate slices inevitably limits the number of pixels that can be incorporated to something like 10^6 (compared to ~10^8 possible on silicon substrates).

This catalogue of negative features may sound somewhat daunting so we should counter it with an example of just what can be achieved in practice—the fabrication of a two-colour IR detector, as described by the Raytheon group, Smith et al. (2006). The concept of a two-colour detector is interesting—it consists of back-to-back diodes in the form of a three-layer n-p$^+$-n structure, the first n-layer absorbing in the 3–5 μm band, and the second, in the 6–11 μm band, as shown in Figure 10.14. (The authors make no comment concerning the dopants they used but standard practice suggests these would be In donors and As acceptors.) The response of the individual diodes can be selected by adjusting the sign of the applied bias across the overall structure. The complete structure was grown by MBE in a single run and the individual 20 μm pixels isolated by dry

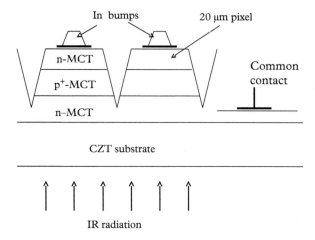

Figure 10.14 *Schematic cross-section of a two-colour IR focal plane array using back-to-back diodes. The compositions of the n-type MCT layers are chosen to absorb in the 3–5 μm and 6–11 μm bands, respectively. The corresponding signals can be selected by varying the sign of the bias voltage across the pixel (Following Smith et al. 2006.)*

mesa etching. Arrays of 640×480 and 1280×720 pixels were formed on 4×4 cm and 6×6 cm wafers, respectively. Quantum efficiencies of 70% and dark currents as low as 300 pA at 78 K were recorded. Readout was by way of silicon ICs, using In bump connections. This represents a good example of a 'third-generation' IRFPA and we should recognise that it has been replicated by numerous commercial establishments around the globe. It suffers mainly from limitation in overall size and the inconvenience of using In bump technology.

Ultimately, one feels, silicon must be the obvious choice of substrate in so far as it is relatively cheap, is available as large, high-quality wafers and possesses a well-established technology for the production of appropriate readout circuitry. On the other hand, it suffers from a lattice mismatch of 19%, a thermal mismatch of ~52% and the problem that it is monatomic—the difficulty of growing films of GaAs on Si, for example, arising from a tendency to form anti-site defects, is well known. The question is, of course, how well can these difficulties be overcome in practice. Perhaps unsurprisingly, a favourable answer involves the use of a suitable buffer layer, commonly CdTe, and careful choice of substrate orientation. Early work made use of (100) substrates tilted 30° away from [011] to avoid anti-site defects but more recently alternative surfaces have been preferred. Garland and Sporken (2010) describe the use of Si (211) on which is grown a CdTe layer with either (111)B or (211)B orientation in order to achieve the necessary crystal quality and surface smoothness, while Yakushev et al. (2011) preferred Si (310). In both cases, a thin layer of ZnTe (~10 nm) was first deposited at a temperature of 200 °C–240 °C, followed by a CdTe layer of the order of 5–10 μm thickness at 280 °C–320 °C. Some care was necessary to obtain a CdTe (111)B surface, the alternative (111)A being extremely rough. The use of an As flux as pretreatment serves to ensure the CdTe film will have the desired (111)B orientation. Nevertheless, in spite of such attention to detail, the resulting MCT layers suffer from a level of threading dislocations of the order of 10^{10} m^{-2} or greater, which has been proved to degrade minority-carrier recombination lifetimes and to short-circuit p–n junctions, thereby seriously increasing dark currents. This proves particularly detrimental to LWIR devices. Much effort is being devoted in attempts to ameliorate these problems but they remain a significant deterrent to the widespread use of silicon substrates for fully integrated FPAs. Compositional uniformity over the area of a slice is important for achieving good image quality, and Yakushev et al. have adopted an unusual approach. They make considerable use of ellipsometry to monitor surface quality and this is difficult to realise while employing the standard rotating substrate so they designed what they refer to as 'coaxial' molecular sources which provide adequate uniformity over 3-inch substrates, without the need for rotation, MWIR MCT samples showing uniformity within ±0.3%. This is of general interest, as the standard method of measuring growth rate by monitoring RHEED oscillations also demands a stationary substrate and is therefore available only in the form of a calibration before or after the growth of 'real' layers.

While GaAs is more expensive, is available as somewhat smaller-sized wafers and is less well developed in regard to readout circuitry, there can be little doubt that, if high-quality films of MCT could be grown on it, it would offer a very tempting alternative to silicon. Considerable effort has therefore been devoted to studying the possibilities, in spite of the obvious mismatch in lattice parameter (~14%) and thermal expansion coefficient (~14%). GaAs has some advantage over Si in so far as it has the same crystal structure as MCT and the readily available (211) substrates allow straightforward choice of the desired (211)B surface. The preparation of a clean surface is also considerably easier, as the native oxide can be removed at relatively low temperature. On the other hand, one should also be aware of the possibility of As diffusion producing p-type doping of the MCT layers. Needless to say, a CdTe buffer layer is required (its thickness (~10 µm) appears to be similar to that used on Si substrates) while there may be some advantage in growing this in a separate MBE chamber, before transferring it to the MCT growth chamber (Jacobs et al. 2012). By careful control of growth temperature, these authors report XRD rocking curve widths as low as ~50 arcsecs (compared with ~20 arcsecs measured on the original GaAs substrates) and which are significantly better than values measured on Si/CdTe substrates (Garland and Sporken 2010). Sadly, however, threading dislocation densities were no better than those measured on Si/CdTe/MCT layers. Vasiliev et al. (2012), demonstrating an apparent Russian preference for the (310) orientation, describe the use of (310) GaAs substrates, together with ZnTe (10 nm), CdTe (6 µm), and $Cd_{0.5}Hg_{0.5}Te$ (1 µm) buffer layers as a basis for LWIR photodiodes ($Cd_{0.2}Hg_{0.8}Te$). They measure XRD rocking curve widths of 170 arcsecs but do not report values of dislocation density. They do, however, report on successful imaging by 288 × 4 linear arrays and by 320 × 256 FPAs using In bump technology on a Si ROIC. MBE-grown 640 × 512 arrays of 15 µm LWIR diodes have also been reported by Wenisch et al. (2012), based on (211) GaAs substrates, which the authors claim show promise for future imaging systems. Taking our scientific reputations in our hands, we might hazard a summary conclusion to the effect that progress with GaAs substrates is currently on a very similar level to that appropriate to silicon substrates! There is still work to be done.

Finally, we should look briefly at the possibilities for InSb as an alternative. It has the obvious advantage of being very nearly lattice-matched and thermally matched to MCT though its quality is far from being as well developed as those of GaAs or Si and we must not overlook the possibility of both In and Sb doping of the MCT layers grown upon it. There is, too, the problem of oxide removal, its being necessary to employ Ar-ion sputtering, with consequent damage which can be only partially removed by thermal annealing. According to Garland and Sporken (2010), both CdTe and MCT have been grown on InSb with encouragingly narrow XRD rocking curves but there is obviously some way to go before the resulting films can be seen as proven LWIR material. Thus, while InSb substrates may look seriously interesting for LWIR devices, it is probably too early to attempt even a tentative summary of their long-term future.

So much for what one may call the 'conventional' approach to detecting infrared radiation; but there has been a great deal of activity in the past two decades concerned to exploit quantum structures. Like so many other innovations, it all began at Bell Labs in 1987 when Levine et al. (1987) demonstrated a device that has come to be known as a QWIP (quantum well infrared photodetector). The idea was to make use of electron or hole states confined within quantum wells and utilise transitions between such states to absorb IR photons (rather than using interband transitions in a narrow gap material such as MCT), such photodetectors being readily tuneable by the simple expedient of varying the width and depth of the wells. What was more, these devices could be fabricated in well-established III-V materials. Levine et al. may or may not have appreciated that they were opening a veritable Aladdin's cave of exciting new device possibilities but such was certainly the case. Looking at the field from the vantage point of the year 2014, the problem for the humble reviewer is one of presenting any kind of balanced account. Not only is there interest in at least five distinct wavebands but there exists a similar number of distinct device structures and distinct materials systems. While recognising the contribution of MOVPE growth technology, we should emphasise that MBE has played a major role in these developments. Our account will, of necessity, be somewhat brief—more detail can be found in Downs and Vandervelde (2013) and Jagadish et al. (2011). The reader can also find a useful comparison between the various types of IR detector in the review by Rogalski (2003).

The 10 μm detector described by Levine et al. made use of an $Al_{0.25}Ga_{0.75}As$/GaAs superlattice in which the GaAs wells were designed to have just two confined states. The 50-period SL was grown by MBE, the 6.5 nm wells being doped with Si only in the central 5 nm region, thus providing a supply of electrons in the well ground state. Photon absorption excited these carriers into the upper level, near the top of the well, from which they could easily tunnel out to produce a photocurrent. Over the temperature range 15 K–80 K, they measured a narrow-band (10%) photoresponse at 10.8 μm with a responsivity of 0.52 AW^{-1}, which compared moderately well with values appropriate to photoconductive MCT detectors. The narrow-band response arises from the fact that

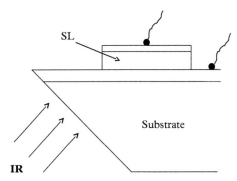

Figure 10.15 *Sample design used by Levine et al. (1987) to allow 45° incidence angle for IR radiation. Similar geometries have been used in many single detector devices.*

the dispersion curves of the two confined states are closely similar and can be seen as a distinct disadvantage for many imaging applications. A further drawback to commercial development of QWIPs is the fact that they are not sensitive to light impinging normally to the plane of the wells—this being a consequence of transition probability between confined states. Levine et al. made use of 45° incidence, as illustrated in Figure 10.15, and this has been employed in many single detectors, though it is not applicable to FPAs. While it is possible to use a scattering filter which redirects the light, it clearly represents a major inconvenience.

Research into the MBE growth of QWIPs has continued right up to the present time (see e.g. the 10 μm detector described by Lee et al. (2012)), though emphasis has undoubtedly shifted to alternative structures based on quantum dots rather than wells, as we shall discuss in a moment. A major area of interest has been the exploration of alternative material systems such as GaAs/InGaAs and InGaAs/GaInP, with the aim of extending detector response out to longer wavelengths, though performance has not been particularly attractive. As we saw in our discussion of IR lasers, there has been a growing interest in the so-called 'terrahertz' spectral region (wavelengths in the range from 30 to 300 μm) and there is an obvious need for photodetectors to function at these wavelengths. An example of attempts to satisfy this particular requirement is provided by the work of Liu et al. (2004b) using AlGaAs/GaAs quantum wells. As they point out, there is a wavelength region between 30 and 40 μm which is inaccessible in this material system on account of GaAs LO phonon absorption (at 36 meV). They demonstrated a detector with responsivity at 8 K of approximately 0.5 AW^{-1} at a wavelength of 42 μm, based on a 50-period Al$_{0.05}$Ga$_{0.95}$As/GaAs MQW structure with 12 nm wells and 40 nm barriers, the central well regions being doped with ~10^{23} m^{-3} Si. Other examples are available but all require similarly low operating temperatures.

At the opposite extreme, there is another obvious requirement for detectors operating in the vicinity of 1.5 μm and this presents a challenge of a different kind, namely, the need for wells of depth greater than 0.8 eV, which stimulated the development of QWIPs based on the AlN/GaN system. Three complications were inherent—the readily available substrates, sapphire and SiC were seriously mismatched from the active material, the constituent lattice constants of the nitrides themselves differed by 3–4%, and the commonly used c-plane orientation was characterised by built-in electric fields, resulting from the associated strain. An early attempt to fabricate a working detector for the 1.3–1.55 μm band, based on plasma-assisted MBE growth on c-plane sapphire substrates (Monroy et al. 2006), dealt with each in turn. First, all device layers were grown on 1 μm-thick buffer layers of AlN to induce compressive strain, thereby avoiding cracking. Second, the active structure, 40 periods of 1.0 nm-thick GaN wells and 2.0 nm-thick AlN barriers, was contained within 30 nm layers of $Al_{0.65}Ga_{0.35}N$ cladding—the Al:Ga ratio matching that of the superlattice. This not only avoids the generation of misfit dislocations but also minimises the electric field in the active region. Finally, the cladding layers were heavily Si doped to reduce series resistance, in the interest of response speed—it should never be forgotten that this particular application implies the detection of ever-increasing bit rates. We should also emphasise the absolute necessity for room-temperature operation. A sample detector showed a narrow-band response at 1.39 μm, though it was straightforward to predict tunability over the range 1.33–1.91 μm by varying the well width from 1 nm to 2.5 nm.

Taking up this vital question of response speed, we note several advances since the early work. In 2007 Giorgetta et al. (2007) reported a room-temperature AlN/GaN QWIP device with a peak response at 1.55 μm functioning at speeds of up to 2.37 GHz, while a year later Vardi et al. (2008) reported operation at 11.4 GHz. More recently, Sakr et al. (2013) increased this to 40 GHz using a quantum cascade photodetector with peak response at 1.55 μm. This device makes use of the large internal electric fields in nitride structures to sweep photogenerated carriers from one quantum section to the next and provides a voltage output at zero applied bias, thus avoiding dark-current noise (the limitation being simply that of Johnson noise). Response speed was found to be RC limited, and detectors with sizes reduced to 7 μm × 7 μm showed 3 dB bandwidths of 42 GHz, which brought them in line with Ge and InGaAs PIN diodes. One may simply 'watch this space' to see whether these MBE-grown nitrides can mount an effective challenge to the incumbent detector species. For a recent review covering a wider range of device possibilities, see Beeler et al. (2013).

From a practical point of view, the future of quantum IR detectors was much brightened by the discovery of self-assembled quantum dots in the middle of the 1980s and, more directly, by their application to laser diodes in the middle of the 1990s. It was clear that dots possessed confined energy states similar to those associated with wells and brought with them an important advantage in so far as optically stimulated transitions between these states were allowed for light falling normally on the surface of the structure. Responsivity was also enhanced by the long lifetimes of carriers in the excited state (due to the phenomenon of phonon bottleneck). The IR dot detector story began in 1997 with several reports of optical absorption measurements on self-organised InAs quantum dots, while the first report of a quantum dot IR photodetector (a QDIP) appeared (from Japan) in 1999 (Horiguchi et al. 1999; Lee et al. 1999). Taking the

Horiguchi paper as example, the sample had ten layers of InAs dots separated by 30 nm GaAs barriers. Dot density was 1×10^{15} m^{-2}, and dot dimensions were 20 nm, diameter, and 5 nm, height. Dots were modulation doped, the centre 10 nm of the GaAs layer being doped with Si to a density of 1×10^{23} m^{-3}, corresponding to one electron per dot. The photocurrent response peaked at approximately 10 μm but was significantly broader than that typical of QWIPs, covering the range from about 7 μm to 14 μm. This was clearly the result of the distribution in dot size already well known from earlier work and, while it may be a distinct disadvantage for QD lasers, it could be seen as a virtue in many examples of IR imaging.

As the subject developed, there were many advances, though we can give only a few examples. One such was concerned with the reduction of dark current. This thermally generated current was more serious in QDIPs on account of the light-generating material (the dots) being only a fraction of the total structure. Due to its thermal nature, dark current increased exponentially with increasing temperature and limited the maximum operating temperature of the detectors. Wang et al. (2001) introduced an $Al_{0.2}Ga_{0.8}As$ current blocking layer, as illustrated in Figure 10.16. This reduced dark current by some three orders of magnitude, while still allowing an acceptable photoresponse. Tang et al. (2001) reported a somewhat different arrangement, employing just two AlGaAs layers, grown either side of the ten-sequence QD structure, and reporting 'near-room-temperature operation' of their InAs/GaAs QDIP. The detectivity at a temperature of 250 K was about an order of magnitude lower than at 77 K but the device was still useful. High-temperature operation was also the target of Chakrabarti et al. (2004), who concentrated on improving the material quality of both dots and barriers; 50 nm GaAs barriers were grown at 600 °C at 2 Ås^{-1}, while dots were grown at 500 °C and 0.3 Ås^{-1} to minimise the density of dislocations and other defects which give rise to dark current. They were able to grow up to 70 QD layers without degradation in material quality, thus enhancing IR photon absorption. Detectors functioned between 2 μm and 6 μm

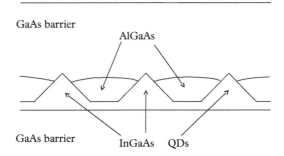

Figure 10.16 *Structure of a QDIP with an AlGaAs current-blocking layer, as introduced by Wang et al. (2001). The AlGaAs blocks dark current flowing from the InAs dots into the GaAs barriers, while allowing photocurrent to flow via the uncovered regions of the dots.*

at temperatures up to 200 K, with performances sufficient to encourage development of FPAs. Another interesting application of QDIPs was to the fibre-optic communication detectors which we discussed above. Hofstetter et al. (2010) compared QDIP detectors with QWIPs and found an increase in responsivity of 60 times for devices with peak response in the vicinity of 1.4 μm. It also allowed performance at significantly higher temperature—160 K, rather than the 80 K typical of QWIPs. However, their suggested reason for the improvement, that the excited-state lifetime is much increased, implies that the response speed must be correspondingly limited and this is anathema to communications engineers, looking for a response to high-speed bit streams.

Yet another advance in detector design was concerned with enhancing the absorption of photons by incorporating the dot layers within a resonant cavity (Asano et al. 2010), thus enabling the dot region to be kept relatively thin and therefore of improved quality. Such a concept brought with it other difficulties, however. An effective match between cavity and absorption wavelengths demanded a narrow absorption band, which had proved difficult in earlier work due to the spread in dot sizes. The cavity also required highly reflecting mirrors which could not be made using the standard AlAs/GaAs DBR process because of the long wavelengths involved (it would have required something like 100–200 μm of material, well beyond the capabilities of the typical MBE process). Asano et al. solved these problems by using short-period $(AlAs)_1(GaAs)_4$ superlattices as barrier material, rather than GaAs and by the use of GaAs/air interfaces as mirrors. These latter were produced by growing AlGaAs/GaAs structures and then selectively etching away the AlGaAs. The resulting QDIPs showed an order of magnitude enhancement in detectivity.

Taking an almost diametrically opposite stance, Lin and Lin (2012) demonstrated the application of variable dot size to achieving a *broader* photoresponse. They grew InAs dots on GaAs and then capped them with 8 nm of InGaAs, followed by 50 nm of GaAs, repeated ten times. The dot size was controlled by a small variation (from 2.0 to 2.8 ML) in the amount of InAs deposited and, somewhat remarkably, samples grown with 2.4 ML spontaneously produced a spread in dot heights which generated a photon response over the range 6–12 μm, compared with the 1–2 μm width of the 2.0 and 2.8 ML samples. (One cannot help wondering whether there is an element here of the 'East' rediscovering what the 'West' knew all too well in 1999!)

Clearly, QDIPs showed promise for a wide range of detector applications but we should take note also of the use of quantum dots within quantum wells (DWELLs), a structure we came across earlier in connection with QD lasers and one which is also grown predominantly by MBE. The first references to DWELL IR detectors appeared about 2002, though at that time there was no clear delineation between them and QDIPs. It is, perhaps, worth emphasising that the optical transition involved in a DWELL detector is frequently one which excites an electron from the ground state in the dot into a confined state within the well and that electronic conduction laterally along the well is a possible mechanism for detecting such a transition. An interesting example of an early DWELL device was described by Krishna et al. (2003). This showed a response at three distinct wavelengths: ~3.8 μm, ~8.5 μm and ~23 μm, the first two involving transitions from dot states into well states, the third (long-wavelength) transition probably within the dots. They grew $In_{0.15}Ga_{0.85}As/GaAs$ wells and InAs dots by solid source MBE, the GaAs being grown at 580 °C, and the well and dot material at 480 °C. The dots were grown very slowly (0.05 MLs^{-1}) and TEM images showed no evidence for threading dislocations (Krishna 2005). Perhaps their most significant observation was that the long-wavelength response was still visible at temperatures up to 80 K, very much higher than similar longwave QWIP detectors, a phenomenon thought to be related to the phonon bottleneck which significantly reduces dark current. Krishna (2005) describes the development of 320 × 256 FPAs based on a 15-layer stack of these DWELL structures and demonstrates images derived from both the 3–5 μm and 8–12 μm IR bands.

The promise of such DWELL imagers was apparent but nevertheless lacked the responsivity needed to challenge well-established devices. As pointed out by Gunapala et al. (2007), it would be necessary to improve quantum efficiency by increasing the number of dots—dot density being very much less than the density of electrons in a QWIP device. They achieved this by increasing the number of dot/well layers to 30 and by the application of a metal reflecting film on top of the structure so as to obtain a double pass for the IR light. Areal dot density was measured as 3 × 10^{14} m^{-2}, which compares well with the typical areal electron density in a QWIP, ~3 × 10^{15} m^{-2}. They also demonstrated the ease of wavelength tuning (over the range 5–12 μm) by varying well width from 5.5 nm to 10 nm, this being very much easier than attempting to control dot dimensions, a clear advantage of DWELL over QDIP devices. Of particular significance in this paper was the observation that, though DWELL structures are, indeed, sensitive to normal incidence radiation, they can be even more sensitive to 45° incident light. Thus, by using a reflection grating they were able to increase responsivity by a factor of 4, achieving a measured responsivity as high as 0.8 AW^{-1} in the wavelength region of 8–9 μm; 640 × 512 pixel LWIR FPAs were made using these structures, and imaging achieved at 60 K. However, this performance still fell short of that available from QWIP structures, on account of their greater sensitivity and uniformity.

The quest for improved quantum efficiency in Dwell detectors took an interesting step forward in 2008 when Ling et al. (2008) reported on the use of an $Al_{0.3}Ga_{0.7}As$ 'confinement enhancing layer'. This layer was 2.5 nm thick, somewhat less than the height of the dots, and therefore looked rather similar to the dark-current blocking layer introduced by Wang et al. (2001) and which we referred to above (see Figure 10.16). What was new was the interpretation of its function—according to Ling et al., it resulted in greater confinement of the energy levels in the dot which, in turn, increased the transition rate between ground and excited states and led to a significant increase in responsivity. Indeed, this was borne out by their experimental data—quantum efficiency was enhanced by as much as 25 times, compared with a similar device in which the AlGaAs layer was replaced with an InGaAs layer, resulting in measured responsivities as high as 8 AW^{-1} at 77 K ($\lambda = 8$ μm). There would appear to be scope for critical discussion on the precise mechanism(s) involved but there can be no argument with the experimental data. A further development saw the application of sub-monolayer deposition to produce rather differently shaped quantum dots (Ting et al. 2009). It follows that if one deposits, say, one-third of a monolayer of InAs on a GaAs surface, this must result in thin isolated regions of InAs (or InGaAs?), and these islands represent a form of quantum dot which can, in principle, be grown at higher density than the more conventional Stranski–Krastanow dots. Albeit, they can be used to form DWELL IR detectors, which can also be made into FPAs, as demonstrated by Ting et al. Alas, these dots also show greater sensitivity to inclined incidence than to normal incidence light and might therefore be improved by better control over their growth dynamics but they clearly offer yet another interesting possibility in the maelstrom of IR detector research.

10.6   Solar Cells

It goes without saying that the utilisation of solar cells to generate electricity for both humdrum domestic and sophisticated satellite applications has been a success story for semiconductor science over the past five decades. Without the solar cell, satellite ventures would have been grounded from the word 'go' and this is clearly a regime where cost is very much of secondary consequence—conversion efficiency being all important. On the other hand, domestic 'PV solar' is ruled by cost, efficiency playing second fiddle to material availability in the form of silicon single-junction devices and, needless to say, MBE has found no role in this particular market. There is, however, an intermediate technology, based on solar concentrators, in which the area of the cell is small enough that it represents no more than 10–15% of the overall system cost and this has encouraged the use of multijunction cells similar to those used on satellites. It is with this aspect of solar cell technology that we shall mainly be concerned, MBE having made a useful, if not dominant, contribution.

The principle behind the use of multijunction cells can be seen as a response to the calculations of Shockley and Queisser (1961). They used detailed balance considerations to show that the maximum possible efficiency for an ideal single-junction cell under 'one-sun' illumination is 31.0% for a material with band gap $E_g = 1.3$ eV, this being limited by, on the one hand, low-energy photons not being absorbed and, on the other, by high-energy photons generating electron–hole pairs which lose energy non-radiatively before reaching the contacts. Ideally, it would be necessary for each group of photons to be absorbed by a material with perfectly matching band gap. The typical three-junction cell (one example of what are generally referred to as a 'third-generation' cells, aimed at reducing the cost of solar electricity to something of the order of 0.3 US$/W) offers a reasonable compromise, given that the additional complexity of multijunction structures adds significantly to production costs. Shockley and Queisser's calculations also showed that, for 'concentrator' applications the corresponding maximum efficiency and band gap for a single-junction cell are 40.8% and 1.1 eV, respectively. It is perhaps worth emphasising that the Shockley–Queisser model assumes electron–hole recombination to be 100% radiative—any non-radiative recombination centres in cell material must inevitably lead to reduced efficiency. A recent article by Conibeer (2007) describes third-generation cells in some detail, listing many of the options and discussing the pros and cons of each in turn. We refer the reader to this for further detail. From the viewpoint of MBE, we shall concentrate mainly on three aspects—the three-junction cell, the use of

Front contact

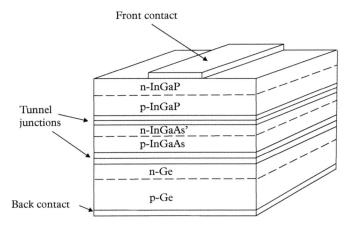

Tunnel
junctions

Back contact

n-InGaP

p-InGaP

n-InGaAs'

p-InGaAs

n-Ge

p-Ge

Figure 10.17 *Structure of a triple junction solar cell, grown epitaxially on a Ge substrate. The Ge junction is formed by diffusion doping, and the InGaAs and InGaP junctions during growth. (Following Conibeer 2007.)*

quantum well structures (which appear to sever the connection between absorption and open-circuit voltage) and the so-called intermediate band cell.

The classic example of a three-junction cell is shown in Figure 10.17, a Ge p–n junction cell forming the substrate $\left(E_g = 0.7\,\text{eV}\right)$, while InGaAs $\left(E_g = 1.4\,\text{eV}\right)$ and InGaP $\left(E_g = 1.9\,\text{eV}\right)$ junctions are grown epitaxially on top, the materials being chosen, of course, so as to achieve lattice-matching, in the interest of minimising dislocation densities. Thus, the incoming radiation meets the widest gap material first, followed by the intermediate and, finally, the narrowest gap material. All three junctions are connected in series via heavily doped tunnel junctions designed to minimise series resistance. For unconcentrated light, the theoretical efficiency is 48.6%, and practical values of slightly more than 40% have been achieved, making such devices useful in space applications. At high concentration ratios, the maximum efficiency is 63.2%, compared with the ideal thermodynamic limit of 86.8%, so there would clearly be scope for further improvement (four or even five junctions?) if financial pressure should make it worthwhile! Until recently, MOVPE was the preferred growth method but it is now recognised that MBE is equally suitable, following the work of Tobin et al. (1990), who compared nominally identical GaAs single-junction cells grown both by MOVPE and by MBE. Record high efficiencies were recorded in both instances, 24.8% and 23.8%, respectively, and the authors concluded that MBE-grown films were 'suitable for high-efficiency solar cells'.

MBE growth certainly came into its own in a significant recent development—a three-junction cell involving dilute nitride material (Sabnis et al. 2012)—and it is worth looking at this in a little more detail. Detailed examination of the classic cell performance reveals that the photocurrents generated by the three junctions are far from identical: ~14 mA for each of the top junctions but ~25 mA from the bottom, Ge junction. As the three are series connected, this means that the excess current from the Ge junction has to be dissipated as heat and the overall efficiency correspondingly reduced. One approach to overcoming this problem is to replace the Ge junction with a slightly higher band-gap material, in this case GaInNAs(Sb), a material that we met earlier, in Sections 10.3 and 10.4. As we remarked in Chapter 7 (Section 7.5), the incorporation of N into GaAs was not only difficult but also led to some unexpected consequences—in particular, it resulted in a reduction in band gap due to the very large bowing effect. What was more (see Box 10.2), it was possible, by balancing the amount of added In and N in the ratio of approximately 2.7:1, to reduce the band gap while maintaining an accurate lattice match to GaAs. Sabnis et al. developed MBE growth techniques, employing Sb as a surfactant which allowed them to obtain band gaps ranging from 0.8 to 1.3 eV and they employed material with $E_g \sim 0.9 - 1.0$ eV as replacement for Ge in the lower junction, thus producing good current matching while obtaining a significantly greater open-circuit voltage. It was also necessary to make slight modifications to the other materials by, for example, using InAlGaP and AlGaInAs to optimise the efficiency of the overall cell. The result was a world-record measured efficiency of 43.5% using concentration factors approaching 1000. Their company, 'Solar Junction', is now marketing such cells and predicting that their use will result in overall efficiencies of concentrator modules as high as 35%.

Good as they obviously are, these tandem cells inevitably suffer from some degree of imperfection. One problem stems from the fact that the solar spectrum changes throughout the day, being red-shifted in the early morning and again in the evening, causing a mismatch with the cell design. Also, the reverse-biased junctions which link the active junctions represent a loss of voltage, with corresponding reduction in efficiency. Several attempts have been made to overcome these limitations and in most cases MBE has been central to the research. First, we make brief mention of an idea originating from former colleagues in Philips (Barnham and Duggan 1990). This involved the incorporation of a MQW structure in the intrinsic region of a single cell p-i-n photodiode, the effective band gap for absorption being set by transitions within the confined states of the quantum wells, while the open-circuit voltage was determined by the band gap of the barriers. There was an obvious question mark over the efficiency of charge transfer from the wells to the contacts but, assuming a value close to 100%, calculations suggested that ultimate solar cell efficiencies of over 40% at one sun and over 50% at concentrations of 1000 times might be anticipated (in both cases ~10% greater than the basic single cell limits), and all this, of course, without the additional complexity inherent in making triple junction cells. Experimental results on test samples grown both by MOVPE and by MBE (Barnham et al. 1991; Paxman et al. 1993) suggested that higher open-circuit voltages could, indeed, be obtained but whether overall efficiency increases are possible still appears to be a matter for debate.

An alternative approach to improved cell efficiency, the so-called intermediate band solar cell was first proposed in 1960 (Wolf 1960) and analysed in detail by Luque and Marti (1997). This depends on introducing a band of impurities within the band gap of material forming the i-region of a p-i-n diode, so as to absorb sub-band-gap photons, thus adding a further contribution to photocurrent. They calculated an optimum efficiency (for a band gap of 1.95 eV) under high concentration of 63.1% (compared to the Shockley–Queiser value of 40.8% for a single-junction device), closely similar to that of a three-junction cell. (Indeed, it can be thought of as two cells in series plus a third in parallel, an arrangement which has the advantage of improved tolerance to variations in the solar spectrum.) However, it was clear that a number of conditions had to be satisfied if such a result was to be achieved in practice. The additional photocurrent had to be generated by the combination of a valence band–intermediate band (VB-IB) transition, followed by an (IB-CB) transition, though not necessarily involving the same electron—which implies electronic conduction in the IB, and this in turn requires that impurities must be present in sufficient quantity to ensure overlap of their electron wavefunctions. In order to balance the two transition rates, the IB had to be doped so that its quasi-Fermi level was at the centre of the band. It was also essential that the IB was not in contact with either contact electrode, so that the cell open-circuit voltage would still be determined by the overall band gap. For maximum efficiency, the IB should lie at an energy one-third of the band gap above the VB. Finally, it was important that the intermediate levels did not function as non-radiative recombination centres—as in the original Shockley–Queisser model, recombination should be purely radiative. While relatively simple in principle, this approach was clearly going to make severe demands on the semiconductor technologist!

Nevertheless, the idea struck a chord with several experimental groups and is still being explored worldwide, MBE growth playing a major role in the various research programmes. While progress has undoubtedly been made, there is yet no clear evidence of improved solar cell efficiency, but hope is still alive and we should look briefly at some of the attempts being made to realise practical success (see e.g. the reviews by Luque and Marti 2010 and by Luque et al. 2012). Three distinct approaches to the formation of the intermediate band are worthy of attention: the use of high-density quantum dots, high-level doping with deep impurity species, and the use of highly mismatched alloys—all making use of MBE growth to obtain the necessary material. In all cases, attention has been concentrated on demonstrating fundamental features such as an increase in photocurrent—meeting all the demands of the theoretical model simultaneously seems likely to require much further time and effort.

An example of research aimed to understand the function of quantum dots as the intermediate band is provided by Marti et al. (2006). They were able to demonstrate photocurrent due to transitions between the IB and CB in a structure where InAs/GaAs dots (grown by the SK technique) were inserted in the i-region of an AlGaAs/GaAs p-i-n diode. Ten layers of dots were grown and the GaAs layers doped with Si to ensure half-filling of the IB. It was probably this proof of principle that encouraged other workers to join the fray, and many groups have observed improvements in photocurrent due to the dot layers, though generally associated with a decrease in open-circuit voltage—sadly, no one appears to have demonstrated any improvement in cell efficiency. There are obvious problems with the dot approach, such as the difficulty in increasing dot density sufficiently to achieve adequate sub-band-gap photon absorption—more dot layers generally leading to increased strain. Though several groups have used 'strain-balanced' or 'strain-compensated' dots (see e.g. Oshima et al. 2008) and have seen an increase in quantum efficiency at longer wavelengths, no overall gain in cell efficiency has resulted. We might also note the fact that, as pointed out in our discussion of IR photodetectors, photon absorption of normally incident radiation is significantly less that that associated with lateral radiation—this is equally a problem for solar cell quantum efficiency. A second complication arises as a consequence of the wetting layer inherent in the SK growth method. Its precise involvement is difficult to quantify but it must inevitably compromise cell performance. Alternative methods of dot growth seem highly desirable. As summarised by Luque et al. (2012), 'The shortcomings of InAs/GaAs QD solar cells are quite well understood, and we believe this knowledge will guide research towards more promising QD structures.' We shall leave the matter there.

An alternative to dot growth which is being explored by several groups is that of doping GaAs or AlGaAs with deep impurity levels such as Fe, Cr or Co but there are severe problems in achieving densities of these impurities adequate to exceed the Mott limit and thus obtain conductivity in the intermediate band to avoid non-radiative recombination. This, in turn, requires non-equilibrium or low temperature growth, so MBE or variants such as MEE are most suitable but, again, there is no evidence yet of improved cell efficiency. Finally, we might make mention of the use of highly mismatched alloys to obtain a suitable intermediate band, as described by Ahsan et al. (2013). As we discussed earlier (Chapter 7, Section 7.5), when N is introduced into GaAs, the band-anti-crossing model predicts that the GaAs CB is split into upper E_+ and lower E_- bands and these two bands may be utilised as CB and IB, respectively, of an IB solar cell, the VB being the unmodified VB of the GaNAs. Ahsan et al. investigated two-photon absorption associated with this material in a solar cell-like structure sandwiched between p-AlGaAs and n-AlGaAs barriers. They made use of Sb as a surfactant to improve the structural quality of the GaNAs layer and demonstrated an increase in photocurrent, though with no overall improvement in cell efficiency. This work was clearly at an early stage of development and further progress can be anticipated in future.

In summary, we should re-emphasise the significance of the record solar cell efficiency achieved by Sabnis et al. (2012) using a three-junction structure with a GaInNAs lower junction. This was not only grown by MBE but represented the result of measurements on a commercial device which is already in production. This leaves alternative approaches to high-efficiency cells a long way behind but we should also accept that such alternatives are still in an early stage of development—there may well be considerable improvements in the coming years and it is clear that MBE will play an important role in the necessary research.

REFERENCES

Ahsan, N, Miyashita, N, Islam, M M, Yu, K M, Walukiewicz, W and Okada, Y (2013) Effect of Sb on GaNAs intermediate band solar cells. IEEE J Photovoltaics 3, 730.
Alferov, Zh I (2001) Nobel Lecture: the double heterostructure concept and its applications in physics, electronics, and technology. Rev Mod Phys 73, 767.
Alferov, Zh I, Andreev, V M, Garbazov, D Z, Zhilyaev, Yu V, Morozov, E P, Portnoi, E L and Trofim, V G (1970) Investigation of the influence of the AlAs-GaAs heterostructure

parameters on the laser threshold current and the realization of continuous emission at the room temperature. Fiz Tekh Poluprovodn 4, 1826 [Sov Phys Semicond 4, 1573 (1971)].

Alferov, Zh I, Andreev, V M, Korol'kov, V I, Trat'yakov, D N and Tuchkevich, V M (1967) High-voltage p-n junctions in $Ga_xAl_{1-x}As$ crystals. Fiz Tekh Poluprovodn 1, 1579 [Sov Phys Semicond 1, 1313 (1968)].

Alferov, Zh I, Vasilev, A M, Ivanov, S V, Kop'ev, P S, Ledentsov, N N, Lutsenko, M E, Metser, B Ya and Ustinov, B M (1988) Reducing of the threshold current in GaAs-AlGaAs DHS SCH quantum well lasers (jth = 52 A/cm^2, T = 300 K) with quantum well restricted by short period superlattice of variable period. Pis'ma Zh Tekh Fiz 14, 1803 [Sov Tech Phys Let 14, 782 (1988)].

Alferov, Zh I and Kazarinov, R F (1963) 'Semiconductor Laser with Electric Pumping', Russian patent application 950840.

Arakawa, Y and Sakaki, H (1982) Multidimensional quantum well laser and temperature dependence of its threshold current. Appl Phys Lett 40, 939.

Asano, T, Hu, C, Zhang, Y, Liu, M, Campbell, J C and Madhukar, A (2010) Design consideration and demonstration of resonant-cavity-enhanced quantum dot infrared photodetectors in mid-infrared wavelength regime (3–5). IEEE J Quantum Electron 46, 1484.

Badcock, T J, Royce, R J, Mowbray, D J, Skolnick, M S, Liu, H Y, Hopkinson, M, Groom, K M and Jiang, Q (2007) Low threshold current density and negative characteristic temperature 1.3 μm InAs self-assembled quantum dot lasers. Appl Phys Lett, 90, 111102.

Bae, H, Bank, S R, Yuen, H B, Sarmiento, T, Pickett, E R, Wistey, M A and Harris, J M (2007) Temperature dependencies of annealing behaviors of GaInNAsSb/GaNAs quantum wells for long wavelength dilute-nitride lasers. Appl Phys Lett 90, 231119.

Bank, S R, Bae, H, Goddard, L L, Yuen, H B, Wistey, M A, Kudrawiec, R and Harris, J S (2007) Recent progress on 1.55-μm dilute-nitride lasers. IEEE J Quantum Electron 43, 773.

Bardeen, J (1963) in 'Collected Works of John Von Neumann'. Ed A H Traub, Vol 5, p 420, Pergamon Press, NY.

Barnham, K W J, Braun, B, Nelson, J, Paxman M, Button, C, Roberts, J S and Foxon, C T (1991) Short-circuit current and energy efficiency enhancement in a low-dimensional structure photovoltaic device. Appl Phys Lett 59, 135.

Barnham, K W J and Duggan G (1990) A new approach to high-efficiency multi-band-gap solar cells. J Appl Phys 67, 3490.

Basov, N G, Krokhin, O N and Popov, Yu M (1961) Production of negative temperature states in pn junctions of degenerate semiconductors. Sov Phys JETP 13, 1320.

Bhattacharya, P, Ghosh, S and Stiff-Roberts, A D (2004) Quantum dot opto-electronic devices. Ann Rev Mater Res 34, 1.

Beck, M, Hofstetter, D, Aellen, T, Faist, J, Oesterle, U, Ilegems, M, Gini, E and Melchior, H (2002) Continuous wave operation of a mid-infrared semiconductor laser at room temperature. Science 295, 301.

Beeler, M, Trichas, E and Monroy, E (2013) III-nitride semiconductors for intersubband optoelectronics: a review. Semicond Sci Technol 28, 074022.

Beere, H E, Fowler, J C, Alton, J, Linfield, E H, Ritchie, D A, Köhler, R, Tredicucci, A, Scalari, G, Ajili, L, Faist, J and Barbieri, S (2005) MBE growth of terahertz quantum cascade lasers. J Cryst Growth 278, 756.

Belkin, M A, Capasso, F, Belyanin, A, Sivco, D L, Cho, A Y, Oakley, D C, Vineis, C J and Turner, G W (2007) Terahertz quantum-cascade-laser source based on intracavity difference-frequency generation. Nature Photon 1, 288.

Bernard, M G A and Duraffourg, G (1961) Laser conditions in semiconductors. Phys Stat Sol (b) 1, 699.

Bimberg, D, Grundmann, M and Ledentsov N N (1999) 'Quantum Dot Heterostructures', Wiley, Chichester.

Bismuto, A, Terazzi, R, Beck, M and Faist, J (2010) Electrically tunable, high performance quantum cascade laser. Appl Phys Lett 96, 141105.

Blood, P (1988) 'Reappraisal Of GaAs-AlGaAs Quantum Well Lasers' in 'Quantum Wells and Superlattices in Optoelectronic Devices and Integrated Optics, Proc SPIE 861' (ed A R Adams), p 34.

Casey, H C Jr, Panish, M B, Schlosser, W O and Paoli, T L (1974) GaAs–Al_xGa_{1-x}As heterostructure laser with separate optical and carrier confinement. J Appl Phys 45, 322.

Casey, H C Jr, Somekh, S and Ilegems, M (1975) Room-temperature operation of low-threshold separate-confinement heterostructure injection laser with distributed feedback. Appl Phys Lett 27, 142.

Chakrabarti, S, Stiff-Roberts, A D, Bhattacharya, P, Gunapala, S, Bandara, S, Rafol, S B and Kennerly, S W (2004) High-temperature operation of InAs-GaAs quantum-dot infrared photodetectors with large responsivity and detectivity. IEEE Photonics Technol Lett 16, 1361.

Chen, H Z, Ghaffari, A, Morkoç, H and Yariv, A (1987) Effect of substrate tilting on molecular beam epitaxial grown AlGaAs/GaAs lasers having very low threshold current densities. Appl Phys Lett 51, 2094.

Cheng, T S, Dawson, P, Lacklison, D E, Foxon, C T, Orton, J W, Hughes, O H and Henini, M (1993) Substrate temperature dependence of the minority carrier lifetime in (AlGa)As/GaAs MQWs grown with As_2 and As_4. J Cryst Growth 127, 841.

Cho, A Y and Casey, H C Jr (1974) GaAs–Al_xGa_{1-x} As double-heterostructure lasers prepared by molecular-beam epitaxy. Appl Phys Lett 25, 288.

Cho, A Y, Dixon, R W, Casey, H C Jr and Hartman, R L (1976) Continuous room-temperature operation of GaAs-Al_xGa_{1-x}As double-heterostructure lasers prepared by molecular-beam epitaxy. Appl Phys Lett 28, 501.

Conibeer, G (2007) Third-generation photovoltaics. Mater Today 10, 42.

Dawson, P and Woodbridge, K (1984) Effects of prelayers on minority-carrier lifetime in GaAs/AlGaAs double heterostructures grown by molecular beam epitaxy. Appl Phys Lett 45, 1227.

Dean, P, Salih, M, Khanna, S P, Li, L H, Saat, N K, Valavanis, A, Burnett, A, Cunningham, J E, Davies, A G and Linfield, E H (2012) Resonant-phonon depopulation terahertz quantum cascade lasers and their application in spectroscopic imaging. Semicond Sci Technol 27, 094004.

Depuis, R D (1987) Preface. IEEE J Quantum Electron 23, 658.

Dingle, R, Wiegmann, W and Henry, C H (1974) Quantum states of confined carriers in very thin Al_xGa_{1-x}As-GaAs-Al_xGa_{1-x} As heterostructures. Phys Rev Lett 33, 827.

Downs, C and Vandervelde, T E (2013) Progress in infrared photodetectors since 2000. Sensors (Basel) 13, 5054.

Egorov, A Yu, Zhukov, A E, Kop'ev, P S, Ledentsov, N N, Maksimov, M V, Ustinov, V M, Tsatsul'nikov, A F, Bert, N A, Kosogov, A O and Alferov, Zh I (1996) Formation of vertically aligned arrays of strained InAs quantum dots in a GaAs(100) matrix. Sov Phys Semicond 30, 879.

Eibelhuber, M, Schwarzl, T, Pichler, S, Heiss, W and Springholz, G (2010) Near room temperature continuous-wave laser operation from type-I interband transitions at wavelengths beyond 4 μm. Appl Phys Lett 97, 061103.

English, J H, Gossard, A C, Störmer, H L and Baldwin, K W (1987) GaAs structures with electron mobility of 5×106 cm^2/V s. Appl Phys Lett 50, 1826.

Faist, J, Beck, M,Aellen, T and Gini, E (2001) Quantum-cascade lasers based on a bound-to-continuum transition. Appl Phys Lett 78, 147.

Faist, J, Capasso, F, Sitori, C, Sivco, D L, Baillargeon, J N, Hutchinson, A L, Chu, S -N G and Cho, A L (1996) High power mid-infrared ($\lambda\sim5$ μm) quantum cascade lasers operating above room temperature. Appl Phys Lett 68, 3680.

Faist, J, Capasso, F, Sivco, D L, Sirtori, C, Hutchinson, A L and Cho, A Y (1994) Quantum cascade laser. Science, 264, 553.

Fathololoumi, S, Dupont, E, Chan, C W I, Wasilewski, Z R, Laframboise, S R, Ban, D, Mátyás, A, Jirauschek, C, Hu, Q and Liu, H C (2012) Terahertz quantum cascade lasers operating up to ~ 200 K with optimized oscillator strength and improved injection tunneling. Opt Express, 20, 3866.

Fathpour, S, Mi, Z, Bhattacharya, P, Kovsh, A R, Mikhrin, S S, Krestnikov, I L, Koshukhov, A V and Ledentsov, N N (2004) The role of Auger recombination in the temperature-dependent output characteristics ($T_0 = \infty$) of p-doped 1.3 μm quantum dot lasers. Appl Phys Lett 85, 5164.

Feit, Z, Mak, P, Woods, R and McDonald, M (1996a) MBE grown buried heterostructure separate confinement multiple quantum well $Pb_{0.9854}Eu_{0.0146}Se_xTe_{1-x}$/$Pb_{0.981}Sn_{0.019}$Te tunable diode lasers for high resolution sepctroscopy. Spectrochim Acta A 52, 851.

Feit, Z, McDonald, M, Woods, R J, Archambault, V and Mak, P (1996b) Low threshold PbEuSeTe/PbTe separate confinement buried heterostructure diode lasers. Appl Phys Lett 68, 738.

Ferguson, J W, Smowton, P M, Blood, P, Bae, H, Sarmiento, T and Harris, J S Jr (2009) Non-radiative recombination in 1.56 μm GaInNAsSb/GaNAs quantum-well lasers. Appl Phys Lett 95, 231104.

Fill, M, Khiar, A, Rahim, M, Felder, F and Zogg, H (2011) PbSe quantum well mid-infrared vertical external cavity surface emitting laser on Si-substrates. Appl Phys Lett 109, 093101.

Fischer, M, Reinhardt, M and Forchel, A (2000) GaInAsN/GaAs laser diodes operating at 1.52 μm. Electron Lett 36, 1208.

Foxon, C T, Clegg, J B, Woodbridge, K, Hilton, D, Dawson, P and Blood, P (1985) The effect of the oxygen concentration on the electrical and optical properties of AlGaAs films grown by MBE. J Vac Sci Technol B 3, 703.

Fujii, T, Hiyamizu, S, Yamakoshi, S and Ishikawa, T (1985) MBE growth of extremely high-quality GaAs–AlGaAs GRIN-SCH lasers with a superlattice buffer layer. J Vac Sci Technol B 3, 776.

Garland, J and Sporken, R (2010) 'Substrates for the Epitaxial Growth of MCT' in 'Mercury Cadmium Telluride: Growth, Properties and Applications' (ed P Capper and J Garland), Wiley, Chichester, p 75.

Geels, R S and Coldren, L A (1990) Submilliamp threshold vertical-cavity laser diodes. Appl Phys Lett 57, 1605.

Geels, R S, Corzine, S W, Scott, J W, Young D B and Coldren, L A (1900) Low threshold planarized vertical-cavity surface-emitting lasers. IEEE Photonics Tech Lett 2, 234.

Giorgetta, F R, Baumann, E, Guillot, F, Monroy, E and Hofstetter, D (2007) High frequency (f=2.37 GHz) room temperature operation of 1.55 μm AlN/GaN-based intersubband detector. Electron Lett, 43, 185.

Grabherr, M, Jäger, R, Miller, M, Thalmaier, C, Heerlein, J, Michalzik, R and Ebeling, K J (1998) Bottom-emitting VCSEL's for high-CW optical output power. IEEE Photonics Tech Lett 10, 1061.

Gunapala, S D, Bandara, S V, Hill, C J, Ting, D Z, Liu, J K, Rafol, S B, Blazejewski, E R, Mumolo, J M, Keo, S A, Krishna, S, Chang, Y-C and Craig, A S (2007) 640 × 512 pixels long-wavelength infrared (LWIR) quantum-dot infrared photodetector (QDIP) imaging focal plane array. IEEE J Quantum Electron 43, 230.

Haase, M A, Qiu, J, DePuydt, J M and Cheng, H (1991) Blue-green laser diodes. Appl Phys Lett 59, 1272.

Hall, R N, Fenner, G E, Kingsley, J D, Soltys, T J and Carlson, R O (1962) Coherent light emission from Ga-As junctions. Phys Rev Lett 9, 366.

Harris, J S Jr (2002) GaInNAs long-wavelength lasers: progress and challenges. Semicond Sci Technol 17, 880.

Hayashi, I (1972) Double heterostructure laser diodes. US Patent No. 3691476.

Hayashi, I, Panish, M B, Foy, P W and Sumski, S (1970) Junction lasers which operate continuously at room temperature. Appl Phys Lett 17, 109.

Hofstetter, D, Beck, M, Aellen, T and Faist, J (2001) High-temperature operation of distributed feedback quantum-cascade lasers at 5.3 μm. Appl Phys Lett 78, 396.

Hofstetter, D, Faist, J, Beck, M, Müller, A and Oesterie, U (1999) Demonstration of high-performance 10.16 μm quantum cascade distributed feedback lasers fabricated without epitaxial regrowth. Appl Phys Lett 75, 665.

Hofstetter, D, Francesco, J D, Kandaswamy, P K, Aparna, D, Valdueza-Felip, S and Monroy, E (2010) Performance improvement of AlN–GaN-based intersubband detectors by using quantum dots. IEEE Photon Technol Lett 22, 1087.

Holonyak, N Jr and Bevacqua, S F (1962) Coherent (visible) light emission from Ga(As$_{1-x}$P$_x$) junctions. Appl Phys Lett 1, 82.

Hooper, S E, Kauer, M, Bousquet, V, Johnson, K, Barnes, J M and Heffernan, J (2004) InGaN multiple quantum well laser diodes grown by molecular beam epitaxy. Electron Lett 40, 33.

Horiguchi, N, Futatsugi, T, Nakata, Y, Yokoyama, N, Mankad, T and Petroff, P M (1999) Quantum dot infrared photodetector using modulation doped InAs self-assembled quantum dots. Jap J Appl Phys 38, 2559.

Huffaker, D L and Deppe, D G (1998) Electroluminescence efficiency of 1.3 μm wavelength InGaAs/GaAs quantum dots. Appl Phys Lett 73, 520.

Huffaker, D L, Deppe, D G, Kumar, R and Rogers, T J (1994) Native-oxide defined ring contact for low threshold vertical-cavity lasers. Appl Phys Lett, 65, 97.

Huffaker, D L, Park, G, Zou, Z, Shchekin, O B and Deppe, D G (1998) 1.3 μm room-temperature GaAs-based quantum-dot laser. Appl Phys Lett 73, 2564.

Hugi, A, Terazzi, R, Bonetti, Y, Wittmann, A, Fischer, M, Beck, M, Faist, J and Gini, E (2009) External cavity quantum cascade laser tunable from 7.6 to 11.4 μm. Appl Phys Lett 95, 061103.

Ibaraki, A, Kawashima, K, Furusawa, K, Ishikawa, T, Yamaguchi, T and Niina, T (1989) Buried heterostructure GaAs/GaAlAs distributed Bragg reflector surface emitting laser with very low threshold (5.2 mA) under room temperature CW conditions. Jap J Appl Phys 28, L667.

Iga, K (2000) Surface-emitting laser-its birth and generation of new optoelectronics field. IEEE J Sel Top Quant Electron 6, 1201.

Iga, K, Nishimura, T, Yagi, K, Yamaguchi, T and Niina, T (1986) Room temperature pulsed oscillation of GaAlAs/GaAs surface emitting junction laser grown by MBE. Jap J Appl Phys 25, 924.

Ilegems, M (1977) Beryllium doping and diffusion in molecular-beam epitaxy of GaAs and Al$_x$Ga$_{1-x}$As. J Appl Phys 48, 1278.

Ishibashi, A (1996) II–VI blue-green light emitters. J Cryst Growth 159, 555.

Ishida, M, Hatori, N, Otsubo, K, Yamamoto, T, Nakata, Y, Ebe, H, Sugawara, M and Arakawa, Y (2007) Low-driving-current temperature-stable 10 Gbit/s operation of p-doped 1.3 μm quantum dot lasers between 20 and 90°C. Electron Lett 43, 219.

Jacobs, R N, Nozaki, C, Almeida, L A, Jaime-Vasquez, M, Lennon, C, Markunas, J K, Benson, D, Smith, P, Zhao, W F, Smith, D J, Billman, C, Arias, J and Pellegrino, J (2012) Development of MBE II–VI epilayers on GaAs(211)B. J Electron Matls 41, 2707.

Jagadish, C, Gunapala, S and Rhiger, D (eds) (2011) 'Semiconductors and Semimetals', Vol 84, Academic Press, New York.

Jager, R, Grabherr, M, Jung, C, Michalzik, R, Reiner, G, Weigl, B and Ebling, K J (1997) 57% wallplug efficiency oxide-confined 850 nm wavelength GaAs VCSELs. Electron Lett 33, 330.

Jäger, R and Riedl, M C (2011) MBE growth of VCSELs for high volume applications. J Cryst Growth 323, 434.

Jewell, J L, Lee, Y H, Harbison, J P, Scherer, A and Florez, L T (1991) Vertical-cavity surface-emitting lasers - design, growth, fabrication, characterization IEEE J Quantum Electron 27, 1332.

Jin, C -Y, Badcock, T J, Liu, H -Y, Groom, K M, Royce, R J, Mowbray, D J and Hopkinson, M (2006) Observation and modeling of a room-temperature negative characteristic temperature 1.3-μm p-type modulation-doped quantum-dot laser. IEEE J Quantum Electron 42, 1259.

Kamath, K, Bhattacharya, P, Sosnowski, T, Norris, T and Philips, J (1996) Room-temperature operation of $In_{0.4}Ga_{0.6}As$/GaAs self-organised quantum dot lasers. Electron Lett 32, 1374.

Kazarinov, R F and Suris, R A (1971) Possibility of amplification of electromagnetic waves in a semiconductor with a superlattice. Sov Phys Semicond 5, 707.

Khiar, A, Rahim, M, Felder, F, Hobrecker, F and Zogg, H (2010) Continuously tunable monomode mid-infrared vertical external cavity surface emitting laser on Si. Appl Phys Lett 97, 151104.

Kirstaedter, N, Ledentsov, N N, Grundmann, M, Bimberg, D, Ustinov, V M, Ruvimov, S S, Maksimov, M V, Kop'ev, P S, Alferov, Zh I, Richter, U, Werner, P, Gosele, U and Heydenreich, J (1994) Low threshold, large T_0 injection laser emission from (InGa) As quantum dots. Electron Lett 30, 1416.

Klopf, F, Krebs, R, Wolf, A, Emmerling, M, Reithmaier, J P and Forchel, A (2001) InAs/GaInAs quantum dot DFB lasers emitting at 1.3 μm. Electron Lett 37, 634.

Köhler, R, Tredicucci, A, Bertram, F, Beere, H E, Linfield, E H, Davies, A G, Ritchie, D A, Lotti, R C and Rossi, F (2002) Terahertz semiconductor-heterostructure laser. Nature 417, 156.

Kondo, M and Ishikawa, F (2012) High-quality growth of GaInNAs for application to near-infrared laser diodes. Adv Opt Technol 2012, ID 754546.

Kondow, M, Kitatani, T, Nakatsuka, S, Larson, M C, Nakahara, K, Yazawa, Y, Okai, M and Uomi, K (1997) GaInNAs: a novel material for long-wavelength semiconductor lasers. IEEE J Sel Top Quant Electron 3, 719.

Kondow, M, Uomi, K, Niwa, A, Kitatani, T, Watahiki, S and Yazawa, Y (1996) GaInNAs: a novel material for long-wavelength-range laser diodes with excellent high-temperature performance. Jap J Appl Phys 35, 1273.

Konttinen, J, Harkonen, A, Tuomisto, P, Guina, M, Rautiainen, J, Pessa, M and Okhotnikov, O (2007) High-power (>1 W) dilute nitride semiconductor disk laser emitting at 1240 nm. New J Phys 9, 140.

Korpijärvi, V -M, Guina, M, Puustinen, J, Tuomisto, P, Rautiainen, J, Härkönen, A, Tukiainen, A, Okhotnikov, O and Pessa, M (2009) MBE grown GaInNAs-based multi-Watt disk lasers. J Cryst Growth, 311, 1868.

Kovsh, A R, Zhukov, A E, Livshits, D A, Egorov, A Yu, Musikhin, Yu G, Ledentsov, N N, Kop'ev P S, Alferov, Zh I and Bimberg, D (1999) 3.5 W CW operation of quantum dot laser. Electron Lett 35, 1161.

Koyama, F, Kinoshita, S and Iga, K (1989) Room-temperature continuous wave lasing characteristics of a GaAs vertical cavity surface-emitting laser. Appl Phys Lett 55, 221.

Krishna, S (2005) Quantum dots-in-a-well infrared photodetectors. J Phys D: Appl Phys 38, 2142.

Krishna, S, Raghaven, S, von Winckel, G, Stintz, A, Ariyawansa, G, Matsik, S G and Perera, A G U (2003) Three-color ($\lambda_{p1} \sim 3.8$ μm, $\lambda_{p2} \sim 8.5$ μm, and $\lambda_{p3} \sim 23.2$ μm) InAs/InGaAs quantum-dots-in-a-well detector. Appl Phys Lett 83, 2745.

Kroemer, H (1963) A proposed class of heterojunction injection lasers. Proc IEEE, 51, 1782.

Kuznetsov, M (2010) 'VECSEL Semiconductor Lasers' in 'Semiconductor Disc Lasers, Physics and Technology' (ed O G Okhotnikov), Wiley–VCH, Weinheim.

Ledentsov, N N, Grundmann, M, Heinrichsdorff, F, Bimberg, D, Ustinov, V M, Zhukov, A E, Maksimov, M V, Alferov, Zh I and Lott, J A (2000) Quantum-dot heterostructure lasers. IEEE J Sel Top Quant Electron 6, 439.

Ledentsov, N N, Shchukin, V A, Grundmann, M, Kirstaedter, N, Böhrer, J, Schmidt, O, Bimberg, D, Ustinov, V M, Egorov, A Yu, Zhukov, A E, Kop'ev, P S, Zaitsev, S V, Gordeev, N Yu, Alferov, Zh I, Borokov, A I, Kosogov, A O, Ruvuimov, S S, Werner, P, Gösele, U and Heydenreich, J (1996) Direct formation of vertically coupled quantum dots in Stranski-Krastanow growth. Phys Rev B 54, 8743.

Ledentsov, N N, Ustinov, V M, Egorov, A Yu, Zhukov, A E, Maksimov, M V, Tabatadze, I G and Kop'ev, P S (1994) Optical properties of heterostructures with InGaAs-GaAs quantum clusters. Semicond 28, 832.

Lee, J -H, Chang, C -Y, Li, C -H, Lin, S -Y and Lee, S -C (2012) Performance improvement of AlGaAs/GaAs QWIP by NH_3 plasma treatment. IEEE J Quantum Electron 48, 922.

Lee, S W, Hirakawa, K and Shimada, Y (1999) Bound-to-continuum intersubband photoconductivity of self-assembled InAs quantum dots in modulation-doped heterostructures. Appl Phys Lett 75, 1428.

Lee, Y H, Jewell, J L, Scherer, A, Tell, B, Brown-Goebeler, K, Harbison, J P and Florez, L T (1989) Electron Lett 25, 225.

Lee, A, Liu, H and Seeds, A (2013) Semiconductor III–V lasers monolithically grown on Si substrates. Semicond Sci Technol 28, 015027.

Levine, B F, Choi, K K, Bethea, C G, Walker J and Malik, R (1987) New 10 μm infrared detector using intersubband absorption in resonant tunneling GaAlAs superlattices. Appl Phys Lett 50, 1092.

Lin, W -H and Lin, S -Y (2012) Broadband InGaAs-capped InAs/GaAs quantum-dot infrared photodetector with Bi-modal dot height distributions. J Appl Phys 112, 034508.

Ling, H S, Wang, S Y, Lee, C P and Lo, M C (2008) High quantum efficiency dots-in-a-well quantum dot infrared photodetectors with AlGaAs confinement enhancing layer. Appl Phys Lett 92, 193506.

Liu, H Y, Childs, D T, Badcock, T J, Groom, K M, Sellers, I R, Hopkinson, M, Hogg, R A, Robbins, D J, Mowbray, D J and Skolnick M S (2006a) High-performance three-layer 1.3-μm InAs-GaAs quantum-dot lasers with very low continuous-wave room-temperature threshold currents. IEEE Photonics Tech Lett 17, 1139.

Liu, H Y, Liew, S L, Badcock, T, Mowbray, D J, Skolnick, M S, Ray, S K, Choi, T L, Groom, K M, Stevens, B, Hasbullah, F, Jin, C Y, Hopkinson, M and Hogg, R A (2006b) p-doped 1.3 μm InAs/GaAs quantum-dot laser with a low threshold current density and high differential efficiency. Appl Phys Lett 89, 073113.

Liu, H Y, Sellers, I R, Badcock, T J, Mowbray, D J, Skolnick, M S, Groom, K M, Gutierrez, M, Hopkinson, M, Ng, J S, David, J P R and Beanland, R (2004a) Improved performance of 1.3 μm multilayer InAs quantum-dot lasers using a high-growth-temperature GaAs spacer layer. Appl Phys Lett 85, 704.

Liu, H C, Song, C Y, SpringThorpe, A J and Cao, J C (2004b) Terahertz quantum-well photodetector. Appl Phys Lett 84, 4068.

Liu, G T, Stintz, A, Li, H, Malloy, K J and Lester, L F (1999) Extremely low room-temperature threshold current density diode lasers using InAs dots in $In_{0.15}Ga_{0.85}As$ quantum well. Electron Lett 35, 1163.

Luque, A and Marti, A (1997) Increasing the efficiency of ideal solar cells by photon induced transitions at intermediate levels. Phys Rev Lett 78, 5014.

Luque, A and Marti, A (2010) The intermediate band solar cell: progress toward the realization of an attractive concept. Adv Mater 22, 160.

Luque, A, Marti, A and Stanley, C (2012) Understanding intermediate-band solar cells. Nature Photon 6, 146.

Manz, C, Yang, Q, Rattunde, M, Schulz, N, Rösener, B, Kirste, L, Wagner, J and Kohler, K (2009) Quaternary GaInAsSb/AlGaAsSb vertical-external-cavity surface-emitting lasers—a challenge for MBE growth. J Cryst Growth 311, 1920.

Marko, I P, Masse, N F, Sweeney, S J, Andreev, A D, Adams, A R, Hatori, N and Sugawara, M (2005) Carrier transport and recombination in p-doped and intrinsic 1.3 μm InAs/GaAs quantum-dot lasers. Appl Phys Let 87, 211114.

Marti, A, Antolin, E, Stanley, C R, Farmer, C D, López, N, Diaz, P, Cánovas, E, Linares, P G and Luque, A (2006) Production of photocurrent due to intermediate-to-conduction-band transitions: a demonstration of a key operating principle of the intermediate-band solar cell. Phys Rev Lett 97, 247701.

Maulini, R, Beck, M, Faist, J and Gini, E (2004) Broadband tuning of external cavity bound-to-continuum quantum-cascade lasers. Appl Phys Lett 84, 1659.

Mi, Z, Bhattacharya, P, Yang, J and Pipe, K P (2005) Room-temperature self-organised $In_{0.5}Ga_{0.5}As$ quantum dot laser on silicon. Electron Lett 41, 742.

Mirin, R, Gossard, A and Bowers, J (1996) Room temperature lasing from InGaAs quantum dots. Electron Lett 32, 1732.

Mirin, R P, Ibbetson, J P, Nishi, K, Gossard, A C and Bowers, J E (1995) 1.3 μm photoluminescence from InGaAs quantum dots on GaAs. Appl Phys Lett 67, 3795.

Monroy, E, Guillot, F, Leconte, S, Bellet-Amalric, E, Baumann, E, Giorgetta, F, Hofstetter, D, Nevou, L, Tchernycheva, M, Doyennette, L, Julien, F H, Remmele, T and Albrecht, M (2006) MBE growth of nitride-based photovoltaic intersubband detectors. Superlattice Microstr 40, 418.

Mowbray, D J and Skolnick, M S (2005) New physics and devices based on self-assembled semiconductor quantum dots. J Phys D: Appl Phys 38, 2059.

Nakamura, S, Senoh, M, Nagahama, S, Iwasa, N, Yamada, T, Matsushita, T, Kiyoku, H and Sugimoto, Y (1996) InGaN-based multi-quantum-well-structure laser diodes. Jap J Appl Phys 35, L74.

Nathan, M I, Dumke, W P, Burns, G, Dill, F H Jr and Lasher, G (1962) Stimulated emission of radiation from GaAs p-n junctions. Appl Phys Lett, 1, 62.

Orton, J W (2004) 'The Story of Semiconductors', Oxford University Press, Oxford.

Orton, J W, Dawson, P, Lacklison, D E, Cheng, T S and Foxon, C T (1994) Recombination lifetime measurements in AlGaAs/GaAs quantum well structures. Semicond Sci Technol 9, 1616.

Orton, J W and Foxon, C T (1998) Group III nitride semiconductors for short wavelength light-emitting devices. Rep Prog Phys 61, 1.

Oshima, R, Takata, A and Okada, Y (2008) Strain-compensated InAs/GaNAs quantum dots for use in high-efficiency solar cells. Appl Phys Lett 93, 083111.

Park, G, Shchekin, O B, Csutak, S, Huffaker, D L and Deppe, D G (1999) Room-temperature continuous-wave operation of a single-layered 1.3 μm quantum dot laser. Appl Phys Lett 75, 3267.

Paxman, M, Nelson, J, Braun B, Connolly, J, Barnham, K W J, Foxon, C T and Roberts, J S (1993) Modeling the spectral response of the quantum well solar cell. J Appl Phys 74, 614.

Peters, F H, Peters, M G, Young, D B, Scott, J W, Thibeault, B J, Corzine, S W and Coldren, L A (1993) High-power vertical-cavity surface-emitting lasers. Electron Lett 29, 200.

Quist, T M, Rediker, R H, Keyes, R J, Krag, W E, Lax, B, McWhorter, A L and Zeigler, H J (1962) Semiconductor maser of GaAs. Appl Phys Lett 1, 91.

Rahim, M, Fill, M, Felder, F, Chappuis, D, Corda, M and Zogg, H (2009a) Mid-infrared PbTe vertical external cavity surface emitting laser on Si-substrate with above 1 W output power. Appl Phys Lett 95, 241107.

Rahim, M, Khiar, A, Felder, F, Fill, M and Zogg, H (2009b) 4.5 μm wavelength vertical external cavity surface emitting laser operating above room temperature. Appl Phys Lett 94, 201112.

Revin, D G, Cockburn, J W, Steer, M J, Airey, R J, Hopkinson, M, Krysa, A B, Wilson, L R and Menzel, S (2007) InGaAs/AlAsSb/InP quantum cascade lasers operating at wavelengths close to 3 μm. Appl Phys Lett 90, 021108.

Rogalski, A (2003) Quantum well photoconductors in infrared detector technology. J Appl Phys 93, 4355.

Ruprecht, H S, Woodall, J M and Pettit, G D (1967) Efficient visible electroluminescence at 300° K from $Ga_{1-x}Al_xAs$ p-n junctions grown by liquid-phase epitaxy. Appl Phys Lett 11, 81.

Sabnis, V, Yuen, H and Wiener, M (2012) High-efficiency multijunction solar cells employing dilute nitrides. AIP Conf Proc 1477 14.

Saito, H, Nishi, K and Sugou, S (1998) Influence of GaAs capping on the optical properties of InGaAs/GaAs surface quantum dots with 1.5 μm emission. Appl Phys Lett 73, 2742.

Sakr, S, Crozat, P, Gacemi, D, Kotsar, Y, Pesach, A, Quach, P, Isac, N, Tchernycheva, M, Vivien, L, Bahir, G, Monroy, E and Julien, F H (2013) GaN/AlGaN waveguide quantum cascade photodetectors at $\lambda \approx 1.55$ μm with enhanced responsivity and ~40 GHz frequency bandwidth. Appl Phys Lett 102, 011135.

Saku, T, Iwamura, H, Hirayama, Y, Suzuki, Y and Okamota, H (1985) Room temperature operation of 650 nm AlGaAs multi-quantum-well laser diode grown by molecular beam epitaxy. Jap J Appl Phys 24, L73.

Sandall, I C, Smowton, P M, Thomson, J D, Badcock, T, Mowbray, D J, Liu, H Y and Hopkinson, M (2006) Temperature dependence of threshold current in p-doped quantum dot lasers. Appl Phys Lett 89, 15111801.

Sandusky, J V and Brueck, S R (1996) A CW external-cavity surface-emitting laser. IEEE Photonics Tech Lett 8, 313.

Schießl, U and Rohr, J (1999) 60°C lead salt laser emission near 5-μm wavelength. Infrared Phys Technol 40, 325.

Schwarzl, T, Heiss, W, Springholz, G, Aigle, M and Pascher, H (2000) 6 μm vertical cavity surface emitting laser based on IV-VI semiconductor compounds. Electron Lett 36, 322.

Sellers, I R, Liu, H Y, Groom, K M, Childs, D T, Robbins, D, Badcock, T J, Hopkinson, M, Mowbray, D J and Skolnick, M S (2004) 1.3 μm InAs/GaAs multilayer quantum-dot laser with extremely low room-temperature threshold current density. Electron Lett 40, 1412.

Shchekin, O B, Ahn, J and Deppe, D G (2002) High temperature performance of self-organised quantum dot laser with stacked p-doped active region. Electron Lett 38, 712.

Shockley, W and Queisser, H J (1961) Detailed balance limit of efficiency of p-n junction solar cells. J Appl Phys 32, 510.

Sirtori, C, Faist, J, Capasso, F, Sivco, D L, Hutchinson, A L and Cho, A L (1997) Mid-infrared (8.5 μm) semiconductor lasers operating at room temperature. IEEE Photon Tech Lett 9, 294.

Sirtori, C, Kruck, P, Barbieri, S, Collot, P, Nagle, J, Beck, M, Faist, J and Oesterie, U (1998) GaAs/Al$_x$Ga$_{1-x}$As quantum cascade lasers. Appl Phys Lett 73, 3486.

Skierbiszewski, C, Siekacz, M, Turski, H, Muziol, G, Sawicka, M, Feduniewicz-Zmuda, A, Smalc-Koziorowska, J, Perlin, P, Grzanka, S, Wasilewski, Z R, Kucharski, R and Porowski, S (2012) InGaN laser diodes operating at 450–460 nm grown by rf-plasma MBE. J Vac Sci Technol B 30, 02B102.

Skierbiszewski, C, Siekacz, M, Turski, H, Sawicka, M, Feduniewicz-Zmuda, A, Perlin, P, Suski, T, Wasilewski, Z, Grzegory, I and Perowski, S (2011) Lith J Physics 51, 276.

Skierbiszewski, C, Wasilewski, Z R, Siekacz, M, Feduniewicz, A, Perlin, P, Wisniewski, P, Borysiuk, J, Grzegory, I, Leszczynski, M, Suski, T and Porowski, S (2005) Blue-violet InGaN laser diodes grown on bulk GaN substrates by plasma-assisted molecular-beam epitaxy. Appl Phys Lett 86, 011114.

Smith, E P G, Patten, E A, Goetz, P M, Venzor, G M, Roth, J A, Nosho, B Z, Benson, J D, Stoltz, A J, Varesi, J B, Jensen, J E, Johnson, S M and Radford, W A (2006) Fabrication and characterization of two-color midwavelength/long wavelength HgCdTe infrared detectors. J Electronic Materials 35, 1145.

Soda, H, Iga, K, Kitahara, C and Suematsu, Y (1979) GaInAsP/InP surface emitting injection lasers. Jap J Appl Phys 18, 2329.

Springholz, G and Bauer, G (2013) 'IV-VI Semiconductors' in 'Semiconductor Quantum Structures: Growth and Structuring. Landolt-Bornstein Group III Vol 34A' (ed C. F. Klingshirn), Springer, New York, p 415.

Springholz, G, Schwarzl, T, Aigle, M, Pascher, H, Heiss, W (2000) 4.8 μm vertical emitting PbTe quantum-well lasers based on high-finesse EuTe/Pb$_{1-x}$Eu$_x$Te microcavities. Appl Phys Lett 76, 1807.

Springholz, G, Schwarzl, T, Heiss, W, Bauer, G, Aigle, M, Pascher, H and Vavra, I (2001) Midinfrared surface-emitting PbSe/PbEuTe quantum-dot lasers. Appl Phys Lett 79, 1225.

SpringThorpe, A J and Majeed, A (1990) Epitaxial growth rate measurements during molecular beam epitaxy. J Vac Sci Technol B 8, 266.

Takada, K, Tanaka, Y, Matsumoto, T, Ekawa, M, Song, H Z, Nakata, Y, Yamaguchi, M, Nishi, K, Yamamoto, T, Sugawara, M and Arakawa, Y (2011) Wide-temperature-range 10.3 Gbit/s operations of 1.3 μm high-density quantum-dot DFB lasers. Electron Lett 47, 206.

Tan, W S, Kauer, M, Hooper, S E, Barnes, J M, Rossetti, M, Smeeton, T M, Bousquet, V and Heffernan, J (2008) High-power and long-lifetime InGaN blue-violet laser diodes grown by molecular beam epitaxy. Electron Lett 44, 351.

Tanabe, K, Rae, T, Watanabe, K and Arakawa, Y (2013) High-temperature 1.3 μm InAs/GaAs quantum dot lasers on Si substrates fabricated by wafer bonding. Appl Phys Express 6, 082703.

Tanaka, H and Mushiage, M (1991) MBE as a production technology for AlGaAs lasers. J Cryst Growth 111, 1043.

Tang, S -F, Lin, S -Y and Lee, S -C (2001) Near-room-temperature operation of an InAs/GaAs quantum-dot infrared photodetector. Appl Phys Lett 78, 2428.

Thompson, G H B and Kirkby, P A (1973) (GaAl)As lasers with a heterostructure for optical confinement and additional heterojunctions for extreme carrier confinement. IEEE J Quantum Electron 9, 311.

Ting, D Z -Y, Bandara, S V, Gunapala, S D, Mumolo, J M, Keo, S A, Hill, C J, Liu, J K, Blazejewski, E R, Rafol, S B and Chang, Y -C (2009) Submonolayer quantum dot infrared photodetector. Appl Phys Lett 94, 111107.

Tobin, S P, Vernon, S M, Bajgar, G, Wojtczuk, S J, Melloch, M R, Keshavarzi, A, Stellwag, T B, Venkatensan, S, Lundstrom, M S and Emery, K A (1990) Assessment of MOCVD- and MBE-growth GaAs for high-efficiency solar cell applications. IEEE Trans Electron Devices 37, 469.

Tokranov, V, Yakimov, M, Katsnelson, A, Lamberti, M and Oktyabrsky, S (2003) Enhanced thermal stability of laser diodes with shape-engineered quantum dot medium. Appl Phys Lett 83, 833.

Tsang, W T (1978) The influence of bulk nonradiative recombination in the wide band-gap regions of molecular beam epitaxially grown GaAs-Al_xGa_{1-x}As DH lasers. Appl Phys Lett 33, 245.

Tsang, W T (1979) Low-current-threshold and high-lasing uniformity GaAs–Al_xGa_{1-x}As double-heterostructure lasers grown by molecular beam epitaxy. Appl Phys Lett 34, 473.

Tsang, W T (1980a) Very low current threshold GaAs-Al_xGa_{1-x}As double-heterostructure lasers grown by molecular beam epitaxy. Appl Phys Lett 36, 11.

Tsang, W T (1980b) Infrared-visible (0.89–0.72 µm) Al_xGa_{1-x}As/Al_yGa_{1-y}As double-heterostructure lasers grown by molecular beam epitaxy. J Appl Phys 51, 917.

Tsang, W T (1981a) High-through-put, high-yield, and highly-reproducible (AlGa)As double-heterostructure laser wafers grown by molecular beam epitaxy. Appl Phys Lett 38, 587.

Tsang, W T (1981b) A graded-index waveguide separate-confinement laser with very low threshold and a narrow Gaussian beam. Appl Phys Lett 39, 134.

Tsang, W T (1981c) Extremely low threshold (AlGa)As modified multiquantum well heterostructure lasers grown by molecular-beam epitaxy. Appl Phys Lett 39, 786.

Tsang, W T (1982) Extremely low threshold (AlGa)As graded-index waveguide separate-confinement heterostructure lasers grown by molecular beam epitaxy. Appl Phys Lett 40, 217.

Tsang, W T (1984) Heterostructure semiconductor lasers prepared by molecular beam epitaxy. IEEE J Quantum Electron 20, 1119.

Tsang, W T, Reinhart, F K and Ditzenberger, J A (1980) The effect of substrate temperature on the current threshold of GaAs-Al_xGa_{1-x}As double-heterostructure lasers grown by molecular beam epitaxy. Appl Phys Lett 36, 118.

Ueno, K, Camargo, E G, Katsumata, T, Goto, H, Kuze, N, Kangawa, Y and Kakimoto, K (2013) InSb mid-infrared photon detector for room-temperature operation. Jap J Appl Phys 52, 092202.

Uesugi, K, Morooka, N and Suemune, I (1999) Reexamination of N composition dependence of coherently grown GaNAs bandgap energy with high resolution X-ray diffraction mapping measurements. Appl Phys Lett 74, 1254.

van der Ziel, J P, Dingle, R, Miller, R C, Wiegmann, W and Nordland, W A (1975) Laser oscillation from quantum states in very thin GaAs–$Al_{0.2}Ga_{0.8}$As multilayer structures. Appl Phys Lett 26, 463.

Vardi, A, Kheirodin, N, Nevou, L, Machhadani, H, Vivien, L, Crozal, P, Tchernycheva, M, Colombelli, R, Julien, F H, Guillot, F, Bougerol, C, Monroy, E, Schacham, S and Bahir, G (2008) High-speed operation of GaN/AlGaN quantum cascade detectors at $\lambda \approx 1.55$ µm. Appl Phys Lett 93, 193509.

Vasiliev, V V, Varavin, V S, Dvoretsky, S A, Marchishin, I M, Mikhailov, N N, Predein, A V, Sabinina, I V, Sidorov, Yu G, Suslyakov, A O and Aseev, A L (2012) 'LWIR photodiodes and focal plane arrays based on novel HgCdTe/CdZnTe/GaAs heterostructures grown by MBE technique' in 'Photodiodes – From Fundamentals to Applications' (ed I Yun), InTech, DOI: 10.5772/50822. < http://www.intechopen.com/books/photodiodes-from-fundamentals-to-applications/lwir-photodiodes-and-focal-plane-arrays-based-on-novel-hgcdte-cdznte-gaas-heterostructures-grown-by->.

Vizbaras, K, Bachmann, A, Arafin, S, Saller, K, Sprengel, S, Boehm, G, Meyer, R and Amann, M -C (2011) MBE growth of low threshold GaSb-based lasers with emission wavelengths in the range of 2.5–2.7 µm. J Cryst Growth 323 446.

Wang, P D, Ledentsov, N N, Sotomayor Torres, C M, Kop'ev, P S and Ustinov, V M (1994) Optical characterization of submonolayer and monolayer InAs structures grown in a GaAs matrix on (100) and high-index surfaces. Appl Phys Lett 64, 1526.

Wang, S Y, Lin, S D, Wu, H W and Lee, C P (2001) Low dark current quantum-dot infrared photodetectors with an AlGaAs current blocking layer. Appl Phys Lett 78, 1023.

Wenisch, J, Eich, D, Lutz, H, Schallenberg, T, Wollrab, R and Ziegler, J (2012) MBE growth of MCT on GaAs substrates at AIM. J Electronic Matls 41, 2828.

Weyers, M, Sato, M and Ando, H (1992) Red shift of photoluminescence and absorption in dilute GaAsN alloy layers. Jap J Appl Phys 31, L853.

Williams, B S (2007) Terahertz quantum-cascade lasers. Nature Photon 1, 517.

Wilson, G C, Hadley, M A, Smith, J S and Lau, K Y (1993) High single-mode output power from compact external microcavity surface-emitting laser diode. Appl Phys Lett 63, 3265.

Wolf, M (1960) Limitations and possibilities for improvement of photovoltaic solar energy converters, Pt. I: considerations for earth's surface operation. Proc IRE 48, 1246.

Woodbridge, K, Blood, P, Fletcher, E D and Hulyer, P J (1984) Short wavelength (visible) GaAs quantum well lasers grown by molecular beam epitaxy. Appl Phys Lett 45, 16.

Yakushev, M, Varavin, V, Vasilyev, V, Dvoretsky, S, Sabinina, I, Sidorov, Y, Sorochidn, A and Aseev, A (2011) 'HgCdTe heterostructures grown by MBE on Si(310) for infra-red photodetectors' in 'Photodiodes – World Activities in 2011' (ed J W Park), InTech, DOI: 10.5772/20587. < http://www.intechopen.com/books/photodiodes-world-activities-in-2011/hgcdte-heterostructures-grown-by-mbe-on-si-310-for-infrared-photodetectors>.

Yang, X, Jurkovic, M J, Heroux, J B and Wang, W I (1999) Molecular beam epitaxial growth of InGaAsN:Sb/GaAs quantum wells for long-wavelength semiconductor lasers. Appl Phys Lett 75, 178.

Yang, H -P D, Lu, C, Hsiao, R, Chiou, C, Lee, C, Huang, C, Yu, H, Wang, C, Lin, K, Maleev, N A, Kovsh, A R, Sung, C, Lai, C, Wang, J, Chen, J, Lee, T and Chi, J Y (2005) Characteristics of MOCVD- and MBE-grown InGa(N)As VCSELs. Semicond Sci Technol 20, 834.

Yu, H C, Wang, J S, Su, Y K, Chang, S J, Lai, F I, Chang, Y H, Kuo, H C, Sung, C P, Yang, H P D, Lin, K F, Wang, J M, Chi, J Y, Hsiao, R S and Mikhrin, S (2006) 1.3-µm InAs-InGaAs quantum-dot vertical-cavity surface-emitting laser with fully doped DBRs grown by MBE. IEEE Photonics Tech Lett 18, 418.

Zhukov, A E, Kovsh, A R, Mikhrin, S S, Maleev, N A, Ustinov, V M, Livshits, D A, Tarasov, A F, Bedarev, D A, Maximov, M V, Tsatsul'nikov, A F, Soshnikov, P S, Kop'ev, P S, Alferov, Zh I, Ledentsov, N N and Bimberg, D (1999) 3.9 W CW power from sub-monolayer quantum dot diode laser. Electron Lett, 35, 1845.

Zogg, H, Alchalabi, K, Zimin, D and Kellermann, K (2002) Lead chalcogenide on silicon infrared sensors: focal plane array with 96×128 pixels on active Si-chip. Infrared Phys Technol 43, 251.

Zogg, H, Arnold, M, Felder, F, Rahim, M, Fill, M and Boye, D (2008) Epitaxial lead-chalcogenides on Si and BaF_2 for mid-IR detectors and emitters including cavities. Proc SPIE 7082, 70820H.

Zogg, H, Fach, A, Maissen, C, Masek, J and Blunier, S (1994) Photovoltaic lead-chalcogenide on silicon infrared sensor arrays. Optical Eng 33, 1440.

Zogg, H, Maissen, C, Masek, J, Hoshino, T, Blunier, S and Tiwara, A N (1991) Photovoltaic infrared sensor arrays in monolithic lead chalcogenides on silicon. Semicond Sci Technol 6, C36.

Companies and Institutions Index

Materials Index

Name Index

General Index

A

acceptor 2, 4, 18, 19, 22, 24–28, 143, 154–159, 191, 214–218, 261, 267, 269, 303, 323–329, 330, 334–336, 340, 341, 347–349, 351, 359, 364, 368, 406, 412, 413, 418, 419, 420, 421, 459, 477
accommodation coefficient 122, 123
activation energy 49, 80, 121, 122, 126, 130, 173, 179, 328, 329, 334, 363
adsorption 121, 126
ammonia MBE (NH3–MBE): *see* section 7.2 254, 284–288, 422, 460
amorphous semiconductor 129, 148, 164, 313, 314, 316, 362, 368, 375
amphoteric doping 16, 18, 24, 25, 153, 155, 156, 403
antiphase domain 300–302
atomic layer epitaxy (ALE): *see* section 5.2 300, 301
Auger electron spectroscopy (AES) 56, 59, 60, 61, 78, 99, 100, 117, 118, 128, 209, 397
Auger recombination 315, 448, 451
AX centre 335

B

bakeout: *see* section 4.7 55, 57, 60, 75, 76, 99, 155, 163
band anti-crossing model 308, 310, 314, 331, 490
beam calibration: *see* section 3.3

bipolar transistor: *see* section 9.3 4, 7, 194, 300, 359, 360–367, 374, 391, 416, 421, 423
bond energies 151, 282, 283
Bragg mirror: *see* section 10.4 461–469
Bridgman growth method 7, 8, 11, 28, 29, 31, 283, 298, 325, 348, 359, 474
buffer layer 27, 63, 164, 185, 246, 255, 294, 304, 312, 336, 338, 345, 346, 358, 370–377, 391–395, 400–415, 418, 426, 441, 452, 468, 476–479, 482

C

Casimir doctrine 59
catalyst 456
chemical bonds 9, 112, 113, 161, 162, 173, 185, 326, 333, 456
chemical beam epitaxy (CBE): *see* section 5.4 101, 186, 190–195, 247, 248, 346
chemisorption 57, 71, 121–124
cleaved edge overgrowth (CEO) 246, 255, 398
coaxial source 397, 479
commercial MBE machines: *see* section 3.7 55, 61, 67, 70, 75, 78, 80, 85, 96, 100, 104–107, 127, 200, 325, 365
computer-control 69, 106, 107, 178, 181, 205, 210, 230, 364, 438
covalent radius 154, 282, 313
critical thickness 103, 249–251, 301, 303, 367–372

cryopanel: *see* section 3.1 76, 77, 80, 118, 163, 285, 436
Czochralski crystal growth 3, 5–8, 27, 107, 283, 299, 474

D

dangling bond 141, 151
delta-doping 219, 220, 230, 267, 335, 337, 380, 402–407
desorption 35, 71, 92, 102, 114–130, 155, 189, 413
diffusion
  gaseous 47, 48
  impurity 9, 22–24, 30, 156, 157, 212, 216–218, 252, 257, 315, 316, 338, 351, 359, 362, 363, 375, 376, 421, 479, 487
  inter-diffusion 30, 31, 158, 159, 175, 179, 182, 210, 211, 235, 325, 343–345, 352, 360, 477
  surface 35, 36, 49, 91, 113, 122, 126, 144, 146, 160, 161, 173, 186, 248, 251, 293–296, 302, 362, 377, 457, 458
digital versatile disc (DVD) 425
dilute nitrides: *see* section 7.5
dimerisation 140, 148, 151, 152, 161
dislocations 102, 103, 185, 247–252, 283, 298, 301, 304, 312, 328, 332, 346, 350, 351, 356–362, 367–369, 373–379, 403–408, 413–416, 422, 446, 447, 450, 452, 476–487

mixer diode 62, 63, 68, 106, 365, 390
modeling: *see* section 4.6
modulated beam growth methods: *see* section 5.2
modulated beam studies: *see* section 4.2
modulation doping: *see* sections 4.5, 6.6 257, 265, 304, 368, 392, 403, 448
modulation doped field effect transistor (MODFET): *see* "high electron mobility transistor" 396
monolayer superlattice 211
Monte Carlo modelling 144, 158–162, 240

N
Nanostructure 380
nanowire (column, whisker) 246–249, 289, 293–296, 356
n-i-p-i superlattice 214–217, 230
Nobel Prize 64, 283, 434
nuclear magnetic resonance 43–45
nucleation 47, 48–50, 102, 143, 144, 173, 179–183, 291–293, 300, 301, 327, 338, 411–415, 426
nucleation-enhanced MBE (NE-MBE): *see* section 5.2 182, 183
nucleation studies: *see* section 2.3

O
operating procedures (MBE): *see* section 4.7

P
parabolic quantum well 224, 230–232
persistant photoconductivity 260
phase-locked epitaxy (PLE): *see* section 5.2 177, 178, 180, 182
phase segregation 456
photo-assisted MBE 346
photoluminescence (PL) 7, 18, 22, 28, 68, 99, 160, 164, 174, 179, 180, 191–194,
210, 217–220, 230–238, 245, 248, 250–255, 307, 339, 379
photoluminescence excitation spectroscopy (PLE) 230–241, 256, 299, 307
photovoltaic solar cell: *see* solar cell
physisorption 57, 71, 121, 124, 125, 160
planar doping 219, 220, 230, 267, 402: *see also* "delta doping"
plasma-assisted MBE (PAMBE): *see* section 7.2 57, 254, 285, 288–294, 312, 341, 346, 411–416, 422, 460, 467, 482
power devices: *see* sections 9.2, 9.3
precursor 71, 117, 121–124, 160
pseudomorphic growth 153, 185, 301, 315, 367, 368, 370–373, 399–402, 410

Q
quadrupole mass spectrometry: *see* section 4.2
quantum dots: *see* sections 6.5 and 10.2 *also* 184, 272, 273, 288, 292, 293, 356, 357, 376–379, 408, 426, 433–435, 464, 465, 474, 481–489
quantum dot ordering 253
quantum Hall effect 7, 201, 258, 262, 263, 265, 269, 274, 424, 425
quantum well: *see* section 6.4, 10.2 and 10.4 *also* 67, 102, 146, 152, 158, 160, 161, 171–181, 189–199, 204, 210–213, 218, 220, 242–246, 251–259, 263, 271, 287, 288, 292, 293, 302–303, 304, 312, 325, 330, 341, 351–359, 377, 379, 380, 396–399, 403–408, 414, 426, 434, 435, 455, 459, 475, 480–483, 487, 488
quantum wire: *see* section 6.5

R
Raman spectroscopy 177, 209, 218, 371, 375, 422
recombination 22, 23, 25, 28, 174, 215, 218–220, 323, 338, 414, 418, 420, 434, 436, 439, 440, 444, 451, 468, 482
Auger 315, 451
interface 440
non-radiative 64, 252, 288, 378, 419, 436, 439, 448, 450, 456, 458, 486, 489, 490
radiative 22, 287, 377, 434, 437, 444
Shockley-Read 444, 458
recombination centre 22, 64, 252, 288, 402, 446, 449, 456, 458, 489
recombination lifetime 213, 214, 217, 220, 255, 341, 391, 419, 423, 440, 444, 448, 449, 477, 478, 484
reflection high energy electron diffraction (RHEED): *see* sections 3.3 and 4.3
oscillations: *see* section 4.3 and 5.2
resistance standard 262, 424, 425
resonant tunnelling 66, 69, 201, 210–213

S
sawtooth doping superlattice 219
scanning electron microscopy (SEM) 65, 66, 103, 294
scanning tunnelling microscopy (STM): *see* section 4.4 103, 114, 367
Shockley-Read recombination: *see* "recombination"
segregation 4, 21, 68, 126, 155, 156, 268, 298, 299, 363, 364, 436, 456
secondary ion mass spectrometry (SIMS) 340, 344, 419, 439, 440
selective area growth 416
self-organised quantum dots: *see* "quantum dots"
solar cells: *see* section 10.6